Recent Advances in Gamma Ray Astrophysics and Future Perspectives

Recent Advances in Gamma Ray Astrophysics and Future Perspectives

Guest Editor

Patrizia Romano

MDPI

Basel • Beijing • Wuhan • Barcelona • Belgrade • Novi Sad • Cluj • Manchester

Guest Editor
Patrizia Romano
Osservatorio Astronomico di Brera
INAF
Merate
Italy

Editorial Office
MDPI AG
St. Alban-Anlage 66
4052 Basel, Switzerland

This is a reprint of the Special Issue, published open access by the journal *Universe* (ISSN 2218-1997), freely accessible at: www.mdpi.com/journal/universe/special_issues/7299902Z97.

For citation purposes, cite each article independently as indicated on the article page online and as indicated below:

Lastname, A.A.; Lastname, B.B. Article Title. *Journal Name* **Year**, *Volume Number*, Page Range.

ISBN 978-3-7258-1244-8 (Hbk)
ISBN 978-3-7258-1243-1 (PDF)
https://doi.org/10.3390/books978-3-7258-1243-1

Contents

About the Editor

Patrizia Romano

Dr. P. Romano obtained their first M.S. at the University of Padova. In their early years, Romano worked with and visited the NFRA in Dwingeloo, the Dept of Physics in Ljubljana, and the Dept of Physics and Astronomy in Tuscaloosa (AL). They obtained an M.S. and a Ph.D. at The Ohio State University, Columbus (OH). After a few years of post doc, Romano became a staff member at INAF-IASF Palermo. They now work as a senior scientist at INAF-Osservatorio Astronomico di Brera in Merate. Romano has been with the Swift Team since launch; they are an expert in long-term multi-wavelength monitoring campaigns. Romano is now also involved with the ASTRI and CTA Projects and has served as coordinator of the CTA Extragalactic Science Working Group. Romano's interests range from HMXBs (especially supergiant fast X-ray transients, SFXTs) to AGN (in particular, NLS1s and other jetted sources) investigating them in optical, UV, X-ray, and gamma-ray energy bands.

universe MDPI

Editorial

Editorial to the Special Issue: "Recent Advances in Gamma Ray Astrophysics and Future Perspectives" †

Patrizia Romano

INAF, Osservatorio Astronomico di Brera, Via E. Bianchi 46, 23807 Merate, LC, Italy; patrizia.romano@inaf.it

† Special Issue web page: https://www.mdpi.com/journal/universe/special$_{i}$*ssues*/7299902Z97 (accessed on 1 May 2024).

This Special Issue is a collection of reviews highlighting the recent progress in the very vast and closely related fields of γ-ray astrophysics and astro-particle physics in recent years, looking toward a very promising future. Unsurprisingly, given that active galactic nuclei (AGN) represent 50% of the sources detected at high energies (100 MeV $<$ E $<$ 100 GeV) [and 3% of those detected at very high energies (E $>$ 100 GeV)] [1], most papers in this Issue deal with cosmic sources of extra-galactic origin. As such, they adhere to my own selection of topics, which is by definition incomplete and understandably affected by personal biases. Nevertheless, I hope that they may be valuable for a broad audience that does not necessarily specialise in γ-ray astrophysics, as they were intended to, and I sincerely thank each author who took the time out their already overcrowded schedules and accepted my invitation to contribute to this effort.

AGNs are galaxies whose centers host an accreting supermassive black hole ($M_{\rm BH} \sim 10^6$–$10^{10}\,{\rm M}_{\odot}$), which generates luminosities of up to $10^{46}\,{\rm erg\,s^{-1}}$ through non-thermal mechanisms. They can be observed throughout the entire electromagnetic spectrum from the radio to γ-rays and are generally classified into different types according to the presence and width of the emission lines in their optical spectra. Blazars, in particular, are recognized as having their jets pointed towards the observers (within 10–15°), and are hence strongly connected to γ-ray emission. These can be subdivided into flat-spectrum radio quasars (FSRQ) and BL Lac objects, with the latter showing almost featureless continua (see [2] for a recent review).

In this context, the paper "Gamma-ray Emission and Variability Processes in High-Energy-Peaked BL Lacertae Objects" by Kapanadze (contribution 1) presents a thorough review of the non-thermal mechanisms that dominate the emissions from high-energy peaked BL Lacs, which can reach the VHE regime, highlighting the fundamental role of γ-ray variability in discriminating the individual processes. The scope of this review is not confined to high-energy peaked BL Lacs, howevers, as the emission mechanisms described are applicable to most jetted extragalactic sources.

One of the main common characteristics of the different types of AGN, which is relevant in order to understand their physics, is indeed the set of properties of their relativistic jets. In "The power of relativistic jets: a comparative study" by Foschini et al. (contribution 2), the authors provide a detailed comparison of different methods to estimate the jet power, and present a set of equations in order to evaluate this important parameter. This set, taken with the proper caveats, also represents a very useful tool-box for young researchers who are starting their studies in the field of jetted sources.

The paper "Highlights of the MAGIC Florian Goebel Telescopes in the study of Active Galactic Nuclei" by Manganaro and Dominis Prester (contribution 3) reviews the main achievements obtained in the field of AGN by the MAGIC [3] Telescopes, one of the sets of the current generation of Imaging Atmospheric Cherenkov Telescopes (IACTs), which includes VERITAS [4] and H.E.S.S. [5]. Indeed, MAGIC discovered six out of the ten VHE-emitting FSRQs, two of which close to a redshift of one, thus enabling fundamental studies of the extra-galactic background light (EBL). Among the tens of AGNs detected

Citation: Romano, P. Editorial to the Special Issue: "Recent Advances in Gamma Ray Astrophysics and Future Perspectives". *Universe* **2024**, *10*, 213. https://doi.org/10.3390/universe10050213

Received: 1 May 2024
Accepted: 8 May 2024
Published: 10 May 2024

by MAGIC, the blazar TXS 0506+056 is worthy of mention, as it was found to be in a flaring state at several wavelengths, almost in coincidence with the neutrino event IceCube-170922A. A lepto-hadronic emission model is favoured in order to explain the full spectral energy distribution.

The paper "A very-high-energy gamma-ray view of the transient sky" by Carosi and López-Oramas (contribution 4) provides an in depth review of the Galactic and extragalactic HE and VHE transients, mostly related to stellar-size compact objects, including novae, microquasars and flaring gamma-ray binaries, supernovae, pulsar-wind nebulae, fast radio bursts and magnetars, and gravitational waves, and also touching upon gamma-ray bursts and tidal disruption events. The observational corpus for each of these, obtained thanks to the current generation of IACTs, is presented and the proposed underlying physical processes are reported. The review provides a broad view of how the physics of these VHE transients is tightly connected with time-domain and multi-messenger astronomy, and what the next generation of IACTs will contribute to this research field.

The next generation of IACTs will commence with the advent of the Cherenkov Telescope Array Observatory (CTAO, [6]) which, with its wide (20 GeV–300 TeV) energy range and unprecedented sensitivity (5–20 times better with respect to the current IACTs), will improve our grasp of the VHE phenomena by leaps and bounds. In the meantime, a few precursors are being built and tested. Among them is the ASTRI Mini-Array, which will be the largest IACT array in operation until CTAO is completed. During the first four years of operation, it will be run in experiment mode, starting in late 2025, and will investigate a few fundamental open VHE questions by performing dedicated deep observations of specific sky regions. In the following phase, the ASTRI Mini-Array will be run as an Observatory, open to proposals from the whole scientific community, and it is expected to investigate both galactic and extra-galactic sources. The two papers "The ASTRI Mini-Array: a new pathfinder for Cherenkov Telescope Arrays" by Scuderi (contribution 5) and "Science with the ASTRI Mini-Array: From Experiment to Open Observatory" by Vercellone (contribution 6) detail the challenges faced and the innovative solutions adopted for the construction, operation, and maintenance of the nine IACTs that will form the array, as well as the expected performance and scientific outcomes that will be afforded based on its wide energy range (1–200 TeV), large field of view ($\sim 10^{\circ}$), angular resolution ($\sim 3'$), and energy resolution ($\sim 10\%$).

AGILE (2007–2024) was one of the major HE space missions in the first two decades of this century, after the Energetic Gamma Ray Experiment Telescope (EGRET, [7]) on board Compton Gamma-Ray Observatory. The review "Scientific Highlights of the AGILE Gamma-ray Mission" by Vercellone, Pittori, and Tavani (contribution 7) presents the instrument and some of the major scientific discoveries and achievements obtained during its seventeen years of operation in both the Galactic and extra-galactic fields. Among those is the discovery of the Crab Nebula variability in the energy range above 100 MeV and the first evidence of hadronic cosmic-ray acceleration in supernova remnants. Both results had a profound impact on the scientific community, especially on theoretical modelling, and they are also important for observations at higher energies with current and future Cherenkov and extended air-shower arrays.

One of the main targets of γ-ray astrophysics, since the beginning of its existence, has been supernova remnants (SNRs) due to their potential link to cosmic ray (CR) sources. The paper "Supernova remnants in gamma rays" by Giuliani and Cardillo (contribution 8) reviews the results obtained regarding SNRs from observations in the GeV, TeV, and PeV bands and the still open questions regarding their contribution to the population of Galactic CRs.

Gamma-ray bursts (GRBs) are often reported as the being the most energetic explosive phenomena in the universe. Understanding GRBs requires the contribution of most astrophysical fields, ranging from stellar evolution to jet formation, from the equation of state of supra-density matter to gravitational waves, helping to explain the scientific community's fascination with them since their discovery. The paper "Gamma-Ray Bursts: 50 Years and Counting!" by Vigliano and Longo (contribution 9) is a riveting historical review of the progress in this study, from GRBs' serendipitous discovery to the modern day, and demonstrates how this progress is tightly bound to the advances in the technological field, driven by the need to answer the questions that continually arise from new discoveries and theoretical models. The paper also summarises the current open questions in the field, and offers a strong motivation for new technological ventures.

As an upbeat conclusion, the paper "Future perspectives for gamma ray bursts investigations from space" by Bozzo et al. (contribution 10) offers a detailed review of many of the future space missions that will be devoted to the investigation of GRBs in the coming years, describing their specifications and their expected scientific impact in the field. It includes the Einstein Probe (launched on 9 January 2024), the enhanced X-ray Timing and Polarimetry (eXTP) mission, the Gamow Explorer, the High-z gamma-ray bursts for unraveling the dark ages mission (HiZ-GUNDAM), the LargE Area burst Polarimeter (LEAP), the Moon Burst Energetics All-sky Monitor (MoonBEAM), POLAR-2, the StarBurst Multimessenger Pioneer, the Spectroscopic Time-Resolving Observatory for Broadband Energy X-rays (STROBE-X), the Space-based Variable astronomical Object Monitor (SVOM), and the Transient High-Energy Sky and Early Universe Surveyor (THESEUS).

In closing, I believe that, as a whole, this volume shows how, from the strong foundations pioneered by the past and pursued by the current generations of ground- and space-based instrumentation, a new era is about to begin that will foster new missions, experiments, and observatories, which will satisfy the scientific curiosity of the γ-ray astrophysics and astro-particle physics communities in the years to come.

Funding: This research received no external funding.

Acknowledgments: I wish to thank Amos for his unwavering support and Heidi for keeping the balls rolling. I acknowledge the constant support from the MDPI *Universe* team from their initial invitation to be Guest Editor for this Special Issue until the completion of this project.

Conflicts of Interest: The author declares no conflict of interest.

Abbreviations

The following abbreviations are used in this manuscript:

AGILE	Astrorivelatore Gamma ad Immagini LEggero
AGN	Active Galactic nuclei
ASTRI	Astrofisica con specchi a tecnologia replicante italiana
CGRO	Compton Gamma-Ray Observatory
CTAO	Čerenkov telescope array Observatory
EGRET	Energetic Gamma Ray Experiment Telescope
EP	Einstein Probe
eXTP	enhanced X-ray Timing and Polarimetry mission
FSRQ	Flat-spectrum radio quasar
GRB	Gamma-ray burst
HE	High-energy
HiZ-GUNDAM	High-z gamma-ray bursts for unraveling the dark ages mission
LEAP	LargE Area burst Polarimeter
MAGIC	Major atmospheric gamma-ray imaging Čerenkov Florian Goebel telescopes
MDPI	Multidisciplinary Digital Publishing Institute
MoonBEAM	Moon Burst Energetics All-sky Monitor
SNR	Supernova remnant
STROBE-X	Spectroscopic Time-Resolving Observatory for Broadband Energy X-rays
SVOM	Space based Variable astronomical Object Monitor
THESEUS	Transient High-Energy Sky and Early Universe Surveyor
VHE	Very high-energy

List of Contributions

1. Kapanadze, B. Gamma-ray Emission and Variability Processes in High-Energy-Peaked BL Lacertae Objects. *Universe* **2023**, *9*, 344. https://doi.org/10.3390/universe9070344.
2. Foschini, L.; Dalla Barba, B.; Tornikoski, M.; Andernach, H.; Marziani, P.; Marscher, A.; Jorstad, S. The power of relativistic jets: A comparative study. *Universe* **2024**, *10*, 156. https://doi.org/10.3390/universe10040156.
3. Manganaro, M.; Dominis Prester, D. Highlights of the MAGIC Florian Goebel Telescopes in the study of Active Galactic Nuclei. *Universe* **2024**, *10*, 80. https://doi.org/10.3390/universe10020080.
4. Carosi, A.; López-Oramas, A. A very-high-energy gamma-ray view of the transient sky. *Universe* **2024**, *10*, 163. https://doi.org/10.3390/universe10040163.
5. Scuderi, S. The ASTRI Mini-Array: A new pathfinder for Cherenkov Telescope Arrays. *Universe* **2024**, *10*, 146. https://doi.org/10.3390/universe10030146.
6. Vercellone, S. Science with the ASTRI Mini-Array: From Experiment to Open Observatory. *Universe* **2024**, *10*, 94. https://doi.org/10.3390/universe10020094.
7. Vercellone, S.; Pittori, C.; Tavani, M. Scientific Highlights of the AGILE Gamma-ray Mission. *Universe* **2024**, *10*, 153. https://doi.org/10.3390/universe10040153.
8. Giuliani, A.; Cardillo, M. Supernova remnants in gamma rays. *Universe* **2024**, *10*, 203. https://doi.org/10.3390/universe10050203.
9. Vigliano, A.; Longo, F. Gamma-Ray Bursts: 50 Years and Counting! *Universe* **2024**, *10*, 57. https://doi.org/10.3390/universe10020057.
10. Bozzo, E.; Amati, L.; Baumgartner, W.; Chang, T.C.; Cordier, B.; De Angelis, N.; Doi, A.; Feroci, M.; Froning, C.; Gaskin, J.; et al. Future perspectives for gamma ray bursts investigations from space. *Universe* **2024**, *10*, 187. https://doi.org/10.3390/universe10040187.

References

1. Abdollahi, S.; Acero, F.; Baldini, L.; Ballet, J.; Bastieri, D.; Bellazzini, R.; Berenji, B.; Berretta, A.; Bissaldi, E.; Blandford, R.D.; et al. Incremental Fermi Large Area Telescope Fourth Source Catalog. *ApJS* **2022**, *260*, 53. [CrossRef]
2. Prandini, E.; Ghisellini, G. The Blazar Sequence and Its Physical Understanding. *Galaxies* **2022**, *10*, 35. [CrossRef]
3. Aleksić, J.; Alvarez, E.A.; Antonelli, L.A.; Antoranz, P.; Asensio, M.; Backes, M.; Barrio, J.A.; Bastieri, D.; Becerra González, J.; Bednarek, W.; et al. Performance of the MAGIC stereo system obtained with Crab Nebula data. *Astropart. Phys.* **2012**, *35*, 435–448. [CrossRef]
4. Holder, J.; Atkins, R.W.; Badran, H.M.; Blaylock, G.; Bradbury, S.M.; Buckley, J.H.; Byrum, K.L.; Carter-Lewis, D.A.; Celik, O.; Chow, Y.C.K.; et al. The first VERITAS telescope. *Astropart. Phys.* **2006**, *25*, 391–401. [CrossRef]
5. Aharonian, F.; Akhperjanian, A.G.; Bazer-Bachi, A.R.; Beilicke, M.; Benbow, W.; Berge, D.; Bernlöhr, K.; Boisson, C.; Bolz, O.; Borrel, V.; et al. Observations of the Crab nebula with HESS. *A&A* **2006**, *457*, 899–915. [CrossRef]
6. Cherenkov Telescope Array Consortium; Acharya, B.S.; Agudo, I.; Al Samarai, I.; Alfaro, R.; Alfaro, J.; Alispach, C.; Alves Batista, R.; Amans, J.P.; Amato, E.; et al. *Science with the Cherenkov Telescope Array*; World Scientific: Singapore, 2019. [CrossRef]
7. Kanbach, G.; Bertsch, D.L.; Fichtel, C.E.; Hartman, R.C.; Hunter, S.D.; Kniffen, D.A.; Hughlock, B.W.; Favale, A.; Hofstadter, R.; Hughes, E.B. The project EGRET (energetic gamma-ray experiment telescope) on NASA's Gamma-Ray Observatory GRO. *SSRv* **1989**, *49*, 69–84. [CrossRef]

universe MDPI

Review

Gamma-ray Emission and Variability Processes in High-Energy-Peaked BL Lacertae Objects

Bidzina Kapanadze [1,2,3]

1 Space Research Center, Department of Astronomy and Astrophysics, School of Natural Sciences and Medicine, Ilia State University, Cholokashvili Av. 3/5, Tbilisi 0162, Georgia; bidzina_kapanadze@iliauni.edu.ge or bkapanadze@abao.ge; Tel.: +995-577413299

2 E. Kharadze National Astrophysical Observatory, Deapartment of Galaxies and Stars, Mt. Kanobili, Abastumani 0803, Georgia

3 Instituto National di Astrofizica e Fizica spatiale, Osservatorio Astronomico di Brera, Via E. Bianchi 46, 23807 Merate, Italy

Abstract: BL Lac objects are active galactic nuclei notable for a beamed nonthermal radiation, which is generated in one of the relativistic jets forming a small angle to the observer's line-of-sight. The broadband spectra of BL Lacs show a two-component spectral energy distribution (SED). High-energy-peaked BL Lacs (HBLs) exhibit their lower-energy (synchrotron) peaks at UV to X-ray frequencies. The origin of the higher-energy SED component, representing the γ-ray range in HBLs, is still controversial and different emission scenarios (one- and multi-zone synchrotron self-Compton, hadronic etc.) are proposed. In γ-rays, HBLs show a complex flaring behavior with rapid and large-amplitude TeV-band variations on timescales down to a few minutes. This review presents a detailed characterization of the hypothetical emission mechanisms which could contribute to the γ-ray emission, their application to the nearby TeV-detected HBLs, successes in the broadband SED modeling and difficulties in the interpretation of the observational data. I also overview the unstable processes to be responsible for the observed γ-ray variability and particle energization up to millions of Lorentz factors (relativistic shocks, magnetic reconnection, turbulence and jet-star interaction). Finally, the future prospects for solving the persisting problems by means of the dedicated gamma-ray observations and sophisticated simulations are also addressed.

Keywords: galaxies; BL Lacertae objects; general

Citation: Kapanadze, B. Gamma-ray Emission and Variability Processes in High-Energy-Peaked BL Lacertae Objects. *Universe* **2023**, *9*, 344. https://doi.org/10.3390/universe9070344

Academic Editor: Binbin Zhang

Received: 26 May 2023
Revised: 3 July 2023
Accepted: 21 July 2023
Published: 24 July 2023

1. Introduction

Blazars are active galactic nuclei (AGNs) which are commonly understood as having the relativistic jets emerging from the central supermassive black holes (SMBH; $M \sim 10^8$–10^{10}) and forming small angles with respect of our line-of-sight ($\theta < 10$–15 deg). Consequently, the relativistic motion of the plasma boosts the non-thermal jet emission into a forward cone pointed to the observer [1]. Owing to such a favourable geometry, the strongly beamed jet radiation often completely outshines the other AGN components [2]. Accreting SMBHs are believed to convert their rotational energy into Poynting flux and power the collimated jets (see [3] and references therein).

BL Lacertae objects (BL Lacs) are a blazar subclass which demonstrate featureless spectra [1] and represent a majority of the AGN detected so far in the TeV band[1] (56 out of the total 89). Their broadband SED consists of two smooth, broad distinct components [4]: the first "hump" extends from the infrared to X-rays (synchrotron emission from relativistic electrons residing the jet emission zone), and a high-energy component having a peak at the MeV–TeV energies. The origin of the latter is still under debate and three fundamentally different approaches have been proposed: *leptonic*, *hadronic* and hybrid *lepto-hadronic* models, based on the particles responsible for the γ-ray emission (see, e.g., [5]).

Depending on the position of the synchrotron SED component, BL Lacs are broadly divided into the low-energy-peaked (LBLs) and high-energy-peaked (HBLs) objects; one additionally considers a group of intermediate-energy-peaked BL Lacs (IBLs; see [4,6]). The sequence LBLs→IBLs→HBLs is characterized by increasing peak frequency in the ν–νF_ν plane, declining dominance of the γ-ray flux over the lower-energy emission, and decreasing bolometric luminosity. In the HBL objects, the lower-energy peak is situated at UV-to-hard X-ray wavelengths; their high-frequency component peak is generally situated beyond ∼100 GeV, and the first SED peak is up to one order higher than the higher-energy one (see Figure 1 and [7]). Moreover, the subclass is widely accepted to possess radiatively inefficient accretion disks (see, e.g., [8]). Note that among the extragalactic TeV sources, the highest energy photons (up to 20 TeV) have been reported for the HBL source Mrk 501 [9].

Generally, the HBL spectra observed above 300 GeV by imaging atmospheric Cherenkov telescopes (IACTs) are frequently quite steep (with the photon index $\Gamma \gtrsim 2.5$), defining an SED turnover [10]. Moreover, very-high-energy γ-rays (VHE, $E > 100$ GeV) emitted by the objects situated beyond $\gtrsim$100 Mpc reach us impacted by significant absorption caused by the extragalactic background light (EBL; via the process $\gamma\gamma \to e^-e^+$). Namely, the γ-ray spectrum undergoes a strong deformation at energies characterized by the optical depth $\tau(E, Z) \gtrsim 1$ [11]. Various studies have revealed that EBL contains two components, namely, at the near- and far-infrared wavelengths, separated by a mid-infrared (MIR) "valley" [12]. Consequently, $\tau(E)$ was found to be strongly dependent of the photon energy below 1 TeV and above 10 TeV, while this dependence is much weaker between 1 and 10 TeV. Therefore, one expects a significant distortion of the VHE spectra of HBLs at energies below 1 TeV and above 10 TeV [11]. Note that the range of 100 MeV–100 GeV covered by the Large Area Telescope (LAT) on board *Fermi* (*Fermi*-LAT; [13]) is characterized only by small $\gamma\gamma$ -attenuation and negligible below 3 GeV [10]. This allow us to discern the underlying particle population more robustly.

Among HBLs, one can additionally discern (i) extreme high-energy peaked BL Lacs (EHBLs), with the synchrotron SED peak $E_p^{sync} \geq 1$ keV and showing a hard X-ray photon index $\Gamma_x < 2$ [14]; (ii) five TeV-detected objects (1ES 0229+200, 1ES 0347−121, 1ES 1101−232, 1ES 1218+304 and RGBJ0710+591) exhibit a higher-energy peak $E_p^{\gamma} \geq 1$ TeV and a hard photon index $\Gamma_\gamma \leq 1.5$–1.9 in the 0.1–10 TeV band, after correction for the EBL [15]. Such sources were classified as ultra-high-energy peaked BL Lacs (UHBLs, see [16]), as well as being EHBLs. The discovery of these UHBLs was a surprise, since the standard emission models yield the higher-energy peak below 1 TeV for HBLs due to the limited maximum energy of the electrons and the Klein-Nishina (KN) effects for X-rays, strongly suppressing the TeV-band emission [8].

This review is focused on the results achieved by the different emission scenarios attempting to explain the origin of γ-ray emission in HBLs. They provide us with an efficient tool to evaluate the physical parameters describing the jet emission zone (by comparing the observed higher-energy SED with those modelled in the framework of the different emission scenarios) and draw conclusions about the jet particle content. In turn, information from the VHE part of the SED is required to constrain model parameters for HBLs, which radiate a significant part of their overall γ-ray emission in that energy range. We briefly review also those acceleration processes which are primary "candidates" for energizing the jet particles up to ultrarelativistic energies required for producing γ-ray photons either by inverse Compton (IC) upscatter or hadronic mechanisms (shocks, magnetic reconnection, magnetohydrodynamic turbulence and magnetospheric vacuum gap), as well as represent the sources of the observed variability on various timescales.

One can not directly resolve the HBL emission zone due to its extremely small angular size. Therefore, a multiwavelength (MWL) variability study represents practically the only way for drawing conclusions about the structure of the jet emission zone. Especially informative is the γ-ray variability study since this emission is produced by the highest energy electrons, which lose energy very quickly and exist only in the vicinity of the acceleration sites. Nevertheless, the VHE emission of HBLs are characterized by the most rapid variability (since the cooling time at these energies are the shortest) and based on light travel time arguments, the corresponding timescales impose constraints on the size of the emission region. The most challenging is an ultra-fast variability shown by some close, bright HBLs on timescales down to a few minutes [17,18]). The latter are significantly shorter than the light-crossing time of the central SMBH and one requires requires extreme physical conditions for their interpretation [2]. Consequently, the γ-ray variability allows us to discern the physical processes operating in the innermost jet area and also represent one of the subjects of the current review.

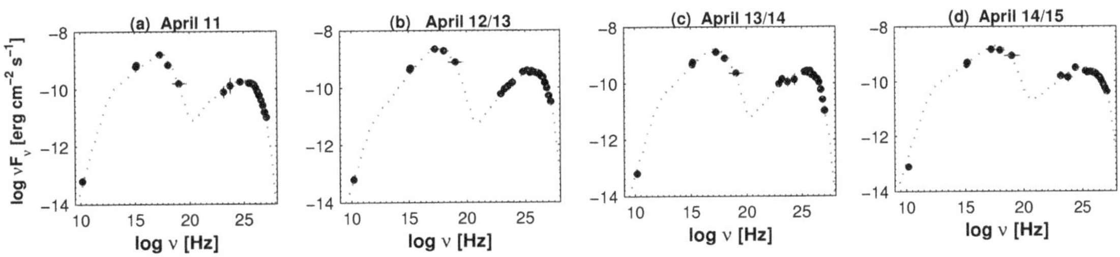

Figure 1. Broadband SEDs of the HBL source Mrk 421 where Panels (**a**–**d**) correspond to the different nights in the epoch of the exceptional X-ray flare in 2013 April. The dashed blue lines represents the one-zone SSC model. Reproduced by permission of AAS from [19].

First of all, we present a review of the γ-ray emission processes in HBLs, their advantages and limitations (Section 2). The γ-ray variability and underlying unstable processes are discussed in Section 3. Finally, we discuss the future prospects of γ-ray observations and the associated simulations for HBLs.

2. Emission Mechanisms

In leptonic models, γ-ray emission is produced by the jet leptonic content (electrons/positrons), while protons do not possess sufficient energies for the photo-pion generation process and significant proton-synchrotron radiation (or such high-energy protons could exist, but not in a high enough number to dominate the electromagnetic emission; see, e.g., [5,20]). The γ-ray emission of HBLs can be produced by the IC upscattering of low-energy photons by the "parent" electron population (*synchrotron self-Compton* model, SSC; [21] and references therein), or "seed" photons can be of external origin (so-called *external inverse-Compton* model, EIC; [7,22]). However, the entire spectrum is likely due to a combination of direct synchrotron and SSC emission in HBLs, without any significant component due to the upscattering of externally produced photons: these sources do not exhibit any significant external radiation fields from the disk, the broad line region (BLR), or the dust torus (e.g., [23]). The EIC process on photons from different parts of the HBL jet seems to be the only possibility [24]. Alternatively, hadronic or lepto-hadronic scenarios have also been considered as the gamma-ray emission mechanism to solve the difficulties with leptonic models [5].

2.1. One-Zone SSC Model

In the framework of the standard SSC scenario, the higher energy gamma-rays detected from HBLs arise from the IC upscattering of radio-to-X-ray photons by the "parent" ultra-relativistic electrons ([22] and references therein) accelerated in a jet which itself moves at relativistic speeds [25]. If external photon fields are neglected (as widely accepted for HBLs), the stationary single-zone SSC model can describe the steady MWL emission. Generally, the SSC models require low magnetic fields for HBLs (0.01–0.1 Gauss), which are significantly different from the equipartition between the magnetic and kinetic energy densities in the γ-ray emitting zone [5].

Within the *homogeneous one-zone* model, the emission zone is generally represented by a spherical "blob" containing a homogeneous magnetic field and a single lepton population. The latter may have five different functional shapes as follows (see, e.g., [20,26]): (1–2) simple and broken power laws; (3) logparabolic; (4) power law with exponential high-energy cutoff and (5) power law at low energies with a log-parabolic high-energy branch. The region moves with constant relativistic velocity $\beta_\Gamma c$ (of the bulk Lorentz factor Γ=$1/\sqrt{1-(V/c)^2}$ with V, the bulk speed) towards the observer, forming a small viewing angle θ. This leads to the boost in the recorded emission by the Doppler factor $\delta = 1/\Gamma(1-\beta cos\theta)$. The electron energy distribution (EED) cools through the synchrotron and IC mechanisms. One considers a temporary equilibrium between particle injection/acceleration, radiative cooling, and escape from a spherical emission region (see, e.g, [5,20] for the corresponding reviews), while an adiabatic expansion of the blob is considered in the one-zone expanding leptonic model discussed below. Generally, the EEDs described by the aforementioned functions can be derived by means of the Fokker–Planck equation, which incorporates the terms corresponding to particle acceleration, radiative (and, possibly, adiabatic) cooling, and particle escape (see, e.g., [27]).

Within the standard one-zone SSC models, the particle spectral index (σ) can be related the photon index (α) as $\sigma = 2\alpha - 1$, if the EED is characterized by insignificant radiation cooling [28]. If there is a strong cooling via the IC-upscattering or the synchrotron mechanism acceleration of the leptons accelerated at the relativistic shock front (see Section 2.1), followed by an escape into the emission zone situated at the shock downstream region, the time-averaged effective EED is given by $\sigma = 2\alpha - 2$ [28].

The position of the IC component peak depends on the upscattering regime. If the electron energy is below $m_e c^2$ in the center-of-momentum frame, the electrons will be non-relativistic and the upscatter is characterized by the Thomson cross-section σ_T (so-called *Thomson limit*; see, e.g., [29]). In the case the particle energy is higher than $m_e c^2$, the KN limit applies and the upscatter cross-section declines with increasing energy. Namely, the Thomson cross-section in the *head-on* approximation $\sigma_T = (8\pi/3)r_e^2$, with r_e^2 to be the classical electron radius. In the KN limit, however, $\sigma = (3/8)\sigma_T\epsilon^{-1}\ln(1+2ln2\epsilon)$, with $\epsilon = h\nu/m_e c^2$ [30]. In the laboratory frame, $\epsilon \ll \gamma^{-1}$ in the Thomson regime and the upscattered photon energy $\epsilon_s \approx \gamma^2\epsilon$, while the KN regime yields $\epsilon_s \approx (1/2)\gamma\epsilon$ with $\epsilon \gg \gamma^{-1}$. In the Thomson regime, the ratio of the SSC peak frequency to the synchrotron 'counterpart' $\nu_{ssc}^p/\nu_{syn}^p = (4/3)\gamma_p^2$, with γ_p, the EED peak [30]. In the case of the KN-upscattering ($\gamma_p h\nu_{syn}^p \gtrsim m_e c^2$) and the SSC peak frequency is given by $\nu_{ssc}^p \approx (2/\sqrt{3})(\gamma_p m_e c^2/h)$. The reduction of the cross section in the KN-regime significantly decreases the IC-upscattering efficiency [20].

In the SSC interpretation, the separation of the γ-ray emission zone from the central SMBH is not strictly constrained [31]. However, there is a constraint on the γ-ray luminosity from the pair opacity whose rate depends on the energy density of the produced radiation in the jet rest frame. This process is characterized by the maximum cross-section $\sigma = 3\sigma_T/16$ when there is a collision between the γ-rays photon of energy ϵ and the target photons with $\epsilon_t = 1/\epsilon$. The minimum energy threshold for this process is $\epsilon_0 = 0.26(E/TeV)^{-1}$ in the case of the head-on collision ($\theta = \pi$). Generally, the TeV opacity is primarily determined by the infrared photons [32]. Therefore, intense infrared fields prevent γ-rays to escape from the emission zone. Moreover, the EBL limits the redshift of HBLs (similar to the other AGN subclasses) from which their TeV-band photons can reach the Earth [32,33]. For the spherical and isotropically emitting jet area, the optical depth of the absorption process is related to the TeV-band luminosity as $\tau_{\gamma\gamma} = (3\sigma_T/16)(L(\epsilon_t)/(4\pi mc^3 R)$. Therefore, the condition $\tau_{\gamma\gamma} \lesssim 1$ requires that the TeV-band luminosity to be constrained as $(L(\epsilon_t \lesssim) \times 10^{43} M_8 R/r_g\,\mathrm{erg\,s^{-1}}$, with M_8 to be the central SMBH mass in units of $10^8 M_\odot$; r_g, the gravitational radius of the central SMBH. Frequently, the observed isotropic TeV-band luminosity of HBLs is one or two orders of magnitude higher compared to this limit (see, e.g., [32]). In combination with the short variability timescales observed for HBLs at the TeV frequencies, this result implies that this emission is strongly Doppler boosted and, correspondingly, generated in a relativistic jet having a large bulk Lorenz factor Γ; a very short timescale implies a compact emission region (see Section 3) which, in turn, can be characterized by a large pair opacity, and the TeV emission could not escape from the jet otherwise.

In turn, this process restricts the location of the γ-ray emission region detected by us [32,34]; it is possible that the synchrotron flares generated at smaller radii than the surface $\tau_{\gamma\gamma}$=1 ("the γ-sphere"; [35]) will have no TeV counterpart. For example, the innermost AGN area up to 40 gravitational radii from the central SMBH was found to be opaque for the TeV-band emission in PKS 2155–304 [32]. However, this opacity is significantly reduced if the disk represents a radiatively inefficient accretion flow (RIAF; see [32,36]).

γ-rays from HBLs may pass through the massive stellar cluster surrounding the jet and luminous stars can emerge close to our line-of-sight. Consequently, the soft radiation field of these stars can absorb the jet-emitted γ-rays. Ref. [37] showed that this process is capable of producing a broad spectral dip in the range of 50–200 GeV, and the time scale of this event was found to a few to tens of days. On the other hand, the jet-surrounding red giants may introduce some large wind-blown "bubbles" into the jet, produce a double-shock structure there. Consequently, the jet particles can be accelerated to (ultra)relativistic energies and contributing to the detected γ-ray emission (see Section 2.4).

The homogeneous one-zone leptonic models were successful in explaining the SEDs and correlated variability in different HBLs. For example,

- Ref. [38] reported strong variations in both X-ray and TeV bands from the MWL observations of Mrk 421 in 1998 April, which were highly correlated and compatible with the standard one-zone SSC model. Similar results were obtained from the MWL campaigns performed in March 2001 [39], January 2006–June 2008 [40], 2009–2012 [41,42], March 2010 [43], January–June 2013 [19], December 2015–April 2018 [44] and for the VHE flares detected with FACT during December 2012–April 2018 [45].
- Ref. [46] modeled the TeV-band variability of Mrk 501 during the MWL campaign in 1994 within the homogeneous SSC model by fitting the quiescent spectrum of the source and then changing the maximum energy of the electron injection spectrum. This produced changes only in the X-ray and TeV bands, leaving all the other bands essentially unaffected. Ref. [47] modeled the April–May 1997 outburst in Mrk 501 by means of the time-dependent SSC model: a steady X-ray emission was combined with a variable SSC component and, moreover, a pre-acceleration of electrons up to γ_{min} = 10^5 was also assumed. The follow-up MWL flare in June 1998 was also modeled by means of one-zone SSC scenario, involving a significant increase in the magnetic field strength and in the electron energy by factors of 3 and 10, respectively [48]. Ref. [49] identified individual TeV and X-ray flares and found a sub-day lag between them (consistent with one-zone SSC model) during the FACT monitoring of the source in 2012 December–2018 April. Mrk 501 showed a low activity during the MWL campaign in 2008 March–May and the one-zone SSC model adequately described the broadband SED [50]. Similarly, the 0.3–10 keV flux was correlated with the HE and VHE emissions during 2017–2020 when the source showed the lowest historical X-ray and γ-ray states [51]. The average SED of Mrk 501 constructed via the data obtained during the MWL campaign performed in March–August 2009, successfully described within the one-zone SSC model with the dominant emission region characterized by the size smaller than 0.1 pc. The total jet power constituted only a very small portion ($\sim 10^{-3}$) of the Eddington luminosity and broken power-law EED was adopted [52].
- Ref. [53] adopted the one-zone SSC model for the broadband SEDs 1ES 1959+650 from the MWL campaign performed in 2012 April–June and deduced that the physical parameters describing the emission zone during the flaring states are significantly different from those corresponding to the low states. The MWL SEDs from the time window 13–14 June 2016 were modeled with the one-zone SSC scenario, requiring relatively large Doppler factors δ = 30–60 [54].

- PKS 2155–304 showed an active γ-ray flaring phase in 1997 November with a similar behavior in X-rays, compatible with the one-zone SSC scenario [55].

The SED "standard" HBLs with steep VHE photon inices ($\Gamma_{\rm VHE} > 2$) can be modeled within the standard one-zone SSC scenario, using only a few physical parameters (redshift, radius of the emission region, magnetic field strength and Doppler factor (see [5,15]). However, to achieve a satisfactory description of UHBL sources within this model, one requires two essential ingredients: (1) even lower magnetic fields compared HBLs ($\lesssim$10 mG). This is required to (1) avoid a softening of the γ-ray spectrum by synchrotron cooling of the ultrarelativistic electrons; (2) explain a large separation between the synchrotron and SSC peaks; (2) a large minimum energy and peculiar EED dominated leptons of large Lorentz factors ($\overline{\gamma_{\rm e}} \sim 10^3$–$10^4$; [15]).

Note that the photon–photon absorption process can yield the arbitrarily hard spectra by assuming that the γ-ray emission passes through the medium containing a hot photon gas with a narrow energy distribution characterized by $E_\gamma \epsilon_0 \gg m_{\rm e}c^2$. In such a situation, the medium becomes optically thick at the lower γ-ray energies and thin at a higher one (due to the decrease in the cross-section of the $\gamma\gamma$ interaction). Consequently, the formation of the intrinsically hard γ-ray spectra can be achieved [11].

While the steady-state lepton models can be used to broadly characterize different activity states of HBLs, a time-dependent description of the electron distribution and/or of the source parameters is required in order to model fastly variable emission: flares can then be modelled through an interplay of particle injection or acceleration with particle cooling and escape, following, e.g., the Fokker–Planck equation as a function of time, or through sharp changes in the magnetic field, Doppler factor or physical extension of the emission region [56].

The main drawback of one-zone SSC models is an assumption that the highest-energy variability of the synchrotron and SSC emissions is produced by the most energetic electrons, the cooling timescale of which is significantly shorter than the light-crossing time of the emission zone. Even if any disturbance in the radiating medium instantaneously passes the emission zone, the observed emission will not contain the information about the fluctuations occurring on timescales shorter than the light-crossing time: they are smeared out owing to light-travel time delays from different source parts [57]. Moreover, representation of the synchrotron component with a single power-law electron distribution is frequently unsuccessful, and broken power-law EED with a much steeper second slope or a logparabolic model are required [5]. This situation indicates that the basic homogeneous one-zone SSC scenario is a simplification and, consequently, a rather complex representation of particle populations and the presence of different inhomogeneities in the emission region are required, in combination with the effects related to the particle injection, acceleration, escape and cooling processes [57]. Moreover, one should account for the differences between the physical conditions within and outside of the emission region (magnetic field strength, particle density etc.).

Moreover, a number of the MWL campaigns challenged the homogeneous one-zone SSC model. For example,

- In June 2004, 1ES 1959+650, underwent a strong "orphan" TeV flare by more than 4 Crab and 7 hr of doubling timescale without simultaneous X-ray event [24,58]. Similarly, strong γ-ray flares in 2009 May and 2012 May were not accompanied by those at synchrotron frequencies. In turn, no significant γ-ray activity was observed during some X-ray flares [59]. A similar behavior was also evident during 2006–2008 [60], January 2016–November 2017 [61–63]. Such events are very difficult to explain within the standard one-zone standard SSC scenarios.
- Mrk 421 underwent a very strong X-ray flare by a factor of 7 within 3 days during the MWL campaign in December 2002–January 2003, which was not accompanied by a comparable TeV-band activity [64]. During the giant flare in 2004, the TeV-band brightness reached its peak several days earlier the X-ray one that was inconsistent with the standard one-zone SSC model, and [65] suggested to be an instance of an

"orphan" TeV flare. [40,66] also found some high X-ray states, not accompanied by TeV flaring and vice versa in 2005–2008. Similar instances were reported by [42,44,67,68] from the periods February 2010–March 2013, November 2015–June 2015 and December 2015–April 2018, respectively. Moreover, there was a quadratic relation between X-ray and VHE variabilities during both the rising and decaying phases of a flare [39]. This is not expected in the KN regime [24]: the γ-ray emission is produced by the electrons having TeV and higher energies, which do not upscatter self-produced synchrotron photons since this is not possible owing to the smaller cross-section typical to the KN regime. However, such particles are capable for upscattering the lower-energy photons (produced by lower-energy electrons) in the Thomson regime. Consequently, the two peaks of the HBL SED are not produced by electrons having the same energy. Consequently, the VHE emission is expected to track the X-ray variability only linearly (instead of quadratically, as shown by Mrk 421). Particularly challenging is to observe a quadratic X-ray–TeV relation in the flare declining phase, owing to the similar energy dependence of both synchrotron and IC cooling ($\propto \gamma^2$) and again, a linear dependence is expected. A quadratic decrease can be achieved even in the Thomson regime, although extremely large beaming factors are required [24,39]. Nevertheless, Mk 421 showed even a super-quadratical X-ray–VHE relation during the fast flare on 19 March 2001 [39].

- During the exceptionally strong X-ray outburst of Mrk 501 in 2014 March–October, the 0.3–10 keV flux was generally correlated with the TeV-band emission, while there was no significant correlation between the 0.3–300 GeV and optical–UV flux variations. Moreover, several cases of the complicated X-ray and γ-ray variabilities were reported, which were inconsistent with the one-zone SSC scenario [69].
- The declining phase of the exceptional TeV flare in PKS 2155−304 exhibited a cubic relation between the VHE and X-ray flux variations, which was even more challenging for one-zone scenarios and showed an inevitable presence of two or more electron populations [24].
- Finally, the recent X-ray polarimetric observations of the nearby bright HBLs with Imaging X-ray Polarimetry Explorer (IXPE; [70–72]) clearly showed a requirement of the inhomogeneous and/or multizone emission region with shock fronts, turbulence and magnetic reconnection (see Sections 3.2–3.4 for the corresponding discussions).

Inhomogeneous SSC variability models of increasing sophistication progressively overcome the problems of one-zone scenarios. For example, an instantaneous particle injection was replaced by variations in the injection rate by [57], which propagate throughout the emission zone and produce variations in the MWL flux, and the light-travel effects on the radiation from the different area were also taken into account. The authors of [73] constructed so-called one-zone expanding leptonic model where the emission region moves through the the jet and undergo a gradual expansion. Consequently, (ultra)relativistic electrons are the subject of the adiabatic, synchrotron and IC losses. A similar scenario was adopted by [74] to model the MWL flaring behavior of Mrk 421 during December 2016–June 2017.

2.2. Multi-Zone SSC Scenarios

Inhomogeneous and *multi-zone* SSC models provide a more realistic representation of the jet zone, which can be significantly extended. However, this is frequently achieved at the cost of a less detailed characterisation of the particle energy distribution [5]. The requirement of such models emerge when the one-zone SSC scenario fails to model the observed SED satisfactorily owing to the reasons as follows: (i) the sources of the MWL emission can be distributed along the jet; (ii) the emission zone propagates along the jet axis; (iii) there might be an electron populations characterized by different acceleration and cooling timescales.

Ref. [75] presented a model based on the SSC scenario by taking into account the time delays with which one observes the variability. Therefore, the source was split into smaller one-zone models which evolve autonomously. Namely, (1) for each zone, the IC upscatter is based only on the synchrotron photons produced locally; (2) the electron energy losses are related only to these photons. This model was satisfactorily adopted for the MWL SED of Mrk 421 constructed from the MWL campaign of 1994 May, but it does not produce the EIC emission based on those synchrotron photons which come from other parts of the source at retarded times and, therefore, can not be valid for any γ-ray flare.

A significant improvement was made by [76], who calculated the SSC emission from a particular area of an inhomogeneous source as follows: one accounts for the synchrotron emission produced in the different source area and reaches the given location at retarded times. This produces more realistic SSC light curves and broadband SED, although in the case when the SSC losses (assumed to be a local process) are negligible. The basic challenge for this inhomogeneous multizone model is neglecting SSC losses due to the time-delayed photons coming from other parts of the emission zone. A further advance was made by [57], which presented such an inhomogeneous model that incorporates the effects of the non-local, time-delayed emission on the SSC losses at the given location. They assumed the presence of a relativistic EED injected in a "pipe" where electrons flow downstream and undergo a radiative cooling. One introduces variations in the injected EED which propagate downstream and are reflected in a frequency-dependent variability. Within this model, the orphan γ-ray flares can be obtained by assuming an increase in the injection of lower-energy electrons (in contrast to the EED's VHE tail).

As noted above, 1ES1959+650 underwent an orphan TeV flare in 2002 June, and [77] explained this event within the SSC scenario incorporating an inhomogenous emission zone. Namely, the primary flare (emerging in both X-ray and TeV γ-ray bands) is due to the injection of nonthermal electrons/positrons. However, when the jet is not uniform and contains different "patchy" area, X-rays produced within the primary flare undergo a scattering at the jet's dense region. This will result in strong increase in the TeV flux by IC upscatter, which can be observed as an orphan TeV flare since there should be a delay with respect to the primary flare.

During the giant 2006 flares of PKS 2155–304, the γ-ray spectra from the non-flaring nights were modeled within one-zone SSC model incorporating only small changes in the physical parameters by [78]. However, the VHE spectral and flux evolution during the flaring time windows were modelled by adopting a multi-zone SSC scenario. The latter succeeded in the interpretation of the hourly VHE variability, and a clear connection between the high activity in γ-rays and long-term increase in the lower-energy bands was deduced. Moreover, Ref. [24] explained the cubic relation between the TeV and X-ray variabilities exhibited by the source during the declining phase of this source as follows: probably, there was the appearance of a new γ-ray flaring component in this phase, which was strongly Compton-dominated and was, therefore, emitting few synchrotron emissions ($L_{\rm IC}/L_{\rm syn} \sim 10$). This component could be also very compact (of the order of several Schwarzschild radii), or dominated by external IC-upscatter of the photon coming from other jet regions.

Ref. [65] obtained a satisfactory fit of the broadband SED of Mrk 421 when introducing additional emission zones for the uncorrelated X-ray and TeV-band flaring activity in 2004. For a similar behavior observed during the TeV outburst in 2010 February, Ref. [79] adopted a two-zone scenario, where the larger zone was responsible for the stationary emission. The smaller emission zone was at the edge of this area, characterized by a transient turbulence and producing the variable emission.

Ref. [63] adopted a two-zone SSC model with different electron densities (outer and inner regions) for six different time windows from the MWL observations of 1ES 1959+650 in 2016. One assumed the presence of stronger magnetic field in the outer blob, amplified by the passing shock. The second, inner blob was characterized by a narrow EED and spectral hardening during the flaring periods, possibly owing to a stochastic acceleration process via Fermi-II process (see Section 2.3).

Mrk 421 underwent a TeV band flare a factor of ten on timescales of several hours in 2017 February, while only a moderate enhancement in the X-rays was observed [80]. The broadband SED from this event was modeled in the framework of a two-zone leptonic scenario, according to which the TeV-flare was explained by introducing a compact second blob, which contained ultrarelativistic electrons characterized by a relatively narrow range of Lorentz factors 2×10^4–6×10^5. This EED was suggested to result from stochastic acceleration in a turbulent jet medium, yielding a quasi-Maxwellian energy distribution (see Section 2.3 for details).

While two-component models frequently show a better fit with the observed SEDs of HBLs, it is problematic to constrain the free parameters due to their large number. Ref. [80] derived such constraints from the VLBI and radio polarization observations of four HBLs (PKS 1424+240, 1ES 1727+502, 1ES 1959+650, 1ES 2344+514) and selected seven epochs from the period 2013–2016 for these objects based on the TeV variability (e.g., low, intermediate and high TeV-states for 1ES 1959+650). The corresponding broadband SEDs were modelled within the two-zone SSC scenario, where the two co-spatial emission zones are situated at the VLBI core (separated by several parsecs from the central SMBH) and the constraints on jet physical factors (magnetic field strengths, Doppler factors etc.) derived from the VLBI observations were used for this purpose.

As noted above, the presence of a higher-energy SED peak above 1 TeV poses a challenge to the one-zone leptonic model. Nevertheless, VHE γ-ray spectra should steepen in the process of the electron acceleration to the ultrarelativistic energies, accompanied by declining in the energy densities of synchrotron seed photons valid to be upscattered in the Thomson regime and increasing dominance of the IC scattering in the KN regime [15]. Ref. [81] proposed a two-component model to explain the extreme high-energy peak of UHBLs: the internal component is produced by the SSC mechanism, while the external one is related to the interaction between the relativistic protons (accelerated within the jet) and the photons from the cosmic microwave background (CMB). Within the latter process, electron-positron pairs will be produced, which upscatter soft photons to γ-ray energies.

Ref. [82] presented a model where particles are accelerated at the recollimation shocks, triggered during the recollimation of the UHBL jet by the external plasma. While the EED generated at the single shock front was sufficient to reproduce the SED of relatively less extreme EHBL sources, the same was not possible for the hardest sources (e.g., in the case of 1ES 0229+200) and the existence of multiple recollimation shocks was proposed. In fact, the latest simulations of the recollimation process in weakly magnetized jets showed that the jet flow is a subject of a rapidly growing instability after the first recollimation shock. Consequently, it becomes highly turbulent and decelerates, preventing the formation of multiple shocks [83]. Based on these findings, Ref. [84] proposed a revised scenario for UHBLs: electrons are accelerated at the recollimation shock font via the Fermi-I mechanism and, subsequently, gain energy through the stochastic acceleration in the turbulent downstream medium. As a result, the observer will record the emission from the entire downstream area, where electrons are at different stages of acceleration. This scenario was applied to 1ES 0229+200 the broadband SED of which was satisfactorily described by using the reasonable values of the jet physical parameters.

Ref. [85] reported the presence of a narrow VHE spectral feature at $\sim$3 TeV for Mrk 501 obtained on 19 July 2014, detected with a significance of higher than 3σ. In order to explain the origin of this feature, a structured jet model was proposed: there could be two jet regions producing the γ-ray emission by means of the SSC mechanism. One region was characterized by an extremely narrow EED, and the second, smaller-size emitting region was additional to the first (larger) one. Two different geometries were considered: (1) these regions were co-spatial, with the second blob is embedded within the first region; (ii) the regions were not co-spatial. In the first case, the photon density within the smaller blob should be sufficiently high, while the external photon field produced by the larger region was negligible for the IC-scattering and for the electron-positron pair creation. Otherwise, the interaction of the relativistic electrons and the emitted gamma rays from the small blob with the synchrotron emission from the large region would broaden and absorb the spectral TeV feature. Within the second scenario, the smaller region should be situated closer to the observer (than the larger one) to avoid the γ-ray absorption by the intense infrared photon field. Even in this situation, a large Doppler factor is required for this blob to produce the aforementioned spectral feature due to the very narrow EED (in contrast to the second region).

A variety of multi-zone lepton scenarios is the so-called spine–sheath model, where the jet contains a quickly moving spine (dominated by the electron-positron plasma) within a less relativistic sheath (possibly, with a significant baryonic content; Ref. [86] and references therein). This scenario is primarily devoted to explain the high bulk Lorentz factors obtained from the SED modelling of HBLs, while the significantly lower values were measured via the radio interferometry. The spine-sheath model assumes a fast variability from a thin spine against slowly variable emission from the sheath. The latter provides a low-energy photon field for the external IC upscatter to the γ-ray energies inside the spine by the local ultrarelativistic leptons.

2.3. *Hadronic and Leptohadronic Processes*

Generally, leptonic models provide a relatively economical approach with respect to the free parameters and the jet energy requirements [5]. However, hadronic models are of particular interest whenever leptonic scenarios face difficulties, and there are various reasons to introduce the hadronic scenarios which are capable of contributing to the observed HBL SEDs. For example, one of the open problems is the origin of ultra high-energy cosmic rays (UHECRs) and high-energy neutrinos. HBLs represent one of the potential emitters of UHCRs; their low-power jets can provide the suitable acceleration [87]. The inclusion of hadronic components in the HBL emission origin is particularly important since it allows to estimate the possible contribution to the flux of neutrinos and UHECRs. (Lepto)hadronic models are capable for discerning potential sites of the UHECR and connect them with the expected emission of the VHE neutrinos and photons [5]. Moreover, there are some evidences from observations and modelling that relativistic blazar jets should contain a significant hadronic component (see, e.g., [88]).

In the framework of the hadronic models, the lower-energy SED component is an electron-synchrotron emission, while the relativistic hadron population contributes to the γ-ray emission of HBLs [8,20]. Both the electron and proton populations are accelerated to ultrarelativistic energies (e.g., at relativistic shock fronts; see Section 3.2), until protons exceed the $p\gamma$ (photo-pion) production threshold on the soft photon field existing in the emission zone [7,82]. There are different possible scenarios for producing higher-energy emission as follows:

- Proton-synchrotron. In the framework of the so-called *synchrotron-proton blazar* (SPB) model ([89,90] and references therein; [7]), a significant portion of the jet kinetic or magnetic power is used to accelerate protons in a strongly magnetized environment to the aforementioned threshold and various synchrotron-emitting pair cascades may develop [8,20]. For this purpose, the acceleration of protons to the energies ($E_p^{\max} \gtrsim 10^{19}$ eV and Lorentz factors $\sim 10^{10}$) is necessary for obtaining a dominant proton-synchrotron emission in the γ-ray energy range. In turn, this requires high magnetic fields of $\sim$1–100 G in order to constrain the Larmor radius smaller than the size of the emission region itself [5,7]. Alternatively, significant hadronic emission can be produced within weaker magnetic fields combined with large particle and/or photon densities [16]. In such a situation, the energy density of relativistic protons needs to largely exceed that of relativistic leptons to contribute significantly to the γ-ray domain. This can be achieved by imposing the specific requirements on the acceleration process [5]. In the case of the aforementioned magnetic field values and Doppler factor δ = 10–50 in the HBL jets, the proton–synchrotron peak frequency is expected in the range of 10–100 GeV [7]. The number of free parameters of the proton–synchrotron scenario is significantly larger than the SSC one (amounting to 14; see, e.g., [7]).

- Modified proton-synchrotron. In the later versions of SPB model, the synchrotron radiation of secondary muons and mesons was also taken into account [91] and references therein; [20]). First of all, one expects a photo-pion production process ($p\pi$) where a photohadronic $p+\gamma$ interaction yields either π^0 or $\pi^\pm$ mesons. For this purpose, the photon energy in the proton frame should be higher than about 145 MeV [7]. Gamma-ray photons can be obtained from the π^0-decay process ("π^0-cascade"), or produced by electrons from the $\pi^\pm \to \mu^\pm \to e^\pm$ decay ("$\pi^\pm$-cascade"). One expects also the proton-synchrotron emission ("p-synchrotron cascade"), as well as the μ-, π-and K-synchrotron photons ("$\mu^\pm$-synchrotron cascad"; [7]). Refs. [16,91] demonstrated that the π^0 and $\pi^\pm$ cascades initiated by ultra-high energy protons generate featureless γ-ray spectra, in contrast to p-synchrotron and $\mu^\pm$-synchrotron processes: the latter produce a two-component γ-ray spectrum, i.e., the muon synchrotron radiation emerges as a third SED component, at higher energies than the synchrotron radiation by the parent protons [7]. Generally, direct proton and $\mu^\pm$ synchrotron radiations are thought to be the main contributors to the higher-energy SED "hump", while the low-energy component is synchrotron radiation from the primary electrons, along with some contribution from the secondary electrons generated by the aforementioned cascades [8,20]. Electrons and positrons produced in the decay of charged pions have extremely high-energy, and their synchrotron radiation can reach even PeV energies [7]. Generally, the jet emission region is "opaque" for first generations of secondary particles and γ-rays, leading to successive reiterations of the above-described cascades [16]. The decay of neutral pions can produce ultra-high-energy (UHE, $E > 100\,\text{TeV}$) γ-rays or so-called PeV-photons [92]. However, these photons do not reach us, being absorbed via pair-production both in the jet, or during the propagation in the (inter)galactic medium [7]. Photo-meson production is characterized by a key property: neutrinos are produced along with photons, escaping the emission zone without any absorption or energy losses and their detection directly indicates the presence of highly-relativistic protons in the jet, as well as is capable for constraining the model key parameters [7]. The proton-proton interactions are thought to be negligible in the SPB models, since this mechanism requires very high particle density and the extreme jet powers for producing a significant γ-ray emission [5].
- Bethe–Heitler pair production. A photohadronic interaction between relativistic protons and photons may also result in the Bethe–Heitler pair production as $p+\gamma \to e^\pm$ ([93] and references therein). This process is in competition with the photo-meson production, although it needs significantly lower lower energies: the threshold for the Bethe–Heitler pair production is lower than the photo-meson one by a factor 0.004 [7]. Consequently, the generated pair produces a lower-energy emission compared to the photo-meson cascades. Namely, the simulations of [94] showed the appearance another higher-energy SED component due to this pair production in the energy range 40 keV–40 MeV (so-called three-hump SED). Although the corresponding peak luminosity can not be always comparable to that emitted above 40 MeV, this keV–MeV SED component may still be observable (not being hidden from other components). Therefore, observation of the three-hump hump SED may indicate a viability of the leptohadronic scenarios.

Another key characteristic of photo-meson interactions is the creation of neutrons, which escape the emitting region without interacting with magnetic fields and Bethe–Heitler pair production [7]. These neutrons can transfer a significant amount of energy at large distances downstream from the jet and decay into protons, radiating synchrotron photons in the presence of magnetic fields. Consequently, the existence of two separate, causally connected hadonic emission zones with a significant separation is possible [7,95].

The spectra of HBLs are relatively well reproduced by proton-synchrotron-dominated SPB models where the intrinsic primary synchrotron photon energy density is small, consistent with the low bolometric luminosity of those objects [5]. As the synchrotron photon energy density increases, relativistic protons undergo increasing energy losses from the $p\gamma$ pion production process. Consequently, the contributions from the $\pi^{\pm}$ and $\mu^{\pm}$ cascades become progressively dominant at higher energies. On the other hand, this process yields a decrease in the peak energy of the γ-ray component [96].

Generally, the hadronic processes are relatively inefficient in point of the produced energy (compared to the leptonic scenarios) and, moreover, require the extreme, super-Eddington jet powers $P_{\rm jet} \sim 100 L_{\rm Edd}$ (so-called energy crisis), where L_{Edd} is the Eddington luminosity [5,97,98]. Consequently, the hybrid *lepto-hadronic* models provide a more reasonable physical approach [8,16,20,82]. Moreover, this energy-crisis hadronic scenario is more inherent to flat-spectrum radio-quasars (FSRQs), while it is less prominent for low-luminosity HBLs and hadronic solutions with $L_{\rm jet} < L_{\rm Edd}$ can be achieved [7]. Nevertheless, the detection of the muon neutrino with the most probable energy of ∼290 TeV from the IBL source TXS 0506+056 (or, more plausibly, a FSRQ object; see [99,100]) revived and deepened interest in these models. Note that [101] modeled the broadband SED of TXS 0506+056 from the neutrino detection epoch by means of a leptohadronic scenario, in which the Bethe–Heitler and pion-decay processes produce the X-rays and VHE γ-ray emissions. The observed neutrino flux was used for the model constraining. Ref. [102] investigated a connection between HBLs, Ice-Cube neutrinos, and UHECRs and found a probability ∼0.18% (2.9σ) after compensation for all the considered trials. Moreover, they deduced that HBLs can account only for ∼10% of the UHECR detections.

Ref. [103] explained a lack of the γ-ray activity along with the X-ray ones in some HBLs by production of the TeV–PeV neutrinos in the case that the X-ray flares are powered by the proton-synchrotron mechanism: neutrinos are expected from the photo-meson interactions of ultrarelativistic protons with their own synchrotron radiation, while the MeV-to-GeV γ-rays emission is produced within the synchrotron-dominated electromagnetic cascades.

Ref. [104] modeled the SEDs of those HBLs thought to be counterparts of the IceCube-detected neutrinos (Mrk 421, PG 1553+113, 1ES 1011+496, H 2359-309, 1RXS J054357.3-553206 and 1H 1914-194), adopting an one-zone leptohadronic model. It was concluded that the model fits with these SEDs by using the reasonable values of the jet physical parameters (e.g., B = 0.05–5 G, δ = 18–31, $\gamma_{\rm e,max} = 8 \times 10^4$–$2 \times 10^6$). In the case of Mrk 421 and 1H 1914−194, a good agreement between the model-predicted and the detected neutrino fluxes (from the events with IDs 9 and 22, respectively) was found. Note also that [105] reported the *AGILE* detection of a candidate γ-ray precursor to the ICECUBE-160731 neutrino event, which was identified with the X-ray source 1RXS J141658.0−001449. Based on the X-ray-to-radio flux ratio, the object was concluded to have properties typical to the HBL sources. However, no further identification of 1RXS J141658.0−001449 was performed (e.g., detection of the featureless optical spectrum). The aforementioned model was adopted for the multi-epoch modeling of TXS 0506+056, including the time windows corresponding to the neutrino detection instances from this object [106]. For the same purpose, [107] developed the `SOPRANO` code which included all hadronic processes yielding high-energy neutrinos. This code (along with the `LeHa` code developed by [16]) was adopted by [51] to model the broadband SEDs of Mrk 501 corresponding to the different time windows of the period 2017–2020 and evaluate the expected neutrino flux from this object.

In the leptohadronic scenarios, the proton–synchrotron component is relatively suppressed via imposing the magnetic field to be not higher than 1 Gauss and the SSC emission dominates in the γ-ray output (in the case of HBLs). The simplest case is the one-zone leptohadronic model: all radiation mechanisms are operating in the same emission zone and external photon fields are negligible [7]. Here, relativistic protons produce secondary leptons via the p-γ interactions over the electron synchrotron photon field. The emission from such leptons can contribute to hard X-rays (as Bethe–Heitler component) and in the TeV band (as photo-meson component; see [104]). Such a model was presented by [108] along with a new extended hadroleptonic code `ExHaLe-jet`, which considers simultaneously the processes related to relativistic protons and electrons. Within a predefined geometry and bulk flow physical parameters, the particle evolution was simulated. Highly relativistic secondary electrons (and positrons) are created through the $\gamma\gamma$ and Bethe–Heitler pair productions, as well as during the pion/muon decay. The ratio of protons to these secondaries was assumed to decrease with distance from the jet base. For particle–photon interactions, all internal and many external photon fields were considered. Note that the external fields were concluded be the more important source for particle–photon interactions leading to the neutrino production. Note that this result is related to the fact that lepto-hadronic solutions also face energetic issues, especially when one tries to maximize the neutrino output and, consequently, the required jet power can quickly become very high [7].

Ref. [109] adopted an one-zone leptohadronic model for the X-ray and γ-ray flares shown by Mrk 421 in 2001 March. First, they performed a preflare SED modeling, using the different leptohadronic scenarios. Afterwards, by introducing small-amplitude variations in the injection rate and in the maximum particle energy, the flaring state SEDs were reproduced. Note that the models incorporating the pion-decay processes successfully reproduced the observed quadratic relation between X-ray and TeV variabilities. Ref. [51] adopted a leptohadronic model for Mrk 501, in which the high-energySED component represents a combination of both leptonic (IC) and hadronic (emission by cascades triggered by hadronic interactions) processes. It was assumed that the bulk of the high-energy SED component is generated by the SSC mechanism, while the hadronic output are subdominant and can emerge (and even dominate the SED) in hard-X-rays, filling in the SED dip, and in the VHE band. In the presented framework, the proton-synchrotron emission is very suppressed owing to the lower magnetization of the emitting region compared to that required otherwise. A similar combination was adopted by [110] to reproduce the broadband SED of the HBL source Mrk 180.

Ref. [111] proposed so-called *hadronic synchrotron mirror* model for explaining the orphan TeV flare of 1ES 1959+65 in 2002 June: X-rays produced in the process of the primary γ-ray flare were reflected by a plasma cloud (situated nearly in the direction of the jet propagation), then collided with the jet protons and, consequently, the pion-poduction cascade was developed which yielded the observed orphan TeV flare. However, this model requires very high proton density in the jet and hadronic jet power. Nevertheless, the model can be physically reasonable, if one takes into account the effects related to the emission zone approaching to the mirror [77]. This scenario was adopted by [53] for the MWL observations of 1ES 1959+65 in 2012 April–June. A leptohadronic model was adopted by [54] for the highest TeV states of 1ES 1959+650 recorded in 13–14 June 2016, although requiring extreme magnetic field ($B \sim 100$ G) and very high values of the jet power ($\sim 10^{46}$ erg s^{-1}).

As noted above, one-zone SSC models are problematic for UHBLs, since they require large Doppler factors as well as extremely high minimum Lorentz factors for the EED. However, the UHBL SEDs were modeled by [16] within a leptohadronic framework, without adopting the extreme Doppler factors the aforementioned extreme minimum Lorentz factors (adopting $\delta = 30$ and $\gamma \sim 10^{2-3}$). In the case of the significant proton-synchrotron radiation, magnetic fields $B{\sim}1$–100 G and maximum proton energies $E_p^{\rm max} \lesssim 10^{19}$ eV were derived. In the case of the synchrotron emission from the $p\gamma$-induced cascades, the range $B \sim 0.1$–1 G was required. Moreover, the deduced jet powers were mostly sub-Eddington, in contrast to previous hadronic modelings (see the corresponding discussion above). A caveat of the [16] model is the very hard spectra of injected particles, required for the co-acceleration of leptons and protons. The hard TeV of 1ES 0229+200 was explained by [112] in the framework of one-zone hadronic model where γ-ray emission is produced via the neutral pion (from proton-proton interaction) decay, but at the cost of adopting a small radius of the radiation zone than the Schwarzschild radius of the central SMBH.

When both leptonic and hadronic radiative models provide similarly good fits to the observed broadband SED, they can be distinguished from the HBL's temporal behavior. While the one-zone SSC model is characterised by correlated variations in both the synchrotron and higher energy ranges, time-dependent hadronic models require the solution of a system of coupled differential equations of the kind $\partial N_X(t,E)/\partial t = Q_X(t,E)$-$L_X(t,E)$ for each particle species X (protons, photons, neutrinos, leptons), with $Q_X(t,E)$ and $L_X(t,E)$ to be the injection and loss terms, respectively [7]. Due to the complexity, the time-dependent hadronic modeling have been used relatively rarely (e.g., [113–116]. For example, Ref. [117] modeled the broadband SED and MWL lightcurves of the HBL source 1ES 1011+496 using a hybrid leptohadronic model, taking all relevant processes into account (acceleration and synchrotron emission of both electrons and protons, IC scattering, photo-hadronic interactions and $\gamma\gamma$ -pair production). This model yielded a more satisfactory representation of the target's VHE flare compared to the pure leptonic modeling. Although the two SED components are produced by two distinct particle populations within the proton–synchrotron models, the observed correlation between the synchrotron and γ-ray variabilities can be achieved by assuming that electrons and protons are energized by the same acceleration mechanism (as done by [109] for Mrk 421; see above). Otherwise, hadronic and hybrid models as useful to reproduce the absent MWL correlation, which is a challenge for the leptonic one-zone scenarios (see the discussion related to the "orphan" flares). For example, ref. [118] proposed that the low energy tail of the SSC photons (1–8 MeV) of 1ES 1959+650 served as the target for the Fermi-accelerated high energy protons of energy $\lesssim$100 TeV, producing the TeV photons through the decay of neutral pions from the Δ-resonance during the orphan TeV flare in 2002 June. Later, this model was adopted for modeling the GeV–TeV flaring episodes of 1ES 0229+200, 1ES 0347−121, 1ES 0806+524, Mrk 501 and HESS 1943+213 [119,120]. This model was expanded into the two-zone photohadronic interpretation, adopting different emission zones below and above the threshold energy of 1 TeV for 1ES 1959+650 (for the time window 2016 November 19–21; [121]), Mrk 501 (2005 May–July and 2012 June; [122]), Mrk 421 (2010 March; [123]) and 1ES 2344+514 (several γ-ray episodes; [124]).

On the other hand, it is generally problematic for the leptohadronic models to deal with a very rapid TeV-band variability of HBLs: the radiative cooling time scales of protons is of the order of several days even in the case of the magnetic fields of $\sim$10 Gauss and typical Doppler factors δ = 10 (adopted for HBLs; see, e.g., [7]). However, the rapid γ-ray variabilities, observed on timescales shorter than the proton cooling time, can be attributed to the geometrical effects (see, e.g., [16,90] and Section 3.3 for the jet-in-jet scenario).

2.4. Magnetospheric Vacuum Gaps, Curvature Emission and EIC Scattering

SMBHs are widely accepted to be the central engines of AGNs (including HBLs), where particles should be accelerated by extracting their rotational energy via the Blandford-Znajek (BZ) process [3]. Moreover, the SMBH magnetospheres could be the sites for the origin of strong and fast VHE flares, depending on the importance of the $\gamma\gamma$ absorption. Note that this effect can be weaker in HBLs which are widely accepted to possess subluminous accretion disks operating in the RIAF regime and sufficiently low magnetic fields (see above).

The BZ-mechanism operates in the force-free magnetosphere containing the high-energy plasma. The latter is continuously removed from there during the jet collimation process and one expects the appearance of an area with a charge deficit, so-called vacuum gaps (or spark gaps). In these gaps, the charged particles can be accelerated up to ultrarelativistic energies along the open magnetic field lines[125]. Vacuum gaps can appear at the jet base, at a few gravitational radii from the horizon and produce the electron-positron plasmoids which are capable of triggering a fastly variable VHE γ-ray emission [5]. Moreover, charged particles can be generated in these gaps via the pair-production cascades [125]. Consequently, vacuum gaps allow us to draw conclusion about the physical conditions in the vicinity of the SMBH horizon.

The TeV-band variability could be enhanced by sharp changes in the physical conditions (e.g., the local accretion rate, abrupt changes in the disc emission, magnetospheric currents) throughout the magnetosphere [126]. Curvature emission is thought to be one of the possible mechanisms producing a fast TeV-band flare [127]. Namely, when electron-positron pairs are created by means of the $\gamma\gamma$-interactions, these particles will be accelerated towards the opposite directions along the field lines (electrons and positrons are accelerated outwards and inwards, respectively) and produce γ-rays via curvature emission, as well as by external IC upscatter of soft photons coming from the inner accretion disc [127].

Note that photons emitted by the innermost disc parts can enter the SMBH magnetosphere and undergo the $\gamma\gamma$ annihilation and supply the vacuum gap with these particles [128,129]. Electron-positron pairs will be accelerated fastly in the gap owing to the large potential drop produced by the rotating SMBH, until reaching those values of the Lorentz factor for which the energy gain is balanced by curvature radiation or EIC losses [129]. The amount of the gap-born TeV-band emission (and, hence, the γ-ray luminosity of the vacuum gap) depends on the gap size along the magnetic field lines, and increases with the declining accretion rate. It was found that the gap width is not smaller than $0.01r_s$ in the case that the density of "seed" charges is below the Goldreich–Julian (GJ) value [127]. When the accretion rate becomes $\dot{m} < 10^{-4}$ ($\dot{m} = \dot{M}/\dot{M}_{\rm Edd}$, with $\dot{M}$, the accretion rate; $\dot{M}_{\rm Edd}$, the Eddington accretion rate), the SMBH magnetosphere becomes charge-starved and a vacuum gap is switched on [130]. In the case of the extremely rotating SMBH, the gap is though to produce a large VHE γ-ray emission and the TeV-band spectrum can extends to higher frequencies with the increasing flux [129]. In the local frame, the peak energy of the curvature spectrum should be limited to $\sim 50(\dot{m})^{3/8}$ TeV, which is below 1 TeV for $\dot{m} \lesssim 10^{-4}$ [129].

However, the TeV luminosity of vacuum gaps is limited by the pair-production process during the interaction of TeV photons with the ambient radiation field. If this occurs outside the gap, the created pairs move away from the gap as the secondary pairs and emit secondary photons via the IC and the synchrotron processes [130]. In the case that these secondary photons materialize within the magnetosphere, the so-called tertiary pairs are produced and, in turn, emit the tertiary photons via IC and synchrotron processes, and so on. The multiplicity of this process was found to depend on the accretion rate [129]. Ref. [130] suggested that such a cascade can propagate up to $60r_g$. First of all, this process is initiated by those TeV photons which have energies higher than 10 TeV (up to $\sim 10^3$ TeV), for which the $\gamma\gamma$-optical depth is much larger than for the lower-energy photons: the pair-production opacity drops with decreasing γ-ray energy and becomes sufficiently small below 10 TeV that allows the photons at this energy range to escape the gap and, eventually, the magnetosphere [129].

The observed γ-ray spectrum depends on the spectrum of soft (scattered) photons and the pair cascade process. Moreover, an increase in the curvature radius of the gap magnetic field lines will boost the maximum energy of accelerated electron-positron pairs. In turn, this will lead to the broadening of the TeV emission spectrum [129]. As noted above, the spark process in the gap can be highly intermittent: the TeV-band luminosity and the variability amplitude depend on the pair-creation opacity (which, in turn, is sensitive to the soft radiation produced by the innermost disc regions). Within this mechanism, a fast TeV variability is expected even in the case of the moderate changes in accretion rate: this will trigger nonlinear fluctuations of the gap potential, induce intermittencies in the pair-production opacity and the strength/geometry of the magnetic field advected by the accretion flow, changing also the co-aligned electric field nonlinearly [129]. The simulations showed that such changes can produce the delayed TeV flares, mainly contributed by the curvature emission. The flare rise duration should be of the order of the light-crossing time of the gap, although the exact shape of the TeV-band light curve depends on the gap's separation from the event horizon: one expects that a strong lensing will significantly affect the observed light curve [127].

As noted above, two mechanisms are thought to be responsible for the γ-ray emission by the electron-positron pairs accelerated in the gap: the EIC scattering and the curvature radiation [127]. The latter represents a synchrotron variant for charged particle moving along curved magnetic field line: the produced radiation is related to the field line curvature, not to the gyro-acceleration. However, there are some distinctions between these mechanisms: (1) for the curvature emission from a single (ultra)relativistic particle, the total emitted power and characteristic frequency $P_{\rm curv} \sim \gamma^4$ and $\nu^c_{\rm curv} \sim \gamma^3$ (versus $P_{\rm syn} \sim \gamma^2$ and $\nu^c_{\rm syn} \sim \gamma^2$ for the synchrotron emission) [127].

An alternative hypothesis to explain the narrow spectral feature at 3 TeV in Mrk 501 (reported by [54]) is based on the gap emission from the electrons accelerated to energies of about 3 TeV in a sporadically active magnetospheric vacuum gap close to the central SMBH [54]: there could be electromagnetic cascades triggered by the interaction of relativistic electrons/positrons with emission line photons coming from the photoionized gas clouds. Even though Mrk 501 is an HBL source and should not have a significant BLR, [54] speculated the possibility that gas clouds from the inner parts of the host galaxy intruded into the AGN. Along with the EIC upscatter, there could be the cascades from the Breit-Wheeler (BWPP; [131]) pair-production process incorporating collisions between the local high-energy and low-energy ambient photons and creating electrons which are capable for the IC-scattering (see [54] for details). Consequently, the corresponding VHE emission manages to escape outward and a narrow TeV-band component was formed, superimposed on the SSC emission from the distinct (larger) jet emission zone.

As discussed in Section 2.2, the uncorrelated VHE flare of Mrk 421 observed in 2017 February was explained within two-zone SSC scenario: there was a compact second blob containing highly energetic electrons which were characterized by a narrow range of Lorentz factors. In turn, this population was suggested to result from an electromagnetic cascade initiated by electrons accelerated in the magnetospheric vacuum gap of Mrk 421 [74].

3. Variability Mechanisms

Because of the very small angular size, it is not possible to directly resolve the emission zone. Therefore, information about the spatial structure of this jet region can be obtained through the MWL variability studies. Particularly important is the variable emission produced by the EED's highest-energy part (X-rays and γ-rays) since these electrons cool very quickly and can exist only close to the site they were produced.

The intense MWL observations of HBLs revealed a complex structure in the γ-ray variability. For example, Mrk 421 underwent two dramatic outbursts of the TeV emission in 1996 May: the first flare showed a flux-doubling time of $\sim$1 h and, eventually, the TeV flux increased by more than a factor of 50, making Mrk 421 the brightest TeV source in the sky. During the second outburst, the brightness boosted by a factor of 20–25 in about 30 min [132].

On some occasions, the TeV variability of HBLs was extremely rapid. For example, Mrk 501 showed flux-doubling time as short as 2 min during the strong flaring activity recorded in 2005 May–July, along with the longer-term VHE variability by an order of magnitude during the entire campaign [18]. A similar range was observed for PKS 2155-304 in 2006 July, when the well-resolved flares on timescales of $\sim$200 s were detected (see [17] and Figure 2). These instances implied highly relativistic sub-parsec scale flows ($\delta \sim$ 50–100; [133]), and the emitting region (constrained by means of the causality relation $R < ct_{\rm var}\delta/(1+z)$; see, e.g., [54]), was comparable or even smaller than the size of the central SMBH horizon (even for high jet bulk Lorentz factors). These observation signatures demonstrate that the HBL jets can be structured on very small spatial scales that are unresolved by the current γ-ray instruments.

The power spectral density (PSD) represents an important tool for characterizing the nature of flux variability: it provides a measure for the contribution of different timescales to the variability power by quantifying the amount of variability power as a function of temporal frequency ($\nu \sim 1/t$; [41,134]). Similar to other AGN subclasses, one of the major observational characteristics of γ-ray variability in HBLs is their power-law type behavior. Namely, $PSD(\nu) \sim \nu^{-\beta}$, with $\beta \geqslant 0$ associated with a random-walk behavior in the time domain (i.e., nonperiodical variability; [134]) as follows:

- white noise (β = 0);

- pink or flicker noise (β = 1);
- red or Brownian noise (β = 2).

In the case that the source is showing a broken power-law PSD with the break frequency ν_b, then the flux variability is characterized by the characteristic timescale $t_{char} \sim 1/\nu_b$ which also can be a variability period ([41,135]; see also Section 3.1). For example, PKS 2155-304 exhibited a red-noise behavior with $\beta \sim 2$ during the aforementioned exceptional VHE flare [136]. The long-term VHE and HE observations of the source (the H.E.S.S. and *Fermi*-LAT data, respectively) demonstrated a a flicker noise with $\beta \sim 1$ [134]. Mrk 421 showed a pink-noise VHE behavior during the intense MAGIC observations during 2009 January–June [41], etc.

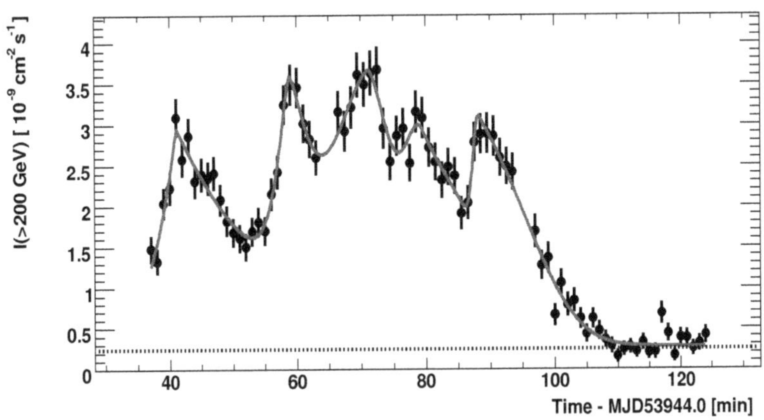

Figure 2. The VHE variability of PKS 2155−304 shown at the energies higher than 200 GeV shown on 28 July 2006. The horizontal line represents the Crab emission in the same energy range. Reproduced by permission of AAS from [17].

3.1. Variability Models and Quasiperiodic Flux Changes in HBLs

To date, several models have been proposed to explain the γ-ray variability of HBLs:

- *Shock-in-jet* scenario (e.g., [21,76]);
- *Jets-in-jet* model and *relativistic magnetic reconnection* [133,137,138];
- *Jet turbulence* [26,139,140];
- Instabilities in the magnetospheric gaps (see Section 2.4);
- Jet precession ([141–143]).

In the latter case, the system consists of a primary SMBH (with the associated accretion disk and jet nearly pointed to the observer), and a smaller-mass, secondary BH orbiting the primary one (see, e.g., [143]). In such a situation, a quasiperiodic flux variability may emerge due to the periodic change of the jet orientation towards the observer. Such a change can be caused by the two different effects [141]:

- The dominant effect (causing the jet angle to vary by the greatest amount) is simply an imprint of the SMBH orbital velocity on the jet: since the jet-carrying primary SMBH is moving along a circular orbit with the velocity V, the highly-relativistic ejected material is expected to have the same velocity component in the observer's rest frame. Consequently, the jet will precess with respect to the distant observer and the γ-ray emitting the region is observed at an angle θ_{obs} oscillating with an amplitude $\theta_{obs} = [2q/(1+q)][GM/Rc^2]^{1/2}$ and period $T = 2\pi(R^3/GM)^{1/2}$, with M—the system's total mass; q, the primary-to-secondary mass ratio; R, the separation between the components, assumed to be larger than a few Schwarzschild radii. Consequently, the jet's instantaneous shape will be helical, where radius of the coils increases linearly with the distance from the primary SMBH and one should observe a quasiperiodic flux variability.

- The second-order effects can be caused by the general-relativistic deflection and Lense–Thirring precession. Namely, a Lense–Thirring precession ([144] and references therein) of the primary SMBHs acretion disc can be triggered by the gravitational field of the secondary SMBH. Consequently, the primary's jet is also expected to precess with the same period. However, the corresponding oscillation angle (and, hence, the amplitude of the periodic flux variability) will be significantly smaller than that caused by orbital movement of the primary SMBH, if these two SMBHs are separated by more than a few Schwarzschild radii. The general-relativistic effect causes a deflection of the relativistic ejecta's trajectory by the gravitational field of the secondary SMBH. Note that the general-relativistic effects are expected to be negligible on the few-years timescale [141].

Note that these variability mechanisms differ from the others in that they produce long-term (quasi)deterministic periodicity and trends, rather than stochastic variability (as discussed below).

Currently, the light curves compiled from the continuous*Fermi*-LAT observations form the most suitable database to search for the quasiperiodic variability in the HE γ-ray band. Among HBLs, PG 1553+113 is the primary candidate among HBLs for hosting a binary SMBH system. Initially, a HE periodicity with $\approx$2.2 yr was reported by [145] by using the Lomb-Scargle Periodogram (LSP) and continuous wavelet transform (CWT) for the *Fermi*-LAT observations of the source. However, the LSP peak is below the 3σ significance and the periodicity detection can not considered as highly-credible (see, e.g., [146,147]). A similar result was reported by [142] from 9-yr *Fermi*-LAT observations of the source, although no detection significance was evaluated. Ref. [146] adopted the PDS method and concluded that the constructed PDS for PG 1553+113 was compatible to noise, i.e., non-periodical variability was singled out with a significance level higher than 95%. Ref. [148] adopted different periodicity searching techniques (LSP, REDFIT, CWT etc.) for the LAT 2008–2017 observations of the source and deduced a period of 2.2 yr by the average significance higher than 4σ. Moreover, the use of these traditional Fourier-like methods for the periodicity search (as done by [148]) was called into question by [149], which adopted the Gaussian process methods—CARMA and Celerite—for the same purpose. For PG 1553+113, possible evidence for the period of $\sim$800 days was found, with local significance of $\gtrsim$95%. However, the global significance was only 50–90% when constructing the LAT-band light curves with different time bins. A further improvement was done by [150] which adopted a Gaussian process modeling (along with the LSP technique) and various tests to conclude a quasiperiodic variability in the optical R- and LAT 0.1–200 GeV bands. The obtained $\sim$2.2 yr period was confirmed also by [151], which performed the analysis of the LAT data by adopting different methods for this purpose. Ref. [143] hypothesized that PG 1553+113 should posses a relativistic jet which rotates with a constant angular velocity around some axis, owing to the central binary SMBH (similar explanations were also presented by authors claiming the periodicity detection for the source). Consequently, the jet Doppler factor (and, hence, the observed brightness) undergoes a periodical variability. Based on this scenario, a light curve with 2.2-yr periodical oscillations was generated and compared to the 0.1–300 GeV light curve constructed from the 13.6-yr LAT observations of PG 1553+113. However, no attempt was made to evaluate the periodicity significance. Finally, Ref. [152] concluded that that periodicity detection significance becomes less than 1σ after the trial correction.

Ref. [148] reported the period of 1.7 yr with a significance $> 3\sigma$ for PKS 2155-304, presented within the previous studies (see, e.g., [153]) but concluded to be non-periodic γ-ray variability by [146]. The later analysis of the optical R- and LAT-band data by [150] did not yield a firm detection of the quasiperiodic variability. Later, such a variability with a period of $\sim$1.7 yr was reported by [151] with a significance of 2.5σ–5σ by different periodicity searching technique from the LAT observations of PKS 2155-304 performed during 2008–2020. However, the period of $\sim$3.4 yr was also obtained from the same data train by means of the phase dispersion minimization (PDM, [154]) method. The authors of [152] did not find a significant periodicity after performing a trial correction.

Moreover, Refs. [155,156] reported quasi-periodicity detections for Mrk 421 and Mrk 501. In addition to the issues related to the analysis of the *Fermi*-LAT data (e.g., the energy range of 0.1–300 GeV instead of 0.3–300 GeV generally adopted for HBLs; see [157]), the reported periods (285 and 330 days, respectively) are detected below the 3σ significance and/or show some changes with time. Consequently, such detections are not robust (according to the criteria of [147]). No significant periodicity was found with a trial correction made by [152].

3.2. Relativistic Shocks and Fermi-I Process

Relativistic shocks are naturally expected in such supersonic outflows as the HBL jets. Shock fronts represent efficient sites for dissipating the bulk kinetic energy for accelerating leptons and hadrons up to the ultrarelativistic energies and produce a flux variability on various timescales [10,158]. A variety of shocks are expected: reconfinement shocks, stationary or moving shocks along the jet [5]. The simulations of [56] demonstrated that mildly-relativistic shocks in weakly magnetized jet flows produce relativistic particle acceleration. These events may result from the intermittent changes in the physical conditions in the innermost AGN area, which can saturate the jet with extremely energetic plasma having a significantly higher pressure than the steady-state jet flow and a forward shock front is formed [76].

Gamma-ray flares of HBLs sometimes show a long-term increase of the flux (weeks to a few months, expected by the shock propagation through the jet), superimposed by shorter-timescale variations (lasting several days to a few weeks; see, e.g., [19,42] for Mrk 421). Such rapid variations could be related to the excitation of the recollimation nozzle by the external perturbation (so-called recollimation shock caused by an external medium; [82]), or by the interaction between the moving shock front and local, jet-inherent inhomogeneities [76]. In order to explain a flaring behavior of Mrk 501, Ref. [34] considered the shock structures as follows: (1) a double shock system with forward and reverse shocks, and (2) a single shock along with a rarefaction wave. The presented model predicts correlated multi-band variability with some cross-band time lags and spectral hysteresis patterns. The average SED of Mrk 501 from the period 2009 March–August, well described by [52] within the standard one-zone SSC model, in which the bulk of the energy was generated within a single emission zone associated with the relativistic, proton-mediated shocks. The multi-zone SSC model of [57] (see Section 2.2) also incorporated a shock acceleration of electrons by a standing or propagating shock in a collimated jet. By assuming that the radio-to-X-ray flux from HBLs is synchrotron radiation of isotropically distributed electrons in the randomly oriented jet magnetic field, Ref. [159] obtained the underlying EED and adopted it to construct the SSC SED as a function of the Doppler factor, magnetic field strength and variability timescale. This method was adopted to model the VHE spectra of PKS 2155-304 and Mrk 421 during the giant outburst on 28–30 July 2006 and during the 2001 March flare, respectively. Temporal variability was assumed to result from Fermi-I mechanism, adiabatic expansion, and radiative cooling.

In HBL jets, one could expect the simultaneous existence of the perturbations having different origin and producing the erratic behavior exhibited by the observed γ-ray light curves of HBLs. Shock interaction with the turbulent jet medium (characterized by the enhanced density and magnetic field) can generate rapid fluctuations with the observed timescale $\Delta t_{\rm obs} = (l_{\rm inh}/V_{\rm sh})(1+z)/\delta$, with $l_{\rm inh}$ to be the inhomogeneity length; $V_{\rm sh}$, the shock speed [56,76]. The observed peak flux and variability amplitude can be strongly enhanced by relativistic effects if the angle between the jet and our line-of-sight is small [76].

The collision between the fast and slower shock fronts (or high-energy plasma blobs) can trigger a system of forward and reverse shocks, which confine two subsequent emission zones and yield complex flare profiles, e.g., double-peaked flares [76,160]. According to [82], the interaction of jet matter with an obstacle can also trigger a double-shock structure depending on the relative momentum fluxes carried by the jet and the obstacle, respectively.

The dominant particle acceleration mechanisms at mildly and non-relativistic shocks are *diffusive shock acceleration* (DSA; [161]) and *shock drift acceleration* (SDA, [162]), which are collectively referred to as *first-order Fermi acceleration* (hereinafter, Fermi-I mechanism; [56]). In DSA, the energization of charged particles is owing to the repeated shock crossings when they interact quasi-elastically with self-generated small length-scale magnetic fluctuations, which are anchored in the converging upstream and downstream plasmas and producing a magnetohydrodynamical (MHD) turbulence [56,163]. That is, energetic particles are confined in the vicinity of the shock front by their scattering on the magnetic turbulence which, in turn, is amplified by these particles; diffusively transported back and forth across the shock, each time achieving an average energy gain $\overline{\Delta E} \sim (\Delta u/c)E \propto [(r-1)/r](u_{\rm sh}/c)$ each cycle, where Δu is the relative velocity between the shock upstream and downstream medium, $u_{\rm sh}$—the shock speed in the frame of the upstream medium, and r—the shock compression ratio, i.e., downstream-to-upstream fluid density ratio (of the order of 4 for strong very supersonic shocks [10]). However, the Fermi-I mechanism will not be efficient in a cold, highly magnetized relativistic plasma dominated by the Poynting flux [164,165].

Fermi acceleration can be efficient and very fast with the acceleration rates $\dot{\gamma}$ of the order of the gyrofrequency $\omega_{\rm g} = eB/mc$. This is due to the gyroresonance-dominated interactions of electrons with the MHD turbulence [10]. In the limit of Bohm diffusion (see below), the mean free paths of accelerating electrons nearly equal to their gyroradii and by achieving the highest possible energy, they can radiate synchrotron photons up to energy $\sim 150\eta(v_s/c)^2)$ MeV, with v_s—the shock velocity and $\eta(\leqslant 1)$—the inverse of the between the diffusion coefficient and its value in the Bohm limit [163].

For example, Ref. [166] presented an one-zone SSC model where a jet blob, containing the separate acceleration and emission zones, is moving relativistically toward us with Doppler factor δ. The acceleration zone (dominated by the DSA mechanism) is represented by a slab containing shock front and is spatially separated from the emission zone. For the stationary emission, the number of electrons injected in the AZ and that of escaped into the emission zone is equal, while X-ray-to-TeV flare was modeled by time variations in the acceleration timescale, yielding more energetic electrons within shorter time intervals leading to the hardening of the γ-ray spectrum.

Ref. [167] presented a time-dependent two-zone SSC model for the MWL observations of Mrk 421 in March 2001, where the second component is (i) pre-existing and co-spatial and participates in the evolution of the active region ("background"), or (ii) spatially separated and independent, only diluting the observed variability ("foreground"). The flux variability was ascribed to the injection of relativistic electrons in the emission zone as a shock front crosses this jet area. However, a quadratic relation between the X-ray and TeV flux variabilities was not reproduced. The authors [168] also adopted a two-zone SSC model for the different MWL states of Mrk 501 during 2011, where the γ-ray emission is produced within two jet blobs containing ultrarelativistic electrons accelerated by the Fermi-I process.

According to [169], the EED established within the Fermi-I process at the relativistic shock front can be represented by a simple power law $N(\gamma) \sim (\gamma/\gamma_0)^{-s+1}$, with the EED spectral index $s = -\log p/\log\epsilon$; p, the probability that electron will undergoes the acceleration step i (characterized by the energy gain ϵ, assumed to be independent from the electron energy as $\gamma_i = \epsilon\gamma_{i-1}$). A log-parabolic EED can be established when the condition that p is energy-independent is broken and the probability of the particle's further acceleration is declining as energy increases, i.e., the probability $p_{\rm i}$ of further acceleration at the step i is given by $p_{\rm i} = g/\gamma_{\rm i}^q$, with g and q to be constants. If $q > 0$, the probability $p_{\rm i}$ decreases with energy (the so-called *energy-dependent acceleration probability* (EDAP) process).

In the case of relativistic shocks, different physical factors (the lifetime of the shock front and spatial extent) can limit the energy to be attained by charge during the Fermi-I process. However, acceleration will eventually cease even in the absence of these factors: when the radiative energy losses (syncrotron plus SSC, inevitably associated with the acceleration) overwhelm the energy gains obtained upon the shock crossings [163]. Moreover, the microphysics of the jet turbulence represents an important factor which determines the value of the power-law photon index and the number of the energy orders passed by particles during the DSA process [28]: when electrons undergo infrequent large-angle scatterings, they produce harder power laws (than in the case of small-angle scatterings) and pass significantly more energy orders before establishing a power-law EED.

Initially, the magnetic field is thought to be random in the jet emission region, but the shock passage can compress it and produce an ordered component [170]. Generally, one assumes during the jet modeling that shocks have a direction transverse to the flow. However, the VLBI observations sometimes show features indicating that shock fronts should be oblique with respect to the jet axis and, consequently, the presence of conical shocks were suggested (e.g., [171]). On some occasions, γ-ray flares can be triggered not by propagating shocks: these events are expected also during the encounter of propagating particle density or magnetic field enhancement and stationary jet inhomogeneity (e.g., a recollimation shock; [170]).

Oblique, relativistic shocks are referred to as "superluminal", implying that they cannot be the sites of first-order Fermi acceleration [162]. As an alternative mechanism for particle acceleration is the shock-drift mechanism. In this scenario, particles can be accelerated by means of a single shock crossing when they undergo a drift parallel (or anti-parallel) to the electric field ([162] and references therein). In turn, this field can be induced when the charged particle is moving towards the shock. This mechanism is mentioned as fast Fermi process: particles are allowed to boost their energy by an order of magnitude even during a single shock encounter [5]. Note that the level of the MHD turbulence should be relatively low for a shock-drift acceleration to be the most efficient (in contrast to DSA and stochastic processes). In the case of weak turbulence, shock-drift acceleration can become dominant in oblique shocks and produce a hard-spectrum EED up to the highest energies (as obtained within the Monte Carlo simulations; see, e.g., [164]). Consequently, the Fermi-I mechanism can produce high-energy electron populations characterized by a large range of power-law indices: from very steep indices down to very hard ones ($p \simeq 1$) depending on the properties of magnetic field and turbulence, shock speed and obliquity in the case of mildly relativistic shocks [28]. Consequently, DSA and SDA complete each other in point of the acceleration capability; the first mechanism is dominating in the case of strong turbulence near the shock front, while the SDA is more efficient when the magnetic field is substantially more laminar on larger spatial scales [28]. This model was adopted by [56] to reproduce the broadband SED and γ-ray light curves presented in [52].

In the case of ultrarelativistic shocks (with Lorentz factor $\Gamma_{sh} \gg 1$), only the particles with $\gamma \gg \Gamma_{sh}$ can manage to cross the shock front from downstream to upstream. In a magnetized medium, such crossing is possible only for parallel or quasi-parallel shocks, in the case of small angles between the magnetic field and the flow direction [28,158]. Consequently, the Fermi-I processes is expected to be much less efficient (especially, in the presence of perpendicular and quasi-perpendicular shocks), and the EED spectral indices are significantly softer than in the case of mildly-relativistic shocks, with a universal value $\sigma \approx 2.2$–2.3 [172].

Ref. [76] modeled a rapid MWL variability by assuming that the γ-ray emission is produced via the SSC mechanism and accounting for (i) the energy stratification established by particle acceleration at shock fronts; (ii) electron cooling by synchrotron emission; and (iii) the effects of light-travel delays for the synchrotron emission providing the seed photons for the IC up-scattering. An MWL flare was produced by the collision between the relativistic shock and jet inhomogeneity, triggering a forward-reverse shock structure. These simulations indicated that relative delays between the γ-ray and synchrotron flares are determined by the energy stratification and geometry of the emitting regions confined between the forward and reverse shocks and yielding both negative and positive time delays depending on the spectral band. Moreover, the light-travel effects related to the seed photons for the EIC upscatter may lead to the delay of the γ-ray variability with respect to those observed at synchrotron frequencies when the jet axis is (nearly) aligned with our line-of-sight.

The 2–8 keV polarimetric observations of Mrk 421 and Mrk 501 with IXPE showed that the X-ray polarization degree was more than a factor of two–three higher than the optical one. These results were explained by the shock presence in the emission zone: higher-energy, hard X-ray emitting particles should populate the magnetically more-ordered region closer to the shock front, and then diffuse away to the area with less-ordered magnetic field, producing optical emission with lower polarization [70,71].

In order to explain the very hard VHE spectra of UHBLs, Ref. [82] revisited the one-zone model by assuming that electrons are co-accelerated with protons by relativistic internal or recollimation shocks in the case of the physical situations as follows: (1) low jet magnetisation and (2) electrons could be preheated in the shock transition layer, yielding large minimum Lorentz factors when involved in the Fermi-I process. While acceleration by a single shock was sufficient for the hardest UHBL SED, re-acceleration on a second shock was considered. The γ-ray emission from the accelerated proton population (with the same number density as the electrons) did not make a significant contribution.

3.3. Jets-in-Jet Model and Relativistic Magnetic Reconnection

In the framework of a jets-in-jet model (e.g., [133,137]), the TeV-band emission is generated in the small-size emitting regions which move relativistically with respect to the main jet. In turn, the latter also is relativistic characterized by the overall bulk Lorentz factor Γ_b. It was concluded that such a geometry is capable for producing a high-amplitude variability on timescales which are significantly shorter than the light-crossing time of the central SMBH. Namely, the emission from a tiny source zone to be beamed through a narrow cone and the observed TeV-band flux is amplified without needing to impose any extreme requirement on the emitting zone.

Moreover, detections of extremely fast flaring TeV emission in Mrk 501 and PKS 2155-304 (see above) impose limits on the spatial scales of the high-energy emission region, which are much shorter than the light-crossing time of the central SMBH (amounting to hours for the blazars with the SMBH masses $\sim 10^9 M_\odot$) and, consequently, it is reasonable to suggest a compact jet emission region to be the source of such TeV flares. On the other hand, the escape of TeV photons from such very compact emission zone implies that the latter should move with a bulk Lorentz factor $\Gamma_{em} \gtrsim 50$ (in order to avoid the annihilation within the soft radiation fields; [133]).

Magnetic reconnection is considered as very efficient and rapid mechanism for using the jet magnetic field energy for accelerating electrons to the energies required, e.g., to upscatter synchrotron photons to gamma-rays. Magnetic field lines in the HBL jets may undergo breaking and reconnection. Consequently, a significant portion of the magnetic energy can be converted into the kinetic energy of jet plasma and accelerate particles. In the particular medium, the time evolution of the magnetic field is given by $\delta \mathbf{B}/\delta t = \nabla \times (u \times \mathbf{B}) - \nabla \times \eta(\nabla \times \mathbf{B})$, where η is a magnetic diffusivity of the medium. In astrophysical plasmas with high magnetic conductivity, the first term on the equations right hand side is generally dominant and, consequently, magnetic field is frozen-in and no reconnection can occur. However, this term may become negligible in some area of the jet plasma, e.g., around the stagnation points, over some lines or surfaces characterized by $u \simeq 0$ [5]. Thereby, strong currents or current sheets with non-zero electric field are induced, leading to the plasma heating and particle acceleration. The relativistic (MHD) and three-dimensional particle-in-cell (PIC) simulations demonstrated that a blazar jet can become unstable to the kink-mode instabilities [173], producing a filamentary current density pattern which is inclined to the magnetic reconnection. Namely, such currents may trigger growing kink instabilities and turbulences which can then lead to the development of an anomalous resistivity. The latter can strongly amplify the magnetic field's dissipation [174].

While the relativistic shocks convert a fraction of the jet kinetic energy of the jet, magnetic reconnection is a highly efficient mechanism for extracting a magnetic field energy and using it for the particle acceleration to ultrarelativistic energies. Namely, this process can rapidly convert a sizeable magnetic energy into the particle kinetic energy via the rearrangement of the field lines [175]. The simulations showed an spontaneous appearance of plasmoids (or magnetic flux tubes) in the sufficiently long and thin current layers, owing to the tearing instability [176]. These plasmoids enhance the overall reconnection rate by trapping the energised particles and evacuating them along with the reconnected magnetic field from the so-called magnetic X-points. They can represent the compact blobs adopted in the different emission models [176].

The reconnection-based mini-jet model was proposed to explain the extremely fast variability shown by the TeV-detected blazars [137]. The model incorporates two wedge-shape regions with relativistically flowing plasma ("mini-jets") and separated by a stationary shock. In such a geometry, mini-jets are perpendicular to the relativistic axis. They are formed in the process of magnetic reconnection and leave the reconnection site in the form of blobs which are moving with relativistic speeds. They produce a fastly variably TeV-band emission within a narrow beam by means of the SSC mechanism (Figure 3). The sequence of the fast TeV-band flares shown by PKS 2155−304 in 2006 July was explained by the existence of the multiple, reconnection-born mini-jets [137].

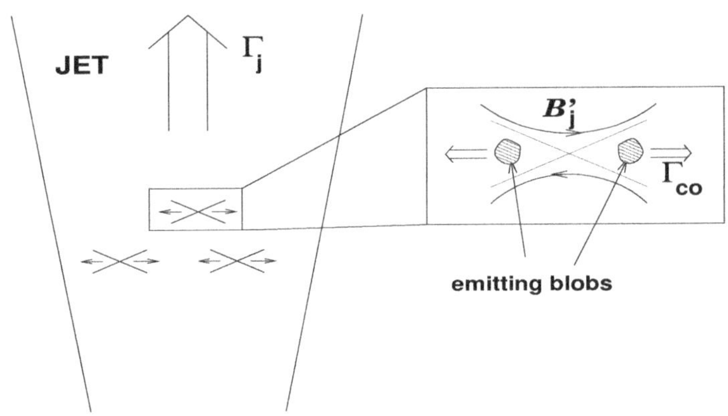

Figure 3. A schematic representation of the jets-in-jet geometry. Reproduced according to the MNRAS guidelines from [137].

Ref. [175] showed that the magnetic reconnection process is often relativistic in the high-energy universe: the energy density of the reconnecting magnetic field B_0 is higher than that of the ambient medium: $\sigma_0 = B_0^2/(4\pi w_0) \gg 1$, with w_0, the relativistic enthalpy including the rest-mass energy. In such a situation, a relativistic magnetic reconnection is the most efficient mechanism to dissipate the magnetic field energy and accelerate particles. In that case, a hard power-law particle energy distribution $N(\gamma \propto \gamma^{-p})$ can be established with $p \to 1$ [177].

Relativistic magnetic reconnection has been studied for blazar jets having an electron–ion or mixed composition. For example, Ref. [138] performed the 2-D PIC simulations of this process for the electron–positron and electron–proton compositions. It was concluded that the reconnection mechanism yields (i) efficient conversion of the magnetic energy into that of accelerated particles; (ii) an extended, non-thermal relativistic energy distribution of particles, and (iii) plasmoids characterized by a rough equipartition between the energies of magnetic fields and that of the relativistic particles. Ref. [178] demonstrated that fast magnetic reconnection can form a self-similar chain of plasmoids which grow in time, while their interiors undergo an compression and amplification of the internal magnetic fields. Consequently, particle energization follows from the conservation of magnetic moment. If particles are injected into plasmoids with a power-law energy distribution, the aforementioned process conserves the original functional shape but adds a nonthermal tail described by $f(E) \propto E^{-3}$ at higher energies, followed by an exponential cutoff with the maximum energy increasing with time as $E_{\rm cut} \propto \sqrt{t}$.

Ref. [179] presented a model for flares produced by individual reconnection plasmoids. In this model, the peak luminosity and flux doubling time-scale were represented as the functions of the plasmoid size and momentum. Ref. [180] interpreted the exceptional X-ray outburst in 2013 April 11–19 and simultaneous MWL behavior in the framework of magnetic reconnection scenario. Here, the multi-hour flux variability is modeled as a combination of the emission from the plasmoids of different size and velocity. As for the sub-hour variability, one a adopted a scenario incorporating a dominant emission from a single small plasmoid which is moving across the magnetic reconnection layer.

An uncorrelated VHE flare of Mrk 421 observed on 4 February 2017, and explained within two-zone SSC scenario by the existence of a compact second blob containing highly energetic electrons with a narrow range of Lorentz factors (see Section 2.2) was suggested to result also from a magnetic reconnection [74]: the blobs containing ultrarelativistic particles could be formed at the jet's reconnection sites and produce a high-energy emission. During the reconnection process, the dissipated magnetic energy is converted into the kinetic energy of nonthermal particles, leading to a decrease in the magnetic field strength with increasing gamma-ray activity. In turn, the ratio $U_B/U_e \sim 10^{-3}$ can be obtained, which is needed to reproduce the observed broadband SED.

One of the explanations of the resent X-ray and optical polarimetric results obtained for Mrk 421 and Mrk 501 was related to the turbulence-induced reconnection in the jet characterized by transverse velocity gradients and, therefore, yielding higher-ordered fields in the jet's transverse direction [70,71].

3.4. Jet Turbulence and Fermi-II Process

Magnetized turbulence is very important for blazar jets in different aspects (see [181] for a review): (a) at the least, it provides scattering agents for DSA; (2) turbulence generates magnetic reconnection, or the converse; (3) it represents an efficient mechanism of particle acceleration, by means of the stochastic or second-order Fermi (Fermi-II) acceleration in a shock downstream region: a particle interacting with randomly moving magnetic inhomogeneities with a typical velocity dispersion $\beta_{\rm m}c$ can gain a large energy stochastically with a rate $\propto(\beta_{\rm m}c)^2$.

As the blazar jets propagate, its interactions with the ambient medium can lead to the different instabilities and mass loading. Consequently, the turbulence responsible for the Fermi-II process can be triggered by (i) a Kelvin–Helmholtz instability (see, e.g., [139]); (ii) a current-driven instability [182]; (iii) a recollimation shock [83].

Refs. [70,72] explained a large change in the polarization degree from X-ray to optical frequencies as follows: in the jet plasma crossing a shock front and having a turbulent magnetic field, particle acceleration is expected to be the most efficient in those cells where the magnetic field is nearly parallel to the shock normal. Consequently, a higher polarization degree and stronger variability should be observed at higher frequencies.

Ref. [183] presented a relativistic turbulence model for the very fast TeV variability of PKS 2155-304, in which a MHD turbulence in the blazar jet generates compact plasma blobs on the spatial scales smaller than the event horizon radius of the central SMBH (similar to the jets-in-jet scenario). These sub-regions move relativistically in random directions and the variability time-scale is determined by the size of each region in their own comoving frames. In the case of the variable orientation during the turbulent blob movement, the observer may receive its radiation only during the short time interval when the beam is pointed to the Earth.

In order to achieve a satisfactory representation of the very hard VHE spectrum of the UHBL source 1ES 1101−232, Ref. [184] used a time-dependent SSC model where extremely hard electron distribution is achieved within the stochastic acceleration yielding a steady-state, relativistic, Maxwellian-type particle distribution peaking at high electron Lorentz factors ∼10^5. This distribution represents a time-dependent solution of the Fokker-Plank equation that incorporates the radiative energy losses of accelerating particles and is capable to reproduce the observed hard TeV-band spectra. Depending on the physical conditions in the jet emission zone, (e.g., if particles undergo cooling beyond the acceleration zone, or the jet medium is clumpy), the combination of different pile-up distributions is capable of interpreting the observed features.

The uncorrelated VHE flare of Mrk 421 observed on 4 February 2017 was explained within the two-zone SSC scenario, incorporating the presence of a compact second blob of highly-energetic electrons with a narrow range of Lorentz factors (see Section 2.2). This event was also explained in the framework of the Fermi-II acceleration: quasi-Maxwellian EEDs could be established in the process of energy exchanges with resonant Alfven waves in a highly turbulent medium [74].

The broadband SED of Mrk 501, constructed by using the MWL data collected on 9 June 2012, showed a transient UHBL nature with the higher-energy peak at ∼2 TeV [185]. A two-zone model was adopted for this case: the first, larger zone dominating in the optical and MeV energy ranges; to be steady or slowly variable. The second, smaller zone, spatially separated from the first one and characterized by a very narrow EED (owing to the stochastic acceleration), was the dominant source of the variable X-rays and VHE emissions, producing also the aforementioned TeV-band high-energy.

Moreover, Ref. [186] reported hard high-energy spectra characterized by the photon index $\Gamma < 1.5$ (down to 0.89 ± 0.29) above 10 GeV on 17 occasions from the *Fermi*-LAT 7-yr data of Mrk 501, each with 30-d integration time. The corresponding SEDs (whenever the VHE spectral points were available) were modeled by using a two-zone SSC scenario: two co-moving blobs (with δ ∼10) characterized by the narrow power-law and relativistic Maxwellian EEDs, established by means of the first and second-order Fermi mechanisms, respectively. We also found a number of the instances of very hard LAT-band photon index for Mrk 501, as well as for another HBL source 1ES 0033+595 ([69,187]; see also Figure 4). Note that these objects showed the features of the efficient stochastic acceleration, as well as very fast X-ray variability explained by presence of small scale jet inhomogeneities with strong turbulent magnetic fields (see [68,187]).

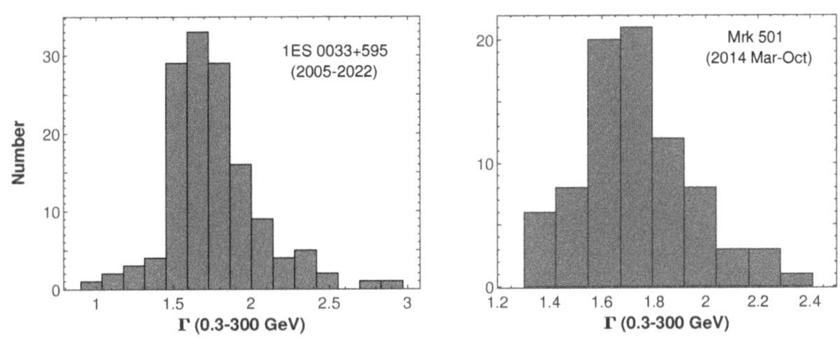

Figure 4. Distribution of the LAT-band photon indices in the HBL sources 1ES 0033+595 (**left**) and Mrk 501 (**right**). The histograms are constructed using the values derived by us from the LAT data analysis according to the recipe provided in [187].

The 2-D MHD modellings of a mildly relativistic shock propagating through a jet inhomogeneous medium showed that the post-shock jet regions may become highly turbulent if there are pre-shock density inhomogeneities. Moreover, magnetic fields can be strongly amplified in these regions due to the stretching and folding of field lines existing in the turbulent velocity field [140,188]: if the initial magnetic field is perpendicular to the shock normal, it will be compressed by the shock front and then undergo an additional amplification by turbulent motions. The amplified magnetic field evolves into a filamentary structure and the turbulence spectrum is flatter than the Kolmogorov function (see below).

Charged particles in turbulent jet plasma are expected to be accelerated via interactions mainly with Alfven waves propagating in the magnetized medium [189]. The astrophysical collisionless turbulence is generally represented as an energy cascade which is spanned over the different orders of spatial scales (from large down to small wavelengths). Generally, most of the fluctuation power (in velocity and electromagnetic fields) are carried by the larger-scales turbulence[181]. The wave energy distribution and intensity are given by $W(k) = (\delta B k_0^2/8\pi)(k/k_0)^{-q}$ and I_k=$I_0(k_0/k)^q$, respectively, with $k = 2\pi/\lambda$ to be the wave number; δB, the turbulent component of the jet magnetic field; q, the turbulent field spectral slope: $q = 3/2$, $q = 5/3$ and q = 2 for the Kraichnan, Kolmogorov and 'hard-sphere' turbulences, respectively. For the wavenumbers below an inverse correlation length k_0, the wave intensity per logarithmic bandwidth is assumed to be equal to the background field intensity; i.e., $I(k)$ = $B_0^2 k^{-1}$ when $k < k_0$ [190].

The Fermi-II process is based on (quasi)elastic reflections (scatterings) of the charged particles by the magnetic inhomogeneities or plasma waves. Consequently, if there are waves propagating towards both directions at a given position, a stochastic acceleration of charged particles can be developed [190,191]: particle gain or lose energy when the "mirror" is approaching or receding, respectively. However, the simulations showed a higher probability of the head-on collisions compared to the rear-on reflections and, on average, it can gain energy. Note that the energy gain per bounce is proportional to the square of the mirror velocity (hence the name of the mechanism: second-order Fermi acceleration). However, the net energy gain depends also on the scattering rate. The Fermi-II mechanism is widely accepted to be a stochastic process and is known also as stochastic acceleration. This process becomes less efficient at the energies $\gamma > \gamma_0$ = $\Omega_{e,0}/k_0 c \gg 1$, with $\Omega_{e,0} = eB/m_e c$; k_0, the inverse correlation length.

Two different types of the stochastic particle transport are considered [192]: (1) in the case of the small spatial scales, charged particles interact with the MHD waves moving in the local magnetic field and undergo a stochastic acceleration. The mean-free path of particles on such scales is equal to the coherence length for the MHD turbulence, $l_{\rm MHD} = c\sigma_{\rm mag}/(3D)$, where $\sigma_{\rm mag} = (V_{\rm A}/c)^2$; (ii) particle transport is implemented via diffusion in the case of large spatial scales. In such a situation, a mean free path is determined by the relativistic electron's Larmor radius, $r_L = E/qB$ with E and q, electron's energy and charge, respectively, and B, the blob's magnetic field. This regime is referred as a Bohm diffusion. When the electron energy becomes sufficiently high, its Larmor radius becomes comparable to the blob radius and this high-energy electron can escape from the system. Therefore, no further acceleration is possible in the case of $r > r_L$ (so-called Hillas condition [193]).

Therefore, a stochastic acceleration can be characterized by a diffusion coefficient in the momentum space. The magnitude and scaling of this coefficient affect the evolution of the EED in the Fokker-Planck equation [194]. In most astrophysical situations, one considers a large-amplitude turbulence, since the stochastic acceleration is found to be very slow within in the case of small-amplitude turbulences: the acceleration timescale (generally referred as the time required on average to double a particle energy), is proportional to $(\delta B/B)^{-2}$. In the case of strong turbulence, particularly important is the relativistic limit where the Alfven speed $v_A \sim c$, where v_A = $B_0 c/sqrt 4\pi hn + B_0^2$; h = $(\rho + P)/n$, n, ρ, and P are the specific enthalpy, number density, total energy density, and gas pressure (measured in the local plasma frame; [195]). Therefore, we are mainly interested in those particles interacting with large-scale (the scattering timescale is a growing function of energy) modes of a fast-moving turbulence spectrum [181].

The simulations of [190] showed that the efficiency of the Fermi-II process can be comparable to that of the shock acceleration. Moreover, it operates over much longer timescales than the Fermi-I mechanism. Stochastic acceleration, on the other hand, may be present on some level in the turbulent downstream of shocks and deliver pre-accelerated particles to the shock front (see, e.g., [26]). Moreover, there can be a combined acceleration process: firstly, particles are efficiently accelerated at the shock front via the Fermi-I mechanism and after the escape into the shock downstream region, they will be involved in the Fermi-II process. Consequently, their energy can be boosted sufficiently to allow particles to return in the shock acceleration zone and repeat the previous acceleration cycle [191].

In the local jet frame, the mean free path represents the spatial scale at which the particle's momentum vector is deflected by $\pi/2$ on average [10]: the wave-particle interaction can be presented by the scattering relation $\lambda_{\|} = \lambda_1(\rho_1/\rho)(r_g/r_{g1})^{\alpha} \equiv \eta_1 r_{g1}(p/p_1)^{\alpha}, \kappa_{\|} = \lambda_{\|}v/3$, with $\lambda_{\|}$ ($\kappa_{\|}$) is the mean free path in the local frame and parallel to the field $\mathbf{B}$, $v = p/m$—particle's velocity in the local frame, $r_g = pc/(QeB)$—gyroradius of a particle carrying a charge Qe, ρ—the plasma density with a far shock upstream value of ρ_1. Note that the condition $\lambda_{\|} \gtrsim r_g$ is the so-called Bohm limit, representing a fundamental bound for the physically meaningful diffusion. The parameter $\eta \equiv \lambda_{\|}/r_g \propto p^{\alpha-1}$ characterizes the scattering strength and, hence, the importance of particle's cross-field diffusion: when $\eta \sim 1$ (i.e., $\lambda \sim r_g$) at the Bohm diffusion limit, $\kappa_{\perp} \sim \kappa_{\|}$ and particles diffuse across magnetic field lines quickly. Note that the Bohm diffusion corresponds to extremely turbulent magnetic fields with fluctuations satisfying $\delta B/B \sim 1$ and $\alpha = 1$. The condition $\lambda \geqslant r_g$ is required for the physically meaningful diffusion resulting from gyroresonant wave–particle interactions. Therefore, the case α =1 is highly important for the different astrophysical situations [10].

The γ-ray SED of Mrk 501 corresponding to the 1–5 May 2009 window was modeled within the one-zone SSC scenario yielding α = 1.5, implying the interactions with weaker turbulence for more energetic particles which undergo a diffusion on larger spatial scales [10]. The modelling indicated that the turbulence strength declines with distance from the shock front in the relativistic jet medium. Consequently, the particle diffusion becomes significantly different from the Bohm limit at all energies, and the diffusion scale λ will increase with the particle momentum. This is required for leptonic models in order to explain very hard HE and VHE spectra, which yielded $\lambda_{\|} \propto p^{\alpha}$ with $\alpha \gtrsim 1.5$, associated with a weaker turbulence. It was concluded the the electron mean free paths should be orders of one magnitude larger than their gyroradii at the Lorentz factors derived from the SED simulations [10].

The stochastic acceleration rate depends on the wave spectrum [190] as follows: for $q = 2$ (hard-sphere turbulence), charged particles have the same, rigidity- and energy-independent mean free paths and, therefore, the Fermi-II mechanism accelerates them at a constant rate. The situation is different within the Kolmogorov turbulence ($q = 5/3$): the mean free paths and acceleration rate of particles decline while the energy increases, and the Fermi-II process becomes gradually inefficient with higher-energy particle distributions. As the parameter q increases and the turbulence spectrum becomes steeper, a larger portion of the turbulence energy is contained in longer waves which, in turn, can interact resonantly with higher-energy particles. Consequently, steeper-spectrum turbulences are capable for producing harder particle distributions that the Kolmogorov turbulence. Consequently, hard-sphere scattering centers are more efficient to accelerating charged particles compared with a Kolmogorov-type wave ensemble. On the other hand, the latter is more efficient compared to the Kraichnan spectrum ($q = 3/2$). The spatial scales for the Kolmogorov turbulence are up to an order of magnitude shorter than in the case of the hard-sphere spectrum [10]. It was concluded that a hardening in the turbulence spectrum shifts the EED cutoffs to higher energies [190].

The EED behavior during the stochastic acceleration depends on the strength of the background magnetic field: the intensity ratio of the Alfven waves in the shock downstream region to those in the upward region is a function of the quasi-Newtonian Alfvenic Mach number $M = u_1/u_{A,1}$, with the shock proper speed $u_1 = c\sqrt{\Gamma_1^2 - 1}$; Γ_1^2, the Lorentz factor of the upstream bulk flow; $u_{A,1}$, Alfven speed in the shock upstream region. In the case of relativistic shocks, the Alfven waves are seen to propagate predominantly backward for relatively low Mach number shocks [196]. Consequently, the compression ratio r_k of the scattering center becomes larger than the gas compression ratio r and, eventually, a significantly harder EED is established (compared to the frozen-in case; [190]). In the cases of weak magnetic fields and a quasi-Newtonian Alfvenic Mach numbers much exceeding the critical Mach number ($M \gg M_c = \sqrt{r}$; with r, the shock compression ratio), the effects of stochastic acceleration are overwhelmed by the much stronger Fermi-I acceleration [190].

Ref. [190] found that the contribution of the Fermi-II mechanism to the particle energy distribution is insignificant compared to that of the Fermi-I acceleration at the shock for high Alfvenic Mach numbers (M = 1000, corresponding to $B_0 \approx 1.4\,\mathrm{G}$ in a Hydrogen plasma); the distribution sustains its shape and energy range unchanged at least for tens of thousands of the electron's mean free paths, regardless of the applied turbulence spectrum. For the stronger magnetic fields (M = 10 and M = 3, corresponding to 0.14 G and 0.46 G, in a Hydrogen plasma and to 4.6 and 15 mG in a electron-positron plasma, respectively), the stochastic acceleration effects are much more significant: the Fermi-II mechanism will further energize particles just after the shock front, and the entire energy spectrum gradually shifts to the higher energies [190].

When the particle energy becomes higher than the turnover energy γ_0, the rate of energization is expected to go down: after this threshold, the particle's mean free path will increase much faster, leading to the decrease in the stochastic acceleration efficiency. Moreover, particles will be able to escape into the shock downstream region and even manage to return back to the shock (at least for $M = 3$). After the turnover energy, particles undergo a pile-up and, consequently, one expects appearance of the narrow distribution bump immediately beyond the γ_0 [190]. Note that one of the alternative scenarios to explain the narrow spectral feature at VHE is pileup in the high-energy range of the relativistic EED due to stochastic acceleration [54].

Mrk 501 showed peculiar high-energy characteristics of during the LAT observations in 2009 May, exhibiting a flaring activity and spectral hardening above 10 GeV while a weak activity was detected at the lower energies. In order to explain this behavior, Ref. [197] adopted a "leading blob" model: the observed radiation was produced within several emitting blobs, where electrons were accelerated to relativistic energies by the Fermi-II mechanism and produced a narrow (piled-up) distribution. All blobs were assumed to have similar physical parameters, with exclusion of the characteristic energies of their EEDs. A TeV-band flare and hard spectral feature was reproduced by dominance of one (or a few) of the radiating components. This so-called leading blob could boost its apparent luminosity by changing its Doppler factor or the injected energy.

Although the spatial scales of stochastic acceleration are enormous compared to the Fermi-I process, they are still orders of magnitude smaller than the spatial scales resolvable by the current VLBI observations [190]. The acceleration timescales can be also very short: the time required to shift the entire EED from the initial energy range to the turnover energy takes from 10 to 50 min in the M = 10 case, and for M = 3 the acceleration times are $\lesssim$1 min in the shock frame.

In the process of the interaction between the HBL jets and the ambient medium, a sharp boundary layer may be formed in the case of the large velocity difference between them. Moreover, a jet may have unequal bulk Lorentz factors at different distances from the axis [5]. Rayleigh–Taylor-type instabilities can be triggered at such spine-sheath interfaces in the shearing layer of the thickness $\sim 0.1 R_{\rm jet}$, where the resulting turbulence can accelerate particles via the Fermi-II mechanism (see, e.g., [198]). Namely, different simulations showed that this process can accelerate electrons up to PeV energies and protons up to EeV energies [5]. The average energy gain per interaction $\overline{\Delta E} \propto (u/c)^2 E \propto (l/c)^2 (\delta V_z/\delta x)^2 E)$, with V_z to be a flow speed along the z-axis (transverse to the x-axis directed along the spine-sheath interface); $u = l(\delta V_z/\delta x)$, the speed change of the scattering centers in the particle's frame, crossing the shearing flow along the x-axis and having the mean free path l. The acceleration time $\tau_{\rm acc} \propto (\delta V_z/\delta x)^2$, i.e. a shear acceleration is faster for those particles which already have high energies. Moreover, this acceleration is more efficient for protons with large l than for electrons [135].

3.5. Jet–Star Interactions

Ref. [199] proposed a possibility of fast TeV-flares by compact magnetized blobs, produced when red giant (RG) stars cross the blazar jet close to the central SMBH. In the framework of jet–star interaction, the RG stars should cross the jet at different distances from its base. In this process, one expects a shocks trigger which will transfer the bulk kinetic energy to charged particles [200]. Stars are expected to traverse faster the innermost jet parts which are thought to be narrower than those with large separations from the central engine and, therefore, are the most plausible sites of the aforementioned fast flares. In the case that the RG is even slightly tidally disrupted by the SMBH, there can be a large amount of stellar material to be blown by the jet. This material ("bubble") will expand quickly, until being shocked by the jet ram pressure [201]. A shock will propagate through the bubble, heats up its material and accelerate particles to relativistic speeds [202]. The shock will propagate until reaching the stagnation radius, where the bubble and jet pressures are equal. At this position, one expects formation of a double bow shock structure, which energizes via the Fermi-I mechanism up to ultrarelativistic energies. The accelerated particles (primarily, leptons) can contribute to the jet's emission [202]. Namely, these particles are expected to radiate in γ-rays predominantly through the proton-synchrotron mechanism or EIC-upscattering by electrons (assuming the jet synchrotron photons as external). Within this scenario, the variable γ-ray emission might be produced during relatively short time interval [203]. Ref. [199] adopted this model to the minute-scale TeV flares superimposed on the longer (daily timescales) γ-ray variability of PKS 2155−304 in 2006 July.

According to [203], advection escape of charged particles dominates their radiation cooling during the star interaction with a moderately powerful jet (as accepted to be the case for HBLs). The produced radiation was found to peak from X-rays to MeV energies in the synchrotron emission (depending on the fraction of energy in magnetic field). Another peak can be situated in the 100–1000 GeV range for the IIC-upscatter, depending on the stellar type: the cooler (either older or less massive) stars are expected to yield the higher SED peak energy (up to $\sim$1 TeV). The radiation spectrum is related to the efficient advection of low-energy electrons even in the case relatively high magnetic fields); Interactions of jets with cold stars may yield even harder IC spectrum owing to the Klein–Nishina effects [203].

The emission generated during the jet–star interaction events can be relatively persistent at high energies, through either IC or synchrotron mechanisms within low magnetic fields (generally expected in HBLs). However, the steady state emission of the whole population seems to be undetectable [202]. Within strong magnetic fields (corresponding the equipartition value; viz., in the small-scale jet region with relativistic magnetic reconnection in HBLs), emission from the jet-star interaction can be seen at the energies $\sim$ 100 MeV as a bright, fast flaring instance superimposed on the persistent, lower-level IC radiation [202].

Note that [44] reported the MeV-excess SEDs from the period 2016 April–August and this result was explained by a possible jet interaction with a wind-blown bubble from a nearby red giant star.

3.6. Impact of Disc Instabilities on the Observed γ-ray Variability

Similar to other spectral ranges, the γ-ray variability of HBL frequently carries out a lognormal character, i.e., the γ-ray fluxes are preferentially log-normally distributed (when the observations are not limited by poor statistics; see [135] for a review). Several scenarios about the lognormality origin are proposed:

- First of all, a lognormality hints at the impact of the accretion disk instabilities on the jet [135,204]: there should be independent density fluctuations in the disk on the local viscous timescale, characterized by negligible damping. They can propagate toward the innermost disc area and couple there producing a multiplicative behavior. If the latter is transferred to the jet flow (e.g., via the jet collimation rate), the γ-ray emission can be modulated accordingly. However, the timescale for particle acceleration and radiative losses within the jet should be correspondingly small for this purpose.

Lognormal variability in the different energy range and over various timescales is then anticipated.

- Cascade-related emission processes (see Section 2.3) are also thought to lead to lognormal flux distributions [129]. However, the latter are expected only in the optical-to-γ-ray ranges. Moreover, there are limited timescales over which log-normality can be detected (i.e., from sub-hour to yearly timescales in the TeV band; [135]). Moreover, there can be some limitations by the gap travel time for the magnetospheric processes and from the dynamical or escape properties of the hadronic cascades, e.g., [135].
- Alternatively, the lognormal variability can be produced in the case of random fluctuations in the particle acceleration rate [205]. However, one should observe an energy-dependent lognormality in this case, to be progressively weakly expressed towards lower energies and disappearing beyond some threshold energy. Moreover, fluctuations in the acceleration rate can be also characterized by the Gaussian distribution of the photon indices along with the lognormal flux distribution [205].

It is possible that the lognormality in different states (long level and short term flares, low and high levels) can be dominated by one out of the aforementioned processes [135], or the observed lognormality stems from their their combination.

The lognormal γ-ray variability has been reported for a number of HBLs. Namely, this was the case for the observations performed with LAT (Mrk 421 [44,206,207]), Mrk 501 ([208]), 1ES 0033+595 ([187]), PKS 2005−489 ([209], RGB 0136+391 ([210]), H1722+119 ([210] etc.) and Cherenkov-type telescopes (Mrk 421 [44,207], Mrk 501 [134,208] and PKS 2155−304 ([134]).

4. Future Prospects

The next generation of the ground-based and space telescopes are crucial for simultaneously accessing the entire γ-ray domain, carrying out the polarization measurements and studying the flux variability down to the shortest timescales. In particular,

- The Cherenkov Telescope Array Observatory (CTAO) is a next-generation IACT array, using telescopes of multiple sizes to achieve a high sensitivity in the 20 GeV–300 TeV energy range [211,212]. The observatory installations in the Southern and Northern Hemispheres will provide visibility of the entire sky and a sensitivity at least an order of magnitude higher than those of the current major Cherenkov telescopes (H.E.S.S., MAGIC, and VERITAS).
- The ASTRI ("Astrofisica con Specchi a Tecnologia Replicante Italiana") mini-array incorporates a technologically innovative solution for small size (about 4 meters diameter) and large field-of-view (more than 10 degrees) IACTs. It is sensitive in the range 1–200 TeV, achieves an angular resolution of a few arcmin and is devoted to study various types relatively bright VHE sources (a few $\times 10^{-12}$ erg cm^{-2}s^{-1} at 10 TeV; including HBLs) at the energies beyond 10 TeV [213,214]. A prototype telescope, deployed on Mt. Etna (Italy), started its scientific operations in 2018.
- The space missions AMEGO (All-sky Medium Energy Gamma-ray Observatory) and AMEGO-X (the funded projects), will detect γ rays through both Compton scattering and pair production, filling a "MeV gap" in sensitivity [215] . They are optimized for continuum sensitivity in the MeV range in different ways. AMEGO-X uses monolithic silicon pixel detectors for a lower energy threshold and higher low-energy effective area than AMEGO. AMEGO also has the Low-Energy Calorimeter that enhances the polarization and narrow-line sensitivity. For blazars, IC scattering is typically unpolarized or has a very low polarization degree (a few to ten percent) in a partially ordered magnetic field, while the hadronic models usually predict at least 20% polarization degree in the MeV band [216]. Consequently, one will be able to discern the underlying emission mechanism directly from observations.
- Southern Wide-field Gamma-ray Observatory (SWGO) as a next-generation Water Cherenkov Detector (WCD) instrument that will provide the observational coverage of the southern sky with nearly continuous up-time and an instantaneous field of view

(FOV) of ~2 sr at energies from 100 GeV to above hundreds of TeV from a site in the Andes mountains. Simultaneous operations with CTA are planned [217].

- e-ASTROGAM is a proposed space mission for measuring γ-ray emission in the range from 300 keV to a few GeV. The e-ASTROGAM is expected to reach a sensitivity by one-two orders of magnitude higher than its predecessors, and offers enhanced capabilities to detect fast transient events in soft γ-rays [218].

These instruments are anticipated to detect much larger samples of HBLs, achieve much higher temporal/spectral coverage and angular resolution. They will be useful to search for the specific spectral features (predicted in the framework of different theories and simulations) and, especially, for the γ-ray SED components anticipated within some hadronic cascades. These instruments will also collect a large number of high-resolution datasets from the γ-ray flares of HBLs; search for (quasi)periodicities, time delays and spectral features; constrain the location, geometry and kinematics of the γ-ray emission zones; allow to select the valid emission scenarios. Thanks to the planned jump in sensitivity, a detailed exploration of very fast sub-minute variability (particularly, in the TeV-band) will be accessible that is crucially important to study the properties of the emission zones and their locations, leading to great progress in our understanding of the HBL jet physics. High-level VHE spectral analysis and long-term monitoring of the HBLs and UHBL sources will allow us to deal with the challenges associated with the current emission scenarios.

Note that the detection of UHE photons from HBLs still has not been reported by means of the current instruments; perhaps they are considered as one candidate class to produce such events [92]. For examples, Ref. [219] analyzed the data obtained for Mrk 421 and Mrk 501 with High Altitude Water Cherenkov (HAWC) Gamma-Ray Observatory during 2015 June and 2018 July. Although this array is sensitive in the 300 GeV to >100 TeV energy range (see [220]), the maximum energies at which the Mrk 421 and Mrk 501 were detected are 9 and 12 TeV, respectively [219]. Note that the detection of 12 sources of the UHE photons up to energies 1.4 PeV with the Large High Altitude Air Shower Observatory (LHAASO; [221]) was reported by [222]; none of these events are expected to have an extragalactic origin: these photons strongly interact with the EBL and their detection poses a challenge to even the next generation of the Cherenkov-type telescopes. Namely, the attenuation length to about 30 kpc around 1 PeV, which increases to the order of 10 Mpc around 10^{19} eV [92]. On the other hand, the expected modification of the reaction $\gamma\gamma \to e^{\pm}$ by Lorentz invariance violation (LIV, if it exists) at energies $E \gtrsim 10$ TeV can lead to the reduction of cosmic opacity. Consequently, this effect should allow photons of this extreme energy range to evade absorption and reach the Earth (see [223]).

The detection of such absorption anomalies is still very problematic: the performance of current TeV-instruments is not capable of obtaining high-quality spectra at energies higher than 10 TeV. Note that [224] did not find any signature of the LIV existence in the energy spectra of two PeV sources, and the lower limits on the LIV energy scale were imposed. Similarly, Ref. [225] failed to detect the LIV features from the H.E.S.S. observations of Mrk 501 during the strong flare recorded in June 2014: the non-detection of energy-dependent time delays and the absence of deviations between the measured spectrum and that of a supposed power-law intrinsic spectrum with standard EBL attenuation were used independently to derive strong constraints on the energy scale of LIV. CTA and ASTRI will be characterized by greatly improved sensitivity beyond this threshold, and allow us to probe the viability of the LIV scenarios. An important point is that prolonged exposures allow to reach the highest TeV-band energies [223]: the detected energy limit is expected to increase linearly with the exposure for photons with $E > 10$ TeV. Moreover, TeV-band observations at high zenith angles (corresponding to large effective areas of the TeV-band instruments), could be particularly favorable for the Lorentz LIV studies [223].

The current rare detections of ultra-fast VHE variability leaves the duty cycle (DC: fraction of the net exposure during which the object showed such a variability) of these events unknown: a proper evaluation of the duty cycle require a dense and frequent monitoring of the most promising targets, along with the enhanced sensitivity of the next-generation instruments. Nevertheless, the DC can test and constrain some of the proposed variability scenarios [226]. Moreover, the events detected so far occurred only during the VHE flaring states of the sources, whereas it would be important to clarify whether such phenomena can also occur in states of low/quiescent activity or in the MeV–GeV range (not reported for HBLs to date).

Similar to other types of high-energy astrophysical systems, magnetic field is important for HBL physics. While the polarimetric studies can directly probe the magnetic field morphology and evolution in these systems, they still have not been carried out in the γ-ray energy range. Since the latter is associated with the most energetic processes, γ-ray polarization can probe more energetic phenomena in more extreme physical environments than X-rays. Gamma-ray polarimetry can directly disentangle the radiation mechanisms in relativistic jets and probe the existence of the anticipated hadronic signatures (e.g., the detection of a the MeV-band polarization will be a direct indication that the proton-synchrotron emission). Although the current hadronic models require higher jet powers and face more challenges in explaining of the very fast variability than leptonic scenarios, they still provide us with an important and physically motivated alternative for interpreting the available observational data. Gamma-ray polarization is also expected in the framework of the SSC and EIC models, but the proton–synchrotron radiation should be significantly more polarized [7,227]. The much higher sensitivity and spectral resolution of CTA compared to the current IACTs and expected coverage of the entire VHE range to above 100 TeV facilitates a search for potential hadronic signatures in the TeV spectrum.

A combination of very high angular resolution of the VLBI observations and very high temporal resolution to be achieved with CTA in the VHE range is crucial, particularly in the framework of Global VLBI Alliance which is under development [5]. The latter will have the power to resolve the inner jet regions, explore the detailed properties and the evolution of the magnetic field, and discern the physical mechanisms responsible for the jet-launching, particle acceleration and energy generation via the fast and multi-frequency VLBI study through the total and polarized light which is generated in the vicinity of the central SMBHs. Combined with the VHE monitoring, this technique is anticipated to firmly and accurately identify the sites of the TeV-band emission and flaring mechanisms; determine the importance of the BH magnetosphere in the generation of fastly variable TeV emission.

As noted above, the γ-ray studies can contribute to solve different fundamental problems of the modern physics and cosmology. For example, the LIV studies are useful for making a progress in the understanding of the intrinsic γ-ray spectra and variability of HBLs. Consequently, they will allow us to explore the nature of space-time via the propagation of their VHE photons, impact on our knowledge about the EBL, intergalactic magnetic fields (IGMF), facilitate axion-like particle searches, etc. Note that a LIV search was performed for Mrk 501 by [225] using the *Fermi*-LAT and H.E.S.S. data collected on 23–24 June 2014 when the source was showing rapid γ-ray variability. Based on the non-detection of energy-dependent time delays, as well as the absence of deviations between the measured spectrum and that of a supposed power-law intrinsic spectrum with standard EBL attenuation (anticipated within the LIV theory), some constraints related to the LIV energy scale were imposed.

Investigating whether relativistic shocks, reconnection zones, MHD turbulence or shear boundaries provide the dominant energization site for ultrarelativistic particles in HBLs represents the major target for future theoretical studies and highest-level simulations. Indeed, a realistic study of the jet internal shocks requires a time-dependent evaluation of the time-delayed radiation fields from all jet regions.

Funding: This research was funded Shota Rustaveli National Science Foundation grant FR-21-307.

Institutional Review Board Statement: Not applicable.

Informed Consent Statement: Not applicable.

Data Availability Statement: Data presented in Figures 1 and 4 are available upon request from the author.

Acknowledgments: The author thanks Shota Rustaveli National Science Foundation and E. Kharadze National Astrophysical Observatory (Abastumani, Republic of Georgia) for the fundamental research grant FR-21–307, Patrizia Romano for proofreading the manuscript and reviewers for their fruitful suggestions.

Conflicts of Interest: The author declares no conflict of interest.

Note

1 See http://tevcat.uchicago.edu (accessed on 25 May 2023).

References

1. Padovani, P.; Alexander, D.M.; Assef, R.J.; De Marco, B.; Giommi, P.; Hickox, R.C.; Richards, G.T.; Smolčić, V.; Hatziminaoglou, E.; Mainieri, V.; et al. Active galactic nuclei: What's in a name? *Astron. Astrophys. Rev.* **2017**, *25*, 2. [CrossRef]
2. Bodo, G.; Tavecchio, F. Recollimation shocks and radiative losses in extragalactic relativistic jets. *Astron. Astrophys.* **2018**, *609*, A122. [CrossRef]
3. Blandford, R.D.; Znajek, R.L. Kerr black holes. *Mon. Not. R. Astron. Soc.* **1977**, *179*, 433–456. [CrossRef]
4. Falomo, R.; Pian, E.; Treves, A. An optical view of BL Lacertae objects. *Astron. Astrophys. Rev.* **2014**, *22*, 73. [CrossRef]
5. Sol, H.; Zech, A. Blazars at Very High Energies: Emission Modelling. *Galax* **2022**, *10*, 105. [CrossRef]
6. Padovani, P.; Giommi, P. The Connection between X-ray– and Radio-selected BL Lacertae Objects. *Astrophys. J.* **1995**, *444*, 567–581. [CrossRef]
7. Cerruti, M. Leptonic and Hadronic Radiative Processes in Supermassive-Black-Hole Jets. *Galax* **2020**, *8*, 72. [CrossRef]
8. Biteau, J.; Prandini, E.; Costamante, L.; Lemoine, M.; Padovani, P.; Pueschel, E.; Resconi, E.; Tavecchio, F.; Taylor, A.; Zech, A. Progress in unveiling extreme particle acceleration in persistent astrophysical jets. *Nat. Astron.* **2020**, *4*, 124–131. [CrossRef]
9. Aharonian, F.A.; Akhperjanian, A.G.; Barrio, J.A.; Bernlöhr, K.; Bolz, O.; Börst, H.; Bojahr, H.; Contreras, J.L.; Cortina, J.; Denninghoff, S.; et al. Reanalysis of the high energy cutoff of the 1997 Mkn 501 TeV energy spectrum. *Astron. Astrophys.* **2001**, *366*, 62–66. [CrossRef]
10. Baring, M.G.; Böttcher, M.; Summerlin, E.J. Probing acceleration and turbulence at relativistic shocks in blazar jets. *Mon. Not. R. Astron. Soc.* **2017**, *464*, 4875–4894. [CrossRef]
11. Aharonian, F.A.; Khangulyan, D.; Costamante, L. Formation of hard very high energy gamma-ray spectra of blazars due to internal photon–photon absorption. *Mon. Not. R. Astron. Soc.* **2008**, *387*, 1206–1214. [CrossRef]
12. Hauser, M.G.; Dwek, E. The Cosmic Infrared Background: Measurements and Implications. *Ann. Rev. Astron. Astrophys.* **2000**, *39*, 248–307. [CrossRef]
13. Atwood, W.B.; Abdo, A.A.; Ackermann, M.; Althouse, W.; Anderson, B.; Axelsson, M.; Baldini, L.; Ballet, J.; Band, D.L.; Barbiellini, G.; et al. The Large Area Telescope on the Fermi Gamma-Ray Space Telescope Mission. *Astrophys. J.* **2009**, *697*, 1071–1102. [CrossRef]
14. Costamante, L.; Ghisellini, G.; Giommi, P.; Tagliaferri, G.; Celotti, A.; Chiaberge, M.; Fossati, G.; Maraschi, L.; Tavecchio, F.; Treves, A. Extreme synchrotron BL Lac objects: Stretching the blazar sequence. *Astron. Astrophys.* **2001**, *371*, 512–526. [CrossRef]
15. Costamante, L.; Bonnoli, G.; Tavecchio, F.; Ghisellini, G.; Tagliaferri, G.; Khangulyan, D. The NuSTAR view on hard-TeV BL Lacs. *Mon. Not. R. Astron. Soc.* **2018**, *477*, 4257–4268. [CrossRef]
16. Cerruti, M.; Zech, A.; Boisson, C.; Inoue, S. A hadronic origin for ultra-high-frequency-peaked BL Lac objects. *Mon. Not. R. Astron. Soc.* **2015**, *448*, 910–927. [CrossRef]
17. Aharonian, F.; Akhperjanian, A.G.; Bazer-Bachi, A.R.; Behera, B.; Beilicke, M.; Benbow, W.; Berge, D.; Bernlöhr, K.; Boisson, C.; Bolz, O.; et al. An Exceptional Very High Energy Gamma-Ray Flare of PKS 2155−304. *Astrophys. J.* **2007**, *664*, L71–L74. [CrossRef]
18. Albert, J.; Aliu, E.; Anderhub, H.; Antoranz, P.; Armada, A.; Baixeras, C.; Barrio, J.A.; Bartko, H.; Bastieri, D.; Becker, J.K.; et al. Variable Very High Energy γ-Ray Emission from Markarian 501. *Astrophys. J.* **2007**, *669*, 862–883. [CrossRef]
19. Kapanadze, B.; Dorner, D.; Vercellone, S.; Romano, P.; Aller, H.; Aller, M.; Hughes, P.; Reynolds, M.; Kapanadze, S.; Tabagari, L. X-ray Flaring Activity of MRK 421 in the First Half of 2013. *Astrophys. J.* **2016**, *831*, 102. [CrossRef]
20. Böttcher, M.; Reimer, A.; Sweeney, K.; Prakash, A. Leptonic and Hadronic Modeling of Fermi-detected Blazars. *Astrophys. J.* **2013**, *768*, 54. [CrossRef]
21. Marscher, A.P.; Gear, W.K. Models for high-frequency radio outbursts in extragalactic sources, with application to the early 1983 millimeter-to-infrared flare of 3C 273. *Astrophys. J.* **1985**, *298*, 114–127. [CrossRef]
22. Dermer, C.D.; Schlickeiser, R. Model for the High-Energy Emission from Blazars. *Astrophys. J.* **1993**, *416*, 458–484. [CrossRef]
23. Plotkin, R.M.; Anderson, S.F.; Brandt, W.N.; Markoff, S.; Shemmer, O.; Wu, J. The Lack of Torus Emission from BL Lacertae Objects: An Infrared View of Unification with WISE. *Astrophys. J.* **2012**, *745*, L27. [CrossRef]

24. Aharonian, F.; Akhperjanian, A.G.; Anton, G.; Barres de Almeida, U.; Bazer-Bachi, A.R.; Becherini, Y.; Behera, B.; Benbow, W.; Bernlöhr, K.; Boisson, C. Simultaneous multiwavelength observations of the second exceptional γ-ray flare of PKS 2155−304 in July 2006. *Astron. Astrophys.* **2009**, *502*, 749–770. [CrossRef]
25. Kirk, J.G.; Rieger, F.M.; Mastichiadis, A. Particle acceleration and synchrotron emission in blazar jets. *Astron. Astrophys.* **1998**, *333*, 452–458.
26. Tramacere, A.; Giommi, P.; Perri, M.; Verrecchia, F.; Tosti, G. Swift observations of the very intense flaring activity of Mrk 421 during 2006. I. Phenomenological picture of electron acceleration and predictions for MeV/GeV emission. *Astron. Astrophys.* **2009**, *501*, 47–56. [CrossRef]
27. Tramacere, A.; Massaro, E.; Taylor, A.M. Stochastic Acceleration and the Evolution of Spectral Distributions in Synchro-Self-Compton Sources: A Self-consistent Modeling of Blazars' Flares. *Astrophys. J.* **2011**, *739*, 66. [CrossRef]
28. Summerlin, E.J.; Baring, M.G. Diffusive Acceleration of Particles at Oblique, Relativistic, Magnetohydrodynamic Shocks. *Astrophys. J.* **2012**, *745*, 63. [CrossRef]
29. Böttcher, M.; Mause, H.; Schlickeiser, R. γ-ray emission and spectral evolution of pair plasmas in AGN jets. I. General theory and a prediction for the GeV–TeV emission from ultrarelativistic jets. *Astron. Astrophys.* **1997**, *324*, 395–409.
30. Chen, L. On the Jet Properties of γ-Ray-loud Active Galactic Nuclei. *Astrophys. J. Suppl. Ser.* **2018**, *235*, 39. [CrossRef]
31. Zacharias, M.; Schlickeiser, R. A new ordering parameter of spectral energy distributions from synchrotron self-Compton emitting blazars. *Mon. Not. R. Astron. Soc.* **2012**, *420*, 84. [CrossRef]
32. Bicknell, G.V.; Wagner, S.J. TeV gamma ray opacity in PKS 2155−304. *Astron. Astrophys.* **2009**, *526*, A61. [CrossRef]
33. Hinton, J.A.; Hofmann, W. Teraelectronvolt Astronomy. *Ann. ReV. Astron. Astrophys.* **2009**, *47*, 523–565. [CrossRef]
34. Bicknell, G.V.; Wagner, S.J. The Evolution of Shocks in Blazar Jets. *Publ. Astron. Soc. Aust.* **2002**, *19*, 129–137. [CrossRef]
35. Blandford, R.G.; Levinson, A. Pair Cascades in Extragalactic Jets. I. Gamma Rays. *Astrophys. J.* **1995**, *441*, 79–95. [CrossRef]
36. White, C.J.; Quataert, E.; Gammie, C.F. The Structure of Radiatively Inefficient Black Hole Accretion Flows. *Astrophys. J.* **2020**, *891*, 63. [CrossRef]
37. Bednarek, W.; Sitarek, J. Absorption effects in the blazar's γ-ray spectra due to luminous stars crossing the jet. *Mon. Not. R. Astron. Soc.* **2021**, *503*, 2423–2431. [CrossRef]
38. Maraschi, L.; Fossati, G.; Tavecchio, F.; Chiappetti, L.; Celotti, A.; Ghisellini, G.; Grandi, P.; Pian, E.; Tagliaferri, G.; Treves, A.; et al. Simultaneous X-ray and TEV Observations of a Rapid Flare from Markarian 421. *Astron. Astrophys.* **1999**, *526*, L81–L84. [CrossRef]
39. Fossati, G.; Buckley, J.H.; Bond, I.H.; Bradbury, S.M.; Carter-Lewis, D.A.; Chow, Y.C.K.; Cui, W.; Falcone, A.D.; Finley, J.P.; Gaidos, J.A.; et al. Multiwavelength Observations of Markarian 421 in 2001 March: An Unprecedented View on the X-ray/TeV Correlated Variability. *Astrophys. J.* **2008**, *677*, 906–925. [CrossRef]
40. Acciari, V.A.; Aliu, E.; Arlen, T.; Aune, T.; Beilicke, M.; Benbow, W.; Boltuch, D.; Bradbury, S.M.; Buckley, J.H.; Bugaev, V.; et al. TeV and Multi-wavelength Observations of Mrk 421 in 2006–2008. *Astrophys. J.* **2011**, *738*, 25. [CrossRef]
41. Aleksić, J.; Ansoldi, S.; Antonelli, L.A.; Antoranz, P.; Babic, A.; Bangale, P.; Barres de Almeida, U.; Barrio, J.A.; Becerra González, J.; Bednarek, W.; et al. The 2009 multiwavelength campaign on Mrk 421: Variability and correlation studies. *Astron. Astrophys.* **2015**, *576*, A126. [CrossRef]
42. Kapanadze, B.; Dorner, D.; Romano, P.; Vercellone, S.; Kapanadze, S.; Tabagari, L. Swift Observations of Mrk 421 in Selected Epochs. II. An Extreme Spectral Flux Variability in 2009–2012. *Astrophys. J.* **2018**, *858*, 68. [CrossRef]
43. Aleksić, J.; Ansoldi, S.; Antonelli, L.A.; Antoranz, P.; Babic, A.; Bangale, P.; Barres de Almeida, U.; Barrio, J.A.; Becerra González, J.; Bednarek, W. Unprecedented study of the broadband emission of Mrk 421 during flaring activity in March 2010. *Astron. Astrophys.* **2015**, *578*, A22. [CrossRef]
44. Kapanadze, B.; Gurchumelia, A.; Dorner, D.; Vercellone, S.; Romano, P.; Hughes, P.; Aller, M.; Aller, H.; Kharshiladze, O. Swift Observations of Mrk 421 in Selected Epochs. III. Extreme X-ray Timing/Spectral Properties and Multiwavelength Lognormality during 2015 December–2018 April. *Astrophys. J. Suppl. Ser.* **2020**, *247*, 27. [CrossRef]
45. Arbet-Engels, A.; Baack, D.; Balbo, M.; Biland, A.; Blank, M.; Bretz, T.; Bruegge, K.; Bulinski, M.; Buss, J.; Doerr, M.; et al. The relentless variability of Mrk 421 from the TeV to the radio. *Astron. Astrophys.* **2015**, *647*, A88. [CrossRef]
46. Mastichiadis, A.; Kirk, J.G. Variability in the synchrotron self-Compton model of blazar emission. *Astron. Astrophys.* **1997**, *320*, 19–25.
47. Krawczynski, H.; Coppi, P.S.; Aharonian, F. Time-dependent modelling of the Markarian 501 X-ray and TeV gamma-ray data taken during 1997 March and April. *Mon. Not. R. Astron. Soc.* **2002**, *336*, 721–735. [CrossRef]
48. Sambruna, R.M.; Aharonian, F.A.; Krawczynski, H.; Akhperjanian, A.G.; Barrio, J.A.; Bernlohr, K.; Bojahr, H.; Calle, I.; Contreras, J.L.; Cortina, J.; et al. Correlated Intense X-ray and TEV Activity of Markarian 501 in 1998 June. *Astrophys. J.* **2000**, *538*, 127–133. [CrossRef]
49. Arbet-Engels, A.; Baack, D.; Balbo, M.; Biland, A.; Bretz, T.; Buss, J.; Dorner, D.; Eisenberger, L.; Elsaesser, D.; Hildebrand, D. et al. Long-term multi-band photometric monitoring of Mrk 501. *Astron. Astrophys.* **2015**, *655*, A93. [CrossRef]
50. Aleksić, J.; Ansoldi, S.; Antonelli, L.A.; Antoranz, P.; Babic, A.; Bangale, P.; Barres de Almeida, U.; Barrio, J.A.; Becerra González, J.; Bednarek, W.; et al. Multiwavelength observations of Mrk 501 in 2008. *Astron. Astrophys.* **2015**, *573*, A50. [CrossRef]

51. MAGIC Collaboration; Abe, H.; Abe, S.; Acciari, V.A.; Agudo, I.; Aniello, T.; Ansoldi, S.; Antonelli, L.A.; Arbet Engels, A.; Arcaro, C.; et al. Multi-messenger characterization of Mrk 501 during historically low X-ray and γ-ray activity. *Astrophys. J. Suppl. Ser.* **2023**, *266*, 37. [CrossRef]
52. Abdo, A.A.; Ackermann, M.; Ajello, M.; Allafort, A.; Baldini, L.; Ballet, J.; Barbiellini, G.; Baring, M.G.; Bastieri, D.; et al. Insights into the High-energy γ-ray Emission of Markarian 501 from Extensive Multifrequency Observations in the Fermi Era. *Astrophys. J.* **2019**, *727*, 129. [CrossRef]
53. Aliu, E.; Archambault, S.; Arlen, T.; Aune, T.; Barnacka, A.; Beilicke, M.; Benbow, W.; Berger, K.; Bird, R.; Bouvier, A.; et al. Investigating Broadband Variability Of The Tev Blazar 1es 1959+650. *Astrophys. J.* **2014**, *797*, 89. [CrossRef]
54. MAGIC Collaboration; Acciari, V.A.; Ansoldi, S.; Antonelli, L.A.; Arbet Engels, A.; Baack, D.; Babić, A.; Banerjee, B.; Barres de Almeida, U.; Barrio, J.A.; et al. Broadband characterisation of the very intense TeV flares of the blazar 1ES 1959+650 in 2016. *Astron. Astrophys.* **2020**, *640*, A132.
55. Chiappetti, L.; Maraschi, L.; Tavecchio, F.; Celotti, A.; Fossati, G.; Ghisellini, G.; Giommi, P.; Pian, E.; Tagliaferri, G.; Treves, A. Spectral Evolution of PKS 2155−304 Observed with BeppoSAX during an Active Gamma-Ray Phase. *Astrophys. J.* **1999**, *521*, 552–560. [CrossRef]
56. Böttcher M.; Baring M.G. Multi-wavelength Variability Signatures of Relativistic Shocks in Blazar Jets. *Astrophys. J.* **2019**, *887*, 133. [CrossRef]
57. Graff, P.B.; Georganopoulos, M.; Perlman, E.S.; Kazanas, D. A Multizone Model for Simulating the High-Energy Variability of TeV Blazars. *Astrophys. J.* **2008**, *689*, 68–78. [CrossRef]
58. Krawczynski, H.; Hughes, S.B.; Horan, D.; Aharonian, F.; Aller, M.F.; Aller, H.; Boltwood, P.; Buckley, J.; Coppi, P.; Fossati, G.; et al. Multiwavelength Observations of Strong Flares from the TeV Blazar 1ES 1959+650. *Astrophys. J.* **2004**, *601*, 151–164. [CrossRef]
59. Kapanadze, B.; Romano, P.; Vercellone, S.; Kapanadze, S.; Mdzinarishvili, T.; Kharshiladze, G. The long-term Swift observations of the high-energy peaked BL Lacertae source 1ES 1959+650. *Mon. Not. R. Astron. Soc.* **2016**, *457*, 704–722. [CrossRef]
60. Tagliaferri, G.; Foschini, L.; Ghisellini, G.; Maraschi, L.; Tosti, G; Albert, J.; Aliu, E.; Anderhub, H.; Antoranz, P.; et al. Simultaneous Multiwavelength Observations of the Blazar 1ES 1959+650 at a Low TeV Flux. *Astrophys. J.* **2008**, *679*, 1029–1039. [CrossRef]
61. Kapanadze, B.; Dorner, D.; Vercellone, S.; Romano, P.; Hughes, P.; Aller, M.; Aller, H.; Reynolds, M.; Kapanadze, S.; Tabagari, L. The second strong X-ray flare and multifrequency variability of 1ES 1959+650 in 2016 January-August. *Mon. Not. R. Astron. Soc.* **2018**, *473*, 2542–2564. [CrossRef]
62. Kapanadze, B.; Dorner, D.; Vercellone, S.; Romano, P.; Hughes, P.; Aller, M.; Aller, H.; Reynolds, M.; Tabagari, L. Strong X-ray and Multiwavelength Flaring Activity for 1ES 1959+650, 2016 August-2017 November. *Astrophys. J. Suppl. Ser.* **2018**, *238*, 13. [CrossRef]
63. Patel, S.R.; Shukla, A.; Chitnis, V.R.; Dorner, D.; Mannheim, K.; Acharya, B.S.; Nagare, B.J. Broadband study of blazar 1ES 1959+650 during flaring state in 2016. *Astron. Astrophys.* **2018**, *611*, A44. [CrossRef]
64. Rebillot, P.F.; Badran, H.M.; Blaylock, G.; Bradbury, S.M.; Buckley, J.H.; Carter-Lewis, D.A.; Celik, O.; Chow, Y.C.; Cogan, P.; Cui, W.; et al. Multiwavelength Observations of the Blazar Markarian 421 in 2002 December and 2003 January. *Astrophys. J.* **2006**, *641*, 740–751. [CrossRef]
65. Błazejowski, M.; Blaylock, G.; Bond, I.H.; Bradbury, S.M.; Buckley, J.H.; Carter-Lewis, D.A.; Celik, O.; Cogan, P.; Cui, W.; Daniel, M. A Multiwavelength View of the TeV Blazar Markarian 421: Correlated Variability, Flaring, and Spectral Evolution. *Astrophys. J.* **2005**, *630*, 130–141. [CrossRef]
66. Kapanadze, B.; Vercellone, S.; Romano, P.; Hughes, P.; Aller, M.; Aller, H.; Kharshiladze, O.; Kapanadze, S.; Tabagari, L. Swift Observations of Mrk 421 in Selected Epochs. I. The Spectral and Flux Variability in 2005–2008. *Astrophys. J.* **2018**, *854*, 66. [CrossRef]
67. Abeysekara, A.U.; Benbow, W.; Bird, R.; Brill, A.; Brose, R.; Buchovecky, M.; Buckley, J.H.; Christiansen, J.L.; Chromey, A.J.; Daniel, M.K.; et al. The Great Markarian 421 Flare of 2010 February: Multiwavelength Variability and Correlation Studies. *Astrophys. J.* **2020**, *890*, 97. [CrossRef]
68. Kapanadze, B.; Dorner, D.; Romano, P.; Vercellone, S.; Kapanadze, S.; Tabagari, L. Mrk 421 after the Giant X-ray Outburst in 2013. *Astrophys. J.* **2017**, *848*, 103. [CrossRef]
69. Kapanadze, B.; Dorner, D.; Romano, P.; Vercellone, S.T.; Mannheim, K.; Lindfors, E.; Nilsson, K.; Reinthal, R.; Takalo, L.; Kapanadze, S.; et al. The prolonged X-ray flaring activity of Mrk 501 in 2014. *Mon. Not. R. Astron. Soc.* **2017**, *469*, 1655–1672. [CrossRef]
70. Liodakis, I.; Marscher, A.P.; Agudo, M.; Berdyugin, A.V.; Bernardos, M.I.; Bonnoli, G.; Borman, G.A.; Casadio, C.; Casanova, V.; Cavazzuti, E.; et al. Polarized blazar X-rays imply particle acceleration in shocks. *Nature* **2022**, *611*, 677–681. [CrossRef]
71. Di Gesu, L.; Donnarumma, I.; Tavecchio, F.; Agudo, I.; Barnounin, T.; Cibrario, N.; Di Lalla, N.; Di Marco, A.; Escudero, J.; Errando, M.; et al. The X-ray Polarization View of Mrk 421 in an Average Flux State as Observed by the Imaging X-ray Polarimetry Explorer. *Astrophys. J.* **2022**, *938*, 7. [CrossRef]
72. Di Gesu, L.; Marshall, H.L.; Ehlert, S.R.; Kim, D.E.; Donnarumma, I.; Tavecchio, F.; Liodakis, I.; Kiehlmann, S.; Agudo, I.; Jorstad, S.G.; et al. Discovery of X-ray polarization angle rotation in active galaxy Mrk 421. *arXiv* **2023**, arXiv:2305.13497.
73. Boula, S.; Mastichiadis, A. Expanding one-zone model for blazar emission. *Astron. Astrophys.* **2021**, *657*, A20. [CrossRef]

74. Acciari, V.A.; Ansoldi, S.; Antonelli, L.A.; Engels, A.A.; Artero, M.; Asano, K.; Babić, A.; Baquero, A.; Barres de Almeida, U.; MAGIC Collaboration; et al. Investigation of the correlation patterns and the Compton dominance variability of Mrk 421 in 2017. *Astron. Astrophys.* **2021**, *655*, A89.
75. Chiaberge, M.; Ghisellini, G. Rapid variability in the synchrotron self-Compton model for blazars. *Mon. Not. R. Astron. Soc.* **1999**, *306*, 551–560. [CrossRef]
76. Sokolov, A.; Marscher, A.P.; McHardy, I.M. Synchrotron Self-Compton Model for Rapid Nonthermal Flares in Blazars with Frequency-dependent Time Lags. *Astrophys. J.* **2004**, *613*, 725–746. [CrossRef]
77. Kusunose, M.; Takahara, F. A structured leptonic jet model of the 'orphan' tev gamma-ray flares in tev blazars. *Astrophys. J.* **2006**, *651*, 113–119. [CrossRef]
78. Abramowski, A.; Acero, F.; Aharonian, F.; Akhperjanian, A.G.; Anton, G.; Balzer, A.; Barnacka, A.; De Almeida, U.B.; Becherini, Y.; Becker, J.; et al. A multiwavelength view of the flaring state of PKS 2155−304 in 2006. *Astron. Astrophys.* **2011**, *539*, 149–170.
79. Dmytriiev, A.; Sol, H.; Zech, A. Connecting steady emission and very high energy flaring states in blazars: The case of Mrk 421. *Mon. Not. R. Astron. Soc.* **2021**, *505*, 2712–2730. [CrossRef]
80. MAGIC Collaboration; Acciari, V.A.; Ansoldi, S.; Antonelli, L.A.; Arbet Engels, A.; Baack, D.; Babić, A.; Banerjee, B.; Barres de Almeida, U.; Barrio, J.A.; et al. Testing two-component models on very high-energy gamma-ray-emitting BL Lac objects. *Astron. Astrophys.* **2020**, *638*, A14.
81. Dong, Q.; Zheng, Y.G.; Yang, C.Y. Two components model for TeVγ-ray emission from extreme high-energy BL Lac objects. *Astrophys. Space Sci.* **2021**, *366*, 36. [CrossRef]
82. Zech, A.; Lemoine, M. Electron-proton co-acceleration on relativistic shocks in extreme-TeV blazars. *Astron. Astrophys.* **2021**, *654*, 96–113. [CrossRef]
83. Matsumoto, J.; Komissarov, S.S.; Gourgouliatos, K.N. Magnetic inhibition of the recollimation instability in relativistic jets. *Mon. Not. R. Astron. Soc.* **2021**, *503*, 4918–4929. [CrossRef]
84. Tavecchio, F.; Costa, A.; Sciaccaluga, A. Extreme blazars: The result of unstable recollimated jets? *Mon. Not. R. Astron. Soc.* **2022**, *517*, L16–L20. [CrossRef]
85. MAGIC Collaboration; Acciari, V.A.; Ansoldi, S.; Antonelli, L.A.; Babic, A.; Banerjee, B.; Barres de Almeida, U.; Barrio, J.A.; Becerra González, J.; Bednarek, W.; et al. Study of the variable broadband emission of Markarian 501 during the most extreme Swift X-ray activity. *Astron. Astrophys.* **2020**, *637*, 86.
86. Sol, H.; Pelletier, G.; Asseo, E. Two-flow model for extragalactic radio jets. *Mon. Not. R. Astron. Soc.* **1989**, *237*, 411–429. [CrossRef]
87. Dermer, C.D.; Razzaque, S. Acceleration of Ultra-high-energy Cosmic Rays in the Colliding Shells of Blazars and Gamma-ray Bursts: Constraints from the Fermi Gamma-ray Space Telescope. *Astrophys. J.* **2010**, *724*, 1366–1372. [CrossRef]
88. Celotti, A.; Ghisellini, G. The power of blazar jets. *Mon. Not. R. Astron. Soc.* **2008**, *385*, 283–300. [CrossRef]
89. Mannheim, K. The proton blazar. *Astron. Astrophys.* **1993**, *269*, 67–76.
90. Aharonian, F. TeV gamma rays from BL Lac Objects due to synchrotron radiation of extremely high energy protons. *New Astron.* **2000**, *5*, 377–395. [CrossRef]
91. Mücke, A.; Protheroe, R.J.; Engel, R.; Rachen, J.P.; Stanev, T. BL Lac objects in the synchrotron proton blazar model. *Astropart. Phys.* **2003**, *18*, 593–613. [CrossRef]
92. Bhatta, G. Blazar Jets as Possible Sources of Ultra-High Energy Photons: A Short Review. *University* **2022**, *8*, 513. [CrossRef]
93. Bethe, H.; Heitler, W. On the Stopping of Fast Particles and on the Creation of Positive Electrons. *Proc. R. Soc. Lond.* **1934**, *134*, 83–112.
94. Petropoulou, M.; Mastichiadis, A. Bethe–Heitler emission in BL Lacs: Filling the gap between X-rays and γ-rays. *Mon. Not. R. Astron. Soc.* **2015**, *447*, 36–48. [CrossRef]
95. Dermer, C.D.; Murase, K.; Takami, H. Variable Gamma-Ray Emission Induced by Ultra-high Energy Neutral Beams: Application to 4C+21.35. *Astrophys. J.* **2012**, *755*, 147. [CrossRef]
96. Böttcher, M. Modeling the Emission Processes in Blazars. *Astrophys. Space Sci.* **2007**, *309*, 95–104. [CrossRef]
97. Zdziarski, A.; Böttcher, M. Hadronic models of blazars require a change of the accretion paradigm. *Mon. Not. R. Astron. Soc.* **2015**, *450*, L21–L25. [CrossRef]
98. Liodakis, I.; Petropoulou, M. Proton Synchrotron Gamma-Rays and the Energy Crisis in Blazars. *Astrophys. J.* **2020**, *893*, 20. [CrossRef]
99. Padovani, P.; Oikonomou, F.; Petropoulou, M.; Giommi, P.; Resconi, E. TXS 0506+056, the first cosmic neutrino source, is not a BL Lac. *Mon. Not. R. Astron. Soc.* **2019**, *484*, L104–L108. [CrossRef]
100. IceCube Collaboration; Aartsen, M.G.; Ackermann, M.; Adams, J.; Aguilar, J.A.; Ahlers, M.; Ahrens, M.; Samarai, I.; Altmann, D.; Andeen, K.; et al. Neutrino emission from the direction of the blazar TXS 0506+056 prior to the IceCube-170922A alert. *Science* **2018**, *361*, 147–151.
101. Acciari, V.A.; Aniello, T.; Ansoldi, S.; Antonelli, L.A.; Arbet Engels, A.; Artero, M.; Asano, K.; Baack, D.; Babić, A.; Baquero, A.; et al. Investigating the Blazar TXS 0506+056 through Sharp Multiwavelength Eyes During 2017–2019. *Astrophys. J.* **2022**, *927*, 197. [CrossRef]
102. Resconi, E.; Coenders, S.; Padovani, P.; Giommi, P.; Caccianiga, L. Connecting blazars with ultrahigh-energy cosmic rays and astrophysical neutrinos. *Mon. Not. R. Astron. Soc.* **2017**, *468*, 597–606. [CrossRef]
103. Mastichiadis, A.; Petropoulou, M. Hadronic X-ray Flares from Blazars. *Astrophys. J.* **2017**, *906*, 131. [CrossRef]

104. Petropoulou, M.; Dimitrakoudis, S.; Padovani, P.; Mastichiadis, A.; Resconi, E. Photohadronic origin of γ-ray BL Lac emission: implications for IceCube neutrinos. *Mon. Not. R. Astron. Soc.* **2015**, *448*, 2412–2429. [CrossRef]
105. Lucarelli, F.; Pittori, C.; Verrecchia, F.; Donnarumma, I.; Tavani, M.; Bulgarelli, A.; Giuliani, A.; Antonelli, L.A.; Caraveo, P.; Cattaneo, P.W.; et al. AGILE Detection of a Candidate Gamma-Ray Precursor to the ICECUBE-160731 Neutrino Event. *Astrophys. J.* **2017**, *846*, 111. [CrossRef]
106. Petropoulou, M.; Murase, K.; Santander, M.; Buson, S.; Tohuvavohu, A.; Kawamuro, T.; Vasilopoulos, G.; Negoro, H.; Ueda, Y.; Siegel, M.H.; et al. Multi-epoch Modeling of TXS 0506+056 and Implications for Long-term High-energy Neutrino Emission. *Astrophys. J.* **2020**, *891*, 115. [CrossRef]
107. Gasparyan, S.; Bégué, D.; Sahakyan, N. Time-dependent lepto-hadronic modelling of the emission from blazar jets with SOPRANO: The case of TXS 0506+056, 3HSP J095507.9+355101, and 3C 279. *Mon. Not. R. Astron. Soc.* **2022**, *509*, 2102–2121. [CrossRef]
108. Zacharias, M.; Reimer, A.; Boisson, C.; Zech, A. EXHALE-JET: An extended hadro-leptonic jet model for blazars - I. Code description and initial results. *Mon. Not. R. Astron. Soc.* **2022**, *512*, 3948–3971. [CrossRef]
109. Mastichiadis, A.; Petropoulou M.; Dimitrakoudis, S. Mrk 421 as a case study for TeV and X-ray variability in leptohadronic models. *Mon. Not. R. Astron. Soc.* **2013**, *434*, 2684–2695. [CrossRef]
110. Mondal, S.K.; Das, S.; Gupta, N. Exploring the Emission Mechanisms of Mrk 180 with Long-term X-ray and γ-ray Data. *Astrophys. J.* **2023**, *948*, 75. [CrossRef]
111. Böttcher, M. A Hadronic Synchrotron Mirror Model for the "Orphan" TeV Flare in 1ES 1959+650. *Astrophys. J.* **2005**, *621*, 176–180. [CrossRef]
112. Li, W.-J.; Xue, R.; Long, G.-B.; Wang, Z.-R.; Nagataki, S.; Yan, D.-H.; Wang, J.-C. Can the one-zone hadronuclear model explain the hard-TeV spectrum of BL Lac objects? *Astron. Astrophys.* **2022**, *659*, 184L. [CrossRef]
113. Mastichiadis, A.; Protheroe, R.J.; Kirk, J.G. Spectral and temporal signatures of ultrarelativistic protons in compact sources. I. Effects of Bethe-Heitler pair production. *Astron. Astrophys.* **2005**, *433*, 765–776. [CrossRef]
114. Dimitrakoudis, S.; Mastichiadis, A.; Protheroe, R.J.; Reimer, A. The time-dependent one-zone hadronic model. First principles. *Astron. Astrophys.* **2012**, *546*, A120. [CrossRef]
115. Diltz, C.; Böttcher, M.; Fossati, G. Time Dependent Hadronic Modeling of Flat Spectrum Radio Quasars. *Astrophys. J.* **2015**, *802*, 133. [CrossRef]
116. Tramacere, A. JetSeT: Numerical modeling and SED fitting tool for relativistic jets. *Astrophys. Sourc. Cod. Libr.* **2020**, *122*, ascl:2009.001.
117. Weidinger, M.; Spanier, F. A self-consistent and time-dependent hybrid blazar emission model. Properties and application. *Astron. Astrophys.* **2015**, *573*, A7. [CrossRef]
118. Sahu, S.; Oliveros, A.F.O.; Sanabria, J.C. Hadronic origin orphan TeV flare from 1ES 1959+650. *Phys. Rev. D* **2013**, *87*, 103015. [CrossRef]
119. Sahu, S.; López F.; Carlos E.; Nagataki, S. Multi-TeV flaring from high energy blazars: An evidence of the photohadronic process. *Astrophys. J.* **2019**, *884*, 17. [CrossRef]
120. Sahu, S.; López Fortín, C.E.; Iglesias Martínez, M.E.; Nagataki, S.; Fernández de Córdoba, P. The VHE SED modelling of Markarian 501 in 2009. *Mon. Not. R. Astron. Soc.* **2020**, *492*, 2261–2267. [CrossRef]
121. Sahu, S.; López Fortín, C.E.; Castañeda Hernández, L.H.; Rajpoot, S.A. A Two-zone Photohadronic Interpretation of the EHBL-like Behavior of the 2016 Multi-TeV Flares of 1ES 1959+650. *Astrophys. J.* **2021**, *906*, 91. [CrossRef]
122. Sahu, S.; López Fortín, C.E.; Castañeda Hernández, L.H.; Nagataki, S.; Rajpoot, S. A Two-zone Photohadronic Scenario for EHBL-like Behavior of Mrk 501. *Astrophys. J.* **2020**, *901*, 132. [CrossRef]
123. Sahu, S.; López Fortín, C.E.; Castañeda Hernández, L.H.; Valadez Polanco, I.A.; Rajpoot, S. Extreme HBL-like Behavior of Markarian 421 and Its Two-zone Photohadronic Interpretation. *Astrophys. J.* **2021**, *914*, 120. [CrossRef]
124. Sahu, S.; López Fortín, C.E.; Valadez Polanco, I.A.; Rajpoot, S. Photohadronic interpretations of the different incarnations of 1ES 2344+514. *Mon. Not. R. Astron. Soc.* **2022**, *515*, 5235–5241. [CrossRef]
125. Parfrey, K.; Philippov, A.; Cerruti, B. First-principles plasma simulations of black hole jet launching. *Phys. Rev. Lett.* **2019**, *122*, 03521. [CrossRef]
126. Crinquand, B.; Cerutti, B.; Dubus, G.; Parfrey, K.; Philippov, A. Synthetic gamma-ray lightcurves of Kerr black hole magnetospheric activity from particle-in-cell simulations. *Astron. Astrophys.* **2021**, *650*, A163. [CrossRef]
127. Kissaka, S.; Levinson, A.; Toma, K.; Niv, I. The Response of Black Hole Spark Gaps to External Changes: A Production Mechanism of Rapid TeV Flares? *Astrophys. J.* **2022**, *924*, 28. [CrossRef]
128. Aleksić, J.; Ansoldi, S.; Antonelli, L.A.; Antoranz, P.; Babic, A.; Bangale, P.; Barrio, J.A.; González, J.B.; Bednarek, W.; Bernardini, E.; et al. Black hole lightning due to particle acceleration at subhorizon scales. *Science* **2014**, *346*, 1080–1084. [CrossRef]
129. Levinson, A.; Rieger, F. Variable TeV Emission as a Manifestation of Jet Formation in M87? *Astrophys. J.* **2011**, *730*, 123. [CrossRef]
130. Hirotani, K.; Pu, H.-Y. Energetic Gamma Radiation from Rapidly Rotating Black Holes. *Astrophys. J.* **2016**, *818*, 50. [CrossRef]
131. Breit, G.; Wheeler, J.A. Collision of Two Light Quanta. *Phys. Rev.* **1934**, *46*, 1087–1091. [CrossRef]
132. Gaidos, J.A.; Akerlof, C.W.; Biller, S.; Boyle, P.J.; Breslin, A.C.; Buckley, J.H.; Carter-Lewis, D.A.; Catanese, M.; Cawley, M.F.; Fegan, D.J.; et al. Extremely rapid bursts of TeV photons from the active galaxy Markarian 421. *Nature* **1996**, *383*, 319–320. [CrossRef]

133. Begelman M.C.; Fabian, A.C.; Rees, M.J. Implications of very rapid TeV variability in blazars. *Mon. Not. R. Astron. Soc.* **2008**, *384*, L19–L23. [CrossRef]
134. Rieger, F.M. Gamma-Ray Astrophysics in the Time Domain. *Galax* **2019**, *7*, 28. [CrossRef]
135. Rieger, F.M. An Introduction to Particle Acceleration in Shearing Flows. *Galax* **2019**, *7*, 78. [CrossRef]
136. H.E.S.S. Collaboration; Abramowski, A.; Acero, F.; Aharonian, F.; Akhperjanian, A.G.; Anton, G.; Barres de Almeida, U.; Bazer-Bachi, A.R.; Becherini, Y.; Behera, B.; et al. VHE γ-ray emission of PKS 2155−304: spectral and temporal variability. *Astron. Astrophys.* **2010**, *520*, A83.
137. Giannios D.; Uzdensky D.A.; Begelman M.C. Fast TeV variability in blazars: Jets in a jet. *Mon. Not. R. Astron. Soc.* **2009**, *395*, L29–L33. [CrossRef]
138. Sironi, L.; Petropoulou, M.; Giannios, D. Relativistic jets shine through shocks or magnetic reconnection? *Mon. Not. R. Astron. Soc.* **2015**, *450*, 183–191. [CrossRef]
139. Mizuno, Y.; Hardee, P.; Nishikawa, K.-Y. Three-dimensional Relativistic Magnetohydrodynamic Simulations of Magnetized Spine-Sheath Relativistic Jets. *Astrophys. J.* **2007**, *662*, 835–850. [CrossRef]
140. Mizuno, Y.; Pohl, M.; Niemiec, J.; Zhang, B.; Nishikawa, K.-I.; Hardee, P.E. Magnetic-field Amplification by Turbulence in a Relativistic Shock Propagating Through an Inhomogeneous Medium. *Astrophys. J.* **2011**, *726*, 62. [CrossRef]
141. Sobacchi, E.; Sormani, M.C.; Stamerra, A. NA model for periodic blazars. *Mon. Not. R. Astron. Soc.* **2017**, *465*, 161–172. [CrossRef]
142. Tavani, M.; Cavaliere, A.; Munar-Adrover, P.; Argan, A. The Blazar PG 1553+113 as a Binary System of Supermassive Black Holes. *Astrophys. J.* **2008**, *854*, 11. [CrossRef]
143. Gao, Q.; Lu, F.-W.; Qin, L.; Gong, Y.-L.; Yu, G.; Li, H.; Yi, T. A Geometric Model to Interpret the γ-Ray Quasiperiodic Oscillation of PG 1553+113. *Astrophys. J.* **2023**, *945*, 146. [CrossRef]
144. Begelman, M.C.; Blandford, R.D.; Rees, M.J. Massive black hole binaries in active galactic nuclei. *Nature* **1980**, *287*, 307–309. [CrossRef]
145. Ackermann, M.; Ajello, M.; Albert, A.; Atwood, W.B.; Baldini, L.; Ballet, J.; Barbiellini, G.; Bastieri, D.; Becerra Gonzalez, J.; Bellazzini, R.; et al. Multiwavelength Evidence for Quasi-periodic Modulation in the Gamma-Ray Blazar PG 1553+113. *Astrophys. J.* **2007**, *813*, L41. [CrossRef]
146. Covino, S.; Sandrinelli, A.; Treves, A. Gamma-ray quasi-periodicities of blazars. A cautious approach. *Mon. Not. R. Astron. Soc.* **2019**, *482*, 1270–1274. [CrossRef]
147. O'Neill, S.; Kiehlmann, S.; Readhead, A.C.S.; Aller, M.F.; Blandford, R.D.; Liodakis, I.; Lister, M.L.; Mróz, P.; O'Dea, C.P.; Pearson, T.J.; et al. The Unanticipated Phenomenology of the Blazar PKS 2131-021: A Unique Supermassive Black Hole Binary Candidate. *Astrophys. J.* **2022**, *926*, L35. [CrossRef]
148. Peñil, P.; Domínguez, A.; Buson, S.; Ajello, M.; Otero-Santos, J.; Barrio, J.A.; Nemmen, R.; Cutini, S.; Rani, B.; Franckowiak, A.; et al. Systematic Search for γ-Ray Periodicity in Active Galactic Nuclei Detected by the Fermi Large Area Telescope. *Astrophys. J.* **2020**, *896*, 134. [CrossRef]
149. Yang, S.; Yan, D.; Zhang, P.; Dai, B.; Zhang, L. Gaussian Process Modeling Fermi-LAT γ-Ray Blazar Variability: A Sample of Blazars with γ-Ray Quasi-periodicities. *Astrophys. J.* **2021**, *907*, 105. [CrossRef]
150. Covino, S.; Landoni, M.; Sandrinelli, A.; Treves, A. Looking at Blazar Light-curve Periodicities with Gaussian Processes. *Astrophys. J.* **2020**, *895*, 122. [CrossRef]
151. Peñil, P.; Ajello, M.; Buson, S.; Domínguez, A.; Westernacher-Schneider, J.R.; Zrake, J. Evidence of Periodic Variability in γ-ray Emitting Blazars with Fermi-LAT. *arXiv* **2022**, arXiv:2211.01894.
152. Ren, H.X.; Cerruti, M.; Sahakyan, N. Quasi-periodic oscillations in the γ-ray light curves of bright active galactic nuclei. *Astron. Astrophys.* **2023**, *672*, 86. [CrossRef]
153. Sandrinelli, A.; Covino, S.; Treves, A.; Holgado, A.M.; Sesana, A.; Lindfors, E.; Ramazani, V.F. Quasi-periodicities of BL Lacertae objects. *Astron. Astrophys.* **2018**, *615*, A118. [CrossRef]
154. Stellingwerf, R.F. Period determination using phase dispersion minimization. *Astrophys. J.* **1978**, *224*, 953–960. [CrossRef]
155. Bhatta, G.; Dhital, N. The Nature of γ-Ray Variability in Blazars. *Astrophys. J.* **2020**, *891*, 120. [CrossRef]
156. Bhatta, G. Blazar Mrk 01 shows rhythmic oscillations in its γ-ray emission. *Mon. Not. R. Astron. Soc.* **2019**, *487*, 3990–3997. [CrossRef]
157. Abdo, A.A.; Ackermann, M.; Ajello, M.; Baldini, L.; Ballet, J.; Barbiellini, G.; Bastieri, D.; Bechtol, K.; Bellazzini, R.; Berenji, B.; et al. Fermi Large Area Telescope Observations of Markarian 421: The Missing Piece of its Spectral Energy Distribution. *Astrophys. J.* **2011**, *736*, 131. [CrossRef]
158. Spitkovsky, A. Particle Acceleration in Relativistic Collisionless Shocks: Fermi Process at Last? *Astrophys. J. Lett.* **2008**, 682, L5–L8. [CrossRef]
159. Finke, J.D.; Dermer, C.D.; Böttcher, M. Synchrotron Self-Compton Analysis of TeV X-ray-Selected BL Lacertae Objects. *Astrophys. J.* **2008**, *686*, 181–194. [CrossRef]
160. Böttcher, M.; Dermer, C.D. Timing Signatures of the Internal-Shock Model for Blazars. *Astrophys. J.* **2010**, *711*, 445–460. [CrossRef]
161. Blandford, R.D.; Eichler, D. Particle acceleration at astrophysical shocks: A theory of cosmic ray origin. *Phys. Rep.* **1987**, *154*, 1–75. [CrossRef]
162. Begelman M.C.; Kirk, J.G. Shock-Drift Particle Acceleration in Superluminal Shocks: A Model for Hot Spots in Extragalactic Radio Sources. *Astrophys. J.* **1990**, *353*, 66–80. [CrossRef]

163. Kirk, J.G.; Reville, B. Radiative Signatures of Relativistic Shocks. *Astrophys. J.* **2010**, *710*, L16–L20. [CrossRef]
164. Sironi, L.; Spitkovsky, A.; Arons, J. The maximum energy of accelerated particles in relativistic collisionless shocks. *Astrophys. J.* **2013**, *771*, 54. [CrossRef]
165. Nalewajko, K.; Giannios, D.; Begelman, M.C.; Uzdensky, D.A.; Sikora, M. Radiative properties of reconnection-powered minijets in blazars. *Mon. Not. R. Astron. Soc.* **2011**, *413*, 333–346. [CrossRef]
166. Kusunose, M.; Takahara, F.; Li, H. Electron Acceleration and Time Variability of High-Energy Emission from Blazars. *Astrophys. J.* **2000**, *636*, 299–307. [CrossRef]
167. Chen, X.; Fossati, G.; Liang, E.P.; Böttcher, M. Time-dependent simulations of multiwavelength variability of the blazar Mrk 421 with a Monte Carlo multizone code. *Mon. Not. R. Astron. Soc.* **2012**, *416*, 2368–2387. [CrossRef]
168. Shukla, A.; Chitnis, V.R.; Singh, B.B.; Acharya, B.S.; Anupama, G.C.; Bhattacharjee, P.; Britto, R.J.; Mannheim, K.; Prabhu, T.P.; Saha, L.; et al. Multi-frequency, Multi-epoch Study of Mrk 501: Hints for a Two-component Nature of the Emission. *Astrophys. J.* **2011**, *798*, 2. [CrossRef]
169. Massaro, E.; Perri, M.; Giommi, P.; Nesci, R. Log-parabolic spectra and particle acceleration in the BL Lac object Mkn 421: Spectral analysis of the complete BeppoSAX wide band X-ray data set. *Astron. Astrophys.* **2004**, *413*, 489–503. [CrossRef]
170. Hughes, P.A.; Aller, M.F.; Aller, H.D. Oblique Shocks as the Origin of Radio to Gamma-ray Variability in Active Galactic Nuclei. *Astrophys. J.* **2011**, *735*, 81. [CrossRef]
171. Marscher, A.P.; Jorstad, S.G. The Megaparsec-scale X-ray Jet of The BL Lac Object OJ287. *Astrophys. J.* **2011**, *729*, 26. [CrossRef]
172. Achterberg, A.; Gallant, Y.A.; Kirk, J.G.; Guthmann, A.W. Particle acceleration by ultrarelativistic shocks: Theory and simulations. *Mon. Not. R. Astron. Soc.* **2001**, *328*, 393–408. [CrossRef]
173. Duran, R.B.; Tchekhovskoy, A.; Giannios, D. Simulations of AGN jets: Magnetic kink instability versus conical shocks. *Mon. Not. R. Astron. Soc.* **2017**, *469*, 4957–4978. [CrossRef]
174. Lazarian, A.; Eyink, G.; Vishniac, E.; Kowal, G. Reconnection in Turbulent Astrophysical Fluids. *ASP Conf. Ser.* **2015**, *488*, 3.
175. Werner G.R.; Uzdensky D.A. Nonthermal Particle Acceleration in 3D Relativistic Magnetic Reconnection in Pair Plasma. *Astrophys. J.* **2017**, *843*, L27. [CrossRef]
176. Giannios, D. Reconnection-driven plasmoids in blazars: fast flares on a slow envelope. *Mon. Not. R. Astron. Soc.* **2013**, *431*, 355–362. [CrossRef]
177. Sironi, L.; Spitkovsky, A. Relativistic Reconnection: An Efficient Source of Non-thermal Particles. *Astrophys. J.* **2014**, *783*, L21. [CrossRef]
178. Hakobyan, H.; Petropoulou, M.; Spitkovsky, A.; Sironi, L. Secondary Energization in Compressing Plasmoids during Magnetic Reconnection. *Astrophys. J.* **2021**, *912*, 48. [CrossRef]
179. Petropoulou, M.; Giannios, D.; Sironi, L. Blazar flares powered by plasmoids in relativistic reconnection. *Mon. Not. R. Astron. Soc.* **2016**, *462*, 3325–3353. [CrossRef]
180. Acciari, V.A.; Ansoldi, S.; Antonelli, L.A.; Arbet Engels, A.; Baack, D.; Babić, A.; Banerjee, B.; Barres de Almeida, U.; Barrio, J.A.; Becerra González, J.; et al. Unraveling the Complex Behavior of Mrk 421 with Simultaneous X-ray and VHE Observations during an Extreme Flaring Activity in 2013 April. *Astrophys. J. Suppl. Ser.* **2013**, *248*, 29. [CrossRef]
181. Demidem, C.; Lemoine, M.; Casse, F. Particle acceleration in relativistic turbulence: A theoretical appraisal. *Phys. Rev. D* **2020**, *102*, 023003. [CrossRef]
182. Narayan, R.; Li, J.; Tchekhovskoy, A. Stability of Relativistic Force-free Jets. *Astrophys. J.* **2009**, *697*, 1681–1694. [CrossRef]
183. Narayan, R.; Piran, T. Variability in blazars: Clues from PKS 2155–304. *Mon. Not. R. Astron. Soc.* **2012**, *420*, 604–612. [CrossRef]
184. Lefa, E.; Rieger, F.M.; Aharonian, F. Formation of Very Hard Gamma-Ray Spectra of Blazars in Leptonic Models. *Astrophys. J.* **2011**, *740*, 64. [CrossRef]
185. Ahnen, M.L.; Ansoldi, S.; Antonelli, L.A.; Arcaro, C.; Babić, A.; Banerjee, B.; Bangale, P.; Barres de Almeida, U.; Barrio, J.A.; Becerra González, J.; et al. Extreme HBL behavior of Markarian 501 during 2012. *Astron. Astrophys.* **2012**, *620*, A181. [CrossRef]
186. Shukla, A.; Mannheim, K.; Chitnis, V.R.; Roy, J.; Acharya, B.S.; Dorner, D.; Hughes, G.; Biland, A. Detection of Very Hard γ-ray Spectrum from the TeV Blazar Mrk 501. *Astrophys. J.* **2016**, 177. [CrossRef]
187. Kapanadze, B.; Gurchumelia, A. Long-term multi-wavelength variability and extreme spectral properties of the TeV-detected blazar 1ES 0033+595. *Astron. Astrophys.* **2022**, *668*, A75. [CrossRef]
188. Mizuno, Y.; Pohl, M.; Niemiec, J.; Zhang, B.; Nishikawa, K.-I.; Hardee, P.E. Magnetic field amplification and saturation in turbulence behind a relativistic shock. *Mon. Not. R. Astron. Soc.* **2014**, *439*, 3490–350. [CrossRef]
189. Becker, P.A.; Le, T.; Dermer, C.D. Time-dependent Stochastic Particle Acceleration in Astrophysical Plasmas: Exact Solutions Including Momentum-dependent Escape. *Astrophys. J.* **2006**, *646*, 539–551. [CrossRef]
190. Virtanen, J.J.P.; Vainio, R. Stochastic Acceleration in Relativistic Parallel Shocks. *Astrophys. J.* **2005**, *621*, 313–323. [CrossRef]
191. Katarzynski, K.; Ghisellini, G.; Mastichiadis, A.; Tavecchio, F.; Maraschi, L. Stochastic particle acceleration and synchrotron self-Compton radiation in TeV blazars. *Astron. Astroph.* **2006**, *453*, 47–56. [CrossRef]
192. Lewis, T.R.; Becker, P.A.; Finke, J.D. Time-dependent Electron Acceleration in Blazar Transients: X-ray Time Lags and Spectral Formation. *Astrophys. J.* **2016**, *824*, 108. [CrossRef]
193. Hillas, A.M. The Origin of Ultra-High-Energy Cosmic Rays. *Ann. Rev. Astron. Astroph.* **1984**, *22*, 422–445. [CrossRef]
194. Stawarz, Ł.; Petrosian, V. On the momentum diffusion of radiating ultrarelativistic electrons in a turbulent magnetic field. *Astrophys. J.* **2008**, *681*, 1725–1744. [CrossRef]

195. Comisso, L.; Sironi, L. Particle Acceleration in Relativistic Plasma Turbulence. *Phys. Rev. Lett.* **2018**, *121*, 255101. [CrossRef]
196. Virtanen, J.J.P.; Vainio, R.; Schlickeiser, R. Alfvén-wave transmission and test-particle acceleration in parallel relativistic shocks. *Astron. Astrophys.* **2003**, *409*, 821–829.
197. Lefa, E.; Aharonian, F.A.; Rieger, F.M. 'Leading Blob' Model in a Stochastic Acceleration Scenario: The Case of the 2009 Flare of Mkn 501. *Astrophys. J.* **2012**, *743*, L19. [CrossRef]
198. Rieger, F.M.; Duffy, P. Turbulence and particle acceleration in shearing flows. *Astrophys. J. Lett.* **2021**, 907, L2. [CrossRef]
199. Barkov, M.V.; Aharonian, F.A.; Bogovalov, S.V.; Kelner, S.R.; Khangulyan, D. Rapid TeV Variability in Blazars as a Result of Jet-Star Interaction. *Astrophys. J.* **2011**, *749*, 119. [CrossRef]
200. Bosch-Ramon, V. Non-thermal emission from standing relativistic shocks: An application to red giant winds interacting with AGN jets. *Astron. Astrophys.* **2015**, *575*, A109. [CrossRef]
201. Barkov, M.V.; Aharonian, F.A.; Bosch-Ramon, V. Gamma-ray Flares from Red Giant/Jet Interactions in Active Galactic Nuclei. *Astrophys. J.* **2011**, *724*, 1517–1523. [CrossRef]
202. Torres-Albà, N.; Bosch-Ramon, V. Gamma rays from red giant wind bubbles entering the jets of elliptical host blazars. *Astron. Astrophys.* **2019**, *623*, A91. [CrossRef]
203. de la Cita, V.M.; Bosch-Ramon, V.; Paredes-Fortuny, X.; Khangulyan, D.; Perucho, M. Coupling hydrodynamics and radiation calculations for star-jet interactions in active galactic nuclei. *Astron. Astrophys.* **2016**, *591*, A15. [CrossRef]
204. Giebels, B.; Degrange, B. Lognormal variability in BL Lacertae. *Astron. Astrophys.* **2009**, *503*, 797–799. [CrossRef]
205. Sinha, A.; Khatoon, R.; Misra, R.; Sahayanathan, S.; Mandal, S.; Gogoi, R.; Bhatt, N. The flux distribution of individual blazars as a key to understand the dynamics of particle acceleration. *Mon. Not. R. Astron. Soc.* **2014**, *480*, L116–L120. [CrossRef]
206. Sinha, A.; Shukla, A.; Saha, L.; Acharya, B.S.; Anupama, G.C.; Bhattacharya, P.; Britto, R.J.; Chitnis, V.R.; Prabhu, T.P.; Singh, B.B.; et al. Long-term study of Mkn 421 with the HAGAR Array of Telescopes. *Astron. Astrophys.* **2016**, *591*, 83. [CrossRef]
207. Acciari, V.A.; Ansoldi, S.; Antonelli, L.A.; Asano, K.; Babić, A.; Banerjee, B.; Baquero, A.; de Almeida, U.B.; Barrio, J.A.; Becerra González, J.; et al. Multiwavelength variability and correlation studies of Mrk 421 during historically low X-ray and γ-ray activity in 2015-2016. *Mon. Not. R. Astron. Soc.* **2021**, *504*, 1427–1451. [CrossRef]
208. Kapanadze, B.; Gurchumelia, A.; Aller M. Long-Term X-ray Outburst in the Tev-Detected Blazar Mrk 501 in 2021–2022: Further Clues for the Emission and Unstable Processes. *Astrophys. J. Suppl. Ser.* 2023, *in press*.
209. Kapanadze, B. The long-term multiwavelength observations of the blazar PKS 2005−489. *Astropart. Phys.* **2021**, *132*, 102620. [CrossRef]
210. Kapanadze, B.; Gurchumelia, A.; Vercellone, S.; Romano, P.; Kapanadze, S.; Kharshiladze, O. Long-term Swift and multiwavelength observations of two TeV-detected blazars with unknown redshifts. *Astrophys. Space Sci.* **2023**, *368*, 23. [CrossRef]
211. Engel, K.; Goodman, J.; Huentemeyer, P.; Kierans, C.; Lewis, T.R.; Negro, M.; San-tander, M.; Williams, D.A.; Allen, A.; Aramaki, T.; et al. [for the CTA Consortium] CTA— The Cherenkov Telescope Array. In The Future of Gamma-Ray Experiments in the MeV–EeV Range. *arXiv* **2022**, arXiv:2203.07360.
212. Acharya, B.S.; Agudo, I.; Al Samarai, I.; Alfaro, R.; Alfaro, J.; Alispach, C.; Alves Batista, R.; Amans, J.P.; Amato, E.; Ambrosi, G.; et al. *Science with the Cherenkov Telescope Array*; CTA Consortium, Ed.; World Scientific Publishing Co. Pte. Ltd.: Singapore, 2018; ISBN 9789813270091.
213. Vercellone, S.; Bigongiari, C.; Burtovoi, A.; Cardillo, M.; Catalano, O.; Franceschini, A.; Lombardi, S.; Nava, L.; Pintore, F.; Stamerra, A.; et al. ASTRI Mini-Array core science at the Observatorio del Teide. *J. High Energy Astrophys.* **2022**, *35*, 1–42. [CrossRef]
214. Scuderi, S.; Giuliani, A.; Pareschi, G.; Tosti, G.; Catalano, O.; Amato, E.; Antonelli, L.A.; Becerra Gonzàles, J.; Bellassai, G.; Bigongiari, C.; et al. The ASTRI Mini-Array of Cherenkov telescopes at the Observatorio del Teide. *J. High Energy Astrophys.* **2022**, *35*, 52–68. [CrossRef]
215. Kierans, C.; Caputo, R.; McEnery, J.; Perkins, J.S. AMEGO and AMEGO-X— The All-sky Medium Energy Gamma-ray Observatory. In The Future of Gamma-Ray Experiments in the MeV–EeV Range. *arXiv* **2022**, arXiv:2203.07360v1.
216. Zhang, H. Gamma-Ray Polarimetry. In The Future of Gamma-Ray Experiments in the MeV–EeV Range. *arXiv* **2022**, arXiv:2203.07360v1.
217. Engel, K.; Harding, P.J.; Brisbois, C. [for The SWGO] Collaboration SWGO— The Southern Wide-field Gamma-ray Observatory. In The Future of Gamma-Ray Experiments in the MeV–EeV Range. *arXiv* **2022**, arXiv:2203.07360.
218. De Angelis, A.; Tatischeff, V.; Tavani, M.; Oberlack, U.; Grenier, I.; Hanlon, L.; Walter, R.; Argan, A.; von Ballmoos, P.; Bulgarelli, A.; et al. Exploring the extreme Universe with gamma rays in the MeV—GeV range. *Exper. Astron.* **2017**, *44*, 25–82. [CrossRef]
219. Albert, A.; Alfaro, R.; Alvarez, C.; Angeles Camacho, J.R.; Arteaga-Velázquez, J.C.; Arunbabu, K.P.; Avila Rojas, D.; Ayala Solares, H.A.; Baghmanyan, V.; Belmont-Moreno, E.; et al. Long-term Spectra of the Blazars Mrk 421 and Mrk 501 at TeV Energies Seen by HAWC. *Astrophys. J.* **2022**, *929*, 129. [CrossRef]
220. Abeysekara, A.U.; Albert, A.; Alfaro, R.; Alvarez, C.; Álvarez, J.D.; Arceo, R.; Arteaga-Velázquez, J.C.; Avila Rojas, D.; Ayala Solares, H.A.; Barber, A.S.; et al. Daily Monitoring of TeV Gamma-Ray Emission from Mrk 421, Mrk 501, and the Crab Nebula with HAWC. *Astrophys. J.* **2017**, *841*, 100. [CrossRef]
221. Cao, Z.; della Volpe, D.; Liu, S.; Bi, X.; Chen, Y.; D'Ettorre Piazzoli, B.; Feng, L.; Jia, H. The Large High Altitude Air Shower Observatory (LHAASO) Science Book (2021 Edition). *Chines Phys. C* **2022**, *47*, 035001–035007.

222. Cao, Z.; Aharonian, F.A.; An, Q.; Axikegu, B.L.X.; Bai, Y.X.; Bao, Y.W.; Bastieri, D.; Bi, X.J.; Bi, Y.J.; Cai, H.; et al. Ultrahigh-energy photons up to 1.4 petaelectronvolts from 12 γ-ray Galactic sources. *Nature* **2021**, *594*, 33. [CrossRef]
223. Tavecchio, F.; Bonnoli, G. On the detectability of Lorentz invariance violation through anomalies in the multi-TeV γ-ray spectra of blazars. *Astron. Astrophys.* **2016**, *595*, A25. [CrossRef]
224. Cao, Z.; Aharonian, F.; An, Q.; Axikegu, B.L.X.; Bai, Y.X.; Bao, Y.W.; Bastieri, D.; Bi, X.J.; Bi, Y.J.; Cai, H. Exploring Lorentz Invariance Violation from Ultrahigh-Energy γ-Rays Observed by LHAASO. *Phys. Rev. Let.* **2022**, *128*, 051102. [CrossRef] [PubMed]
225. Abdalla, H.; Aharonian, F.; Ait Benkhali, F.; Angüner, E. O.; Arakawa, M.; Arcaro, C.; Armand, C.; Arrieta, M.; Backes, M.; Barnard, M.; et al. The 2014 TeV γ-Ray Flare of Mrk 501 Seen with H.E.S.S.: Temporal and Spectral Constraints on Lorentz Invariance Violation. *Astrophys. J.* **2019**, *870*, 93. [CrossRef]
226. Tavecchio, F. Gamma rays from blazars. *AIP Conf. Proc.* **2017**, *1792*, 020007.
227. Peirson, A.L.; Liodakis, I.; Romani, R.W. Testing High-energy Emission Models for Blazars with X-ray Polarimetry. *Astrophys. J.* **2022**, *931*, 59. [CrossRef]

universe

MDPI

Article

The Power of Relativistic Jets: A Comparative Study

Luigi Foschini [1,*], Benedetta Dalla Barba [1,2], Merja Tornikoski [3], Heinz Andernach [4,†], Paola Marziani [5], Alan P. Marscher [6], Svetlana G. Jorstad [6], Emilia Järvelä [7], Sonia Antón [8] and Elena Dalla Bontà [9]

1 Osservatorio Astronomico di Brera, Istituto Nazionale di Astrofisica (INAF), 23807 Merate, Italy
2 Dipartimento di Scienza e Alta Tecnologia (DiSAT), Università degli Studi dell'Insubria, 22100 Como, Italy
3 Metsähovi Radio Observatory, Aalto University, 02540 Kylmälä, Finland
4 Thüringer Landessternwarte, D-07778 Tautenburg, Germany
5 Osservatorio Astronomico di Padova, Istituto Nazionale di Astrofisica (INAF), 35122 Padova, Italy
6 Department of Astronomy, Boston University, Boston, MA 02215, USA
7 Homer L. Dodge Department of Physics and Astronomy, University of Oklahoma, Norman, OK 73019, USA
8 CFisUC, Departamento de Física, Universidade de Coimbra, 3004-516 Coimbra, Portugal
9 Dipartimento di Fisica e Astronomia, Università di Padova, 35122 Padova, Italy
* Correspondence: luigi.foschini@inaf.it
† On leave of absence from: Departamento de Astronomía, Universidad de Guanajuato, Guanajuato 36023, GTO, Mexico.

Abstract: We present the results of a comparison between different methods to estimate the power of relativistic jets from active galactic nuclei (AGN). We selected a sample of 32 objects (21 flat-spectrum radio quasars, 7 BL Lacertae objects, 2 misaligned AGN, and 2 changing-look AGN) from the very large baseline array (VLBA) observations at 43 GHz of the Boston University blazar program. We then calculated the total, radiative, and kinetic jet power from both radio and high-energy gamma-ray observations, and compared the values. We found an excellent agreement between the radiative power calculated by using the Blandford and Königl model with 37 or 43 GHz data and the values derived from the high-energy γ-ray luminosity. The agreement is still acceptable if 15 GHz data are used, although with a larger dispersion, but it improves if we use a constant fraction of the γ-ray luminosity. We found a good agreement also for the kinetic power calculated with the Blandford and Königl model with 15 GHz data and the value from the extended radio emission. We also propose some easy-to-use equations to estimate the jet power.

Keywords: relativistic jets; active galactic nuclei; Seyfert galaxies; BL Lac objects; flat-spectrum radio quasars

Citation: Foschini, L.; Dalla Barba, B.; Tornikoski, M.; Andernach, H.; Marziani, P.; Marscher, A.P.; Jorstad, S.G.; Järvelä, E.; Antón, S.; Dalla Bontà, E. The Power of Relativistic Jets: A Comparative Study. *Universe* **2024**, *10*, 156. https://doi.org/10.3390/universe10040156

Academic Editor: Ioana Dutan

Received: 16 February 2024
Revised: 25 March 2024
Accepted: 26 March 2024
Published: 27 March 2024

1. Introduction

Accreting compact objects can emit powerful relativistic jets (see [1,2] for recent reviews on jets from active galactic nuclei (AGN)). One key quantity to understand the physics of jets and its impact on the nearby environment (the host galaxy and/or the intergalactic medium) is the power—both radiative and kinetic—that is dissipated in these structures. There are many ways to estimate the power based on different observational quantities, but the results are generally not consistent, with differences of one or more orders of magnitude. Despite this clear mismatch between the various methods, very few works have been published to understand the origin of this problem.

Pjanka et al. [3] compared four methods: a one-zone leptonic model by Ghisellini et al. [4,5], radio core shifts [6,7], extended radio emission (radio lobes or steep radio spectrum) [8], and high-energy gamma-ray luminosity [5,9]. They found that the powers estimated according to the leptonic model and the radio core shifts are almost consistent, while the value derived from the γ-ray luminosity is about half, and that from the radio lobes is about one order of magnitude smaller. These calculations can be reconciled by taking into account the source variability across time (the power derived from extended radio emission is an average over the source lifetime) or a change in the ratio between the number of leptons to hadrons (at least 15 to 1) or in the magnetization of the jet (and giving up the ideal magnetohydrodynamic theory). However, Pjanka et al. concluded that they are unable to decide which option is best.

We too made a preliminary study by comparing the interpretations of the same method by different authors [10]. Therefore, we compared the radiative power derived from the γ-ray luminosity and the Lorentz and Doppler factors from radio observations at different frequencies from [11–14]. We compared the broad-band spectral modeling by [5,15,16] and the observation of extended radio emission by [17,18]. We also compared the relationships by [19], based on the 15 GHz radio luminosity, with radiative power from the γ-ray luminosity. Although the models by Ghisellini [5] and Paliya [15,16] were described as the same (one-zone leptonic model), their results are systematically different toward low power values, with Paliya's values being about one order of magnitude greater than Ghisellini's. Something similar was found also by comparing the power calculated from the extended radio emission with differences of one to two orders of magnitudes at low powers. In this case, the reason was likely the different approaches: while Nokhrina [18] directly considered observations at 326 MHz, Meyer [17] started from 1.4 GHz observations and extrapolated to 300 MHz. The latter is not suitable for estimating the steep-spectrum radio emission from lobes because it fades as the frequency increases, and extended emission might not be detected already at GHz. The comparison of Foschini's relationship between jet power and 15 GHz radio core luminosity (see Equation (1) in [19]), with the radiative power from γ rays plus Lorentz and Doppler factors from radio observations at 43 GHz by Jorstad [11], resulted in a fair agreement, although with significant dispersion.

One major limitation of our previous work was to compare published works. Therefore, we could not select the epochs of observations, change models, or reanalyze data. In the present work, we overcome these limitations and address, in some more detail, the estimation of the jet power from a small but reliable sample of jetted AGN. Our aim is to understand the reasons of discrepancies and, if possible, to propose solutions. We also search for easy-to-use solutions, which might be of great value for the analysis of large samples of objects. A simple relationship between the power and an observed quantity or an equation linking a few observed quantities is easier to use than a detailed but complex numerical model. Obviously, the discrepancies have to be smaller than one order of magnitude to be acceptable.

We adopted the most recent value of the Hubble constant for the local Universe, $H_0 = 73.3$ km s^{-1} Mpc^{-1} from [20], and calculated the luminosity distance $d_{\rm L}$ by using the simplified equation:

$$d_{\rm L} \sim \frac{cz}{H_0}(1+\frac{z}{2}) \text{ [Mpc]} \tag{1}$$

where c is the speed of light in vacuum and z is the redshift.

Since we are comparing different methods based on the same data, we did not consider measurement errors, which are often quite large, but we focused on the dispersion σ of the values.

2. Sample Selection

We selected the sample of the very long baseline array (VLBA) Boston University (BU) blazar program (now BEAM-ME, https://www.bu.edu/blazars/BEAM-ME.html, accessed on 27 March 2024) [11,21]. It is composed of 36 objects observed with VLBA at 43 GHz between June 2007 and January 2013. We cross-matched this sample with the catalog of revised classifications and redshifts for the jetted AGN sample in the fourth *Fermi* Large Area Telescope (LAT) catalog (4FGL) as published by [22]. We thus removed four objects: 3C 66A, S5 0716 + 71, and PKS 0735 + 17, because they have no spectroscopic redshift (only estimates from photometry or the imaging of the host galaxy), and 3C 111, because its galactic latitude is $|b| \leq 10^{\circ}$ and therefore not included in the above-cited work.

The remaining 32 objects are listed in Table 1, and were classified in [22] as follows: 21 flat-spectrum radio quasars (FSRQs), 7 BL Lac objects (BLLAC), 2 misaligned AGN (MIS, also known as radio galaxies), and 2 changing-look AGN (CLAGN). The latter type has different meanings, depending on the authors. It was originally introduced by Matt et al. [23] to indicate AGN switching from Compton-thin to Compton-thick obscuration. In more recent years, also changes in the accretion were considered (e.g., [24]). In the present case, CLAGN indicates jetted AGN with optical spectra showing dramatic changes, from a featureless continuum to a line-dominate spectrum, or vice versa, thus moving from one class to another (for example, from BLLAC to FSRQ and/or vice versa; cf. [22]). We kept in the sample both MIS and CLAGN to avoid reducing a small sample too much and to have some insight on how large viewing angles and dramatic changes in the electromagnetic emission can affect the jet power.

Table 1. Sample of jetted AGN derived from [11]. Column explanation: (1) IAU source name referred to J2000, (2) a more common alias, (3) right ascension ([deg], J2000), (4) declination ([deg], J2000), (5) classification (BLLAC: BL Lac object; MIS: misaligned AGN; FSRQ: flat-spectrum radio quasar; CLAGN: changing-look AGN), and (6) redshift. Information for columns (5) and (6) was taken from [22].

Name (1)	Alias (2)	RA (3)	Dec (4)	Class (5)	z (6)
J0238 + 1636	PKS 0235 + 164	39.66	+16.62	BLLAC	0.940
J0319 + 4130	NGC 1275	49.95	+41.51	MIS	0.0176
J0339 − 0146	PKS 0336 − 01	54.88	−1.78	FSRQ	0.852
J0423 − 0120	PKS 0420 − 01	65.82	−1.34	FSRQ	0.915
J0433 + 0521	3C 120	68.30	+5.35	MIS	0.0336
J0530 + 1331	PKS 0528 + 134	82.73	+13.53	FSRQ	2.07
J0830 + 2410	S3 0827 + 24	127.72	+24.18	FSRQ	0.941
J0831 + 0429	PKS 0829 + 046	127.95	+4.49	BLLAC	0.174
J0841 + 7053	4C +71.07	130.35	+70.89	FSRQ	2.17
J0854 + 2006	OJ 287	133.70	+20.11	BLLAC	0.306
J0958 + 6533	S4 0954 + 65	149.70	+65.56	BLLAC	0.368
J1058 + 0133	4C +01.28	164.62	+1.57	FSRQ	0.892
J1104 + 3812	Mkn 421	166.11	+38.21	BLLAC	0.0308
J1130 − 1449	PKS 1127 − 145	172.53	−14.82	FSRQ	1.19
J1159 + 2915	Ton 599	179.88	+29.24	FSRQ	0.725
J1221 + 2813	W Comae	185.38	+28.23	BLLAC	0.102
J1224 + 2122	4C +21.35	186.23	+21.38	FSRQ	0.434
J1229 + 0203	3C 273	187.28	+2.05	FSRQ	0.158
J1256 − 0547	3C 279	194.05	−5.79	FSRQ	0.536
J1310 + 3220	OP 313	197.62	+32.34	FSRQ	0.996
J1408 − 0752	PKS B1406 − 076	212.24	−7.87	FSRQ	1.49
J1512 − 0905	PKS 1510 − 089	228.21	−9.10	FSRQ	0.360
J1613 + 3412	OS 319	243.42	+34.21	FSRQ	1.40
J1626 − 2951	PKS B1622 − 297	246.52	−29.86	FSRQ	0.815
J1635 + 3808	4C +38.41	248.81	+38.13	FSRQ	1.81
J1642 + 3948	3C 345	250.74	+39.81	FSRQ	0.593
J1733 − 1304	PKS 1730 − 13	263.26	−13.08	FSRQ	0.902
J1751 + 0939	OT 081	267.89	+9.65	CLAGN	0.320
J2202 + 4216	BL Lac	330.68	+42.28	BLLAC	0.0686
J2225 − 0457	3C 446	336.45	−4.95	CLAGN	1.40
J2232 + 1143	CTA 102	338.15	+11.73	FSRQ	1.04
J2253 + 1608	3C 454.3	343.49	+16.15	FSRQ	0.858

It is worth noting that there are some slight differences in the values of the redshift with respect to [11]. Therefore, we recalculated the affected quantities (e.g., the brightness temperature) to take into account these changes. This is mostly for the sake of consistency, rather than a significant change in the affected quantities.

In addition to the Boston University blazar program, there is also another excellent VLBA program: the Monitoring of Jets in Active Galactic Nuclei with VLBA Experiments

(MOJAVE (https://www.cv.nrao.edu/MOJAVE/, accessed on 27 March 2024), [25]). We cross-matched the above-cited sample with the larger sample (447 AGN) of the MOJAVE program [26], which offers a comparable set of physical quantities measured from radio observations at 15 GHz or derived from them. All 32 objects from the BU blazar program have been observed in the MOJAVE program. Additionally, in this case, we found some cases of a slightly different redshift, and we corrected the affected quantities.

3. The Blandford and Königl Model

The first step is to use the simplified and evergreen model by Blandford and Königl [27] to estimate the jet power. For the sake of simplicity, we shortly recall the main concepts and refer to the above-cited work [27] for more details. Blandford and Königl considered a conical jet, with an opening semiangle ϕ, and the axis inclined by an angle θ with respect to the line of sight to the observer, so that the observed opening angle is $\phi_{\rm obs} = \phi / \sin\theta$. The jet is a stream of relativistic electrons with distribution:

$$N(\gamma_{\rm e}) = K\gamma_{\rm e}^{-2} \tag{2}$$

where K is a normalization constant, and $\gamma_{\rm e}$ is the random Lorentz factor of the electrons in the range $\gamma_{\rm e,min} < \gamma_{\rm e} < \gamma_{\rm e,max}$. The magnetic field B is tangled with the plasma, and the bulk motion of the electrons has a constant speed β (in units of c), linked to the measured apparent speed $\beta_{\rm app}$ via the following:

$$\beta = \frac{\beta_{\rm app}}{\beta_{\rm app}\cos\theta + \sin\theta} \tag{3}$$

The electron energy density is as follows:

$$u_{\rm e} = Km_{\rm e}c^2 \log(\frac{\gamma_{\rm e,max}}{\gamma_{\rm e,min}}) \tag{4}$$

where $m_{\rm e}$ is the electron rest mass, while the energy density of the magnetic field is as follows:

$$u_{\rm B} = \frac{B^2}{8\pi} \tag{5}$$

Equipartition between $u_{\rm e}$ and $u_{\rm B}$ is assumed via the constant $k_{\rm eq}$, generally smaller than 1 (Blandford and Königl assumed $k_{\rm eq} = 0.5$ in their example [27]).

Blandford and Königl then calculated the expected flux density at radio frequencies, given the power of the jet (Equation (29) in [27]):

$$S_\nu \sim \frac{1}{2}(1+z)k_{\rm eq}^{\frac{5}{6}}\Delta^{-\frac{17}{12}}(1+\frac{2}{3}k_{\rm eq}\Lambda)^{-\frac{17}{12}}\Gamma^{-\frac{17}{6}}\beta^{-\frac{17}{12}}\delta^{\frac{13}{6}}(\sin\theta)^{-\frac{5}{6}}\phi_{\rm obs}^{-1}P_{44}^{\frac{17}{12}}d_{\rm L,9}^{-2}\ [\rm Jy] \tag{6}$$

where S_ν is the observed flux density at the frequency ν, $\Delta = \log(r_{\rm max}/r_{\rm min})$, where $r_{\rm min}$ and $r_{\rm max}$ refer to the size of the emission region, $\Lambda = \log(\gamma_{\rm e,max}/\gamma_{\rm e,min})$; Γ is the bulk Lorentz factor; δ is the Doppler factor; $d_{\rm L,9}$ is the luminosity distance in units of Gpc; and P_{44} is the total jet power in units of 10^{44} erg s^{-1}. We can rearrange Equation (6) to calculate the jet power as a function of the observed radio flux density and the other observed physical quantities:

$$P_{44} \sim k_1 k_2 \left(\frac{S_\nu d_{\rm L,9}^2}{1+z}\right)^{12/17} \tag{7}$$

where the factor k_1 depends on the electron random Lorentz factors and the size of the emitting region:

$$k_1 = \left(\frac{1}{2}\right)^{-12/17} k_{\rm eq}^{-10/17} \Delta(1 + \frac{2}{3} k_{\rm eq} \Lambda) \tag{8}$$

while k_2 depends on the observed quantities:

$$k_2 = \Gamma^2 \beta \delta^{-26/17} (\sin\theta)^{10/17} \phi_{\rm obs}^{12/17} \tag{9}$$

The synchrotron radiative power is as follows:

$$P_{\rm rad,syn,44} \sim \frac{k_{\rm eq}}{2(1 + \frac{2}{3} k_{\rm eq} \Lambda)} P_{44} \tag{10}$$

By adopting the typical values suggested by Blandford and Königl [27] for $k_{\rm eq} = 0.5$, $\Delta = 5$, and $\Lambda = 3$, we obtain $k_1 \sim 24.5$. Therefore,

$$P_{44} \sim 24.5 k_2 \left(\frac{S_\nu d_{\rm L,9}^2}{1+z}\right)^{12/17} \tag{11}$$

$$P_{\rm rad,syn,44} \sim \frac{1}{8} P_{44} \tag{12}$$

It immediately follows that the jet kinetic power is as follows:

$$P_{\rm kin,44} \sim \frac{7}{8} P_{44}. \tag{13}$$

4. Very Long Baseline Array (VLBA) Observations

4.1. All Epochs

The data collected by the BU blazar program span from June 2007 to January 2013, while the MOJAVE program covers the years from 1994 to 2019. The first check was based on all the data available (Table 2). We noted that one object in the MOJAVE sample (J0238 + 1636) has no measurement of $\beta_{\rm app}$; therefore, we adopted the value from Jorstad et al. [11]. The MOJAVE program has also no measurement of $\phi_{\rm obs}$, and therefore, we calculated it by means of the relationship $\Gamma\phi \sim 0.1$–0.2 [3,28]. We adopted $\Gamma\phi \sim 0.11$ as suggested by [3], but tested also the case of $\Gamma\phi \sim 0.2$, resulting in no significant changes. Since we need the observed opening angle $\phi_{\rm obs}$ in Equation (6), the above-cited relationship can be rewritten as follows:

$$\phi_{\rm obs} = \frac{0.11}{\Gamma \sin\theta} \tag{14}$$

This equation was used also to calculate $\phi_{\rm obs}$ at 43 GHz for J0238 + 1636 because this measurement was missing.

We distinguished two cases. In *case 1*, the Doppler factor at 43 GHz was derived from the flux density variability of the jet knots, according to Equation (3) in [11]:

$$\delta = \frac{15.8 s d_{\rm L,9}}{\tau(1+z)} \tag{15}$$

where s is the angular size of the knot (mas), and τ is the variability time scale (years). Therefore, we corrected Equation (15) to take into account the slightly different values of z and $d_{\rm L,9}$.

Then, Γ and θ were also updated according to the well-known equations (e.g., [11,26]):

$$\Gamma = \frac{\beta_{\rm max}^2 + \delta^2 + 1}{2\beta_{\rm max}} \tag{16}$$

$$\theta = \arctan \frac{2\beta_{\max}}{\beta_{\max}^2 + \delta^2 - 1} \tag{17}$$

In *case 2*, we tested the effect of recalculating the Doppler factor at 43 GHz by using the brightness temperature ratio:

$$\delta = \frac{T_{\mathrm{b},43}}{T_{\mathrm{b,int}}} \tag{18}$$

where $T_{\mathrm{b},43}$ is the observed brightness temperature [K] (see Table 2), and $T_{\mathrm{b,int}} = 5 \times 10^{10}$ K is the theoretical intrinsic value [29]. We underline that case 1 and case 2 differ in the calculation of δ at 43 GHz (Equation (15) vs. Equation (18)). The Doppler factor at 15 GHz is always derived from the brightness temperature. It is also worth noting that we adopt $\beta_{\max}$ as the reference apparent speed because it correlates better with T_b, as suggested by [26].

Table 2. Input data corrected for different redshifts and H_0 (all epochs). Columns description: (1) source name (J2000), (2) median flux density at 15 GHz [Jy], (3) 15 GHz brightness temperature [K], (4) maximum apparent speed as measured from 15 GHz observations [*c*], (5) median flux density at 43 GHz [Jy], (6) 43 GHz brightness temperature [K], (7) maximum apparent speed as measured from 43 GHz observations [*c*], (8) Doppler factor as measured according to Equation (15), and (9) observed jet opening semiangle [deg]. Original data at 15 and 43 GHz are taken from [26] and [11], respectively. To avoid reducing the small sample too much, we considered the few cases of lower limits as detections.

Name (1)	$S_{15\,\mathrm{GHz}}$ (2)	$\log T_{\mathrm{b},15}$ (3)	$\beta_{\max,15}$ (4)	$S_{43\,\mathrm{GHz}}$ (5)	$\log T_{\mathrm{b},43}$ (6)	$\beta_{\max,43}$ (7)	δ_{43} (8)	ϕ_{obs} (9)
J0238 + 1636	1.33	11.70	26.27 [1]	1.77	10.97 [2]	26.27	48.75	5.84 [3]
J0319 + 4130	2.84	11.25	0.41	15.51	10.79	0.36	10.37	22.1
J0339 − 0146	1.56	12.03	24.5	1.68	11.19 [2]	31.42	14.39	7.2
J0423 − 0120	4.80	12.49	5.46	4.74	11.90	15.53	15.56	23.4
J0433 + 0521	0.958	11.36	6.28	1.67	11.46	8.7	4.33	6.6
J0530 + 1331	2.18	12.14	18.41	2.02	11.91	77.94	20.79	22.8
J0830 + 2410	1.19	11.78	19.8	1.28	11.47	18.04	21.03	24.0
J0831 + 0429	0.525	11.39	10.2	0.57	10.97	7.23	12.33	7.9
J0841 + 7053	1.50	12.14	21.51	1.73	11.20	25.15	16.80	6.8
J0854 + 2006	2.74	12.27	15.14	4.68	11.88	8.6	7.9	33.0
J0958 + 6533	0.903	11.76	14.8	1.05	11.45	17.58	7.78	21.0
J1058 + 0133	3.57	12.50	6.61	4.02	11.61	14.14	18.42	24.8
J1104 + 3812	0.319	11.14	0.218	0.28	10.18	1.07	23.42	55.2
J1130 − 1449	1.12	11.80	19.8	1.76	11.36	23.37	19.68	15.4
J1159 + 2914	1.57	11.95	24.6	1.40	11.59	15.47	10.75	13.6
J1221 + 2813	0.226	11.31	8.2	0.25	10.79	4.76	8.96	9.2
J1224 + 2122	1.40	11.83	21.8	1.19	11.66	13.81	6.72	16.2
J1229 + 0203	3.52	11.95	14.91	11.88	12.51	11.83	3.97	6.6
J1256 − 0547	11.94	12.76	20.5	18.05	11.92	16.01	16.54	47.4
J1310 + 3220	1.55	11.99	27.5	2.14	10.95	13.73	19.37	58.4
J1408 − 0752	0.802	12.06	22.77	0.59	11.32	29.42	10.89	16.4
J1512 − 0905	1.87	11.95	28.0	2.44	11.15	29.6	31.98	11.4
J1613 + 3412	2.73	12.25	31.1	1.53	11.09	9.82	7.29	20.8
J1626 − 2951	0.959	12.01	12.0	1.35	11.34	11.04	8.68	30.8
J1635 + 3808	2.02	12.46	30.8	2.93	11.86	10.17	12.71	41.2
J1642 + 3948	3.27	12.29	19.37	4.47	11.63	19.45	10.77	18.6
J1733 − 1304	3.07	12.29	27.3	3.31	11.92	23.52	7.27	16.2
J1751 + 0939	3.52	12.62	6.85	3.60	11.65	17.66	15.97	26.2
J2202 + 4216	2.28	11.87	10.0	4.21	11.99 [2]	11.89	7.00	6.0
J2225 − 0457	4.75	12.30	17.7	3.82	11.62	22.20	13.19	22.0
J2232 + 1143	2.04	12.38	20.0	2.71	11.59	27.93	28.49	23.8
J2253 + 1608	3.53	12.26	17.0	14.44	12.22	9.06	22.35	22.6

[1] Missing. Value from 43 GHz measurements. [2] Lower limit. [3] Missing. Value from Equation (14).

Figure 1 shows the comparison of δ as measured with the two cited methods. We noted some cases with extreme differences: J0238 + 1636, $\delta_1 \sim 49$, $\delta_2 \sim 2$; J1104 + 3812, $\delta_1 \sim 23$, $\delta_2 \sim 0.30$; J1229 + 0203, $\delta_1 \sim 4$, $\delta_2 \sim 65$. The reasons might be that, e.g., J0238 + 1636 has no measured ϕ_{obs}, and its $T_{b,43}$ is a lower limit, and J1229 + 0203 has the highest $T_{b,43}$. This might imply the breakdown of the equipartition assumption for the observed brightness temperature $T_{b,obs} > 10^{13}$ K, as already noted in [11,29]. The case of J1104 + 3812 might be due to the so-called Doppler crisis in BL Lacs [30,31].

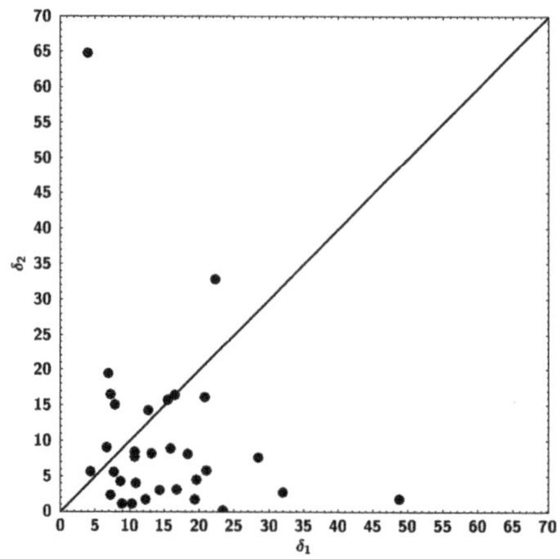

Figure 1. Doppler factor δ_1 (case 1) estimated from 43 GHz data and Equation (15) with the cosmology adopted in the present work vs. δ_2 (case 2) calculated from the brightness temperature. The continuous line indicates the equality of the two values.

Figure 2 displays the total jet power in the two cases and compared with the values derived from 15 GHz data. It is also worth studying the distribution of the coefficients k_2 (see Equation (9)), which is shown in Figure 3 for case 1.

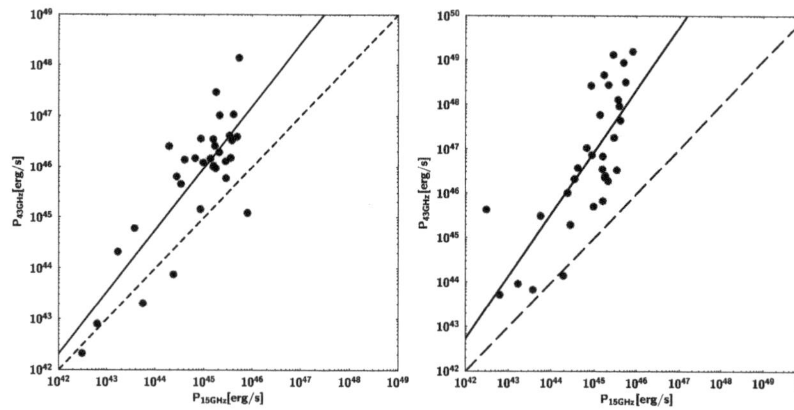

Figure 2. Total jet power calculated from Equation (11) and all the data from 15 and 43 GHz observations: (**left panel**) case 1; (**right panel**) case 2. The dashed line represents the equality of the two powers, while the continuous line is the linear fit to the data.

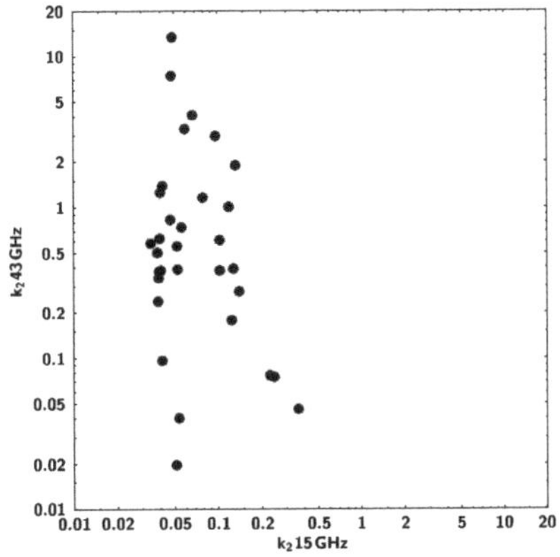

Figure 3. Distribution of k_2 values at 15 and 43 GHz for case 1.

The mean value of k_2 is 0.088 ($\sigma \sim 0.072$), and 1.4 ($\sigma \sim 2.6$) for 15 and 43 GHz, respectively (please note that $k_2 > 0$ by definition, so the dispersion is mostly toward values greater than the average). In case 2 (not shown), the dispersion increases to $\sim$131, while the mean value rises to $\sim$61. It is worth noting that the distribution of k_2 for 15 GHz data is quite narrow, with all the values between $\sim$0.034 and $\sim$0.36.

It is evident that the derivation of the Doppler factor from the brightness temperature at 43 GHz (case 2) leads to a more pronounced divergence at higher powers and a larger dispersion. The linear fit in the following form:

$$\log P_{43\,\text{GHz}} = m \log P_{15\,\text{GHz}} + C \quad (19)$$

gives the following values: $m \sim 1.22$ and $C \sim -8.9$ for case 1; $m \sim 1.40$ and $C \sim -16$ for case 2. The correlation factor ρ is 0.82 and 0.77 for case 1 and case 2, respectively, while the dispersion σ is 0.72 and 0.96.

Nonetheless, the relatively small range of k_2 values (particularly at 15 GHz, see Figure 3) offers an interesting possibility to derive the jet power only on the basis of the flux density at radio frequencies, although some caveats must be taken into account (see Section 8).

4.2. Overlapping Epochs

We remind the reader that 43 GHz data span from June 2007 to January 2013 [11], while 15 GHz data cover the years from 1994 to 2019 [26]. In the previous subsection, we considered all the available epochs, but now we want to study the case of overlapping epochs. Therefore, we collected 15 GHz data only if observed between 1 June 2007, and 31 January 2013 (Table 3). The results are shown in Figure 4.

There are no significant changes with respect to the previous cases. The linear fit gives these parameters: $m \sim 1.26$, $C \sim -11$, $\rho \sim 0.82$, $\sigma \sim 0.71$. However, we note an increase in the mean value of k_2 at 15 GHz and its dispersion, from $\sim$0.088 ($\sigma \sim 0.072$) to $\sim$0.18 ($\sigma \sim 0.26$). Since k_2 is a function of β, Γ, δ, θ, and ϕ_{obs} (see Equation (9)), a change in the observing epochs results in a different median $T_{\text{b},15\,\text{GHz}}$, which in turn affects δ and all the other parameters of k_2. The possibility of having a greater or smaller mean value and dispersion depends on the activity of the objects during the selected time interval. Anyway, in the present case, the distribution is still narrow, with only two values greater than the previous limit of $\sim$0.36. The two objects are J0831 + 0429 ($k_2 \sim 0.57$) and J1221 + 2813 ($k_2 \sim 1.5$). Given the lack of significant changes with respect to $P_{43\,\text{GHz}}$, we concluded that changes in k_2 were partially compensated by changes in flux density.

Table 3. Input data corrected for different redshifts and H_0 (overlapping epochs, from June 2007 to January 2013). Columns description: (1) source name (J2000), (2) median flux density at 15 GHz [Jy], (3) 15 GHz brightness temperature [K], and (4) median flux density at 37 GHz [Jy]. Original data at 15 GHz from [26]. See Section 5 for 37 GHz data of the Metsähovi Radio Observatory.

Name (1)	$S_{15\,GHz}$ (2)	$\log T_{b,15}$ (3)	$S_{37\,GHz}$ (4)
J0238 + 1636	3.37	12.31	1.50
J0319 + 4130	3.26	11.27	17.34
J0339 − 0146	1.58	12.09	2.33
J0423 − 0120	4.65	12.27	5.18
J0433 + 0521	0.675	11.22	1.95
J0530 + 1331	1.69	12.26	1.69
J0830 + 2410	1.23	11.74	1.42
J0831 + 0429	0.434	11.32	0.724
J0841 + 7053	2.10	12.56	2.20
J0854 + 2006	3.99	12.27	5.01
J0958 + 6533	1.07	11.72	1.19
J1058 + 0133	4.37	12.56	4.25
J1104 + 3812	0.292	11.12	0.428
J1130 − 1449	1.32	11.94	–
J1159 + 2914	1.53	11.79	1.61
J1221 + 2813	0.216	11.06	0.363
J1224 + 2122	1.68	12.21	1.69
J1229 + 0203	3.66	11.86	16.49
J1256 − 0547	9.82	12.72	18.67
J1310 + 3220	2.44	12.09	2.20
J1408 − 0752	0.708	11.75	0.812
J1512 − 0905	2.38	12.00	2.62
J1613 + 3412	1.61	12.07	2.35
J1626 − 2951	0.959	12.02	–
J1635 + 3808	2.27	12.26	3.62
J1642 + 3948	4.86	12.37	5.69
J1733 − 1304	3.32	12.30	3.69
J1751 + 0939	4.70	12.72	3.38
J2202 + 4216	3.44	11.98	4.50
J2225 − 0457	4.56	11.89	3.43
J2232 + 1143	2.11	12.46	2.79
J2253 + 1608	9.16	12.76	7.21

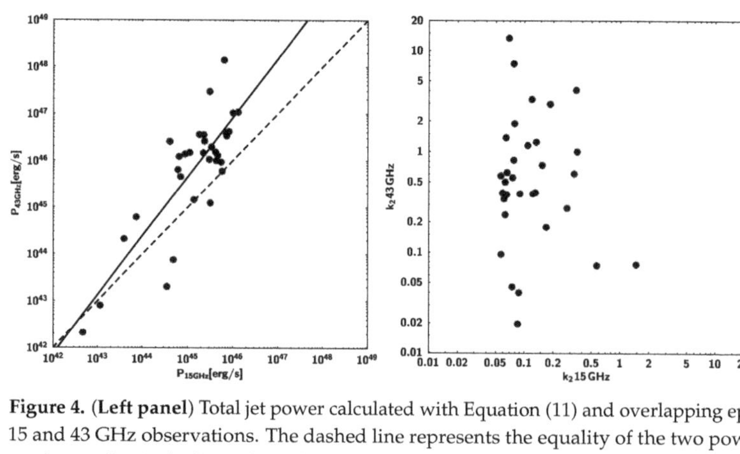

Figure 4. (**Left panel**) Total jet power calculated with Equation (11) and overlapping epoch data from 15 and 43 GHz observations. The dashed line represents the equality of the two powers, while the continuous line is the linear fit to the data. (**Right panel**) Distribution of k_2 values at 15 and 43 GHz.

5. Single-Dish Observations

The next test is to use the above calculated k_2 factors to estimate the jet power from single-dish observations. This type of observations does not allow for measuring or deriving all the quantities necessary to calculate k_2 (which are $\delta, \Gamma, \beta, \theta, \phi_{\rm obs}$, cf. Equation (9)): it is possible to measure only δ from the brightness temperature [13,32], but then it is necessary to take $\beta_{\rm app}$ from VLBA observations to derive the other quantities according to Equations (14), (16), and (17). Therefore, we can try to use k_2 as measured from the above-cited VLBA observations coupled to the flux density as measured from single-dish observations.

Data from the Metsähovi Radio Observatory (https://www.metsahovi.fi/opendata/, accessed on 27 March 2024) (MRO) of Aalto University (Finland) were used. MRO is a $\sim$14 m single dish equipped with a 1 GHz-band dual-beam receiver centered at 36.8 GHz. The high electron mobility pseudomorphic transistor (HEMPT) front end operates at ambient temperature. The observations, with typical exposures of $\sim 10^3$ s, are Dicke-switched ON–ON observations, alternating between the source and the sky in each feed horn. The detection threshold is $\sim$0.2 Jy in the best case. Calibration sources were the HII regions DR 21, NGC 7027, 3C 274, and 3C 84. More information about data reduction and analysis can be found in [33].

All the objects in Table 1 were monitored for more than 30 years, with the exception of J1130 − 1449 and J1626 − 2951. For the sake of simplicity, we considered only the case of overlapping epochs (see Table 3).

The results are displayed in Figure 5. We note a very good correlation between the new values of jet power from MRO at 37 GHz and the values of MOJAVE (15 GHz) and BU (43 GHz), with a best result if k_2 is measured from VLBA observations at the closer frequency (43 GHz), as expected. The linear fit (cf. Equation (19)) gives the following results:

- k_2 from MOJAVE (15 GHz): $m \sim 1.08$, $C \sim -3.8$, $\rho \sim 0.99$, $\sigma \sim 0.14$;
- k_2 from BU (43 GHz): $m \sim 1.00$, $C \sim -0.46$, $\rho \sim 1.00$, $\sigma \sim 0.067$;

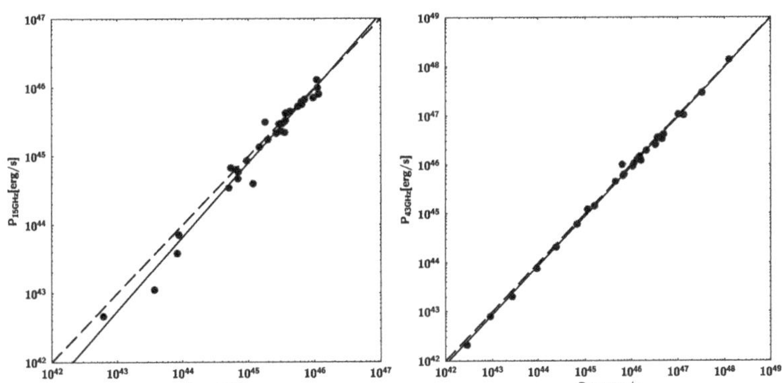

Figure 5. (**Left panel**) Total jet power derived from 37 GHz flux density and k_2 from 15 GHz observations vs. jet power from the same observations. (**Right panel**) Total jet power calculated by using 37 GHz flux density and k_2 from 43 GHz observations vs. jet power from the same observations. The dashed line represents the equality of the two powers, while the continuous line is the linear fit to the data.

6. Kinetic Power Estimated from the Extended Emission

The extended radio emission offers the opportunity to estimate the kinetic power of the jet. McNamara et al. [34] found a deficit of X-ray emission from the surrounding cluster at the location of the radio lobes of Hydra A, indicating that the jet had excavated cavities in the intergalactic medium. Then, by studying these X-ray cavities of a sample of radio galaxies in clusters, Bîrzan et al. [35] found a correlation between the jet kinetic power and its extended radio emission at 327 MHz:

$$\log \frac{P_{\mathrm{kin}}}{10^{42}} = 0.51 \log \frac{P_{327\,\mathrm{MHz}}}{10^{40}} + 1.51 \tag{20}$$

where P_{kin} is the jet kinetic power [erg/s], while $P_{327\,\mathrm{MHz}}$ is the radio power as measured at 327 MHz [erg/s]. Later, Cavagnolo et al. [36] enlarged the sample by adding also isolated giant elliptical galaxies, and proposed a new relationship based on the extended radio emission measured at 200–400 MHz:

$$\log \frac{P_{\mathrm{kin}}}{10^{42}} = 0.64 \log \frac{P_{200-400\,\mathrm{MHz}}}{10^{40}} + 1.54 \tag{21}$$

where $P_{200-400\,\mathrm{MHz}}$ is the radio power as measured at 200–400 MHz [erg/s]. The authors also proposed a relationship with the radio power as measured at 1.4 GHz, but this is less reliable [10,35,36], and therefore, we do not consider it.

To measure the radio power, we followed the procedure outlined in [36], and extracted the radio data from the CATS database (https://www.sao.ru/cats/, accessed on 27 March 2024) [37]. As noted by Cavagnolo [36], it is difficult to find 327 MHz data for all the objects, and therefore, the search was extended to the range 200–400 MHz. In the case of our sample, we found 327 MHz data for 21/32 objects. To avoid reducing our small sample too much, we used radio data at close frequencies (227, 318, 325, 333 MHz) when 327 MHz data were not available. In the case of 200–400 MHz, we also considered the cited frequency range with a tolerance of $\pm 10\%$. We performed the K-correction of the radio fluxes by adopting an average spectral index $\alpha = 0.8$ ($S_\nu \propto \nu^{-\alpha}$), as performed by [36]. We did not restrict the selected data from observations in the period 2007–2013 because the time necessary to excavate cavities in the intergalactic medium is of the order of several 10^8 years (e.g., [34]). Therefore, the measure of the kinetic power refers to an average over a very long time scale. The flux densities are displayed in Table 4.

Figure 6 shows comparisons of the jet kinetic power as calculated with Equations (20) and (21). The two values are well correlated ($\rho \sim 0.99$, $\sigma \sim 0.084$), but there is an evident divergence at high radio powers ($m \sim 1.24$, $C \sim -10.6$). This is somehow expected, given the different slopes of the two relationships ($0.64/0.51 \sim 1.25$, cf. Equations (20) and (21)). The reason for this divergence might be the different samples adopted by Bîrzan [35] and Cavagnolo [36]: while the former built the correlation by selecting a sample of radio galaxies in clusters (where, given the density and temperature of the intergalactic gas, it is easier to detect X-ray cavities), the latter added also a group of isolated giant elliptical galaxies (where X-ray cavities might be more difficult to detect).

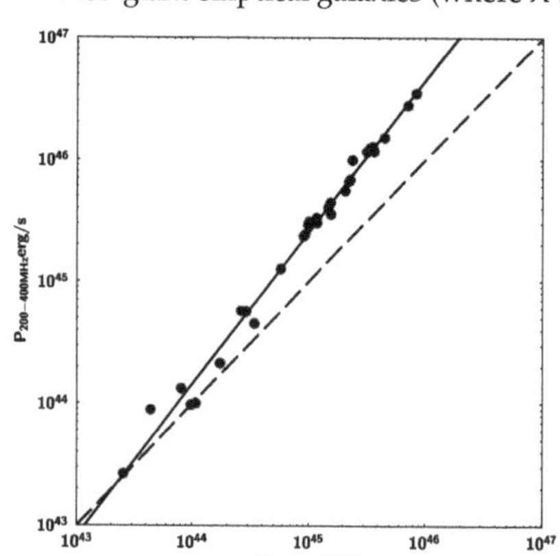

Figure 6. Comparison of the jet kinetic power as estimated from Equations (20) and (21). The dashed line represents the equality of the two powers, while the continuous line is the linear fit to the data.

Table 4. Input data for the kinetic power. Column description: (1) source name (J2000), (2) median flux density at 327 MHz [Jy], and (3) median flux density at 200–400 MHz [Jy]. All the data were extracted from the CATS database [37].

Name (1)	$S_{327\,\mathrm{MHz}}$ (2)	$S_{200-400\,\mathrm{MHz}}$ (3)
J0238 + 1636	1.04	1.26
J0319 + 4130	42.8 [1]	27.06
J0339 − 0146	0.943	1.33
J0423 − 0120	0.820	1.20
J0433 + 0521	2.37	6.33
J0530 + 1331	1.13 [1]	1.05
J0830 + 2410	0.660 [2]	0.770
J0831 + 0429	1.19 [2]	0.837
J0841 + 7053	5.07	5.07
J0854 + 2006	0.790	1.15
J0958 + 6533	0.624 [1]	0.742
J1058 + 0133	4.39	4.42
J1104 + 3812	0.961	1.14
J1130 − 1449	4.51 [3]	5.35
J1159 + 2914	3.52	2.71
J1221 + 2813	1.45	0.790
J1224 + 2122	3.98 [2]	4.80
J1229 + 0203	62.89	64.0
J1256 − 0547	14.79	14.58
J1310 + 3220	1.43	1.42
J1408 − 0752	0.535 [4]	0.584
J1512 − 0905	2.51	2.73
J1613 + 3412	2.55	3.11
J1626 − 2951	2.37 [4]	2.46
J1635 + 3808	2.51	2.31
J1642 + 3948	9.93	8.70
J1733 − 1304	4.66	7.61
J1751 + 0939	1.17 [2]	0.720
J2202 + 4216	1.82 [1]	2.77
J2225 − 0457	12.71	12.15
J2232 + 1143	6.99	7.88
J2253 + 1608	11.67	12.44

[1] From 325 MHz observations. [2] From 318 MHz observations. [3] From 333 MHz observations. [4] From 227 MHz observations.

Figure 7 displays the four comparisons between the kinetic power calculated with Equation (13) and data from 15 or 43 GHz observations and the values calculated with Equation (20) or Equation (21) with the measurements of the extended radio emission at MHz frequencies.

The linear fits give these values:

- 327 MHz vs. 15 GHz: $m \sim 1.14$, $C \sim -6.2$, $\rho \sim 0.89$, $\sigma \sim 0.37$;
- 327 MHz vs. 43 GHz: $m \sim 1.65$, $C \sim -28$, $\rho \sim 0.84$, $\sigma \sim 0.68$;
- 200–400 MHz vs. 15 GHz: $m \sim 0.91$, $C \sim 3.9$, $\rho \sim 0.89$, $\sigma \sim 0.38$;
- 200–400 MHz vs. 43 GHz: $m \sim 1.35$, $C \sim -15.5$, $\rho \sim 0.86$, $\sigma \sim 0.64$.

All the powers are well correlated ($\rho \sim 0.84$–0.89), showing a smaller dispersion when using 15 GHz data. In all cases, we noted a systematic underestimation of the power as calculated with Equation (13) for weak sources, with $P_{\mathrm{kin}} \lesssim 10^{44}$ erg/s (or an overestimation of the relationships based on the extended radio emission). The comparison with 43 GHz data shows a clear divergence toward higher radio powers. One source of bias is the fact that we used the integrated flux density, while we should have taken only the steep spectrum emission of the lobes. However, since our sources have a moderate to high redshift (with a few exceptions), it is not possible to disentangle the core from the lobes.

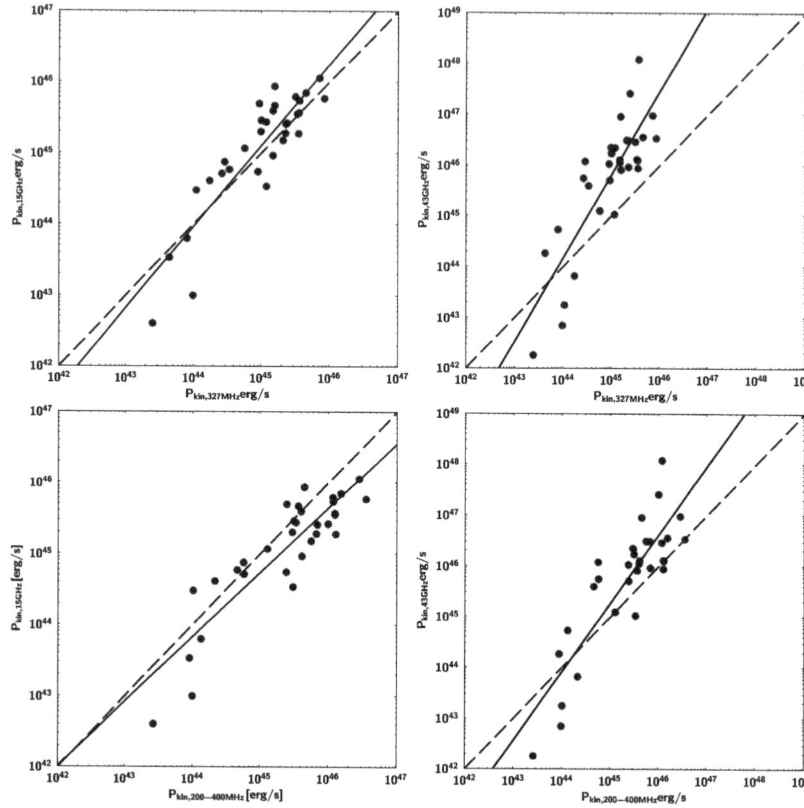

Figure 7. Kinetic jet power. (**Upper panels**) Comparison of Equations (13) and (20) with 15 GHz data (**left**) and 43 GHz data (**right**). (**Lower panels**) Comparison of Equations (13) and (21) with 15 GHz data (**left**) and 43 GHz data (**right**). The dashed line represents the equality of the two powers, while the continuous line is the linear fit to the data.

We would also like to note that Equations (20) and (21) are not the result of a theoretical calculation, but are correlations derived from observed quantities. Therefore, as is well known that correlation is not causation, the above-cited relationships heavily rely on the adopted samples, as also shown by the change in the slope from Equation (20) to Equation (21) displayed in Figure 6.

7. Radiative Power

The last test is with the radiative power as measured at high-energy γ rays by the *Fermi* Large Area Telescope (LAT) [38]. Since all the versions of the LAT catalogs cover a time span greater than the Boston University program [11], we extracted the data from the *Fermi* LAT Light Curve Repository (https://fermi.gsfc.nasa.gov/ssc/data/access/lat/LightCurveRepository/index.html, accessed on 27 March 2024) [39] covering only the epoch of the Boston University program (2007–2013). This web site is an automatic generator of light curves based on the likelihood with a power-law model and with a limited selection of parameters. We selected a 1-month time bin and left the photon index free to vary. We extracted the light curves starting from the beginning of LAT operations (August 2008) until January 2013, and then calculated the weighted mean of the observed 0.1–100 GeV flux F_γ and the spectral index α_γ (Table 5).

Table 5. *Fermi*/LAT data in the period August 2008–January 2013. Column description: (1) source name (J2000), (2) 0.1 − 100 GeV flux [10^{-11} erg cm^{-2} s^{-1}], and (3) spectral index α_γ. All the data were downloaded from the *Fermi* LAT Light Curve Repository [39].

Name (1)	$F_{0.1-100\,\mathrm{GeV}}$ (2)	α_γ (3)
J0238 + 1636	10.0	1.20
J0319 + 4130	22.0	1.07
J0339 − 0146	4.6	1.50
J0423 − 0120	5.5	1.40
J0433 + 0521	1.5	1.70
J0530 + 1331	3.5	1.60
J0830 + 2410	3.2	1.70
J0831 + 0429	4.0	1.20
J0841 + 7053	3.4	1.80
J0854 + 2006	6.4	1.20
J0958 + 6533	1.6	1.40
J1058 + 0133	8.2	1.20
J1104 + 3812	44.0	0.73
J1130 − 1449	2.1	1.60
J1159 + 2914	8.2	1.30
J1221 + 2813	4.0	1.20
J1224 + 2122	30.0	1.60
J1229 + 0203	18.0	2.00
J1256 − 0547	23.0	1.40
J1310 + 3220	2.8	1.50
J1408 − 0752	2.1	1.40
J1512 − 0905	52.0	1.46
J1613 + 3412	1.3	1.40
J1626 − 2951	2.7	1.70
J1635 + 3808	20.0	1.40
J1642 + 3948	4.6	1.20
J1733 − 1304	6.0	1.50
J1751 + 0939	4.4	1.30
J2202 + 4216	17.0	1.28
J2225 − 0457	2.1	1.60
J2232 + 1143	14.0	1.50
J2253 + 1608	174	1.50

From these values, we calculated the 0.1–100 GeV luminosity:

$$L_\gamma = 4\pi d_\mathrm{L}^2 \frac{F_\gamma}{(1+z)^{1-\alpha_\gamma}} \tag{22}$$

The minimum radiative power $P_{\mathrm{rad},\gamma}$ from high-energy γ rays (i.e., via inverse-Compton scattering) can be estimated as follows [40]:

$$P_{\mathrm{rad},\gamma} \sim \frac{\Gamma^2}{\delta^4} L_\gamma \tag{23}$$

The values of Γ and δ can be taken from the VLBA observations at 15 and 43 GHz. Then, $P_{\mathrm{rad},\gamma}$ can be compared with the synchrotron radiative power calculated according to Equation (12). The results are displayed in Figure 8.

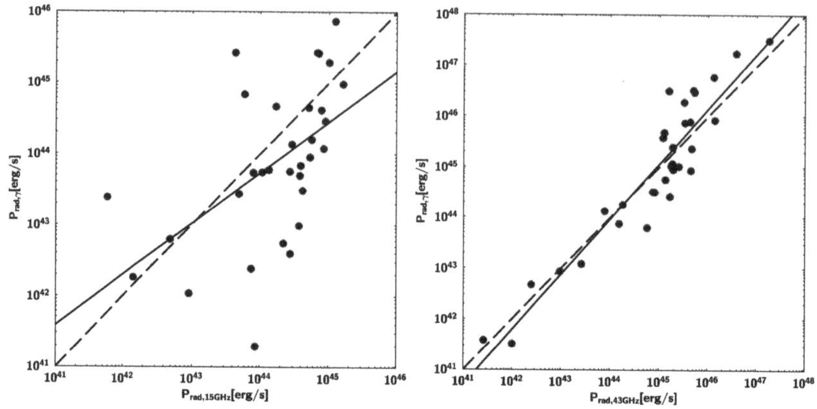

Figure 8. Radiative jet power. Comparison of values from high-energy γ rays and radio observations at 15 GHz (**left panel**) and at 43 GHz (**right panel**). The dashed line represents the equality of the two powers, while the continuous line is the linear fit to the data.

We note a good agreement, with a smaller dispersion when using 43 GHz data. The results of the linear fits are as follows:

- γ rays vs. 15 GHz: $m \sim 0.71$, $C \sim 12$, $\rho \sim 0.54$, $\sigma \sim 0.92$;
- γ rays vs. 43 GHz: $m \sim 1.09$, $C \sim -3.9$, $\rho \sim 0.94$, $\sigma \sim 0.51$.

It is worth noting that Equation (12) calculates the radiative power emitted via the synchrotron process, while the radiative power measured at high-energy γ rays can have a significant contribution from the external Compton process in FSRQs. It is known (e.g., [41]) that the total power radiated by relativistic electrons is as follows:

$$P_{\rm rad,tot} = P_{\rm rad,syn} + P_{\rm rad,\gamma} = \frac{4}{3}\sigma_{\rm Th} c \gamma_e^2 u_{\rm B}(1 + k_{CD}) \tag{24}$$

where $P_{\rm rad,syn}$ is the power dissipated via synchrotron radiation, $P_{\rm rad,\gamma}$ is the power due to the inverse-Compton process, and $\sigma_{\rm Th} \sim 0.66 \times 10^{-28}$ m^2 is the Thompson cross section. The Compton dominance parameter $k_{\rm CD}$ is defined as follows:

$$k_{\rm CD} = \frac{u_{\rm seed}}{u_{\rm B}} \tag{25}$$

where $u_{\rm seed}$ is the energy density of the seed photon field (from accretion disk, broad-line region, molecular torus, etc.). The Compton dominance can be measured from the observations of the peaks of synchrotron and inverse-Compton emissions:

$$k_{\rm CD} \sim \frac{\nu F_\nu^{\rm IC}}{\nu F_\nu^{\rm syn}} \tag{26}$$

From the inspection of a large sample of spectral energy distributions (SEDs) of blazars (e.g., [42]), it is possible to estimate $k_{\rm CD} \sim 1$ for BL Lac objects and $k_{\rm CD} \sim 10$ for FSRQs. Therefore, we applied this correction to FSRQs, and the results are displayed in Figure 9.

The comparison of the powers from 43 GHz and $\gamma-$ray observations does not change, with the linear fit giving these values: $m \sim 0.85$, $C \sim 6.3$, $\rho \sim 0.93$, $\sigma \sim 0.51$. However, the comparison with 15 GHz data is not so good: $m \sim 0.47$, $C \sim 22$, $\rho \sim 0.51$, $\sigma \sim 0.94$).

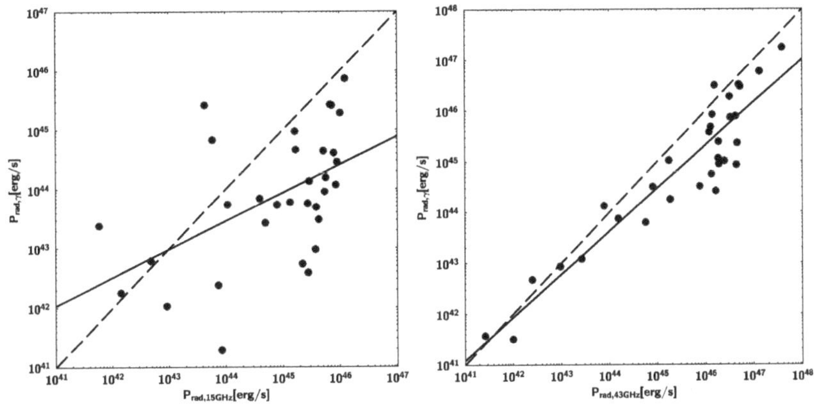

Figure 9. Radiative jet power corrected for the Compton dominance. Comparison of values from high-energy γ rays and radio observations at 15 GHz (**left panel**) and at 43 GHz (**right panel**). The dashed line represents the equality of the two powers, while the continuous line is the linear fit to the data.

The reason seems to be the use of the brightness temperature to estimate the Doppler factor, as shown already in Section 4. As a matter of fact, if we adopt the same method also for 43 GHz data, the consistency with the radiative power from γ-ray observations is lost (Figure 10). The linear fit is still acceptable, but with a large dispersion: $m \sim 1.10$, $C \sim -4.5$, $\rho \sim 0.83$, and $\sigma \sim 1.29$. Another source of bias is the use of a single value of $k_{\rm CD}$ for all FSRQs. This quantity depends on the characteristics of the source and its activity (an outburst can result in a greater value of $k_{\rm CD}$).

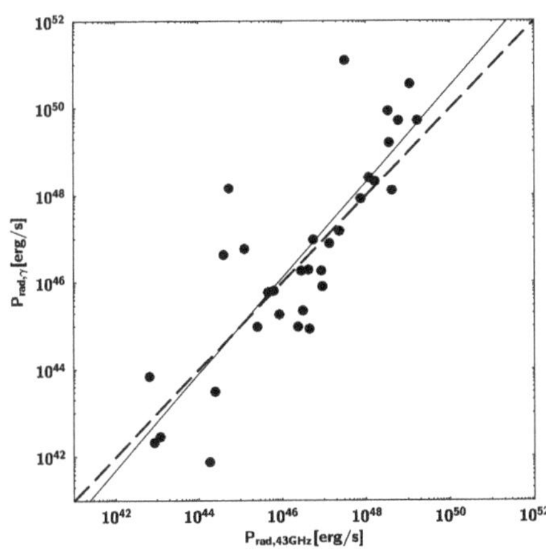

Figure 10. Radiative jet power corrected for the Compton dominance. Comparison of values from high-energy γ rays and radio observations at 43 GHz, with the Doppler factor calculated by using the brightness temperature (case 2, Section 4). The dashed line represents the equality of the two powers, while the continuous line is the linear fit to the data.

8. Fudge Factors

As noted in Section 4, the value of k_2 (see Equation (9)) is within a small range, particularly for 15 GHz data, with some exceptions. Therefore, we can try estimating the jet power by setting k_2 equal to a constant value (mean, median, etc.). We selected

$k_2 = 0.183$, which is the median value calculated by selecting all the available epochs. Therefore, Equation (11) becomes the following:

$$P_{44} \sim 4.5\left(\frac{S_\nu d_{L,9}^2}{1+z}\right)^{12/17} \tag{27}$$

We then consider as reference the total jet power at 43 GHz, calculated with Equation (11), and compare it with the power at 15 and 37 GHz calculated with Equation (27). The only variable is now the flux density at the selected frequency (15 or 37 GHz). The results are shown in Figure 11.

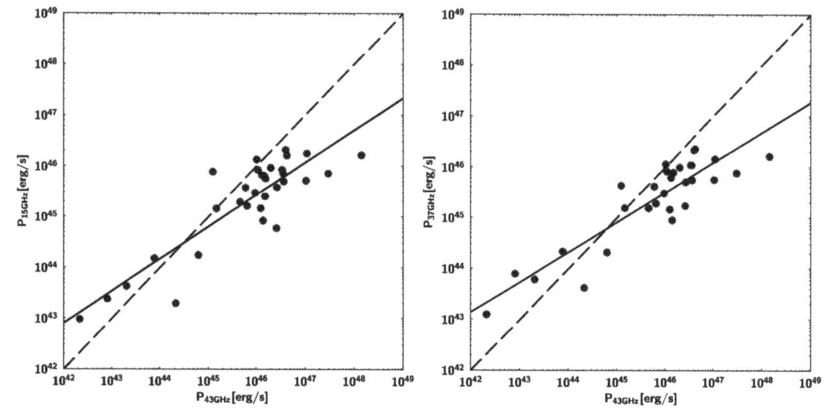

Figure 11. Total jet power calculated with a constant k_2 (Equation (27)) and flux densities at 15 (**left panel**) and 37 GHz (**right panel**) compared with the power at 43 GHz. The dashed line represents the equality of the two powers, while the continuous line is the linear fit to the data.

The linear fit gives the following results:

- 43 vs. 15 GHz: $m \sim 0.63$, $C \sim 16$, $\rho \sim 0.87$, $\sigma \sim 0.44$;
- 43 vs. 37 GHz: $m \sim 0.59$, $C \sim 18$, $\rho \sim 0.89$, $\sigma \sim 0.38$,

with slightly better values for 37 GHz, as expected. However, the slope $\sim$0.6 indicates a divergence toward low and high powers. We note that selecting another value for k_2 (median or average from another data set) will change only the value of C, but not all the others. The dispersion is contained within $\sim$0.4.

We also studied the distributions of the correction factor Γ^2/δ^4 to be applied to the γ-ray luminosity to estimate the radiative power (cf. Equation (23)). We adopted the median calculated from all data, which is $\Gamma^2/\delta^4 \sim 0.0027$. We adopted the latter value as constant in Equation (23) and compared the radiative power estimated with the proper value for each source (Figure 12).

The result of the linear fit is now as follows:

- γ vs. 15 GHz: $m \sim 0.97$, $C \sim 0.43$, $\rho \sim 0.90$, $\sigma \sim 0.56$;
- γ vs. 43 GHz: $m \sim 0.64$, $C \sim 15$, $\rho \sim 0.81$, $\sigma \sim 0.75$.

This time, there is a better agreement with 15 GHz data, but it is worth reminding that a good correlation does not imply a causation. This agreement is likely to be a chance coincidence because the previous tests (see Section 7, Figure 8, left panel, and Figure 9, left panel) do not display any hint of such agreement ($\rho \sim 0.51$–0.54). The only suitable explanation is that, by using constant fudge factors, most of fluctuations have been smoothed out by a mere chance coincidence. Taking constant average values for k_2 and Γ^2/δ^4 has no physical reason and is only for our convenience to get rid of the lack of adequate measurements.

The comparison with the radiative power estimated from 43 GHz data is still acceptable, but with a larger dispersion and a divergence at high powers.

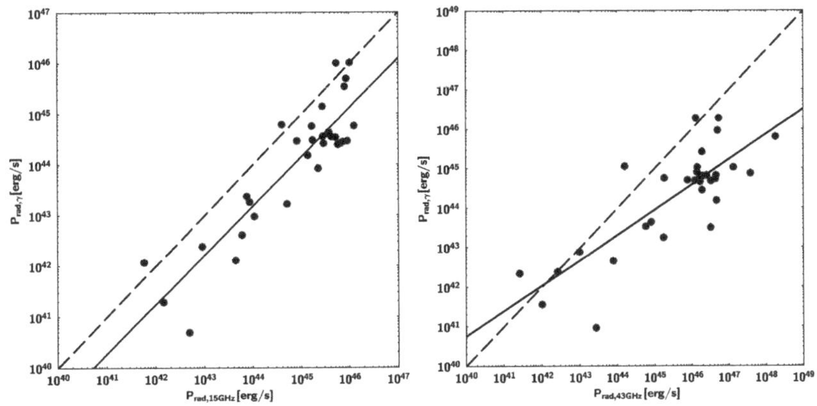

Figure 12. Radiative jet power calculated with a constant Γ^2/δ^4 compared with the power estimated with Γ^2/δ^4 from 15 GHz (**left panel**) and 43 GHz data (**right panel**). The dashed line represents the equality of the two powers, while the continuous line is the linear fit to the data.

9. Discussion and Conclusions

We compared the jet power as measured by different methods mostly based on radio observations. We can summarize the main results as follows:

- The jet power estimates based on the Blandford and Königl model [27] plus VLBA data at 15 and 43 GHz are in good agreement (Section 4). The almost simultaneity of observations does not imply significant changes in the calculated jet power, at least with the present data set (Section 4.2). One source of bias is the measurement of the Doppler factor δ via the brightness temperature (see Equation (18) and Figure 1). This problem has already been noted by several authors (e.g., [11,13,43], and particularly see the extensive discussion in [26]), and is related to both the physics of the jets (opacity, absorption, activity of the jet, etc.) and the instrumental/observational issues (frequency, cadence of observations, etc.). We do not know the intrinsic brightness temperature for any source and cannot measure it. Therefore, we need either to make theoretical hypotheses [29] or to follow a statistical approach by assuming that every jetted AGN has more or less the same T_b equal to the median or the mean of the sample [26]. The approach proposed by Jorstad et al. [11,43] to calculate δ (cf. Equation (15)) based on the flux variability is much more reliable, as shown by the excellent agreement with the radiative power measured from high-energy γ rays (see Section 7, particularly Figure 9, right panel). This approach seems to be not suitable for 15 GHz data, as radio observations at this frequency are sampling the jet downstream, where the flux variability is affected by effects other than radiative losses only [26].
- The use of single-dish flux densities at 37 GHz (Section 5), with k_2 calculated from 15 and 43 GHz observations (see Equation (9)), is consistent with the power derived from VLBA observations. The best result is with 43 GHz data, as expected, because of the smaller difference in frequency.
- The kinetic power calculated on the basis of the extended radio emission at MHz frequencies and the relationships by [35,36] (Section 6) gives better results when compared with the power estimated from the Blandford and Königl [27] model and 15 GHz data. However, we noted a systematic disagreement of the power for weak sources ($P_{kin} \lesssim 10^{44}$ erg/s).
- The comparison of the radiative power estimated from the Blandford and Königl [27] model and high-energy γ-ray observations from *Fermi*/LAT (Section 7) resulted in an

excellent agreement, particularly with 43 GHz data, and when taking into account the Compton dominance. The larger dispersion in the comparison with 15 GHz data seems to be due to the above-cited limitations of δ calculated via T_b (Figure 10). However, a quite good agreement with 15 GHz data is recovered when using a constant value for Γ^2/δ^4 to estimate the radiative power, even though it is systematically lower than the value from radio observations and is likely to be a chance coincidence (Section 8).

- Searching for an easy-to-use equation to estimate the jet power, we proposed Equation (27), based on the limited range of values of k_2, particularly from 15 GHz data. The comparison of power derived from 15, 37, and 43 GHz data is fairly correlated ($\rho \sim 0.9$) with an acceptable dispersion $\sigma \sim 0.4$. The use of a constant Γ^2/δ^4 to estimate the radiative power from the γ-ray luminosity resulted in a slightly greater dispersion ($\sigma \sim$ 0.6–0.7).

For the sake of simplicity, we recall in Table 6 the proposed easy-to-use equations to estimate the jet power, with the caveat of divergence at low and high powers.

Table 6. Jet power in [erg s^{-1}] calculated with our proposed easy-to-use equations based on fudge factors described in Section 8. We remind that the radio flux density S_ν is measured in [Jy], the luminosity distance $d_{L,9}$ is in [Gpc], and L_γ is in [erg s^{-1}].

Jet Power	Equation	Notes
Total	$(4.5 \times 10^{44})\left(\frac{S_\nu d_{L,9}^2}{1+z}\right)^{\frac{12}{17}}$	From Equation (11)
Kinetic	$(3.9 \times 10^{44})\left(\frac{S_\nu d_{L,9}^2}{1+z}\right)^{\frac{12}{17}}$	From Equation (13)
Radiative (synchrotron)	$(5.6 \times 10^{43})\left(\frac{S_\nu d_{L,9}^2}{1+z}\right)^{\frac{12}{17}}$	From Equation (12)
Radiative (Compton)	$0.0027 L_\gamma$	From Equation (23)

We want to stress that equations in Table 6 must be used with great care because the fudge factors are affected by the variability of the source and the uncertainties in the measurement or derivation of the physical quantities $\beta, \Gamma, \delta, \theta$, and ϕ (that we did not consider in this work). However, given the difficulty of measuring or inferring all these quantities without dedicated VLBA observations (preferably at high frequencies, such as 43 GHz), these equations can offer a useful first estimate of the jet power, being careful when dealing with extremely weak or extremely powerful jets.

Before concluding, some more words of caveat should be written, which are also the points to be addressed to improve our methods to estimate the jet power. The possible sources of bias in the present work are as follows:

- The sample is composed mostly of blazars (30/32 objects), whose electromagnetic emission is dominated by relativistic beaming, because of the small viewing angle. Only two objects are misaligned AGN (radio galaxies), and there are no jetted Seyferts. It is necessary to expand the sample to cover all types of jetted AGN, beamed or not.
- To convert redshifts into luminosity distances, we employed the simplified Equation (1). This resulted in an overestimation of the luminosity distance of ~10% for the farthest object (J0841 + 7053, $z = 2.71$), which quickly decreases to ~4% for objects at $z \sim 1$. This is not a problem in the present work, since we compared the jet power of the same object calculated with different methods, but a comparison with values from other works should be dealt with care in the case of high-redshift objects.
- The Blandford and Königl [27] model is for flat-spectrum radio sources. Deviation from a flat radio spectrum, such as in cases of steep spectra of misaligned AGN, might imply large errors. In our sample, we have only two radio galaxies, too few to draw useful conclusions.
- The extended radio emission to estimate the kinetic power (Section 6) should be only due to radio lobes, with a steep spectrum. However, for the sake of simplic-

ity, we considered the whole integrated flux. As a matter of fact, the typical resolution at 200–400 MHz is about one arcminute, which is equivalent to ∼0.1 Mpc at $z \sim 0.1$. Therefore, most of the objects in our sample are pointlike at MHz frequencies, and it is not possible to isolate the steep-spectrum extended emission from the core. Anyway, at MHz frequencies, the core contribution should be less important than the lobes. The low-frequency array (LOFAR) might be a viable solution for a better angular resolution (∼0.21″ at 240 MHz for a 1000 km baseline (https://science.astron.nl/telescopes/lofar/lofar-system-overview/observing-modes/lofar-imaging-capabilities-and-sensitivity/, accessed on 27 March 2024)), but it is necessary to recalibrate Equations (20) and (21) because the maximum frequency of LOFAR is 250 MHz.

- In this work, we always used median or weighted mean values calculated over long periods. The shortest period is 2007–2013, about 5.5 years. Given the strong variability of jetted AGN, the use of values from single-epoch observations or from only one VLBA knot might result in significant deviations. For example, we considered J0433 + 0521 with VLBA data at 43 GHz: the total jet power with the data used in this work results to be $\sim 2.1 \times 10^{44}$ erg/s. We want to compare with the most recent data from [44], which extended the work in [11] to December 2018. By using the median values, we calculate $\sim 4.1 \times 10^{44}$ erg/s, consistent within a factor 2 with the present work. If we calculate the jet power by using the data, for example, of the component C15 only, we obtain $\sim 5.2 \times 10^{43}$ erg/s, about one order of magnitude smaller.
- We also need to underline that this work was conducted by considering the same physical factors $\Delta = \log(r_{\rm max}/r_{\rm min})$ and $\Lambda = \log(\gamma_{\rm e,max}/\gamma_{\rm e,min})$ for all the sources. Therefore, a part of the dispersions in the comparisons is surely due to this assumption. For example, an outburst changing the electron distribution will alter Λ, which in turn will change the coefficient k_1 of Equation (8). Therefore, it is necessary to also address the microphysics of the jet and, particularly, the particle content (leptons vs. hadrons), the energy distribution of electrons, the size of the emission region vs. opacity, and the equipartition hypothesis.

Author Contributions: Conceptualization, L.F.; writing—original draft, L.F.; writing—review and editing, L.F., B.D.B., M.T., H.A., P.M., A.P.M., S.G.J., E.J., S.A. and E.D.B. All authors have read and agreed to the published version of the manuscript.

Funding: This research received no external funding.

Data Availability Statement: This research made use of publicly available data from the VLBA Boston University blazar program [11,21], the MOJAVE program [25], the Metsähovi Radio Observatory (https://www.metsahovi.fi/opendata/, accessed on 27 March 2024), the CATS database [37], and the *Fermi* LAT Light Curve Repository [39]. Recalculated or new values resulting from this work are available upon reasonable requests.

Acknowledgments: L.F. would like to thank (in alphabetical order) Stefano Ciroi, Maria J. M. Marchã, Patrizia Romano, and Stefano Vercellone for their valuable comments on the draft.

Conflicts of Interest: The authors declare no conflicts of interest.

References

1. Blandford, R.; Meier, D.; Readhead, A. Relativistic Jets from Active Galactic Nuclei. *Annu. Rev. Astron. Astrophys.* **2019**, *57*, 467–509. [CrossRef]
2. Foschini, L. Some notes about the current researches on the physics of relativistic jets. *Front. Astron. Space Sci.* **2022**, *8*, 794891. [CrossRef]
3. Pjanka, P.; Zdziarski, A.A.; Sikora, M. The power and production efficiency of blazar jets. *Mon. Not. R. Astron. Soc.* **2017**, *465*, 3506–3514. [CrossRef]
4. Ghisellini, G.; Tavecchio, F. Canonical high-power blazars. *Mon. Not. R. Astron. Soc.* **2009**, *397*, 985–1002. [CrossRef]
5. Ghisellini, G.; Tavecchio, F.; Maraschi, L.; Celotti, A.; Sbarrato, T. The power of relativistic jets is larger than the luminosity of their accretion disks. *Nature* **2014**, *515*, 376–378. [CrossRef] [PubMed]
6. Lobanov, A. Ultracompact jets in active galactic nuclei. *Astron. Astrophys.* **1998**, *330*, 79–89.

7. Zdziarski, A.A.; Sikora, M.; Pjanka, P.; Tchekhovskoy, A. Core shifts, magnetic field and magnetization of extragalactic jets. *MNRAS* **2015**, *451*, 927–935. [CrossRef]
8. Willot, C.J.; Rawlings, S.; Blundell, K.M.; Lacy, M. The emission line-radio correlation for radio sources using the 7C redshift survey. *Mon. Not. R. Astron. Soc.* **1999**, *309* 1017–1033. [CrossRef]
9. Nemmen, R.S.; Georganopoulos, M.; Guiriec, S.; Meyer, E.T.; Gehrels, N.; Sambruna, R.M. A Universal Scaling for the Energetics of Relativistic Jets from Black Hole Systems. *Science* **2012**, *338*, 1445–1448. [CrossRef]
10. Foschini, L.; Lister, M.; Hovatta, T.; Kovalev, Y.; Romano, P.; Vercellone, S.; Lähteenmäki, A.; Savolainen, T.; Tornikoski, M.; Angelakis, E.; et al. Calibrating The Power Of Relativistic Jets. In Proceedings of the High Energy Phenomena in Relativistic Outflows VII ((HEPRO VII)), Barcelona, Spain, 9–12 July 2019; Volume 354.
11. Jorstad, S.G.; Marscher, A.P.; Morozova, D.A.; Troitsky, I.S.; Agudo, I.; Casadio, C.; Foord, A.; Gómez, J.L.; MacDonald, N.R.; Molina, S.N.; et al. Kinematics of parsec-scale jets of gamma-ray blazars at 43 GHz within the VLBA-BU-BLAZAR program. *Astrophys. J.* **2017**, *846*, 98. [CrossRef]
12. Pushkarev, A.B.; Kovalev, Y.Y.; Lister, M.L.; Savolainen, T. MOJAVE—XIV. Shapes and opening angles of AGN jets. *Mon. Not. R. Astron. Soc.* **2017**, *468*, 4992–5003. [CrossRef]
13. Liodakis, I.; Hovatta, T.; Huppenkothen, D.; Kiehlmann, S.; Max-Moerbeck, W.; Readhead, A.C.S. Constraining the Limiting Brightness Temperature and Doppler Factors for the Largest Sample of Radio-bright Blazars. *Astrophys. J.* **2018**, *866*, 137. [CrossRef]
14. Finke, J.D. The properties of parsec-scale blazar jets. *Astrophys. J.* **2019**, *870*, 28. [CrossRef]
15. Paliya, V.S.; Marcotulli, L.; Ajello, M.; Joshi, M.; Sahayanathan, S.; Rao, A.R.; Hartmann, D. General Physical Properties of CGRaBS Blazars. *Astrophys. J.* **2017**, *851*, 33. [CrossRef]
16. Paliya, V.S.; Parker, M.L.; Jiang, J.; Fabian, A.C.; Brenneman, L.; Ajello, M.; Hartmann, D. General Physical Properties of Gamma-Ray-emitting Narrow-line Seyfert 1 Galaxies. *Astrophys. J.* **2019**, *872*, 169. [CrossRef]
17. Meyer, E.T.; Fossati, G.; Georganopoulos, M.; Lister, M.L. From the Blazar Sequence to the Blazar Envelope: Revisiting the Relativistic Jet Dichotomy in Radio-loud Active Galactic Nuclei. *Astrophys. J.* **2011**, *740*, 98. [CrossRef]
18. Nokhrina, E.E.; Beskin, V.S.; Kovalev, Y.Y.; Zheltoukhov, A.A. Intrinsic physical conditions and structure of relativistic jets in active galactic nuclei. *Mon. Not. R. Astron. Soc.* **2015**, *447*, 2726–2737. [CrossRef]
19. Foschini, L. The Unification of Relativistic Jets. *Int. J. Mod. Phys. Conf. Ser.* **2014**, *28*, 1460188. [CrossRef]
20. Riess, A.G.; Yuan, W.; Macri, L.M.; Scolnic, D.; Brout, D.; Casertano, S.; Jones, D.O.; Murakami, Y.; Anand, G.S.; Breuval, L.; et al. A Comprehensive Measurement of the Local Value of the Hubble Constant with 1 km s^{-1} Mpc^{-1} Uncertainty from the Hubble Space Telescope and the SH0ES Team. *Astrophys. J.* **2022**, *934*, L7. [CrossRef]
21. Jorstad, S.; Marscher, A. The VLBA-BU-BLAZAR Multi-Wavelength Monitoring Program. *Galaxies* **2016**, *4*, 47. [CrossRef]
22. Foschini, L.; Lister, M.L.; Andernach, H.; Ciroi, S.; Marziani, P.; Antón, S.; Berton, M.; Dalla Bontà, E.; Järvelä, E.; Marchã, M.J.M.; et al. A New Sample of Gamma-Ray Emitting Jetted Active Galactic Nuclei. *Universe* **2022**, *8*, 587. [CrossRef]
23. Matt, G.; Guainazzi, M.; Maiolino, R. Changing look: From Compton-thick to Compton-thin, or the rebirth of fossil active galactic nuclei. *Mon. Not. R. Astron. Soc.* **2003**, *342*, 422–426. [CrossRef]
24. Ricci, C.; Trakhtenbrot, B. Changing-look Active Galactic Nuclei. *arXiv* **2022**, arXiv:2211.05132.
25. Lister, M.L.; Aller, M.F.; Aller, H.D.; Hodge, M.A.; Homan, D.C.; Kovalev, Y.Y.; Pushkarev, A.B.; Savolainen, T. MOJAVE XV. VLBA 15 GHz Total Intensity and Polarization Maps of 437 Parsec-scale AGN Jets from 1996 to 2017. *Astrophys. J. Suppl. Ser.* **2018**, *234*, 12. [CrossRef]
26. Homan, D.C.; Cohen, M.H.; Hovatta, T.; Kellermann, K.I.; Kovalev, Y.Y.; Lister, M.L.; Popkov, A.V.; Pushkarev, A.B.; Ros, E.; Savolainen, T. MOJAVE XIX. Brightness Temperatures and Intrinsic Properties of Blazar Jets. *Astrophys. J.* **2021**, *923*, 67. [CrossRef]
27. Blandford, R.D.; Königl, A. Relativistic Jets as Compact Radio Sources. *Astrophys. J.* **1979**, *232*, 34–48. [CrossRef]
28. Clausen-Brown, E.; Savolainen, T.; Pushkarev, A.B.; Kovalev, Y.Y.; Zensus, J.A. Causal connection in parsec-scale relativistic jets: results from the MOJAVE VLBI survey. *Astron. Astrophys.* **2013**, *558*, A144. [CrossRef]
29. Readhead, A.C.S. Equipartition Brightness Temperature and the Inverse Compton Catastrophe. *Astrophys. J.* **1994**, *426*, 51–59. [CrossRef]
30. Edwards, P.G.; Piner, B.G. The Subluminal Parsec-Scale Jet of Markarian 501. *Astrophys. J.* **2002**, *579*, L67–L70. [CrossRef]
31. Piner, B.G.; Edwards, P.G. The Parsec-Scale Structure and Jet Motions of the TeV Blazars 1ES 1959 + 650, PKS 2155 − 304, and 1ES 2344 + 514. *Astrophys. J.* **2004**, *600*, 115–126. [CrossRef]
32. Hovatta, T.; Valtaoja, E.; Tornikoski, M.; Lähteenmäki, A. Doppler factors, Lorentz factors, and viewing angles for quasars, BL Lacertae objects and radio galaxies. *Astron. Astrophys.* **2009**, *494*, 527–537. [CrossRef]
33. Teräsranta, H.; Tornikoski, M.; Mujunen, A.; Karlamaa, K.; Valtonen, T.; Henelius, N.; Urpo, S.; Lainela, M.; Pursimo, T.; Nilsson, K.; et al. Fifteen years monitoring of extragalactic radio sources at 22, 37 and 87 GHz. *Astron. Astrophys. Suppl.* **1998**, *132*, 305–331. [CrossRef]
34. McNamara, B.R.; Wise, M.; Nulsen, P.E.J.; David, L.P.; Sarazin, C.L.; Bautz, M.; Markevitch, M.; Vikhlinin, A.; Forman, W. R.; Jones, C.; et al. Chandra X-ray Observations of the Hydra A Cluster: An Interaction between the Radio Source and the X-ray-emitting Gas. *Astrophys. J.* **2000**, *534*, L135-L138. [CrossRef] [PubMed]
35. Bîrzan, L.; McNamara, B.R.; Nulsen, P.E.J.; Carilli, C.L.; Wise, M.W. Radiative efficiency and content of extragalactic radio sources: Toward a universal scaling relation between jet power and radio power. *Astrophys. J.* **2008**, *686*, 859–880. [CrossRef]

36. Cavagnolo, K.W.; McNamara, B.R.; Nulsen, P.E.J.; Carilli, C.L.; Jones, C.; Bîrzan, L. A relationship between AGN jet power and radio power. *Astrophys. J.* **2010**, *720*, 1066–1072. [CrossRef]
37. Verkhodanov, O.V.; Trushkin, S.A.; Andernach, H.; Chernenkov, V.N. Current status of the CATS database. *arXiv* **2005**, arXiv:0705.2959.
38. Atwood, W.B. et al. [Fermi LAT Collaboration]. The Large Area Telescope on the Fermi Gamma-Ray Space Telescope Mission. *Astrophys. J.* **2009**, *697*, 1071–1102. [CrossRef]
39. Abdollahi, S. et al. [Fermi LAT Collaboration]. The Fermi-LAT Lightcurve Repository. *Astrophys. J. Suppl.* **2023**, *265*, 31. [CrossRef]
40. Maraschi, L.; Tavecchio, F. The jet-disk connection and blazar unification. *Astrophys. J.* **2003**, *593*, 667–675. [CrossRef]
41. Böttcher, M.; Harris, D.E.; Krawczynski, H. (Eds.) *Relativistic Jets from Active Galactic Nuclei*; Wiley: Weinheim, Germany, 2012.
42. Ghisellini, G.; Tavecchio, F.; Foschini, L.; Ghirlanda, G.; Maraschi, L.; Celotti, A. General physical properties of bright Fermi blazars. *Mon. Not. R. Astron. Soc.* **2010**, *402*, 497–518. [CrossRef]
43. Jorstad, S.G.; Marscher, A.P.; Lister, M.L.; Stirling, A.M.; Cawthorne, T.V.; Gear, W.K.; Gómez, J.L.; Stevens, J.A.; Smith, P.S.; Forster, J.R.; et al. Polarimetric Observations of 15 Active Galactic Nuclei at High Frequencies: Jet Kinematics from Bimonthly Monitoring with the Very Long Baseline Array. *Astron. J.* **2005**, *130*, 1418–1465. [CrossRef]
44. Weaver, Z.R.; Jorstad, S.G.; Marscher, A.P.; Morozova, D.A.; Troitsky, I.S.; Agudo, I.; Gómez, J.L.; Lähteenmäki, A.; Tammi, J.; Tornikoski, M. Kinematics of Parsec-scale Jets of Gamma-Ray Blazars at 43 GHz during 10 yr of the VLBA-BU-BLAZAR Program. *Astrophys. J. Suppl. Ser.* **2022**, *260*, 12. [CrossRef]

universe

MDPI

Review

Highlights of the Magic Florian Goebel Telescopes in the Study of Active Galactic Nuclei

Marina Manganaro † and Dijana Dominis Prester *,†

Faculty of Physics, University of Rijeka, Radmile Matejčić 2, 51000 Rijeka, Croatia; marina.manganaro@uniri.hr
* Correspondence: dijana@uniri.hr
† These authors contributed equally to this work.

Abstract: The MAGIC (Major Atmospheric Gamma-ray Imaging Cherenkov) Florian Goebel telescopes are a system of two Cherenkov telescopes located on the Canary island of La Palma (Spain), at the Roque de Los Muchachos Observatory, which have been operating in stereo mode since 2009. Their low energy threshold (down to 15 GeV) allows the investigation of Active Galactic Nuclei (AGNs) in the very-high-energy (VHE, E > 100 GeV) gamma-ray range with a sensitivity up to the redshift limit of the existing IACT (Imaging Atmospheric Cherenkov Telescopes) systems. The MAGIC telescopes discovered 36 extragalactic objects emitting VHE gamma-rays and performed comprehensive studies of galaxies and their AGNs, also in a multi-wavelength (MWL) and multi-messenger (MM) context, expanding the knowledge of our Universe. Here, we report on the highlights achieved by the MAGIC collaboration since the beginning of their operations.

Keywords: γ-ray astrophysics; active galactic nuclei; very-high-energy gamma rays; non-thermal; high energy astrophysics

check for updates

Citation: Manganaro, M.; Dominis Prester, D. Highlights of the Magic Florian Goebel Telescopes in the Study of Active Galactic Nuclei. *Universe* **2024**, *10*, 80. https://doi.org/10.3390/universe10020080

Academic Editor: Pat Romano

Received: 8 December 2023
Revised: 25 January 2024
Accepted: 26 January 2024
Published: 6 February 2024

1. Introduction

The main targets of high-energy extragalactic astrophysics are Active Galactic Nuclei (AGNs), galaxies powered by a central supermassive black hole ($M_\bullet \gtrsim 10^6\ M_\odot$, where $M_\bullet$ is the mass of the black hole and $M_\odot \approx 2 \times 10^{30}$ Kg is the solar mass) emitting non-thermal photons across the electromagnetic spectrum. AGNs whose radiation reaches the gamma-ray energy range account for only 10% of the AGNs observed in our Universe, and they exhibit collimated jets of relativistic plasma. The presence of the jet is closely linked to gamma-ray emission, and when such a jet is oriented in the direction of the observer, AGNs are called blazars and are further subdivided into BL Lac-type objects (BL Lacs) and Flat Spectrum Radio Quasars (FSRQs). AGNs can show variability in all energy bands, and due to the many studies and observations that have been conducted in each band, categorizing such objects can be challenging. When AGNs emit from the radio to the VHE (very-high-energy, E > 100 GeV) gamma-ray band, they present a double-peaked spectral energy distribution (SED) (e.g., [1]) that can be studied to identify and predict the emission mechanism of the radiation. The first bump, which extends between radio and X-ray energies, is universally attributed to synchrotron radiation emanating from a population of electrons rotating in the magnetic field of the AGN, while the second bump, which can reach VHE gamma-ray energies, can be explained, depending on the position and on its shape, by different theoretical models which can be fully leptonic or including an hadronic component. In the synchrotron self-Compton model (SSC, [2,3]), the high-energy bump is attributed to the inverse Compton scattering of lower-energy photons by relativistic electrons and is therefore often referred to as the IC bump. The SSC model is a leptonic model that successfully describes the MWL SED of most blazars, but in some cases it is necessary to consider an external component responsible for the emission of high-energy (HE, 100 MeV < E < 100 GeV) or VHE gamma rays, as in leptonic EC (external Compton) models (i.e., [4]). In the case of FSRQs, for example, the external radiation fields can be the accretion disk, broad-line region (BLR) or a dusty torus. In hadronic models, the high-energy gamma rays can also be produced by relativistic protons, by proton-synchrotron radiation or by photopion production. For more details on the theoretical models, see for example a review of the theoretical challenges in the study of AGNs [5,6] for a review of the results of high energy gamma-ray observations in the context of radiative mechanisms and models.

The subdivision of blazars into the subclasses BL Lacs and FSRQs is based on the absence or presence of emission/absorption lines in their optical spectra. BL Lacs have faint lines while the optical spectra of FSRQs have strong resolved lines that suggest the presence of radiatively efficient accretion disks and that can be measured providing a value for the luminosity and consequently the mass of their black-hole [7]. BL Lacs can be further categorized according to the position of the peak of their synchrotron bump, which can be measured by their broadband SED. They are defined as LBL (low-energy peaked), HBL (high-energy peaked) or IBL (intermediate) if their synchrotron peaks are in the submillimeter to infrared energy bands, in the ultraviolet to X-ray energy bands, or in the middle between the two aforementioned cases, respectively.

An effective recent categorization consists of dividing AGNs into two main classes, jetted and non-jetted AGNs [8]. This approach is very effective as it decouples the study of AGNs from features attributed to a single energy band and projects them in a multi-wavelength (MWL) and multi-messenger (MM) context. It is therefore of utmost importance to collect simultaneous data in a MWL context and investigate the time scale of the variability at different energies, correlations between the energy bands, polarization and periodicities, in order to derive a theoretical interpretation of the dataset and shed light on the emission scenarios of AGNs. In this context, the most important goals are:

- to determine the location of the gamma-ray and VHE gamma-ray emission region in blazars;
- to understand the composition of blazar jets and the particle population responsible for the observed high-energy emission;
- to study the temporal evolution of MWL SEDs;
- the search for a comprehensive explanation for the correlations and anti-correlations observed between different wavebands;
- the investigation of the causes of variability.

The observation of AGNs has always been one of the priorities in the physics programme of the MAGIC Florian Goebel telescopes (Figure 1).

Figure 1. The MAGIC Florian Goebel telescopes at the Roque de Los Muchachos Observatory, Spain. Image credit: Chiara Righi.

An overview of the latest highlights of the MAGIC Florian Goebel telescopes, which concern not only AGNs but also other targets of interest in this field (such as binary star systems, gamma-ray bursts and pulsars), can be found in [9].

In this work, Section 2 is devoted to MAGIC observations of AGNs, first presenting the peculiarities and characteristics of the MAGIC system (see Section 2.1) and then the observational strategies used (see Section 2.2). Section 3 collects the results highlights obtained with the MAGIC telescopes concerning the study of AGNs and Section 4 gives a brief overview of the results of other IACTs (Imaging Atmospheric Cherenkov Telescopes). In Section 5 we summarise the results presented.

2. AGNs Studies with MAGIC

2.1. MAGIC Characteristics

The MAGIC collaboration began its activities and data taking with a single telescope, now called MAGIC I, which has been in operation at the Roque de Los Muchachos Observatory on the Canary Island of La Palma since 2004. Since 2009, a second telescope, MAGIC II, has been part of the system and the two telescopes operate simultaneously (stereo mode). MAGIC telescopes are IACTs, ground-based instruments that detect VHE gamma rays thanks to the extensive air showers they generate in the atmosphere. The Cherenkov light emitted by the charged particles in the extensive air showers is captured by the segmented mirror dishes of the IACTs and imaged by their cameras. The IACT technique is described in detail in [10–12]. The MAGIC Florian Goebel telescopes were built with the aim of lowering the energy threshold for the detection of VHE gamma rays, in order to explore more distant sources in this energy range and thus study earlier parts of the Universe. At the time MAGIC I was built, the existing IACTs operated with an energy threshold $\geq$ 300 GeV [13,14]. A lower energy threshold was also necessary to close the gap to the HE gamma-ray detectors, as the latter reach an energy of a few tenths of a GeV. The connection to HE gamma-ray data constrains the HE bump of broadband SEDs and helps to effectively determine the emission scenarios of jetted AGNs. This goal has been successfully achieved, making MAGIC the IACT system with a lower energy threshold, 50 GeV, but capable of going down to 15 GeV using a dedicated stereoscopic analog trigger (SumTriggerII, see [15]), as was the case in MAGIC's recent work on Geminga observations [16]. Important innovations were introduced in the design and construction of the telescopes compared to the other IACTs. Techniques known in accelerator particle physics experiments were used, such as fast electronics and automatic control of the instruments, as well as computers and networks capable of recording and reconstructing large volumes of data and performing interrelations. The main characteristics of MAGIC telescopes (see for details [17]) are as follows:

- Active mirror surface of 236 m^2, made of square elements 49.5 $\times$ 49.5 cm or 99 $\times$ 99 cm; f/D (focal length to diameter ratio) = 1.03;
- Support frame made of reinforced carbon fibre tubes (<70 tons);
- Approximately hexagonal camera with a diameter of 1.05 m, with 1039 PMTs of 1 inch diameter each (some PMTs of 2 inches diameter in the MAGIC I camera, see [17] for details); all PMTs have an effective quantum efficiency of 25 to 35%, depending on the wavelength; The camera is kept as light as possible and is held by an aluminium support arch, stiffened by a net of thin steel cables;
- The maximum repositioning speed is more than 7 degrees per second, which means that the telescopes can be pointed to any point on the observable sky in less than 25 s;
- Analogue signals are transmitted from the camera to the counting house via optical fibres; only the amplifiers and laser diode modulators for the transmission are located in the camera housing;
- Digitization is performed by the Domino Ring Sampler (DRS4) chip with a sampling frequency of 1.64 GHz to use the timing information in the pulse.

The performance of the MAGIC Florian Goebel telescopes is described in detail in [17,18].

2.2. Observational Strategies

Sophisticated observation strategies are used to make the observation time of the MAGIC telescopes efficient. The aim is to maximise the observing time assigned to the various scientific projects within the collaboration and to fully exploit the potential of the telescopes. The MAGIC observational cycles have a duration of one year and the allocation of observing time is based on the submission of observational proposals, which can also be carried out by external scientists (https://magic.mpp.mpg.de/public/magicop/, accessed on 12 January 2024). As far as the observation of AGNs is concerned, MAGIC strategies can be summarised in two types of approaches:

- ToO (Target of Opportunity) observations;
- monitoring with short or long cadence.

MAGIC has a long history of ToO observations that began with the start of operations [19]. ToOs are quickly organised to respond to alerts reporting increased flux from other instruments, for example from optical or HE gamma-ray telescopes. Collaboration with other instruments and facilities is key to the prompt issuance of ToO observations and the resulting rapid re-planning of targets. The collection of MWL simultaneous data allows the target sources to be investigated in a broadband context, and in the case of a non-detection, the upper limits obtained by MAGIC can still help to describe the HE part of the broadband SED and facilitate the theoretical interpretation. Most of MAGIC's detections in the VHE gamma-ray range, listed in Table 1, were made within the ToO MAGIC programme. Especially in the case of rapid variability, the timely response to alerts and the rapid rescheduling play a fundamental role.

The monitoring of specific targets has also yielded several results, as indicated in Section 3.8. The cadence of observation may vary depending on the characteristics of the target. Some AGNs are bright enough in the VHE gamma-ray range to be detected even when in a quiescent or low state, such as Mrk 421 and Mrk 501, allowing long-term monitoring with regularly repeated quick snapshots of observations, others are faint in the VHE gamma-ray range and require longer time windows for each scheduled observation. In any case, MWL campaigns are foreseen and organised in collaboration with other telescopes and facilities to obtain simultaneous data.

The observational proposals also consider observations at high zenith [20,21] and observations under moon conditions [22] for specific targets.

To reduce systematic uncertainties arising from atmospheric conditions, the MAGIC collaboration operates an elastic LIDAR (LIght Detection And Ranging) system [23]. This enables data corrections based on atmospheric transmission profiles [24].

3. Results

3.1. MAGIC Discoveries in the VHE Gamma-Ray Range

Since the beginning of operations, MAGIC has detected many AGNs emitting in the VHE gamma-ray range, 26 BL Lacs, six FSRQs (see Section 3.3), one radio galaxy (see Section 3.6) and three blazars (no final classification yet). Table 1 lists the AGNs discovered by MAGIC in the VHE gamma-ray range, ordered by their redshift.

Table 1. List of AGNs which were discovered to emit VHE gamma rays by the MAGIC telescopes. Data retrieved from TeVCat [1].

Name	Type	Redshift	Date of Announcement	References
RGB J2042+244	HBL	0.104	2019.11	[25]
Mrk 180	HBL	0.045	2006.09	[26]
TXS 0210+515	HBL	0.049	2019.01	[25]
1ES 2037+521	HBL	0.053	2016.10	[25]
1ES 1727+502	HBL	0.055	2011.11	[27]
2WHSP J073326.7+515354	HBL	0.065	2018.04	[28]

Table 1. *Cont.*

Name	Type	Redshift	Date of Announcement	References
1ES 1741+196	HBL	0.084	2011.08	[29]
B2 1811+31	IBL	0.117	2020.10	[30]
B3 2247+381	HBL	0.1187	2010.10	[31]
TXS 1515-273	HBL	0.1284	2019.02	[32]
1ES 1215+303	HBL	0.131	2011.01	[33]
RX J1136.5+6737	HBL	0.1342	2014.04	[34]
1RXS J081201.8+023735	HBL	0.1721	2021.02	([35] (video))
MAGIC J2001+435	IBL	0.1739	2010.07	[36]
1ES 1218+304	HBL	0.182	2006.05	[37]
IC 310	AGN (radio galaxy)	0.0189	2010.03	[38]
RBS 0723	HBL	0.198	2014.01	[25]
1ES 1011+496	HBL	0.212	2007.09	[39–42]
MS 1221.8+2452	HBL	0.218	2013.05	[43]
RGB J0136+391	HBL	>0.27	2012.07	[44]
H 1722+119	HBL		2013.05	[45]
1ES 0647+250	HBL	>0.29	2010.07	[46]
PKS 1413+135	Blazar	$0.247 < z < 0.5$ [47]	2022.01	[48]
S5 0716+714	IBL	0.26 [49]	2008.04	[50,51]
OT 081	LBL	0.322	2016.07	[52]
TXS 0506+056	Blazar	0.3365	2017.10	[53,54]
S2 0109+22	IBL	0.36	2015.07	[55]
S4 0954+65	Blazar	0.3694	2015.02	[56]
PKS 1222+216	FSRQ	0.432	2010.06	[57]
1ES 0033+595	HBL	0.467	2011.10	[58]
GB6 J1058+2817	BL Lac (class unclear)	0.4793 [59]	2021.04	[60]
3C 279	FSRQ	0.5362	2008.06	[61–63]
B2 1420+32	FSRQ	0.682	2020.01	[64]
TON 0599	FSRQ	0.7247	2017.12	[65]
PKS 1441+25	FSRQ	0.939	2015.04	[66]
QSO B0218+357	FSRQ	0.954	2014.07	[67,68]

[1] The TeVCat online source catalog, http://tevcat.uchicago.edu; (accessed on 12 January 2024).

3.2. Sources at High Redshift and EBL Studies

VHE gamma rays can interact with EBL photons by pair production. This process leads to an attenuation of the flux of VHE photons, which becomes more significant with increasing distance. For this reason, IACTs have a limit to the redshift they can achieve. This depends on the EBL properties and the energy of the VHE gamma rays emitted by the source under investigation. This limit is represented by the so-called gamma-ray horizon, which is described in detail in [69,70]. Theoretical predictions made in [69] clearly show the importance of a low energy threshold for IACTs to reach sources that are as far away as possible considering EBL attenuation. Given these predictions and the measurements of the EBL made with UV and the mid-infrared telescopes [71], the detection of AGNs near redshift 1 was considered extremely difficult. Nevertheless, thanks to their particularly low energy threshold, the MAGIC telescopes succeeded in detecting to AGNs close to redshift 1 in the VHE gamma-ray range, thus pushing the limits of the gamma-ray horizon. These two discoveries were particularly important for the study of EBL with IACTs. Both sources were detected in 2014 and were both FSRQs. Only recently an even more distant FSRQ, OP 313 with a redshift of 0.99, was detected by LST-1 (Large-Sized Telescope [72,73]).

3.2.1. QSO B0218+357

QSO B0218+357, with a redshift of $z = 0.94$, was also the first gravitationally lensed blazar detected in the VHE gamma ray range, triggered by increased activity in the HE gamma-ray range. Its discovery by MAGIC occurred during the arrival time of the delayed component of the emission [67].

The MWL SED challenged a simple leptonic model and a two-zone leptonic model with an external component (emission region located inside or outside the broad line region) was used to interpret the broadband emission. This model is described in [74] (third scenario). After this peculiar detection, a MWL campaign was organized to collect more information about the emission scenario, even if during a quiescent state. Also in this case, a model with an external Compton component and two zones was necessary to interpret the data [68]. It is interesting to note the different MWL SEDs observed in the first detection case (Figure 2a) and in the quiescent state monitoring (Figure 2b).

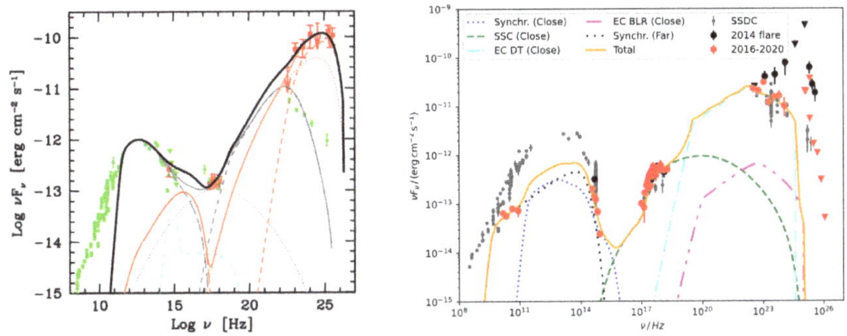

(a) (b)

Figure 2. (**a**) Broadband SED of QSO B0218+357 during the 2014 flare. Green markers show historical data while the black solid line represents the assumed two-zone model. The red markers represent the MWL dataset. Reprinted with permission from [67]. (**b**) MWL SED of QSO B0218+357 during the monitoring campaign (red markers) compared to the 2014 flare (black markers). Reprinted with permission from [68] (Figure 11).

3.2.2. PKS 1441+25

PKS 1441+25 is a FSRQ with a redshift of z = 0.94 and was detected by MAGIC during a flaring state, triggered by high activity in the HE gamma-ray range. Its MWL SED could also not be interpreted by a simple leptonic model and an external Compton was considered, with the location of the emitting region explained as originating from the jet outside the broad line region [66]. With these distant sources, it was possible to study and measure the EBL for the first time at such a high redshift with VHE gamma-ray data, constraining the EBL density between 0.21 and 1.13 μm, as reported in [75].

3.3. *Flat Spectrum Radio Quasars*

FSRQs are highly luminous blazars that show strong emission lines in their optical spectra. This feature indicates the presence of a radiatively efficient accretion disk [6]. The first classification of blazars based on their gamma-ray emission, where one of the properties of FSRQs is their soft gamma index ($\Gamma > 2$), is the famous blazar sequence concept presented in [76]. Subsequent discoveries in the VHE gamma-ray range provided more material to investigate the classification of blazars and consolidate aspects of the blazar sequence while critically reconsidering and extending some of the initial considerations. A full overview on these topics can be found in [77]. Only 10 FSRQs have been confidently detected in the VHE gamma-ray range, six of which were discovered by MAGIC. MAGIC detected in 2006 the absolute first FSRQs in VHE energy, 3C 279, at a redshift of z = 0.53 [61,62]. This result was of utmost importance as it demonstrated the ability of IACTs to reach regions of the Universe where EBL attenuation is not negligible. Moreover, the emission of FSRQs in the VHE gamma-ray range leads to an MWL SED that exhibits a high Compton Dominance (CD, ratio between the so-called IC peak and the synchrotron peak). This latter property leads to a more sophisticated modelling approach with respect to BL Lacs. The high CD observed in the FSRQs detected in VHE gamma rays, was explored for the first time in detail with 3C 279. As shown in Figure 3, the MWL SED exhibits a high CD and models with an external component were required to describe the data set [62].

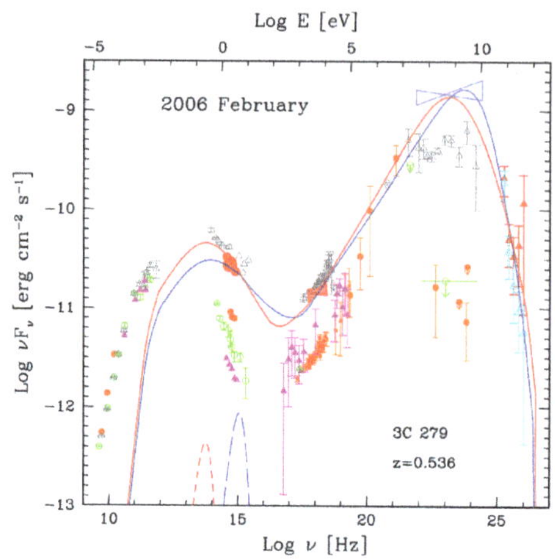

Figure 3. MWL SED of 3C 279 in 23 February 2006. The blue solid line represents the modelling under the assumption of that the external component is inside the BLR, and the red solid line when the external component is outside the BLR. Reprinted with permission from [62], Figure 8.

The second FSRQ detected by MAGIC was PKS 1222+216 in 2010 [57], and the VHE gamma-ray emission was found to be variable on a timescale of 10 min. This was an important result that made it possible to constrain the size of the gamma-ray emission region and to strongly constrain the emission models of blazar jets. The detection of PKS 1222+216 by MAGIC in fact challenged the models of jet emission and opened the discussion between different possible scenarios (far dissipation [78]; recollimation [79]; small compact emission regions in the jet [80]; reconfinement shocks [81]; relativistic filamentation [82]). The most distant FSRQs detected by MAGIC, already mentioned in Section 3.2, were detected in 2014 (QSO B0218+357) and 2015 (PKS 1441+25). After them, MAGIC detected another FSRQ in 2017, TON 0599, which lies at a redshift 0.72 [65]. This source also shows a MWL SED of complex interpretation which is the subject of a paper in preparation for the MAGIC collaboration. In 2020, MAGIC detected a new FSRQ, B2 1420+32 [64], adding another piece of the puzzle to the interpretation of such rare and powerful sources in the VHE gamma-ray sky.

3.4. *Transitional Blazars*

In some cases the usual categorization of blazars based on the properties of their optical spectra is complicated. BL Lacs are supposed to present an optical spectrum with very weak or absent emission lines. Instead in some cases (even for BL Lacertae), the optical spectra measured in different activity states can vary, making the categorization uncertain. Some blazars can be detected in the VHE gamma-ray range even when they are in quiescent state, while others can only be detected during flaring states of activity. Thus, it happens that some sources categorized as BL Lacs based on their optical spectra, suddenly show strong VHE gamma-ray emission in a flaring state that cannot be explained by the simplest leptonic models, which are often precise enough to describe the MWL emission of BL Lacs. Another categorization is related to the position of the synchrotron peak, which can also vary depending on the activity of the source. These sources could be called transitional sources [83], although it is not clear whether their behavior is simply related to exceptional gamma-ray activity in or to a real change in properties over time. To better understand these transitional AGNs, it is important to obtain a long-term dataset in

MWL. MAGIC observed and detected for the first time in the VHE gamma-ray range two AGNs categorized as BL Lacs, but from the MWL dataset collected simultaneously with the MAGIC observations, they were found to have properties close to those of FSRQs. The first, S4 0954+65, was observed by MAGIC in 2015 and studied in an MWL context in [56]. It was found that the MWL SED of S4 0954+65 is reproduced by a leptonic model typical of FSRQs with an external Compton component (dusty torus). Similarly, the blazar OT 081, observed and detected in 2016 by MAGIC together with H.E.S.S. shows features of FSRQs rather than BL Lacs, in particular the high CD [52].

3.5. Extreme Sources

The BL Lacs that exhibit a synchrotron peak at unusually high energies, above 10^{17} Hz, belong to the EHBL class (extreme HBLs, [84]). EHBLs are expected to be very faint in the VHE gamma-ray range and very difficult to detect by IACTs. Nevertheless, their study is appealing as they can be used for testing the gamma-ray propagation at high energy and in particular to probe EBL and derive limits on the IGMF (inter galactic magnetic field, see [85] for a review). Ref. [86] reports in detail on the progress made in studying such AGNs. MAGIC has devoted many observations to the hunting and study of EHBLs.

In 2018, MAGIC observed and detected the EHBL 2WHSP J073326.7+515354 for the first time in the VHE gamma-ray range [28]. The observations were scheduled as part of MAGIC's hunting strategy for such extreme sources, and this source was chosen primarily because of its high synchrotron peak frequency ($\nu_{synch} = 10^{17.9}$ Hz). The successful strategy led to the VHE detection and to an in-depth study of the MWL properties of the SED. The theoretical interpretation of the broadband emission was challenging as expected for this type of AGNs, and four different theoretical models were used to test this particular case. The scenario that best described the dataset was a spine-layer two-zone leptonic model, as described in [87]. Many other EHBLs were observed by MAGIC as part of the EHBL hunting programme, and are collected in [25]. This work also includes the archetypal EHBL 1ES 0229+20 and the results of its observation from 2013 to 2017. Also included in [25] are three EHBLs detected for the first time in the VHE gamma-ray range: 1ES 2037+521, RBS 0723, and TXS 0210+515. Ref. [86] describes and emphasises the existence of two different types of EHBLs: extreme-synchrotron sources (synchrotron peak energy $= h\nu_{synch} \geq 1$ keV) and extreme-TeV sources (gamma-ray peak energy $= h\nu_{\gamma} \geq 1$ TeV). This difference corresponds to a hard spectrum in the soft X-ray band (photon index $\Gamma_{X-ray} < 2$), or in the TeV band ($\Gamma_{\gamma} < 2$). Blazars belonging to the HBL class can take on extreme properties and exhibit extreme behavior. This has been established on two occasions by MAGIC observations, for Mrk 501 [88] and for 1ES 2344+514 [89,90]. EHBLs, and in particular extreme-TeV sources, are perfect candidates for the study of IGMF, since their high energy emission and hard spectrum can be used to constrain the presence of cascades in the IGMF. For this reason, an in-depth study using gamma-ray observations (by MAGIC, H.E.S.S., VERITAS and the *Fermi*-LAT telescopes) of the archetypal EHBL 1ES 0229+20 was performed with the specific aim of detecting or constraining the IGMF-dependent secondary produced during the propagation of TeV gamma rays through the intergalactic medium. The results presented in [91] set robust limits consisting of a lower bound of $B > 1.8 \times 10^{-17}$ G for the long-correlation-length IGMF and $B > 10^{-14}$ G for an IGMF of cosmological origin.

3.6. Black-Hole Lightening: IC310

A very interesting result from MAGIC concerns the radio galaxy IC 310, which is powered by a supermassive black hole ($M = 3 \times 10^8 M_{\odot}$). In 2012, a high amount of activity of the source was observed in the VHE gamma-rays, characterized by a very fast variability (doubling time scales faster than 4.8 min). This result challenged the existing theoretical models of the variability and suggested a new interpretation of the sub-horizon scale variability consisting of a pulsar-like particle acceleration by the electric field across a magnetospheric gap at the base of the radio jet [92]. This interpretation is shown in Figure 4.

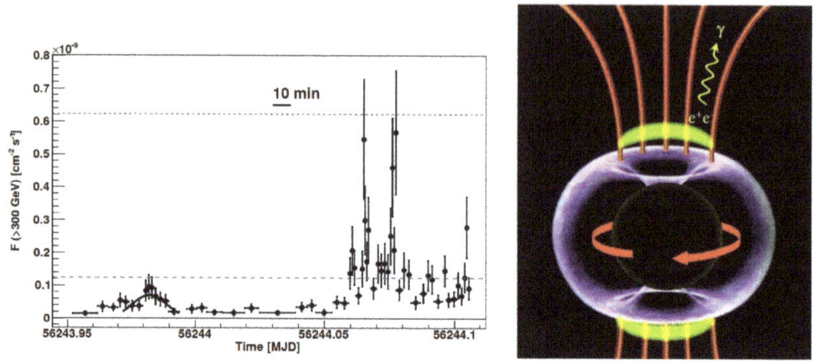

(**a**) (**b**)

Figure 4. (**a**) Light curve of IC 310, observed by the MAGIC telescopes in 12–13 November 2012. Reprinted from [92], Figure S4. (**b**) Emission scenario for the origin of the highly variable gamma-ray emission observed in IC 310 during the 2012 activity: the black sphere represents the rotating black hole with its event horizon accreting plasma from the center of IC 310. The apple-shaped blue form represents the ergosphere surrounding the black hole. The magnetosphere is shown in red and the polar vacuum gap regions in yellow. In the gaps, the electric field of the magnetosphere has a component that runs parallel to the magnetic field accelerating particles to ultra-relativistic energies. The observed gamma rays arise from inverse-Compton scattering and copious pair production due to interactions with low-energy thermal photons from the plasma accreted by the black hole. Reprinted with permission from [92], Figure S5.

3.7. Markarian Galaxies at VHE Gamma-Rays: Mrk 421 and Mrk 501

Two AGNs emitting in the VHE gamma-ray range, Mrk 421 and Mrk 501, belonging to the category of Markarians (AGNs named after Benjamin Markarian, who brought them into attention because of their uncommon excess of emission in the UV band), have been monitored by MAGIC since the beginning of the operations. Those close-by blazars (redshift $z = 0.031$ and $z = 0.034$, respectively) are perfectly suited to study the mechanism of acceleration and broadband emission of blazars for the following reasons: their proximity implies a negligible effect of EBL absorption on their VHE gamma-ray spectra and also very precise VLBI (Very Large Baseline Interferometry) studies that can be used to constrain the emission scenarios; their brightness allows monitoring in different activity states (quiescent, flaring and intermediate states) and consequently a study of the temporal evolution of their broadband SEDs; their variability can be used to break degeneracies between emission models.

MWL campaigns were organized to obtain simultaneous MWL data from Mrk 421 and Mrk 501. This has led to several works over the years describing the broadband emission of such powerful blazars over the years, accompanied by detailed studies of the broadband correlations and variability in different activity states, as shown in Tables 2 and 3.

All publications by MAGIC (in collaboration with many other instruments) on Mrk 421 and Mrk 501, together with the main results and the corresponding references, are listed in Tables 2 and 3. Another Markarian, Mrk 180, was discovered in the VHE gamma-rays by MAGIC in 2006 [26] during an optical outburst. This source is very faint has only been observed in the VHE gamma-ray range by MAGIC on this occasion.

Table 2. List of the main results obtained by MAGIC on Mrk 421.

Mrk 421			
Observational Period	**Main Results**	**Theor. Model**	**Ref.**
November 2004–April 2005	γ-ray/X-ray corr., IC peak $\sim$100 GeV	one-zone SSC	[93]
22–30 April + 14 June 2006	intra-night var. (29 April, $\sim$36 min)	leptonic	[94]
5 August 2008–12 March 2010	MWL SED in a quiescent state characterization	one-zone SCC, proton-synch	[95]
March 2010	γ-ray/X-ray corr., γ-ray and X-ray var.	one-zone SCC, two-zones SSC	[96]
January–June 2009	quiescent state characterization, X-ray harder-when-brighter, γ-ray/X-ray corr., optical/X-ray anti-corr.	one-zone SCC	[97]
January–March 2013	γ-ray/X-ray corr., double-bumped frac. var., low state characterization	one-zone SCC, suggestion of multi-zone leptonic	[98]
March 2007–June 2009	X-ray/soft X-ray corr., frac. var. increasing with energy, different levels of activity	suggested SSC, or generic hadronic scenarios	[99]
28 April–4 May 2014	X-ray spectrum variability	one-zone SSC	[100]
November 2014–June 2016	X-ray and γ-ray harder-when brighter, double-bumped frac. var., X-ray/γ-ray/ corr., VHE intra-night var. (27 January + 12 March 2015)	suggesting that the emission is powered by a multiplicative process	[101]
11–19 April 2013	intra-night var. of X-ray and VHE γ-ray bands, VHE γ-ray/X-ray corr.	magnetic reconnection in a multi-zone scenario	[102]
February 2010	limits on the Doppler factor and size of the emission region, time-lagged corr. optical/VHE	one-zone SSC excluded	[103]
December 2016–June 2017	VHE/X-ray corr., orphan γ-ray activity, intra-night VHE var., UV/X-ray anti-corr.	two-zone leptonic	[104]
December 2007–February 2009	upper limits on extended emission	possible constraints on EGMF	[105]

Table 3. List of the main results obtained by MAGIC on Mrk 501.

Mrk 501			
Observational Period	**Main Results**	**Theor. Model**	**Ref.**
May–July 2005	VHE intra-night var., spectra hardening when increasing flux, var. increasing with energy	one-zone SSC	[106]
July 2006	low state in VHE steep VHE photon index spectral hardening with flux (VHE)	one-zone SSC	[107]
15 March–1 August 2009	low activity characterization	one-zone SSC	[108]
March 2009	quiescent state characterization, X-ray peak shift of two orders of magnitude	one-zone SSC	[109]
1 April–10 August 2013	hard X-ray var. on hour timescales, five MWL SEDs	one-zone SSC	[110]
March–May 2008	low state characterization, hint of X-ray-to-VHE correlation	one-zone SSC	[111]
15 March–1 August 2009	frac. var. increasing with energy, flaring activity coincident with EVPA rotation (1 May)	two-zones SSC	[112]
March–July 2012	hard X-ray and VHE spectral indexes, extreme behaviour, VHE/X-ray corr., frac. var. increasing with energy	one-zone SSC	[88]
16–31 July 2014	frac. var. increasing with energy, VHE/X-ray corr., narrow feature in the VHE spectrum at 3 TeV (19 July)		[113]
February 2017–December 2020	X-ray/VHE corr., HE/radio corr.,	one-zone leptonic, two-zone leptonic, but also hadronic and lepto-hadronic are considered	[114]
May and April 2008	upper limits on extended emission	possible constraints on EGMF	[105]

3.8. Long-Term Monitoring Campaigns

Long-term monitoring of specific AGN targets has been a successful strategy that has yielded many results.

For example, the giant radio galaxy M 87 has been studied by MAGIC over the years allowing in-depth characterization of its quiescent states [115,116], and investigation of its flare activity [117,118]. Since M 87 is a very important target for radio observation, a collaboration in MWL with many telescopes, including EHT (Event Horizon Telescope, ref. [119]), enabled an in-depth MWK study of this close radio galaxy [116,120].

The blazar PG 1553+113 was also the subject of long monitoring campaigns in MAGIC. This blazar has been observed by MAGIC from the beginning of the operations in mono mode [121,122] and as part of MWL monitoring campaigns. Ref. [123] reports on 5 years of observations and the associated results. Recently, this source, already known for its variability in different energy bands, has been found to exhibit a quasi-periodicity in gamma-rays [124–126], making its monitoring even more important to shed light on the complex MWL emission mechanism.

The BL Lac object 1ES 0647+250 was discovered in 2012 as emitting VHE gamma-rays [127] by MAGIC. Following this finding, a monitoring campaign was launched to collect MWL data and investigate this HBL in detail. The campaign resulted in an in-depth study of the correlations between the wavebands and the characterization of the broadband emission in four different activity states [46]. Long-term variability was found in the X-ray and VHE gamma-rays band, as well as correlations between radio, optical and HE gamma-ray fluxes.

The blazar 1ES 1959+650 was detected by MAGIC in 2004 [128]. Blazar 1ES 1959+650 is a nearby AGN (z = 0.048) and an HBL that showed high activity in various energy bands in 2015, especially in the optical, but also in the gamma-ray range. The results of MAGIC and MWL monitoring of this source in recent years are summarised in [129]. These include the detection of intra-night flux variability in the VHE gamma-ray range and the interpretation of the broadband emission within a one-zone SSC model.

3.9. Multi-Messenger Studies

The first MM event in which photons and neutrinos were detected simultaneously from an astrophysical source was the famous explosion of supernova 1987A [130].

In the following years, the development of neutrino experiments made it easier to detect neutrinos from astrophysical objects and to identify their origin. This enabled a complex system of alerts in cooperation with other observatories and telescopes at different photon energies. The efforts of the astronomical community were rewarded when in 2018 when the blazar TXS 0506+056 was found to be flaring in many wavelengths in coincidence with the neutrino event IceCube-170922A. This important detection was made possible by the coordination between the instruments and the prompt response to the IceCube alert. MAGIC was involved in the MM observations and in this discovery, reported in [53]. Several theoretical models have been tested to explain the MM SED and in [54] the dataset is interpreted using a one-zone lepto-hadronic scenario in which the gamma-ray emission is produced by the inverse Compton up-scattering of external photons (see Figure 5).

Thanks to the close collaboration between neutrino experiments and other instruments across the electromagnetic spectrum, MM observations are carried out regularly to be prepared for the next MM event of a blazar.

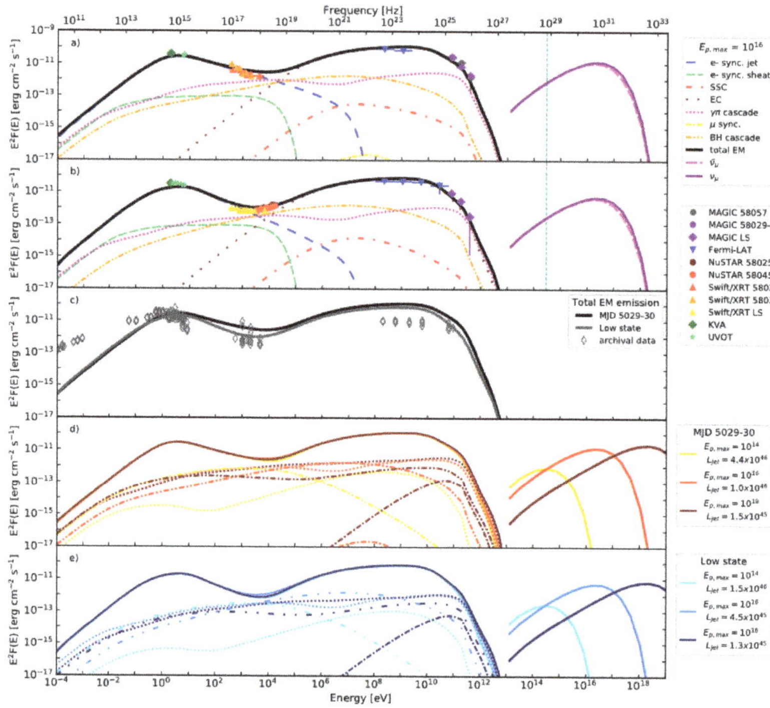

Figure 5. MM SEDs of the blazar TXS 0506+056. Panel (**a**) shows the MM SED for the enhanced VHE gamma-ray state, while panel (**b**) shows the MM SED for the lower state. Panel (**c**) shows a comparison with archival data, while panels (**d**,**e**) show results for different values of the maximum proton energy. Reprinted with permission from [54].

4. Broader Context: Other IACTs

Since MAGIC began the operations, other IACTs have also been active in the VHE gamma-ray field and have collected crucial results on AGNs and other targets. The present work is devoted only to the highlights of MAGIC, but it is important to consider the broader context of telescopes observing in a very similar energy range.

The VERITAS (Very Energetic Radiation Imaging Telescope Array System, ref. [131]) telescope array, which has been active since 2007, has observed and studied several AGNs as part of a dedicated AGN observing program [132]. The energy range of the VERITAS array extends from 85 GeV to 30 TeV. Scientific highlights of VERITAS are presented in [133] and listed on the VERITAS webpage (https://veritas.sao.arizona.edu/the-science-of-veritas/veritas-results; accessed on 12 January 2024). The VERITAS array is located in the Northern Hemisphere, like MAGIC, so many targets are common, allowing for joint publications and scientific collaboration in several cases (e.g., [88,97,100,103]).

The H.E.S.S. (High Energy Stereoscopic System) telescope array has been in operation since 2003. It is located in the Southern Hemisphere and has also studied several AGNs in the VHE gamma-ray range, from 30 GeV to 30 TeV. The highlights of their observations of AGNs are presented for example in [134,135]. A full list of scientific results can be found on their website (https://www.mpi-hd.mpg.de/HESS/pages/publications/, accessed on 12 January 2024). MAGIC has also collaborated with H.E.S.S. in the study of AGNs (e.g., [52,136]).

The FACT (First G-APD Cherenkov Telescope [137]), located at the same site of MAGIC, has been in operation since 2011 and its scientific program focuses mainly on the observations of extragalactic objects and in particular on the long-term monitoring of AGNs [138].

FACT observations can trigger follow-up observations with other IACTS with higher sensitivity, such as MAGIC, H.E.S.S. and VERITAS [88,89,104,113,139].

The LST-1 [140], which was inaugurated in 2018, is located at the same site of MAGIC (ORM observatory). This telescope, together with three others of the same design to be put in operation in the coming years, will form the group of LSTs [72] that will be part of the future CTAO (Cherenkov Telescope Array Observatory [141]). LST-1 has already achieved results in the observation of the gamma-ray source LHAASO J2108+5157 [142]. As for the study of AGNs, it has recently discovered the FSRQ OP 313 [73] in the VHE gamma-ray range. Collaboration with MAGIC in the observation of AGNs and other targets is planned, and joint data analysis will be possible thanks to a dedicated analysis pipeline [143].

5. Conclusions

The study of extragalactic sources has been one of the main targets of the MAGIC Florian Goebel telescopes for the last 20 years. The low energy threshold, which was the main goal behind their constructions (see Section 2.1), allowed a close connection to the other gamma-ray experiments and a smooth reconstruction of the high-energy bump of the MWL SEDs of many AGNs. In addition, it enabled the detection and discovery in the VHE gamma-ray range of two FSRQs close to redshift 1 (QSO B0218+357, Section 3.2.1 and PKS 1441+25, Section 3.2.2), at the limit of the IACT detection capacity, allowing the EBL to be studied for the first time at this distance with VHE gamma-ray measurements. Remarkably, of the only 10 FSRQs detected to date in the VHE gamma-ray range, six were discovered by MAGIC (See Section 3.3).

Thanks to the prompt response to MWL alerts under a dedicated ToO program, MAGIC discovered 36 AGNs in the VHE gamma-ray range (see Section 3.1) and characterized several types of AGNs in detail over the years, focusing on MWL and MM studies, which are of utmost importance for the identification of the broadband emission scenario. MAGIC was involved in the first MM observation of a neutrino blazar, TXS 0506+056, in which VHE gamma-rays were detected in spatial coincidence with the neutrino observed by IceCube (Section 3.9). The latter result suggested that blazars can be sources of high energy neutrinos and stimulated the development of theoretical models that can reproduce the MM SED.

The monitoring of AGNs such as Mrk 421 and Mrk 501, PG 1553+113, M 87, 1ES 0647+250, 1ES 1959+650 enabled long-term studies of correlations between different energy bands and variability (Sections 3.7 and 3.8).

The results obtained by MAGIC during its long operation lifetime will certainly be complemented and extended by the new generation of IACTs—the CTAO (Cherenkov Telescope Array Observatory [141]).

Author Contributions: Conceptualization, M.M. and D.D.P.; Investigation, M.M. and D.D.P.; Writing—original draft preparation, M.M.; writing—review and editing, D.D.P. All authors have read and agreed to the published version of the manuscript.

Funding: This work was funded by the Croatian Science Foundation (HrZZ) Project IP-2022-10-4595 and the University of Rijeka grant number uniri-prirod-18-48.

Informed Consent Statement: Not applicable.

Conflicts of Interest: The authors declare no conflict of interest.

Abbreviations

The following abbreviations are used in this manuscript:

AGN	Active Galactic Nucleus
AGNs	Active Galactic Nuclei
EBL	Extragalactic Background Light
LIDAR	LIght Detection And Ranging
MAGIC	Major Atmospheric Gamma-ray Imaging Cherenkov (telescopes)

IACTs	Imaging Atmospheric Cherenkov Telescopes
HE	High-energy
VHE	Very-high-energy
MWL	Multi-wavelength
MM	Multi-messenger
SED	Spectral Energy Distribution
FSRQ	Flat Spectrum Radio Quasar
BL Lacs	BL Lacertae type objects
f.o.v	Field of view
MJD	Modified Julian Date
ToO	Target of Opportunity
SSC	Synchrotron self-Compton

References

1. Ghisellini, G.; Righi, C.; Costamante, L.; Tavecchio, F. The Fermi blazar sequence. *Mon. Not. R. Astron. Soc.* **2017**, *469*, 255–266. [CrossRef]
2. Konigl, A. Relativistic jets as X-ray and gamma-ray sources. *Astrophys. J.* **1981**, *243*, 700–709. [CrossRef]
3. Maraschi, L.; Ghisellini, G.; Celotti, A. A Jet Model for the Gamma-Ray—Emitting Blazar 3C 279. *Astrophys. J. Lett.* **1992**, *397*, L5–L9. [CrossRef]
4. Dermer, C.D.; Schlickeiser, R. Model for the High-Energy Emission from Blazars. *Astrophys. J.* **1993**, *416*, 458. [CrossRef]
5. Böttcher, M. Progress in Multi-Wavelength and Multi-Messenger Observations of Blazars and Theoretical Challenges. *Galaxies* **2019**, *7*, 20. [CrossRef]
6. Madejski, G.G.; Sikora, M. Gamma-Ray Observations of Active Galactic Nuclei. *Annu. Rev. Astron. Astrophys.* **2016**, *54*, 725–760. [CrossRef]
7. Vestergaard, M.; Peterson, B.M. Determining Central Black Hole Masses in Distant Active Galaxies and Quasars. II. Improved Optical and UV Scaling Relationships. *Astrophys. J.* **2006**, *641*, 689–709. [CrossRef]
8. Padovani, P. On the two main classes of active galactic nuclei. *Nat. Astron.* **2017**, *1*, 0194. [CrossRef]
9. Loporchio, S.; MAGIC Collaboration. Highlights of the very-high-energy gamma-ray sky as seen by MAGIC. *Nucl. Instrum. Methods Phys. Res.* **2023**, *1055*, 168441. [CrossRef]
10. Bose, D.; Chitnis, V.R.; Majumdar, P.; Acharya, B.S. Ground-based gamma-ray astronomy: history and development of techniques. *Eur. Phys. J. Spec. Top.* **2022**, *231*, 3–26. [CrossRef]
11. de Naurois, M.; Mazin, D. Ground-based detectors in very-high-energy gamma-ray astronomy. *Comptes Rendus Phys.* **2015**, *16*, 610–627. [CrossRef]
12. Aharonian, F.; Buckley, J.; Kifune, T.; Sinnis, G. High energy astrophysics with ground-based gamma ray detectors. *Rep. Prog. Phys.* **2008**, *71*, 096901. [CrossRef]
13. Pühlhofer, G.; Bolz, O.; Götting, N.; Heusler, A.; Horns, D.; Kohnle, A.; Lampeitl, H.; Panter, M.; Tluczykont, M.; Aharonian, F.; et al. The technical performance of the HEGRA system of imaging air Cherenkov telescopes. *Astropart. Phys.* **2003**, *20*, 267–291. [CrossRef]
14. Rovero, A.C.; Buckley, J.H.; Fleury, P.; Jiang, Y.; Pare, E.; Sarazin, X.; Urban, M.; Weekes, T.C. Calibration of the Whipple atmospheric Čerenkov telescope. *Astropart. Phys.* **1996**, *5*, 27–34. [CrossRef]
15. Dazzi, F.; Schweizer, T.; Ceribella, G.; Corti, D.; Dettlaff, A.; Garcia, J.R.; Hafner, D.; Herranz, D.; Lopez-Moya, M.; Mariotti, M.; et al. The Stereoscopic Analog Trigger of the MAGIC Telescopes. *IEEE Trans. Nucl. Sci.* **2021**, *68*, 1473–1486. [CrossRef]
16. MAGIC Collaboration; Acciari, V.A.; Ansoldi, S.; Antonelli, L.A.; Arbet Engels, A.; Asano, K.; Baack, D.; Babić, A.; Baquero, A.; Barres de Almeida, U.; et al. Detection of the Geminga pulsar with MAGIC hints at a power-law tail emission beyond 15 GeV. *Astron. Astrophys.* **2020**, *643*, L14. [CrossRef]
17. Aleksić, J.; Ansoldi, S.; Antonelli, L.A.; Antoranz, P.; Babic, A.; Bangale, P.; Barceló, M.; Barrio, J.A.; Becerra González, J.; Bednarek, W.; et al. The major upgrade of the MAGIC telescopes, Part I: The hardware improvements and the commissioning of the system. *Astropart. Phys.* **2016**, *72*, 61–75. [CrossRef]
18. Aleksić, J.; Ansoldi, S.; Antonelli, L.A.; Antoranz, P.; Babic, A.; Bangale, P.; Barceló, M.; Barrio, J.A.; Becerra González, J.; Bednarek, W.; et al. The major upgrade of the MAGIC telescopes, Part II: A performance study using observations of the Crab Nebula. *Astropart. Phys.* **2016**, *72*, 76–94. [CrossRef]
19. Mazin, D.; Lindfors, E. Successful ToO triggers on the extragalactic sources with the MAGIC telescope. *arXiv* **2008**, arXiv:0709.1694. https://doi.org/10.48550/arXiv.0709.1694.
20. Aleksić, J.; Alvarez, E.A.; Antonelli, L.A.; Antoranz, P.; Asensio, M.; Backes, M.; Barres de Almeida, U.; Barrio, J.A.; Bastieri, D.; Becerra González, J.; et al. High zenith angle observations of PKS 2155-304 with the MAGIC-I telescope. *Astron. Astrophys.* **2012**, *544*, A75. [CrossRef]
21. MAGIC Collaboration; Acciari, V.A.; Ansoldi, S.; Antonelli, L.A.; Arbet Engels, A.; Baack, D.; Babić, A.; Banerjee, B.; Barres de Almeida, U.; Barrio, J.A.; et al. MAGIC very large zenith angle observations of the Crab Nebula up to 100 TeV. *Astron. Astrophys.* **2020**, *635*, A158. [CrossRef]

22. Ahnen, M.L.; Ansoldi, S.; Antonelli, L.A.; Arcaro, C.; Babić, A.; Banerjee, B.; Bangale, P.; Barres de Almeida, U.; Barrio, J.A.; Becerra González, J.; et al. Performance of the MAGIC telescopes under moonlight. *Astropart. Phys.* **2017**, *94*, 29–41. [CrossRef]
23. Fruck, C.; Gaug, M.; Hahn, A.; Acciari, V.; Besenrieder, J.; Dominis Prester, D.; Dorner, D.; Fink, D.; Font, L.; Mićanović, S.; et al. Characterizing the aerosol atmosphere above the Observatorio del Roque de los Muchachos by analysing seven years of data taken with an GaAsP HPD-readout, absolutely calibrated elastic LIDAR. *Mon. Not. R. Astron. Soc.* **2022**, *515*, 4520–4550. [CrossRef]
24. Schmuckermaier, F.; Gaug, M.; Fruck, C.; Moralejo, A.; Hahn, A.; Dominis Prester, D.; Dorner, D.; Font, L.; Mićanović, S.; Mirzoyan, R.; et al. Correcting Imaging Atmospheric Cherenkov Telescope data with atmospheric profiles obtained with an elastic light detecting and ranging system. *Astron. Astrophys.* **2023**, *673*, A2. [CrossRef]
25. Acciari, V.A.; Ansoldi, S.; Antonelli, L.A.; Engels, A.A.; Asano, K.; Baack, D.; Babić, A.; Banerjee, B.; Barres de Almeida, U.; Barrio, J.A.; et al. New Hard-TeV Extreme Blazars Detected with the MAGIC Telescopes. *Astrophys. J. Suppl. Ser.* **2020**, *247*, 16. [CrossRef]
26. Albert, J.; Aliu, E.; Anderhub, H.; Antoranz, P.; Armada, A.; Asensio, M.; Baixeras, C.; Barrio, J.A.; Bartko, H.; Bastieri, D.; et al. Discovery of Very High Energy γ-Rays from Markarian 180 Triggered by an Optical Outburst. *Astrophys. J. Lett.* **2006**, *648*, L105–L108. [CrossRef]
27. Aleksić, J.; Antonelli, L.A.; Antoranz, P.; Asensio, M.; Backes, M.; Barres de Almeida, U.; Barrio, J.A.; Becerra González, J.; Bednarek, W.; Berger, K.; et al. Discovery of very high energy gamma-ray emission from the blazar 1ES 1727+502 with the MAGIC Telescopes. *Astron. Astrophys.* **2014**, *563*, A90. [CrossRef]
28. MAGIC Collaboration; Acciari, V.A.; Ansoldi, S.; Antonelli, L.A.; Arbet Engels, A.; Baack, D.; Babić, A.; Banerjee, B.; Barres de Almeida, U.; et al. Testing emission models on the extreme blazar 2WHSP J073326.7+515354 detected at very high energies with the MAGIC telescopes. *Mon. Not. R. Astron. Soc.* **2019**, *490*, 2284–2299. [CrossRef]
29. Ahnen, M.L.; Ansoldi, S.; Antonelli, L.A.; Antoranz, P.; Arcaro, C.; Babic, A.; Banerjee, B.; Bangale, P.; Barres de Almeida, U.; Barrio, J.A.; et al. MAGIC detection of very high energy γ-ray emission from the low-luminosity blazar 1ES 1741+196. *Mon. Not. R. Astron. Soc.* **2017**, *468*, 1534–1541. [CrossRef]
30. Blanch, O. Detection of very-high-energy gamma-ray emission from B2 1811+31 with the MAGIC telescopes. *Astron. Telegr.* **2020**, *14090*, 1–11.
31. Aleksić, J.; Alvarez, E.A.; Antonelli, L.A.; Antoranz, P.; Asensio, M.; Backes, M.; Barrio, J.A.; Bastieri, D.; Becerra González, J.; Bednarek, W.; et al. Discovery of VHE γ-ray emission from the BL Lacertae object B3 2247+381 with the MAGIC telescopes. *Astron. Astrophys.* **2012**, *539*, A118. [CrossRef]
32. Acciari, V.A.; Ansoldi, S.; Antonelli, L.A.; Arbet Engels, A.; Artero, M.; Asano, K.; Baack, D.; Babić, A.; Baquero, A.; Barres de Almeida, U.; et al. First detection of VHE gamma-ray emission from TXS 1515-273, study of its X-ray variability and spectral energy distribution. *Mon. Not. R. Astron. Soc.* **2021**, *507*, 1528–1545. [CrossRef]
33. Aleksić, J.; Alvarez, E.A.; Antonelli, L.A.; Antoranz, P.; Ansoldi, S.; Asensio, M.; Backes, M.; Barres de Almeida, U.; Barrio, J.A.; Bastieri, D.; et al. Discovery of VHE γ-rays from the blazar 1ES 1215+303 with the MAGIC telescopes and simultaneous multi-wavelength observations. *Astron. Astrophys.* **2012**, *544*, A142. [CrossRef]
34. Mirzoyan, R. Discovery of Very High Energy Gamma-Ray Emission from BL Lac object RX J1136.5+6737 by the MAGIC Telescopes. *Astron. Telegr.* **2014**, *6062*, 1.
35. Ventura, S.; Prandini, E.; Fallah Ramazani, V.; Jormanainen, J.; Bonnoli, G.; Da Vela, P. Very-High-Energy Detection of the Extreme Blazar 1RXS J081201.8+023735 with the MAGIC Telescopes. In Proceedings of the 43rd COSPAR Scientific Assembly, Sydney, Australia, 28 January–4 February 2021; Volume 43, p. 1281.
36. Aleksić, J.; Ansoldi, S.; Antonelli, L.A.; Antoranz, P.; Babic, A.; Bangale, P.; Barres de Almeida, U.; Barrio, J.A.; Becerra González, J.; Bednarek, W.; et al. First broadband characterization and redshift determination of the VHE blazar MAGIC J2001+439. *Astron. Astrophys.* **2014**, *572*, A121. [CrossRef]
37. Albert, J.; Aliu, E.; Anderhub, H.; Antoranz, P.; Armada, A.; Asensio, M.; Baixeras, C.; Barrio, J.A.; Bartelt, M.; Bartko, H.; et al. Discovery of Very High Energy Gamma Rays from 1ES 1218+30.4. *Astrophys. J. Lett.* **2006**, *642*, L119–L122. [CrossRef]
38. Aleksić, J.; Antonelli, L.A.; Antoranz, P.; Babic, A.; Barres de Almeida, U.; Barrio, J.A.; Becerra González, J.; Bednarek, W.; Berger, K.; Bernardini, E.; et al. Rapid and multiband variability of the TeV bright active nucleus of the galaxy IC 310. *Astron. Astrophys.* **2014**, *563*, A91. [CrossRef]
39. Albert, J.; Aliu, E.; Anderhub, H.; Antoranz, P.; Armada, A.; Baixeras, C.; Barrio, J.A.; Bartko, H.; Bastieri, D.; Becker, J.K.; et al. Discovery of Very High Energy γ-Rays from 1ES 1011+496 at z = 0.212. *Astrophys. J. Lett.* **2007**, *667*, L21–L24. [CrossRef]
40. Ahnen, M.L.; Ansoldi, S.; Antonelli, L.A.; Antoranz, P.; Babic, A.; Banerjee, B.; Bangale, P.; Barres de Almeida, U.; Barrio, J.A.; Becerra González, J.; et al. MAGIC observations of the February 2014 flare of 1ES 1011+496 and ensuing constraint of the EBL density. *Astron. Astrophys.* **2016**, *590*, A24. [CrossRef]
41. Ahnen, M.L.; Ansoldi, S.; Antonelli, L.A.; Antoranz, P.; Babic, A.; Banerjee, B.; Bangale, P.; Barres de Almeida, U.; Barrio, J.A.; Becerra González, J.; et al. Multiwavelength observations of the blazar 1ES 1011+496 in Spring 2008. *Mon. Not. R. Astron. Soc.* **2016**, *459*, 2286–2298. [CrossRef]
42. Aleksić, J.; Ansoldi, S.; Antonelli, L.A.; Antoranz, P.; Arcaro, C.; Babic, A.; Bangale, P.; Barres de Almeida, U.; Barrio, J.A.; Becerra González, J.; et al. Insights into the emission of the blazar 1ES 1011+496 through unprecedented broadband observations during 2011 and 2012. *Astron. Astrophys.* **2016**, *591*, A10. [CrossRef]

43. Cortina, J. Discovery of Very High Energy Gamma-Ray Emission from MS1221.8+2452 with the MAGIC telescopes. *Astron. Telegr.* **2013**, *5038*, 1.
44. Mazin, D.; MAGIC Collaboration. Highlights from the MAGIC telescopes. In *High Energy Gamma-Ray Astronomy: 5th International Meeting on High Energy Gamma-Ray Astronomy*; Aharonian, F.A.; Hofmann, W.; Rieger, F.M., Eds.; American Institute of Physics Conference Series; AIP Publishing: Melville, NY, USA, 2012; Volume 1505, pp. 186–193. [CrossRef]
45. Ahnen, M.L.; Ansoldi, S.; Antonelli, L.A.; Antoranz, P.; Babic, A.; Banerjee, B.; Bangale, P.; Barres de Almeida, U.; Barrio, J.A.; Becerra González, J.; et al. Investigating the peculiar emission from the new VHE gamma-ray source H1722+119. *Mon. Not. R. Astron. Soc.* **2016**, *459*, 3271–3281. [CrossRef]
46. MAGIC Collaboration; Acciari, V.A.; Aniello, T.; Ansoldi, S.; Antonelli, L.A.; Arbet Engels, A.; Arcaro, C.; Artero, M.; Asano, K.; Baack, D.; et al. Long-term multi-wavelength study of 1ES 0647+250. *Astron. Astrophys.* **2023**, *670*, A49. [CrossRef]
47. Readhead, A.C.S.; Ravi, V.; Liodakis, I.; Lister, M.L.; Singh, V.; Aller, M.F.; Blandford, R.D.; Browne, I.W.A.; Gorjian, V.; Grainge, K.J.B.; et al. The Relativistic Jet Orientation and Host Galaxy of the Peculiar Blazar PKS 1413+135. *Astrophys. J.* **2021**, *907*, 61. [CrossRef]
48. Blanch, O.; Sitarek, J.; Striskovic, J. First detection of very-high-energy gamma-ray emission from PKS1413+135 with the MAGIC telescopes. *Astron. Telegr.* **2022**, *15161*, 1.
49. Dorigo Jones, J.; Johnson, S.D.; Muzahid, S.; Charlton, J.; Chen, H.W.; Narayanan, A.; Schaye, J.; Wijers, N.A. Improving blazar redshift constraints with the edge of the Ly α forest: 1ES 1553+113 and implications for observations of the WHIM. *Mon. Not. R. Astron. Soc.* **2022**, *509*, 4330–4343. [CrossRef]
50. Anderhub, H.; Antonelli, L.A.; Antoranz, P.; Backes, M.; Baixeras, C.; Balestra, S.; Barrio, J.A.; Bastieri, D.; Becerra González, J.; Becker, J.K.; et al. Discovery of very High Energy γ-Rays from the Blazar S5 0716+714. *Astrophys. J. Lett.* **2009**, *704*, L129–L133. [CrossRef]
51. MAGIC Collaboration; Ahnen, M.L.; Ansoldi, S.; Antonelli, L.A.; Arcaro, C.; Baack, D.; Babić, A.; Banerjee, B.; Bangale, P.; Barres de Almeida, U.; et al. Multi-wavelength characterization of the blazar S5 0716+714 during an unprecedented outburst phase. *Astron. Astrophys.* **2018**, *619*, A45. [CrossRef]
52. Manganaro, M.; Seglar-Arroyo, M.; Becerra-González, J.; Sanchez, D.; Cerruti, M.; Tavecchio, F.; Fallah-Ramazani, V.; Agudo, I.; Ciprini, S.; Filippenko, A.V.; et al. MAGIC and H.E.S.S. detect VHE gamma rays from the blazar OT081 for the first time: A deep multiwavelength study. In Proceedings of the 37th International Cosmic Ray Conference, Berlin, Germany, 12–23 July 2021; p. 815.
53. IceCube Collaboration; Aartsen, M.G.; Ackermann, M.; Adams, J.; Aguilar, J.A.; Ahlers, M.; Ahrens, M.; Al Samarai, I.; Altmann, D.; Andeen, K.; et al. Multimessenger observations of a flaring blazar coincident with high-energy neutrino IceCube-170922A. *Science* **2018**, *361*, eaat1378. [CrossRef]
54. Ansoldi, S.; Antonelli, L.A.; Arcaro, C.; Baack, D.; Babić, A.; Banerjee, B.; Bangale, P.; Barres de Almeida, U.; Barrio, J.A.; Becerra González, J.; et al. The Blazar TXS 0506+056 Associated with a High-energy Neutrino: Insights into Extragalactic Jets and Cosmic-Ray Acceleration. *Astrophys. J. Lett.* **2018**, *863*, L10. [CrossRef]
55. MAGIC Collaboration; Ansoldi, S.; Antonelli, L.A.; Arcaro, C.; Baack, D.; Babić, A.; Banerjee, B.; Bangale, P.; Barres de Almeida, U.; et al. The broad-band properties of the intermediate synchrotron peaked BL Lac S2 0109+22 from radio to VHE gamma-rays. *Mon. Not. R. Astron. Soc.* **2018**, *480*, 879–892. [CrossRef]
56. MAGIC Collaboration; Ahnen, M.L.; Ansoldi, S.; Antonelli, L.A.; Arcaro, C.; Baack, D.; Babić, A.; Banerjee, B.; Bangale, P.; Barres de Almeida, U.; et al. Detection of the blazar S4 0954+65 at very-high-energy with the MAGIC telescopes during an exceptionally high optical state. *Astron. Astrophys.* **2018**, *617*, A30. [CrossRef]
57. Aleksić, J.; Antonelli, L.A.; Antoranz, P.; Backes, M.; Barrio, J.A.; Bastieri, D.; Becerra González, J.; Bednarek, W.; Berdyugin, A.; Berger, K.; et al. MAGIC Discovery of Very High Energy Emission from the FSRQ PKS 1222+21. *Astrophys. J. Lett.* **2011**, *730*, L8. [CrossRef]
58. Aleksić, J.; Ansoldi, S.; Antonelli, L.A.; Antoranz, P.; Babic, A.; Bangale, P.; Barres de Almeida, U.; Barrio, J.A.; Becerra González, J.; Bednarek, W.; et al. Discovery of very high energy γ-ray emission from the blazar 1ES 0033+595 by the MAGIC telescopes. *Mon. Not. R. Astron. Soc.* **2015**, *446*, 217–225. [CrossRef]
59. Massaro, F.; Masetti, N.; D'Abrusco, R.; Paggi, A.; Funk, S. Optical Spectroscopic Observations of Blazars and γ-Ray Blazar Candidates in the Sloan Digital Sky Survey Data Release Nine. *Astron. J.* **2014**, *148*, 66. [CrossRef]
60. Blanch, O. Detection of very-high-energy gamma-ray emission from GB6 J1058+2817 with the MAGIC telescopes. *Astron. Telegr.* **2021**, *14506*, 1.
61. MAGIC Collaboration; Albert, J.; Aliu, E.; Anderhub, H.; Antonelli, L.A.; Antoranz, P.; Backes, M.; Baixeras, C.; Barrio, J.A.; Bartko, H.; et al. Very-High-Energy gamma rays from a Distant Quasar: How Transparent Is the Universe? *Science* **2008**, *320*, 1752. [CrossRef]
62. Aleksić, J.; Antonelli, L.A.; Antoranz, P.; Backes, M.; Barrio, J.A.; Bastieri, D.; Becerra González, J.; Bednarek, W.; Berdyugin, A.; Berger, K.; et al. MAGIC Observations and multiwavelength properties of the quasar 3C 279 in 2007 and 2009. *Astron. Astrophys.* **2011**, *530*, A4. [CrossRef]
63. Aleksić, J.; Ansoldi, S.; Antonelli, L.A.; Antoranz, P.; Babic, A.; Bangale, P.; Barres de Almeida, U.; Barrio, J.A.; Becerra González, J.; Bednarek, W.; et al. MAGIC observations and multifrequency properties of the flat spectrum radio quasar 3C 279 in 2011. *Astron. Astrophys.* **2014**, *567*, A41. [CrossRef]

64. MAGIC Collaboration; Acciari, V.A.; Ansoldi, S.; Antonelli, L.A.; Arbet Engels, A.; Artero, M.; Asano, K.; Baack, D.; Babić, A.; Baquero, A.; et al. VHE gamma-ray detection of FSRQ QSO B1420+326 and modeling of its enhanced broadband state in 2020. *Astron. Astrophys.* **2021**, *647*, A163. [CrossRef]
65. Mirzoyan, R. Detection of very-high-energy gamma-ray emission from the FSRQ Ton 0599 with the MAGIC telescopes. *Astron. Telegr.* **2017**, *11061*, 1.
66. Ahnen, M.L.; Ansoldi, S.; Antonelli, L.A.; Antoranz, P.; Babic, A.; Banerjee, B.; Bangale, P.; Barres de Almeida, U.; Barrio, J.A.; Bednarek, W.; et al. Very High Energy γ-Rays from the Universe's Middle Age: Detection of the z = 0.940 Blazar PKS 1441+25 with MAGIC. *Astrophys. J. Lett.* **2015**, *815*, L23. [CrossRef]
67. Ahnen, M.L.; Ansoldi, S.; Antonelli, L.A.; Antoranz, P.; Arcaro, C.; Babic, A.; Banerjee, B.; Bangale, P.; Barres de Almeida, U.; Barrio, J.A.; et al. Detection of very high energy gamma-ray emission from the gravitationally lensed blazar QSO B0218+357 with the MAGIC telescopes. *Astron. Astrophys.* **2016**, *595*, A98. [CrossRef]
68. Acciari, V.A.; Ansoldi, S.; Antonelli, L.A.; Arbet Engels, A.; Artero, M.; Asano, K.; Baack, D.; Babić, A.; Baquero, A.; Barres de Almeida, U.; et al. Multiwavelength study of the gravitationally lensed blazar QSO B0218+357 between 2016 and 2020. *Mon. Not. R. Astron. Soc.* **2022**, *510*, 2344–2362. [CrossRef]
69. Blanch, O.; Martinez, M. Exploring the gamma ray horizon with the next generation of gamma ray telescopes. Part 1: Theoretical predictions. *Astropart. Phys.* **2005**, *23*, 588–597. [CrossRef]
70. Blanch, O.; Martinez, M. Exploring the gamma-ray horizon with the next generation of gamma-ray telescopes. Part 2: Extracting cosmological parameters from the observation of gamma-ray sources. *Astropart. Phys.* **2005**, *23*, 598–607. [CrossRef]
71. Domínguez, A.; Primack, J.R.; Rosario, D.J.; Prada, F.; Gilmore, R.C.; Faber, S.M.; Koo, D.C.; Somerville, R.S.; Pérez-Torres, M.A.; Pérez-González, P.; et al. Extragalactic background light inferred from AEGIS galaxy-SED-type fractions. *Mon. Not. R. Astron. Soc.* **2011**, *410*, 2556–2578. [CrossRef]
72. Pierro, F.D. Status of the Large-Sized Telescope of the Cherenkov Telescope Array. *J. Phys. Conf. Ser.* **2023**, *2429*, 012020. [CrossRef]
73. Cortina, J.; CTAO LST Collaboration. First detection of VHE gamma-ray emission from FSRQ OP 313 with LST-1. *Astron. Telegr.* **2023**, *16381*, 1.
74. Tavecchio, F.; Becerra-Gonzalez, J.; Ghisellini, G.; Stamerra, A.; Bonnoli, G.; Foschini, L.; Maraschi, L. On the origin of the γ-ray emission from the flaring blazar PKS 1222+216. *Astron. Astrophys.* **2011**, *534*, A86. [CrossRef]
75. Acciari, V.A.; Ansoldi, S.; Antonelli, L.A.; Arbet Engels, A.; Baack, D.; Babić, A.; Banerjee, B.; Barres de Almeida, U.; Barrio, J.A.; et al. Measurement of the extragalactic background light using MAGIC and Fermi-LAT gamma-ray observations of blazars up to z = 1. *Mon. Not. R. Astron. Soc.* **2019**, *486*, 4233–4251. [CrossRef]
76. Fossati, G.; Maraschi, L.; Celotti, A.; Comastri, A.; Ghisellini, G. A unifying view of the spectral energy distributions of blazars. *Mon. Not. R. Astron. Soc.* **1998**, *299*, 433–448. [CrossRef]
77. Prandini, E.; Ghisellini, G. The Blazar Sequence and Its Physical Understanding. *Galaxies* **2022**, *10*, 35. [CrossRef]
78. Tavecchio, F.; Ghisellini, G.; Bonnoli, G.; Ghirlanda, G. Constraining the location of the emitting region in Fermi blazars through rapid γ-ray variability. *Mon. Not. R. Astron. Soc.* **2010**, *405*, L94–L98. [CrossRef]
79. Marscher, A.P. Relativistic jets and the continuum emission in QSOs. *Astrophys. J.* **1980**, *235*, 386–391. [CrossRef]
80. Ghisellini, G.; Tavecchio, F. Rapid variability in TeV blazars: the case of PKS2155-304. *Mon. Not. R. Astron. Soc.* **2008**, *386*, L28–L32. [CrossRef]
81. Nalewajko, K.; Sikora, M. A structure and energy dissipation efficiency of relativistic reconfinement shocks. *Mon. Not. R. Astron. Soc.* **2009**, *392*, 1205–1210. [CrossRef]
82. Frederiksen, J.T.; Haugbølle, T.; Medvedev, M.V.; Nordlund, Å. Radiation Spectral Synthesis of Relativistic Filamentation. *Astrophys. J. Lett.* **2010**, *722*, L114–L119. [CrossRef]
83. Ghisellini, G.; Tavecchio, F.; Foschini, L.; Ghirlanda, G. The transition between BL Lac objects and flat spectrum radio quasars. *Mon. Not. R. Astron. Soc.* **2011**, *414*, 2674–2689. [CrossRef]
84. Costamante, L.; Ghisellini, G.; Giommi, P.; Tagliaferri, G.; Celotti, A.; Chiaberge, M.; Fossati, G.; Maraschi, L.; Tavecchio, F.; Treves, A.; et al. Extreme synchrotron BL Lac objects. Stretching the blazar sequence. *Astron. Astrophys.* **2001**, *371*, 512–526. [CrossRef]
85. Durrer, R.; Neronov, A. Cosmological magnetic fields: their generation, evolution and observation. *Astron. Astrophys. Rev.* **2013**, *21*, 62. [CrossRef]
86. Biteau, J.; Prandini, E.; Costamante, L.; Lemoine, M.; Padovani, P.; Pueschel, E.; Resconi, E.; Tavecchio, F.; Taylor, A.; Zech, A. Progress in unveiling extreme particle acceleration in persistent astrophysical jets. *Nat. Astron.* **2020**, *4*, 124–131. [CrossRef]
87. Ghisellini, G.; Tavecchio, F.; Chiaberge, M. Structured jets in TeV BL Lac objects and radiogalaxies. Implications for the observed properties. *Astron. Astrophys.* **2005**, *432*, 401–410. [CrossRef]
88. Ahnen, M.L.; Ansoldi, S.; Antonelli, L.A.; Arcaro, C.; Babić, A.; Banerjee, B.; Bangale, P.; Barres de Almeida, U.; Barrio, J.A.; Becerra González, J.; et al. Extreme HBL behavior of Markarian 501 during 2012. *Astron. Astrophys.* **2018**, *620*, A181. [CrossRef]
89. MAGIC Collaboration; Acciari, V.A.; Ansoldi, S.; Antonelli, L.A.; Arbet Engels, A.; Babić, A.; Banerjee, B.; Barres de Almeida, U.; Barrio, J.A.; Becerra González, J.; et al. An intermittent extreme BL Lac: MWL study of 1ES 2344+514 in an enhanced state. *Mon. Not. R. Astron. Soc.* **2020**, *496*, 3912–3928. [CrossRef]
90. Abe, H.; Abe, S.; Acciari, V.A.; Agudo, I.; Aniello, T.; Ansoldi, S.; Antonelli, L.A.; Arbet Engels, A.; Arcaro, C.; Artero, M.; et al. Multi-year characterisation of the broad-band emission from the intermittent extreme BL Lac 1ES~2344+514. *arXiv* **2023**, arXiv:2310.03922. https://doi.org/10.48550/arXiv.2310.03922.

91. Acciari, V.A.; Agudo, I.; Aniello, T.; Ansoldi, S.; Antonelli, L.A.; Arbet Engels, A.; Artero, M.; Asano, K.; Baack, D.; Babić, A.; et al. A lower bound on intergalactic magnetic fields from time variability of 1ES 0229+200 from MAGIC and Fermi/LAT observations. *Astron. Astrophys.* **2023**, *670*, A145. [CrossRef]
92. Aleksić, J.; Ansoldi, S.; Antonelli, L.A.; Antoranz, P.; Babic, A.; Bangale, P.; Barrio, J.A.; González, J.B.; Bednarek, W.; Bernardini, E.; et al. Black hole lightning due to particle acceleration at subhorizon scales. *Science* **2014**, *346*, 1080–1084. [CrossRef]
93. Albert, J.; Aliu, E.; Anderhub, H.; Antoranz, P.; Armada, A.; Asensio, M.; Baixeras, C.; Barrio, J.A.; Bartko, H.; Bastieri, D.; et al. Observations of Markarian 421 with the MAGIC Telescope. *Astrophys. J.* **2007**, *663*, 125–138. [CrossRef]
94. Aleksić, J.; Anderhub, H.; Antonelli, L.A.; Antoranz, P.; Backes, M.; Baixeras, C.; Balestra, S.; Barrio, J.A.; Bastieri, D.; Becerra González, J.; et al. MAGIC TeV gamma-ray observations of Markarian 421 during multiwavelength campaigns in 2006. *Astron. Astrophys.* **2010**, *519*, A32. [CrossRef]
95. Abdo, A.A.; Ackermann, M.; Ajello, M.; Baldini, L.; Ballet, J.; Barbiellini, G.; Bastieri, D.; Bechtol, K.; Bellazzini, R.; Berenji, B.; et al. Fermi Large Area Telescope Observations of Markarian 421: The Missing Piece of its Spectral Energy Distribution. *Astrophys. J.* **2011**, *736*, 131. [CrossRef]
96. Aleksić, J.; Ansoldi, S.; Antonelli, L.A.; Antoranz, P.; Babic, A.; Bangale, P.; Barres de Almeida, U.; Barrio, J.A.; Becerra González, J.; Bednarek, W.; et al. Unprecedented study of the broadband emission of Mrk 421 during flaring activity in March 2010. *Astron. Astrophys.* **2015**, *578*, A22. [CrossRef]
97. Aleksić, J.; Ansoldi, S.; Antonelli, L.A.; Antoranz, P.; Babic, A.; Bangale, P.; Barres de Almeida, U.; Barrio, J.A.; Becerra González, J.; Bednarek, W.; et al. The 2009 multiwavelength campaign on Mrk 421: Variability and correlation studies. *Astron. Astrophys.* **2015**, *576*, A126. [CrossRef]
98. Baloković, M.; Paneque, D.; Madejski, G.; Furniss, A.; Chiang, J.; Ajello, M.; Alexander, D.M.; Barret, D.; Blandford, R.D.; Boggs, S.E.; et al. Multiwavelength Study of Quiescent States of Mrk 421 with Unprecedented Hard X-Ray Coverage Provided by NuSTAR in 2013. *Astrophys. J.* **2016**, *819*, 156. [CrossRef]
99. Ahnen, M.L.; Ansoldi, S.; Antonelli, L.A.; Antoranz, P.; Babic, A.; Banerjee, B.; Bangale, P.; Barres de Almeida, U.; Barrio, J.A.; Becerra González, J.; et al. Long-term multi-wavelength variability and correlation study of Markarian 421 from 2007 to 2009. *Astron. Astrophys.* **2016**, *593*, A91. [CrossRef]
100. Abeysekara, A.U.; Archambault, S.; Archer, A.; Benbow, W.; Bird, R.; Buchovecky, M.; Buckley, J.H.; Bugaev, V.; Cardenzana, J.V.; Cerruti, M.; et al. A Search for Spectral Hysteresis and Energy-dependent Time Lags from X-Ray and TeV Gamma-Ray Observations of Mrk 421. *Astrophys. J.* **2017**, *834*, 2. [CrossRef]
101. Acciari, V.A.; Ansoldi, S.; Antonelli, L.A.; Asano, K.; Babić, A.; Banerjee, B.; Baquero, A.; de Almeida, U.B.; Barrio, J.A.; Becerra González, J.; et al. Multiwavelength variability and correlation studies of Mrk 421 during historically low X-ray and γ-ray activity in 2015-2016. *Mon. Not. R. Astron. Soc.* **2021**, *504*, 1427–1451. [CrossRef]
102. Acciari, V.A.; Ansoldi, S.; Antonelli, L.A.; Arbet Engels, A.; Baack, D.; Babić, A.; Banerjee, B.; Barres de Almeida, U.; Barrio, J.A.; Becerra González, J.; et al. Unraveling the Complex Behavior of Mrk 421 with Simultaneous X-Ray and VHE Observations during an Extreme Flaring Activity in 2013 April. *Astrophys. J. Suppl. Ser.* **2020**, *248*, 29. [CrossRef]
103. Abeysekara, A.U.; Benbow, W.; Bird, R.; Brill, A.; Brose, R.; Buchovecky, M.; Buckley, J.H.; Christiansen, J.L.; Chromey, A.J.; Daniel, M.K.; et al. The Great Markarian 421 Flare of 2010 February: Multiwavelength Variability and Correlation Studies. *Astrophys. J.* **2020**, *890*, 97. [CrossRef]
104. MAGIC Collaboration; Acciari, V.A.; Ansoldi, S.; Antonelli, L.A.; Arbet Engels, A.; Artero, M.; Asano, K.; Babić, A.; Baquero, A.; Barres de Almeida, U.; et al. Investigation of the correlation patterns and the Compton dominance variability of Mrk 421 in 2017. *Astron. Astrophys.* **2021**, *655*, A89. [CrossRef]
105. Aleksić, J.; Antonelli, L.A.; Antoranz, P.; Backes, M.; Baixeras, C.; Barrio, J.A.; Bastieri, D.; Becerra González, J.; Bednarek, W.; Berdyugin, A.; et al. Search for an extended VHE γ-ray emission from Mrk 421 and Mrk 501 with the MAGIC Telescope. *Astron. Astrophys.* **2010**, *524*, A77. [CrossRef]
106. Albert, J.; Aliu, E.; Anderhub, H.; Antoranz, P.; Armada, A.; Baixeras, C.; Barrio, J.A.; Bartko, H.; Bastieri, D.; Becker, J.K.; et al. Variable Very High Energy γ-Ray Emission from Markarian 501. *Astrophys. J.* **2007**, *669*, 862–883. [CrossRef]
107. Anderhub, H.; Antonelli, L.A.; Antoranz, P.; Backes, M.; Baixeras, C.; Balestra, S.; Barrio, J.A.; Bastieri, D.; Becerra González, J.; Becker, J.K.; et al. Simultaneous Multiwavelength Observation of Mkn 501 in a Low State in 2006. *Astrophys. J.* **2009**, *705*, 1624–1631. [CrossRef]
108. Abdo, A.A.; Ackermann, M.; Ajello, M.; Allafort, A.; Baldini, L.; Ballet, J.; Barbiellini, G.; Baring, M.G.; Bastieri, D.; Bechtol, K.; et al. Insights into the High-energy γ-ray Emission of Markarian 501 from Extensive Multifrequency Observations in the Fermi Era. *Astrophys. J.* **2011**, *727*, 129. [CrossRef]
109. Acciari, V.A.; Arlen, T.; Aune, T.; Beilicke, M.; Benbow, W.; Böttcher, M.; Boltuch, D.; Bradbury, S.M.; Buckley, J.H.; Bugaev, V.; et al. Spectral Energy Distribution of Markarian 501: Quiescent State Versus Extreme Outburst. *Astrophys. J.* **2011**, *729*, 2. [CrossRef]
110. Furniss, A.; Noda, K.; Boggs, S.; Chiang, J.; Christensen, F.; Craig, W.; Giommi, P.; Hailey, C.; Harisson, F.; Madejski, G.; et al. First NuSTAR Observations of Mrk 501 within a Radio to TeV Multi-Instrument Campaign. *Astrophys. J.* **2015**, *812*, 65. [CrossRef]
111. Aleksić, J.; Ansoldi, S.; Antonelli, L.A.; Antoranz, P.; Babic, A.; Bangale, P.; Barres de Almeida, U.; Barrio, J.A.; Becerra González, J.; Bednarek, W.; et al. Multiwavelength observations of Mrk 501 in 2008. *Astron. Astrophys.* **2015**, *573*, A50. [CrossRef]

112. Ahnen, M.L.; Ansoldi, S.; Antonelli, L.A.; Antoranz, P.; Babic, A.; Banerjee, B.; Bangale, P.; Barres de Almeida, U.; Barrio, J.A.; Becerra González, J.; et al. Multiband variability studies and novel broadband SED modeling of Mrk 501 in 2009. *Astron. Astrophys.* **2017**, *603*, A31. [CrossRef]
113. MAGIC Collaboration.; Acciari, V.A.; Ansoldi, S.; Antonelli, L.A.; Babić, A.; Banerjee, B.; Barres de Almeida, U.; Barrio, J.A.; Becerra González, J.; Bednarek, W.; et al. Study of the variable broadband emission of Markarian 501 during the most extreme Swift X-ray activity. *Astron. Astrophys.* **2020**, *637*, A86. [CrossRef]
114. Abe, H.; Abe, S.; Acciari, V.A.; Agudo, I.; Aniello, T.; Ansoldi, S.; Antonelli, L.A.; Arbet-Engels, A.; Arcaro, C.; Artero, M.; et al. Multimessenger Characterization of Markarian 501 during Historically Low X-Ray and γ-Ray Activity. *Astrophys. J. Suppl. Ser.* **2023**, *266*, 37. [CrossRef]
115. Aleksić, J.; Alvarez, E.A.; Antonelli, L.A.; Antoranz, P.; Asensio, M.; Backes, M.; Barrio, J.A.; Bastieri, D.; Becerra González, J.; Bednarek, W.; et al. MAGIC observations of the giant radio galaxy M 87 in a low-emission state between 2005 and 2007. *Astron. Astrophys.* **2012**, *544*, A96. [CrossRef]
116. MAGIC Collaboration.; Acciari, V.A.; Ansoldi, S.; Antonelli, L.A.; Arbet Engels, A.; Arcaro, C.; Baack, D.; Babić, A.; Banerjee, B.; Bangale, P.; et al. Monitoring of the radio galaxy M 87 during a low-emission state from 2012 to 2015 with MAGIC. *Mon. Not. R. Astron. Soc.* **2020**, *492*, 5354–5365. [CrossRef]
117. Albert, J.; Aliu, E.; Anderhub, H.; Antonelli, L.A.; Antoranz, P.; Backes, M.; Baixeras, C.; Barrio, J.A.; Bartko, H.; Bastieri, D.; et al. Very High Energy Gamma-Ray Observations of Strong Flaring Activity in M87 in 2008 February. *Astrophys. J. Lett.* **2008**, *685*, L23. [CrossRef]
118. Abramowski, A.; Acero, F.; Aharonian, F.; Akhperjanian, A.G.; Anton, G.; Balzer, A.; Barnacka, A.; Barres de Almeida, U.; Becherini, Y.; Becker, J.; et al. The 2010 Very High Energy γ-Ray Flare and 10 Years of Multi-wavelength Observations of M 87. *Astrophys. J.* **2012**, *746*, 151. [CrossRef]
119. Event Horizon Telescope Collaboration.; Akiyama, K.; Alberdi, A.; Alef, W.; Asada, K.; Azulay, R.; Baczko, A.K.; Ball, D.; Baloković, M.; Barrett, J.; et al. First M87 Event Horizon Telescope Results. II. Array and Instrumentation. *Astrophys. J. Lett.* **2019**, *875*, L2. [CrossRef]
120. EHT MWL Science Working Group.; Algaba, J.C.; Anczarski, J.; Asada, K.; Baloković, M.; Chandra, S.; Cui, Y.Z.; Falcone, A.D.; Giroletti, M.; Goddi, C.; et al. Broadband Multi-wavelength Properties of M87 during the 2017 Event Horizon Telescope Campaign. *Astrophys. J. Lett.* **2021**, *911*, L11. [CrossRef]
121. Albert, J.; Aliu, E.; Anderhub, H.; Antoranz, P.; Baixeras, C.; Barrio, J.A.; Bartko, H.; Bastieri, D.; Becker, J.K.; Bednarek, W.; et al. MAGIC observations of PG 1553+113 during a multiwavelength campaign in July 2006. *Astron. Astrophys.* **2009**, *493*, 467–469. [CrossRef]
122. Aleksić, J.; Anderhub, H.; Antonelli, L.A.; Antoranz, P.; Backes, M.; Baixeras, C.; Balestra, S.; Barrio, J.A.; Bastieri, D.; Becerra González, J.; et al. Simultaneous multi-frequency observation of the unknown redshift blazar PG 1553+113 in March-April 2008. *Astron. Astrophys.* **2010**, *515*, A76. [CrossRef]
123. Aleksić, J.; Alvarez, E.A.; Antonelli, L.A.; Antoranz, P.; Asensio, M.; Backes, M.; Barrio, J.A.; Bastieri, D.; Becerra González, J.; Bednarek, W.; et al. PG 1553+113: Five Years of Observations with MAGIC. *Astrophys. J.* **2012**, *748*, 46. [CrossRef]
124. Ackermann, M.; Ajello, M.; Albert, A.; Atwood, W.B.; Baldini, L.; Ballet, J.; Barbiellini, G.; Bastieri, D.; Becerra Gonzalez, J.; Bellazzini, R.; et al. Multiwavelength Evidence for Quasi-periodic Modulation in the Gamma-Ray Blazar PG 1553+113. *Astrophys. J. Lett.* **2015**, *813*, L41. [CrossRef]
125. Covino, S.; Landoni, M.; Sandrinelli, A.; Treves, A. Looking at Blazar Light-curve Periodicities with Gaussian Processes. *Astrophys. J.* **2020**, *895*, 122. [CrossRef]
126. Peñil, P.; Domínguez, A.; Buson, S.; Ajello, M.; Otero-Santos, J.; Barrio, J.A.; Nemmen, R.; Cutini, S.; Rani, B.; Franckowiak, A.; et al. Systematic Search for γ-Ray Periodicity in Active Galactic Nuclei Detected by the Fermi Large Area Telescope. *Astrophys. J.* **2020**, *896*, 134. [CrossRef]
127. De Lotto, B.; Magic Collaboration. The MAGIC telescopes: performance, results and future perspectives. *J. Phys. Conf. Ser.* **2012**, *375*, 052021. [CrossRef]
128. Albert, J.; Aliu, E.; Anderhub, H.; Antoranz, P.; Armada, A.; Asensio, M.; Baixeras, C.; Barrio, J.A.; Bartko, H.; Bastieri, D.; et al. Observation of Very High Energy Gamma-Ray Emission from the Active Galactic Nucleus 1ES 1959+650 Using the MAGIC Telescope. *Astrophys. J.* **2006**, *639*, 761–765. [CrossRef]
129. MAGIC Collaboration.; Acciari, V.A.; Ansoldi, S.; Antonelli, L.A.; Arbet Engels, A.; Baack, D.; Babić, A.; Banerjee, B.; Barres de Almeida, U.; Barrio, J.A.; et al. Broadband characterisation of the very intense TeV flares of the blazar 1ES 1959+650 in 2016. *Astron. Astrophys.* **2020**, *638*, A14. [CrossRef]
130. Bionta, R.M.; Blewitt, G.; Bratton, C.B.; Casper, D.; Ciocio, A.; Claus, R.; Cortez, B.; Crouch, M.; Dye, S.T.; Errede, S.; et al. Observation of a neutrino burst in coincidence with supernova 1987A in the Large Magellanic Cloud. *Phys. Rev. Lett.* **1987**, *58*, 1494–1496. [CrossRef] [PubMed]
131. Weekes, T.C.; Badran, H.; Biller, S.D.; Bond, I.; Bradbury, S.; Buckley, J.; Carter-Lewis, D.; Catanese, M.; Criswell, S.; Cui, W.; et al. VERITAS: The Very Energetic Radiation Imaging Telescope Array System. *Astropart. Phys.* **2002**, *17*, 221–243. [CrossRef]
132. Benbow, W. Highlights from the VERITAS AGN Observation Program. *Int. Cosm. Ray Conf.* **2019**, *36*, 632. [CrossRef]
133. McGrath, C. VERITAS Highlights 2022. *J. Phys. Conf. Ser.* **2023**, *2429*, 012015. [CrossRef]

134. Wagner, R. Observations of AGN at very-high energy gamma rays with the H.E.S.S. telescopes. In Proceedings of the Active Galactic Nuclei: What's in a Name? (AGN 2016), Garching, Germany, 27 June–1 July 2016; p. 58. [CrossRef]
135. Taylor, A.M.; Sanchez, D.; Cerruti, M. Extragalactic Observations with HESS: Past and Future. *arXiv* **2017**, arXiv:1708.00775. https://doi.org/10.48550/arXiv.1708.00775.
136. H. E. S. S. Collaboration.; Abdalla, H.; Adam, R.; Aharonian, F.; Ait Benkhali, F.; Angüner, E.O.; Arcaro, C.; Armand, C.; Armstrong, T.; Ashkar, H.; et al. H.E.S.S. and MAGIC observations of a sudden cessation of a very-high-energy γ-ray flare in PKS 1510−089 in May 2016. *Astron. Astrophys.* **2021**, *648*, A23. [CrossRef]
137. Anderhub, H.; Backes, M.; Biland, A.; Boller, A.; Braun, I.; Bretz, T.; Commichau, S.; Commichau, V.; Domke, M.; Dorner, D.; et al. FACT—The first Cherenkov telescope using a G-APD camera for TeV gamma-ray astronomy. *Nucl. Instrum. Methods Phys. Res.* **2011**, *639*, 58–61. [CrossRef]
138. Dorner, D.; Arbet-Engels, A.; Baack, D.; Balbo, M.; Biland, A.; Bretz, T.; Buss, J.; Eisenberger, L.; Elsaesser, D.; Hildebrand, D.; et al. FACT—Highlights from more than Eight Years of Unbiased TeV Monitoring. In Proceedings of the 37th International Cosmic Ray Conference, Online, 18 March 2022; p. 851. [CrossRef]
139. Romoli, C.; Chakraborty, N.; Dorner, D.; Taylor, A.M.; Blank, M. Flux Distribution of Gamma-Ray Emission in Blazars: The Example of Mrk 501. *Galaxies* **2018**, *6*, 135. [CrossRef]
140. Abe, H.; Abe, K.; Abe, S.; Aguasca-Cabot, A.; Agudo, I.; Alvarez Crespo, N.; Antonelli, L.A.; Aramo, C.; Arbet-Engels, A.; Arcaro, C.; et al. Observations of the Crab Nebula and Pulsar with the Large-sized Telescope Prototype of the Cherenkov Telescope Array. *Astrophys. J.* **2023**, *956*, 80. [CrossRef]
141. Actis, M.; Agnetta, G.; Aharonian, F.; Akhperjanian, A.; Aleksić, J.; Aliu, E.; Allan, D.; Allekotte, I.; Antico, F.; Antonelli, L.A.; et al. Design concepts for the Cherenkov Telescope Array CTA: An advanced facility for ground-based high-energy gamma-ray astronomy. *Exp. Astron.* **2011**, *32*, 193–316. [CrossRef]
142. Abe, S.; Aguasca-Cabot, A.; Agudo, I.; Alvarez Crespo, N.; Antonelli, L.A.; Aramo, C.; Arbet-Engels, A.; Artero, M.; Asano, K.; Aubert, P.; et al. Multiwavelength study of the galactic PeVatron candidate LHAASO J2108+5157. *Astron. Astrophys.* **2023**, *673*, A75. [CrossRef]
143. Abe, H.; Abe, K.; Abe, S.; Acciari, V.A.; Aguasca-Cabot, A.; Agudo, I.; Alvarez Crespo, N.; Aniello, T.; Ansoldi, S.; Antonelli, L.A.; et al. Performance of the joint LST-1 and MAGIC observations evaluated with Crab Nebula data. *Astron. Astrophys.* **2023**, *680*, A66. [CrossRef]

universe

Review

A Very-High-Energy Gamma-Ray View of the Transient Sky

Alessandro Carosi [1,*,†] and Alicia López-Oramas [2,*,†]

1 INAF—Osservatorio Astronomico di Roma, Via Frascati 33, I-00040 Monte Porzio Catone, Italy
2 Instituto de Astrofísica de Canarias and Departamento de Astrofísica, Universidad de La Laguna, 38200 La Laguna, Tenerife, Spain
* Correspondence: alessandro.carosi@inaf.it (A.C.); alicia.lopez@iac.es (A.L.-O.)
† These authors contributed equally to this work.

Abstract: The development of the latest generation of Imaging Atmospheric Cherenkov Telescopes (IACTs) over recent decades has led to the discovery of new extreme astrophysical phenomena in the very-high-energy (VHE, E > 100 GeV) gamma-ray regime. Time-domain and multi-messenger astronomy are inevitably connected to the physics of transient VHE emitters, which show unexpected (and mostly unpredictable) flaring or exploding episodes at different timescales. These transients often share the physical processes responsible for the production of the gamma-ray emission, through cosmic-ray acceleration, magnetic reconnection, jet production and/or outflows, and shocks interactions. In this review, we present an up-to-date overview of the VHE transients field, spanning from novae to supernovae, neutrino counterparts or fast radio bursts, among others, and we outline the expectations for future facilities.

Keywords: gamma-ray astrophysics; gamma-ray instrumentation; transients; novae; supernovae; fast radio bursts; magnetars; neutrinos; tidal disruption events; gravitational waves

Citation: Carosi, A.; López-Oramas, A. A Very-High-Energy Gamma-Ray View of the Transient Sky. *Universe* **2024**, *10*, 163. https://doi.org/10.3390/universe10040163

Academic Editor: Fridolin Weber

Received: 31 January 2024
Revised: 13 March 2024
Accepted: 15 March 2024
Published: 29 March 2024

1. Introduction

The very-high-energy (VHE, E > 100 GeV) gamma-ray regime is of the utmost importance in studying extreme astrophysical processes. Transient phenomena, located at the crossroads of time-domain and multi-messenger astronomy, have revealed a plethora of new emitters at VHE. During the last twenty years, IACT experiments have proven to be suitable instruments to perform fast follow-up of transient events, with 3–4 times improved sensitivities at short time scales compared to space-based instruments [1]. At the same time, some IACTs are optimized for a swift reaction and repositioning, see e.g., [2], allowing for the study of short-lived signals during their initial phase. Understanding the recent advancements and open issues in transient and multi-messenger astrophysics at VHE is the key for the science to be developed with current IACTs and with future instrumentation, such as the Cherenkov Telescope Array (CTA) Observatory [3].

In this contribution, we review the phenomenology of transient events of both Galactic and extragalactic origin, which are (mostly) related to stellar-size compact objects and that are shock-powered and/or accretion-powered. Shocks power several transient phenomena, such as novae, supernovae, binary neutron star mergers, or tidal disruption events. The shocks and ejecta in (at least some of) these systems can show morphological resemblances and have similar characteristics, although at different scales and displaying, e.g., a broad range of various velocities and densities. Other sources such as magnetars, even if rotationally powered, can also generate blast waves and shocks.

We review the state of the art in the detection and search for emission of transient events in the GeV-TeV regime for different types of astrophysical sources in Sections 2–8, namely novae (Section 2), microquasars, and flaring gamma-ray binaries (Section 3), supernovae (Section 4), pulsar-wind nebulae (Section 5), fast radio bursts and magnetars (Section 6), and gravitational waves (Section 7). We then briefly mention other transient sources in Section 8, such as gamma-ray bursts (GRBs) and tidal disruption events (TDEs), which are more extensively covered in a dedicated review of this Special Issue [4]. We finally discuss the latest advances and future perspectives in Section 9.

2. Novae

Novae outbursts are thermonuclear explosions that take place on the surface of a white dwarf (WD) accreting material from a companion star. The transferred material is mostly hydrogen; however, helium accretion can also occur in some systems (see [5]). The material accumulates on a layer on the surface of the WD, where hydrogen is burning in degenerate conditions, increasing the temperature and density. Once a critical mass is reached, the system undergoes an unstable burning, provoking a thermonuclear runaway. The ejecta expand at velocities reaching thousands of km s^{-1} and can interact with the surrounding material, if any. Depending on the type of companion star, novae can be classified into *classical* and *symbiotic* systems. Classical novae are cataclysmic variables in which the companion is a main sequence (or slightly evolved) star. The mass-transfer onto the WD happens via Roche-lobe overflow. These systems are characterized for having short orbital periods lasting from hours to days [6]. Symbiotic systems are composed of a red giant (RG) companion and the have longer orbital periods and show larger component separations [7]. The binary is embedded in the RG wind and the WD accretes directly from this wind.

Novae explosions do not disrupt the binary system and hence the cycle of accretion can start again. After enough material is accumulated, another thermonuclear runaway can happen again. The recurrence timescale of these outburst is defined as $\tau_{rec} = M_{acc}/\dot{M}$, M_{acc} being the critical mass to initiate the nuclear burning and $\dot{M}$ the accretion rate. High recurrence times are then associated to more massive WDs (with mass close to the Chandrasekhar limit, $M_{Ch} \approx 1.4 M_{\odot}$) accreting at high rates. The typical recurrence times for classical novae are 10^4–10^5 years. However, some systems known as *recurrent novae* have displayed more than one eruption in a human lifetime ($\tau_{rec} \leq 100$ years). For this to happen, the WD should be close to the Chandrasekhar limit (with at least $M > 1.2 M_{\odot}$) and the system shall have high mass-accretion rates ($\approx 10^{-7} M_{\odot}$ y^{-1}) (see [8]). There are 10 recurrent systems known in the Galaxy up to date which can be further classified into two groups [8] depending on the mechanism which leads to the short recurrence: long-period systems (eight in total, period > one-third of a day) hosting a giant companion (also known as *symbiotic recurrent novae*) in which the accretion is driven by the RG wind and the evolution of the companion. These symbiotic recurrent systems can indeed be the progenitors of type Ia supernovae; and short-period ones (two systems) in which the accretion is driven by the heating of the WD.

The discovery rate of novae is about 5–15 events per year [5], although the estimated number of eruptions in the Galaxy is much larger: 20–70 per year. The lower detection rate could be due to dust obscuration, since many novae happen in the Galactic plane, or simply due to a scarce monitoring; see, e.g., [8], and references therein for a more detailed discussion on novae rates.

The first evidence of non-thermal emission due to particle acceleration up to TeV energies in the blast wave of (recurrent) novae was suggested by [9]. The discovery of symbiotic novae as high energy (HE; E > 100 MeV) emitters was performed by *Fermi*-LAT in 2010 [10] with the detection of V407 Cyg. The HE emission lasted for about two weeks. Only four years later, *Fermi*-LAT also established classical novae as HE sources [11] with the discovery of three systems (V959 Mon, V1324 Sco, and V339 Del). The spectral energy distribution (SED) of these four LAT-detected novae is rather soft, mostly described with power laws with exponential cutoff and with energies up to a maximum of ∼10 GeV. Both hadronic and leptonic scenarios can fit the observed emission and could not be ruled out.

Since then, the satellite has been detecting an average of ∼1 nova per year.[1] Most of the *Fermi*-LAT novae are located in the Galactic disk, although some have been discovered in the Galactic bulge, implying detection up to distances of ∼8 kpc. By studying different classical novae, ref. [12] suggests an inverse relationship between the HE emission duration and the total emitted energies. This could possibly indicate that the presence of more compact and high-density ejecta produces a higher particle acceleration, which leads to stronger emissions and shorter duration.

The detection of HE emission from novae clearly demonstrated that non-thermal mechanisms operate in these cataclysmic binaries. The evident question to pose was whether novae could accelerate particles to sufficiently high energies to produce VHE gamma rays. These particles (leptons and protons) are accelerated at the nova shock and could eventually produce emission at higher energies. In the case of protons, they could reach high energies and emit TeV gamma rays [13]. Since the discovery of HE gamma rays in novae, searches for a VHE component were performed by IACTs for over a decade, without achieving any significant detection. VERITAS observed the 2010 outburst of the symbiotic nova V407 Cyg on days 9 to 16 after the eruption, leading to no signal [14]. MAGIC observed the classical nova V339 Del on the night of the optical peak (although under poor-quality weather conditions) and a few days after the *Fermi*-LAT emission, setting upper limits (ULs) to the VHE emission [15]. In the same work, MAGIC reported no signal from the symbiotic nova YY Her (taken a week after the optical maximum) and the dwarf nova ASASSN-13ax, a system in which the outburst are due to accretion disk instabilities (instead of a thermonuclear runaway).

The first nova for which VHE gamma-ray emission was discovered is RS Oph, a recurrent symbiotic system composed of a massive $M_{WD} \approx 1.2 \div 1.4\ M_{\odot}$) carbon–oxygen WD [16] accreting from a M0-2 III RG star [17]. It shows an orbital period of (453.6 ± 0.4) days [18] and displays major outbursts with a recurrence time of 14.7 years [8]. The fact that the mass of the WD is so close to the Chandrasekhar limit suggest that RS Oph is a possible type Ia SN (see Section 4) progenitor candidate [19–21].

The VHE observations were triggered on 9 August, after optical [22] and HE [23] alerts. A clear gamma-ray signal at VHE was then detected by H.E.S.S [24], MAGIC [25], and confirmed by the LST-1 telescope [26] during the 2021 outburst that started on 8 August 2021 (MJD 59435). The multi-wavelength lightcurve of the RS Oph emission is shown in Figure 1). The VHE gamma-ray signal is significantly detected up to five days after the nova eruption. Observations after the full Moon break revealed no significant signal, with a maximum of 3.3 σ hint integrated over 14.6 h between 25 August and 7 September [24].

The lightcurve reported by IACTs varies in shape depending on the energy range (see Figure 1). MAGIC observed RS Oph between 9 August and 1 September, for a total of 21.4 h. The signal detected in the VHE regime measured during the first 4 days corresponds to the optical and HE maxima. However, the 4-day binned emission >100 GeV is best fit to a constant flux [25], which suggest a migration of the gamma-ray emission toward higher energies, implying an increase in the energies of the parent particle population. A constant flux compatible with that measured by MAGIC has been reported by the LST-1 during the first nights [27]. On the other hand, H.E.S.S. observed this between 9 August and 7 September. The signal was detected by H.E.S.S. at >250 GeV peaks a day after the HE maximum, with a temporal decay of $t^{-(1.43 \pm 0.48)}$ compatible with what observed at HE, explaining the similarities in the lightcurve due to a common origin of the emission, in which the particles are accelerated at the external shock [24].

The VHE component of the SED measured by MAGIC and H.E.S.S. expands from 60 GeV up to 1 TeV [24,25]. Both collaborations performed a joint analysis of the HE and VHE data and suggest that the combined *Fermi*-LAT + MAGIC and *Fermi*-LAT + H.E.S.S. spectra can be described as a single component spanning from 50 MeV to VHE. The emission would be due to a shock created by the ejecta which expand into the surrounding medium and the wind of the RG companion, creating a single shock where particles are accelerated. Both experiments suggest that the gamma-ray emission is best fit by a hadronic scenario, in which protons are accelerated in the shock wave formed by the interaction of the novae ejecta with with the interstellar medium with some contribution of the RG wind. The daily SEDs (Figure 2 for the MAGIC sample) are also best adjusted to a hadronic case, with evidence of increase in the energy cutoff, implying and acceleration of protons and the absence of strong cooling processes. The leptonic scenario does not properly fit the obtained spectra. A lepto-hadronic scenario is also tested by [25], providing a poor fit. In the case of protons, the injected particle spectrum also follows a canonical distribution ($\Gamma = -2$), while the leptonic and lepto-hadronic cases assume more complicated injection models with some strong ad hoc spectral breaks which cannot be fully explained by cooling, and still leading to a poorer fit of the SED. All this together favored the hadronic scenario as mechanism for the VHE gamma-ray production and the settlement of novae as proton accelerators.

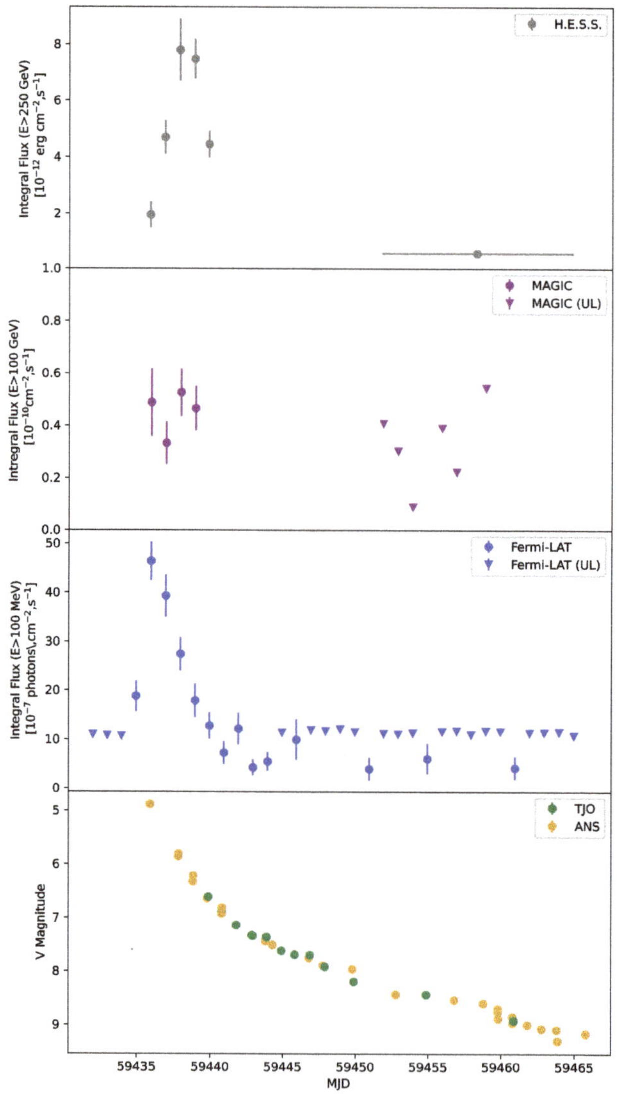

Figure 1. Multi-wavelength lightcurve of the 2021 eruption of RS Oph as seen by H.E.S.S (top panel), MAGIC (second panel), *Fermi*-LAT, (third panel), and optical V magnitude (bottom) observations by ANS (golden points) and TJO (green points). ULs for the late MAGIC emission and *Fermi*-LAT are indicated as inverted triangles. Optical, *Fermi*-LAT, and MAGIC data points taken from the supplementary material in [25], H.E.S.S. fluxes from auxiliary material from [24].

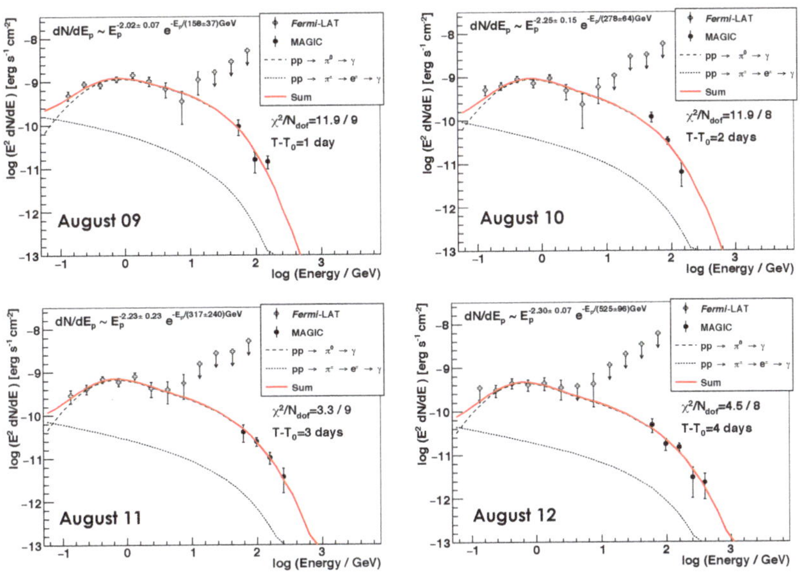

Figure 2. Daily SEDs of RS Oph as seen by *Fermi*-LAT and MAGIC adjusted to an hadronic scenario from the first night on 9 August 2021 to the fourth night, 12 August. A hint of spectral hardening is observed with increasing time, with increased cut-off energies. Reprinted with permission from [25].

Ref. [28] elaborated a model to explain both the gamma-ray and radio emissions assuming a single shock, multi-population (lepto-hadronic) scenario for the first four nights of the outburst. The authors suggest a possible different origin for the HE and VHE components, the HE one being of leptonic origin and the VHE hadronic-dominated. This scenario would also explain the temporal dependence of the measured emissions.

Novae ejecta are not spherical and show some asymmetries whose shape depends mainly on the densities of the surroundings. It has been proposed that, at least in classical novae, two components may be at work: a slow (∼hundreds of km/s) dense flow (from binary motion) that moves in the equatorial plane and a faster (∼few thousand of km/s) less dense isotropic outflow (wind from the WD) that propagates in the polar direction [5,29]. This creates a forward shock that is driven into the slow outflow, while a reverse shock will interact with the faster component. It is still unknown whether this scenario is universal to all types of novae or not.

The 2021 eruption of RS Oph could be a first step to answer this enigma. The 2006 outburst already showed some asymmetric structures in the ejecta with reported extended emission in the east–west direction [30–32]. A similar asymmetry has been reported during the latest outburst, displaying a bipolar structure with a predominant orientation in the same direction [33–35]. Both in the 2006 and 2021 eruptions, a slow moving equatorial ring and a faster bipolar ejecta expanding in the polar (east–west) direction have been reported [30,34]. Ref. [34] claims the formation of a ring-like structure in the orbital plane due to the interaction between the ejecta and an enhanced ambient medium in the equatorial plane and argue that similar torus-like structures are observed in classical novae, although with a different origin (due to the orbital motion on the ejecta, see [29]).

The recent detection of the RS Oph during the 2021 outburst in the GeV-TeV domain together with the multi-wavelength data have led some authors to suggest the presence of multiple shocks (polar and equatorial) in this system. Recently, ref. [36] argued that the presence of multiple ejecta components can explain both the gamma-ray SEDs and the shapes of the lightcurves as seen by *Fermi*-LAT and H.E.S.S., which are consistent with the combination of two shocks: a low-velocity shock which expands in a dense medium and another faster one which expands in a less-dense environment. This scenario suggests that RS Oph (and probably RS Oph-like symbiotic systems) show similar properties to classical novae.

The protons that are accelerated in (symbiotic recurrent) novae ejecta can eventually escape onto the interstellar medium and contribute to the Galactic cosmic ray sea. Novae are less energetic that other events such as supernovae (SNe) ($\sim 10^{43}$ erg vs. $\sim 10^{50}$ erg) but they occur at a higher rate (5–15 detected novae in the Galaxy per year vs. 1–2 core-collapse SNe); hence, their contribution to the cosmic ray budget could be noticeable. However, considering the energetics of RS Oph as detected at VHE and its recurrence time of $\sim$15 years, the overall contribution to the cosmic ray spectrum is negligible, only 0.1% of that of SNe [25]. Nevertheless, novae do create bubbles of enhanced cosmic ray density in their close environment. In the case of RS Oph-like systems, these bubbles can extend up to $\sim$10 pc.

3. Flaring Binary Systems and Microquasars

Different types of gamma-loud binaries have been found to emit both in the HE and VHE regime, such as gamma-ray binaries and microquasars [37]. They represent a good opportunity to study particle acceleration in shocks and jets at relatively short timescales. The so-called gamma-ray binaries are systems that display the peak of their non-thermal emission above 1 MeV and are composed by massive stars (O or Be type) being orbited by a compact object (either NS or BH). Out of eight known gamma-ray binaries, three of them host a pulsar: PSR B1259–63, PSR 2032+4127, and LS I +61 303. Even if the powering engine remains unknown for the others, it could be that all gamma-ray binaries host an NS, due to similarities in the SEDs and flux patterns, although there are differences in a case-to-case basis. Even though the emission in gamma-ray binaries is modulated with the orbital period and some even display super-orbital modulation [38], the systems LS I +61 303 and HESS J0632+057 have shown enhanced transient episodes [39,40]. Since both systems are composed by massive Be stars with a circumstellar disk, the origin of this transient emission is suggested to be associated to clumps or inhomogeneities in the stellar wind or in the interaction region between the stellar and pulsar winds (assuming that HESS J0632+057 also host a NS). In the case of long period binaries such as PSR 2032+4127 (50 years) and PSR B1259−63 (1237 days), their gamma-ray emission is detected during the periastron passage, where VHE signal is detected [41,42]. In the case of PSR B1259−63, some additional HE flares with no flux increase in the TeV counterpart are detected during the periastron passage.

Microquasars are binary systems composed of a compact object accreting material from a companion star, generating accretion disks and jets. In the HE regime, two microquasars, both of them hosting a massive donor star, have been identified to emit transient emission: Cygnus X-1 [43–45] and Cygnus X-3 [46,47]. No HE emission from microquasars with a low-mass companion (so-called low-mass X-ray binaries, LMXBs) has yet been detected in the MeV range, with the strongest hint being that at $\sim 4\sigma$ level of V404 Cyg during the major outburst of 2015 [48,49]. No transient emission from microquasars has been detected in the TeV domain; see, e.g., [50–54]. The strongest TeV hint of emission up to now is that of Cygnus X-1 reported by [55] during a contemporaneous X-ray flare, reaching a 4.1σ (post-trial) signal in an 80 min observation. Only two microquasars have been discovered to emit persistent TeV emission: SS433, a microquasar with two persistent jets that interact with a surrounding nebula, being the interaction regions between the jet and the nebula the TeV-bright region [56], and the microblazar V4641 Sgr [57]. Both were discovered by particle detectors after accumulating few years of data. Only SS433 has been detected by an IACT [58] after accumulating more than 200 h of data.

4. Supernovae

SNe are explosive energetic events that result from a stellar death. They can generally be classified into two large groups depending on whether their spectra is hydrogen-poor (type I SNe) or if they do show Balmer lines (type II SNe). More refined classifications have been appearing in terms of additional spectral features, see e.g., [59] or even depending on whether the optical light curve decays (in magnitudes) linearly (II-L) or forming a plateau (II-P).

These SNe also differ in the nature and structure of the stellar progenitor. Type Ia SNe result from the thermonuclear runaway of a WD in a binary system. The WD has been accreting material from its companion star and it has probably been producing nova eruptions throughout its life. Due to accretion, the WD increases its mass until the Chandrasekhar limit ($M_{Ch} \approx 1.4 M_{\odot}$) is reached and it will explode as a SNe. As mentioned in Section 2, symbiotic systems such as RS Oph are type Ia SNe progenitors. Core-collapse SNe (CCSNe) are the consequence of the death of a massive star ($M > 8\ M_{\odot}$) that has exhausted its fuel, producing a violent explosive release of the external shells and leading to the collapse of the nucleus. Type Ib and Ic are *stripped envelope* SNe, in which the progenitors were Wolf Rayet stars stripped of their H (Ib) and He (Ic) layers [60]. The progenitors of II-P SNe are generally red-supergiant (RSG) stars and those of II-n are luminous blue variables (LBVs). These LBVs are very massive ($M > 25\ M_{\odot}$) and show high eruptive mass-loss processes during their lifetime, leading to the appearance of strong narrow H emission lines due to the interaction of the SNe with the surrounding circumstellar medium (CSM) in the spectra of II-n. RSGs can also show some smaller eruptive events and hence also show early CSM interaction *flash* features in their early spectra. The progenitors of IIb SNe are supergiants that were partially stripped from their H envelope via binary interaction during the pre-SN phase.

CCSNe are of interest due to many aspects. They are the precursors of compact objects (BHs and NS, depending of the initial mass of the progenitor star), they help disperse the heavy elements that have been created by the progenitor star onto the interstellar medium (ISM), and they are sources of cosmic rays, neutrinos, and (likely) GWs. Finally, some supernova remnants (SNRs) have been suggested as counterparts of sources detected up to 100 TeV [61,62] which can contribute to the Galactic cosmic-ray spectrum. The most common SNe are type II-P, representing 57% of the population [63].

4.1. Expected Gamma-Ray Emission from SNe

SNe are expected gamma-ray production sites due to the acceleration of protons in the SNe blastwave [64–66]. They would then contribute to the Galactic cosmic-ray budget. To account for the measured CR spectrum, about 10% of the ejecta energy ($\sim 10^{51}$ erg) shall be converted into kinetic energy.

4.2. Type Ia SNe

Type Ia SNe are thermonuclear explosions generated by carbon–oxygen WDs in binary systems once they exceed their Chandrasekhar's limit. These SNe are used as standard candles since they display similar lightcurves and homogeneous absolute magnitudes, which are used to estimate cosmological parameters. We now know that novae, which are the progenitors of this type of event, are HE (classical and symbiotic) and VHE (recurrent symbiotic novae) gamma-ray emitters. However, no type Ia SNe has been identified as a gamma-ray source. The only observation of this kind was performed by the MAGIC telescopes of SN 2014J [67], setting the firsts and only ULs to the VHE gamma-ray emission of these explosive events. SN 2014J was discovered in 21 January 2014 in M 82 at 3.6 Mpc and was observed by MAGIC for about 5.4 h starting 6 days after explosion and over a total of four nights. The integral flux UL set at 300 GeV is 1.3×10^{-12} photons cm^{-2} s^{-1}. The expected gamma-ray emission in type Ia SNe should be of hadronic nature (which is in line with the hadronic origin of the VHE signal detected in the nova RS Oph) as described in [68]. This model suggests that the hadronic emission shall increase with time (for a constant density medium). The gamma-ray emission should have come from the interaction of the protons accelerated in the SN shock with the surrounding medium. Adopting this model, ref. [67] calculated that the putative gamma-ray flux shall be at the level of $\sim 1.3 \times 10^{-24}$ photons cm^{-2} s^{-1}, well below the sensitivity of IACTs.

4.3. Core-Collapse SNe

CCSNe are considered the best candidates for gamma-ray factories. The interaction of the SN ejecta with the surrounding CSM will produce the GeV-TeV gamma-ray emission via proton–proton interaction with the ambient matter. Hence, CCSNe with strong CSM interaction (type II-n, Ibn, or near II-P with early CSM interaction) are the best candidates for gamma-ray production. The CSM density decreases with increasing distance; hence, the expected GeV-TeV emission shall take place during the first days of the SNe explosion. However, eruptive phases of the progenitor star before during the pre-eruption can cause the CSM to be layered in shells. These shells could then enhance the expected gamma-ray radiation at later times once the ejecta reaches them. However, the gamma photons can also interact with the low-energy photons from the photosphere, producing pair production and hence leading to the strong absorption of the GeV-TeV signal during the first days after explosion [65,66].

No transient emission from SNe has been confirmed in the HE regime. Two candidate sources have been observed in *Fermi*-LAT data, corresponding to the position of two CCSNe [69,70], but due to the large uncertainties in the localization regions which overlap with other gamma-ray sources, their confirmation is challenging. Variable gamma-ray emission has been detected to correspond with the peculiar luminous type II SN iPTF14hls located at 150 Mpc [69]. It shows similar spectra to H-rich SNe but with a different lightcurve and it is located at the same position of another eruption detected in 1954, which is difficult to explain in an SN context. Also, there is a blazar inside the error box. The HE source is detected between days 300 and 850 after the explosion. However, if the association between the HE source and SN iPTF14hls is real, it would be the first SN to be detected in the gamma-ray domain. Although it is challenging to explain the gamma-ray emission via particle acceleration in shocks, since the efficiency should be too high. In the case of Type II-P SN 2004dj located in the galaxy NGC 2403 (3.5 Mpc), HE gamma-ray emission from the young SNR has been detected after the accumulation of 11.4 years of data [70]. The authors suggest that this source, whose emission is banishing over time, is the result of the interaction of the SN ejecta with a high-density shell. Two more candidates have been reported by [71] associated with SN AT2018iwp and SN AT2019bvr, with transient HE signals starting 3 and 6 months after the SN explosion respectively.

No signal from CCSNe has been detected in the VHE regime. The authors in [72] reported ULs on ten different CCSNe observed within a year of the explosion. Nine of them where serendipitously observed, most of them type II-P and located at distances 4–54 Mpc, while ToO observations were performed on SN 2016adj, a type IIb SN located in Cen A galaxy at 3.8 Mpc. The exposure time is also different depending on the target: for four sources including SN 2016adj, observations started around or shortly after the discovery date, while the rest of the observations happened months later (up to 272 days after the explosion). The flux ULs above 1 TeV are of the order of 10^{-13} TeV cm^{-2} s^{-1}. The non-detection of this sample may simply indicate that the putative gamma-ray fluxes are below the sensitivity of current IACTs or that the CSM is not dense enough for particle acceleration, but do not rule out the possibility of SNe being VHE emitters. Most recently [73] observed the superluminous (SLSN) type I SN2015bn and SN2017egm. SLSNe are characterized for displaying luminosities 10 to 100 times larger than ordinary CCSNe and for their lightcurves reaching the peak emission at later times. The VHE observations happened 135 days (49 days from the peak magnitude) after explosion for SN2015bn (serendipitously observed) and 670 days from explosion for SN2017egm, targeted due to the predicted gamma-ray emission derived from the optical lightcurve. No TeV counterpart was detected and the first ULs on type I SLSNe in this regime are set (see Figure 3). Although these ULs do not help constrain the scenarios of a magnetar as central engine or a shock-acceleration they do discard a jet model powered due to fallback accretion onto a black hole (model L_{BH} as seen in Figure 3).

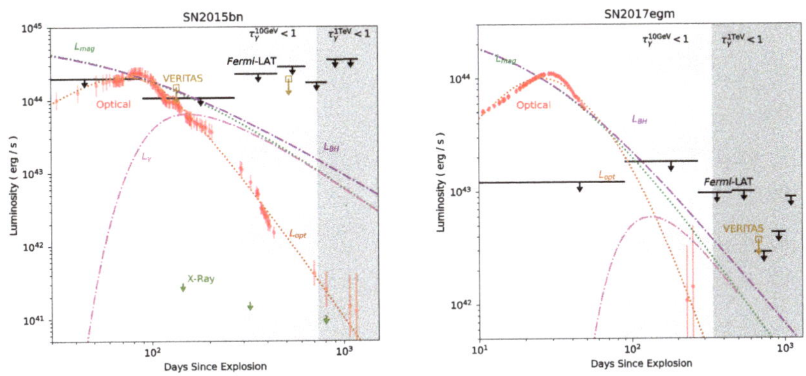

Figure 3. Multi-wavelength lightcurves of SN2015bn from days 30–1500 after explosion (**left** panel) and SN2017egm spanning 10–1300 days after explosion (**right** panel). The gamma-ray ULs from VERITAS (orange arrows) and *Fermi*-LAT are shown. Figure adapted and reprinted with permission from [73].

The most famous CCSNe is probably SN 1987A, the closest explosion in over 300 yr, located in the Large Magellanic Cloud (LMC), and that reached a peak magnitude of 2.9, visible to the naked eye. It is classified as type-II peculiar, since the progenitor was not a RSG but a blue supergiant. Neutrinos were detected a few hours prior to the arrival of electromagnetic emission, likely happening during the collapse of the progenitor's nucleus [74,75]. The central compact object is a neutron star powering a pulsar-wind nebula (PWN) [76,77]. The evolution of the remnant has been studied over decades in which rings have been formed which are the result of the interaction of the ejecta with pre-eruption material ejected by the progenitor star; see, e.g., [78], and references therein for a review. This interaction shall be capable of producing gamma-ray signal via CR acceleration and magnetic field amplification. However, no VHE signal has detected on the remnant after a intensive campaign of 210 h [79].

An important parameter to take into account for CCSNe is that the gamma photons can also interact with the low-energy photons from the photosphere producing pair production and hence leading to strong absorption of the GeV-TeV signal during the first days after explosion. Ref. [65] created a time-dependent model to estimate the gamma-ray emission from type II-P CCSNe (the most abundant type of SNe) during the first month after the explosion, taking into account the expected attenuation. By studying the evolution and dependence of different parameters such as photosphere temperature, the authors probe that the expected signal for type II-P CCSNe at distances >1 Mpc is below the sensitivity of current generation of IACTs, but close-by systems located in the Galaxy or Magellanic Clouds-located SNe could be detectable now and will undoubtedly be by future-generation of IACTs. Regarding $\gamma\gamma$ effects, the expected signal shall be strongly absorbed during the first 10 days approximately (see Figure 4). Ref. [66] developed a model for Type II-n associated to an LBV progenitor and Type II-P associated to a RSG, accounting also for the strong $\gamma\gamma$ absorption expected during the first days after the explosion. Assuming high mass-loss rates of the progenitor before the eruption, the maximum energies reached by protons can reach up to 600 TeV, which could be compatible with the knee feature of the CR spectrum. However, moderate mass-loss rates show lower values for these energies, between 70 TeV (type II-P) and 200 TeV (type II-n). Considering the absorption effects, the expected gamma-ray peak should happen 12 to 30 days after the explosion. The models suggest that current-generation instrumentation should be able to detect nearby events, up to ∼60 kpc for type II-P and ∼1 Mpc for type II-n. Future instruments (such as the southern array of the CTA observatory) shall detect type II-P up to 200 kpc and type II-n up to 3 Mpc. The values obtained by [65,66] are in agreement with the lack of detection of a VHE counterpart.

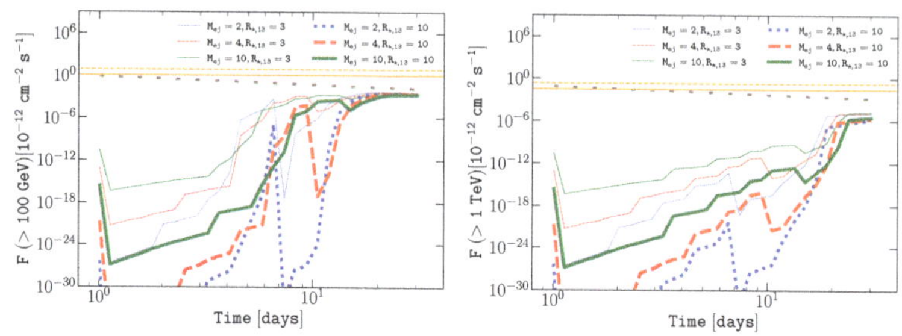

Figure 4. Temporal evolution of the integrated fluxes for type II-P CCSNe at E > 100 GeV (**left**) and E > 1 TeV (**right**) at a distance of 1 Mpc and a mass-loss of 10^{-6} $M_{\odot}$ yr^{-1} for different mass ejecta (blue dotted, red dashed, and green solid) and progenitor radius (thin and thick lines). Figure adapted and reprinted with permission from [65].

The second brightest type II CCSNe since the discovery of SN 1987A is SN 2023ixf which went off on May 2023 in M101 galaxy at ∼6.8 Mpc and reached a peak magnitude of about B = 10.6. It is a type II-P SNe with a RSG progenitor that showed strong interaction with the CSM, revealed via flash spectroscopy. No gamma-ray emission has been detected at HE [80] and none has been reported at VHE. However, considering the type of explosion and its distance, a non-detection is compatible with respect to current models.

5. Flaring Pulsar-Wind Nebulae

Pulsars are highly magnetized neutron stars which are the aftermath of a massive star death. They accelerate leptons (electrons and positrons) in a relativistic wind that halts at the termination shock, creating diffuse structures known as pulsar-wind nebulae (PWNe). In early times, as pulsars are the result of a SN explosion, both the pulsar and its associated nebula are initially surrounded by an SNR. PWNe represent one of the largest VHE source population in the Galaxy. The recent detection (and highly probable association with PWNe counterparts in other cases) of several systems at >100 TeV (ultra high energies, UHEs) has revealed that (at least) a fraction of PWNe are leptonic PeV accelerators or leptonic PeVatrons [61,62].

PWNe show different evolutionary stages; see, e.g., [81–83]: (i) an early phase (typical time t ≤ 10 kyr) of free–free expansion in which the PWNe are contained inside the SNR and there is no interaction yet with the inward SNR reverse shock. At this early stage, the pulsar is located near the SNR center close to its birthplace. The TeV emission should come, in this case, from within the nebula itself; (ii) the *reverberation* or second phase begins once the reverse shock collides with the PWN forward shock (t∼few tens of years kyr), creating a compression on the PWNe that then leads to an expansion, creating oscillations or reverberations. At this point, the PWN becomes disrupted, provoking that the electrons that produce the TeV emission start to escape from the PWNe onto the SNR and possibly into the ISM. The PWNe start suffering morphological changes and the pulsar can start moving from its birthplace, but it is still contained within the nebula; (iii) the final post-reverberation or bow-shock phase is reached once the pulsar abandons the SNR onto the ISM (at least t ≥ 40 kyr), creating a bow-shock structure. At this stage, the escaped leptons can propagate further into the ISM in a region larger than the PWN, creating extended TeV halos. Two TeV halos were initially discovered by HAWC around the evolved pulsars Geminga and Monogem [84] and several more have been identified by the Large High

Altitude Air Shower Observatory (LHAASO) as counterparts for some of the sources on their first catalog [62].

The Crab Nebula is probably the most studied PWNe in the VHE regime. It was the first TeV source ever detected [85] and since then it has served as standard candle for VHE astronomy. It is the result of an SN explosion that happened in 1054 CE; hence, it is powered by a young 33-ms pulsar (PSR J0534+220). This central pulsar has largely been studied by IACTs and its pulsations have been detected from as low as 25 GeV [86] up to 1.5 TeV energies [87]. It is the most powerful pulsar in the Galaxy with a spin-down luminosity of 4.6×10^{38} erg s^{-1}. The multi-wavelength emission of the Crab Nebula is described by synchrotron radiation detected from radio to HE gamma rays, while the TeV component is produced by inverse Compton up-scattering of those synchrotron photons by the relativistic electrons accelerated in the pulsar wind. The TeV PWNe has been resolved to an extension of $\sim$52 arcsec [88]. Its spectrum is measured over 22 decades in energy, described as leptonic emission. The existence of photons at energies >100 TeV was first evidenced by the UHE detectors Tibet ASγ and HAWC [89,90] and by the MAGIC telescopes via very large zenith angle (VLZA) observation techniques [91]. LHAASO has further established the Crab Nebula as a leptonic PeVatron with the detection of photons up to 1.1 PeV [61,62], implying that the parent electrons shall have energies of $\sim$2 PeV.

For a long time, the Crab Nebula was considered a steady source with a stable flux both in the HE and VHE gamma-ray regimes. However, strong flaring emission was discovered at energies >100 MeV by the space-borne instruments AGILE and *Fermi*-LAT [92,93]. These flux enhancements show a short few-hour timescales duration. The flux increase has been found to vary for a factor 3–30 with respect to the nebula average value, as seen in [94], and references therein. These flaring episodes can last for several days/weeks and they show shorter-scale structures. Also, the spectral index shows differences among flares. During these flaring events, no variability in the pulsar flux or significant glitch is detected. Also, no variability in the synchrotron component is detected in the radio, infrared of X-ray bands [92,94]. These flares have been repeatedly appearing at rates of about $\sim$1–2 per year.

This enhanced emission could be extended up to TeV emission and be potentially observable by IACTs. Two scenarios are possible: the detection of the synchrotron tail at the low-end of the VHE regime (up to few tens of GeV) or that the enhanced emission is transferred to the IC component and detected deep in the TeV range (in Klein–Nishina regime) due to synchrotron self Compton process, since the electrons that produce the enhanced MeV emission should upscatter the photons in the nebula to produce TeV emission.However, the IC component seems to remain stable during the HE flares, since IACTs have searched for variability in the TeV component, with no significant flux enhancement detected. Both MAGIC and VERITAS observed during the 2010 HE flare (58 min in one night and 120 min over four nights, respectively), with no VHE enhancement reported [95,96]. The HE flare of 2013 [97], which lasted for about 2 weeks at HE, was observed by H.E.S.S. for five consecutive nights and by VERITAS during a period of about three weeks (see Figure 5), with similar results [98,99]. Whether a flux enhancement deep in the TeV component exists remains yet unknown. A search in the TeV end with the VLZA technique with MAGIC revealed that the stereoscopic system should be able to detect fluctuations 2.25 times brighter that the constant PWNe value [100]; hence, given the right conditions, these flares could potentially be detected by the current generation of IACTs.

It is not trivial to understand the undergoing mechanism of these rapid flares. In the PWNe scenario, the MeV-GeV component is described as synchrotron produced by electrons and positrons in a shocked pulsar wind and hence with energy limited by the synchrotron burn-off (assuming an MHD outflow). The flares surpass this value, hence excluding this ideal scenario. On the other hand, rapid flares cannot be explained in an IC context. Different scenarios have been proposed to account for the origin of these HE gamma-ray flares: (explosive) magnetic reconnection events in a highly magnetized plasma [101] or inductive spikes [102], among others. The absence of flux enhancement at other wavelengths could indicate that the HE flares are produced by a single population of electrons. However, it is the lack of multi-wavelength detections and possible correlations that make the study of the origin of this flaring emission challenging. The fast variability and rapid enhancement at HE implies that the emission should come from a compact region in the PWNe of $\sim 10^{-4}$ pc [94].

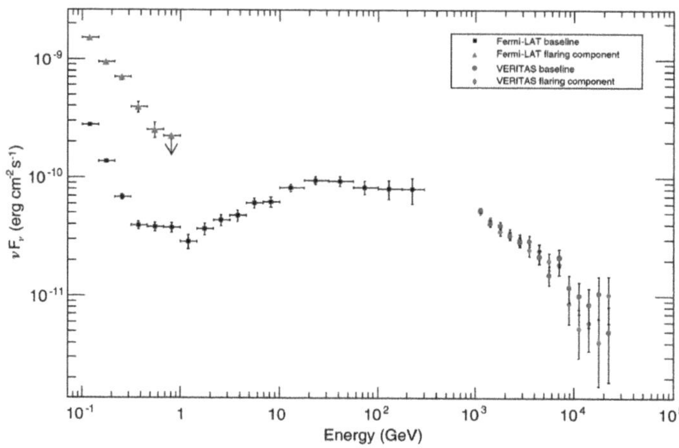

Figure 5. SED of the Crab Nebula during quiescence (black squares for *Fermi*-LAT data and blue dots for VERITAS data) and during the 2013 flare (magenta triangles for *Fermi*-LAT and red dots for VERITAS). While the synchrotron component detected by *Fermi*-LAT shows hardening and enhanced flux, the IC spectrum shows no deviation. The y-axis error bars represent the statistical uncertainties in the flux. The downside arrow in the *Fermi*-LAT flaring component is a flux UL. Reprinted with permission from [99].

PSR J0534+220, powering the Crab Nebula, is the most energetic pulsar in the Galaxy (4.8×10^{38} erg s^{-1}) and it is among the youngest ones. Up to now, the Crab Nebula is the only PWNe showing variable HE gamma-ray emission in the Milky Way. However, two young pulsars in the LMC show similar spin-down powers to Crab: PSR J0537−6910 (1.5×10^{38} erg s^{-1}) and PSR J0540−6919 (4.9×10^{38} erg s^{-1}). Flaring-like activity has been detected with *Fermi*-LAT in different bands: 100–300 MeV, 100–300 MeV, and 1–10 GeV [103]. Due to the spatial closeness of the two pulsars, it is not possible to identify which of them is responsible for the flares in the first two energy regimes. However, since the angular resolution improves at higher energies, it has been found that both pulsars flare at GeV. Gamma-ray flares from these pulsars were already predicted by [102]. The detection of flaring gamma-ray emission from other PWNe apart from the Crab Nebula could then indicate that this type of variability (e.g., inductive spikes) is common to young powerful pulsars.

6. Fast Radio Bursts and Magnetars

Fast Radio Bursts (FRBs) are a relatively newly-discovered Jy-level-class of $\sim$millisecond radio transient events of (mainly) extragalactic origin. At present, since the first discovery of the so-called *Lorimer burst* [104], about 1000 FRBs have been detected. However, once taken into account, various factors such as the sky coverage of the different instruments, survey threshold, and selection effects, it is possible to conclude that FRBs occur at an extraordinary rate, up to $\sim 10^4$ per day distributed over the entire sky. This correspond to a rate of 10^{-3} yr^{-1} per galaxy, much larger than the GRB rate. Nonetheless, only few tens of FRBs have been associated to their host galaxy (with kpc precision) and only a handful have been localized with enough accuracy to be associated with specific regions within those galaxies. Although from the observational point of view FRBs are similar to the pulses detected from Galactic radio pulsars, the observed flux density coupled with their extragalactic origin indicates a total emitted isotropical equivalent energy ranging from $\approx 10^{35}$ to $\approx 10^{43}$ erg, extremely high compared to the radio pulsar standard, but several orders of magnitude smaller than GRBs; for a general review, see, e.g., [105,106].

Most FRBs are one-off events. However, within the population of FRBs detected so far, around 50 events have been observed to produce multiple bursts, the so-called *repeating FRBs* [107]. Although sporadic, the repeating behavior of some FRBs allowed for the first time to perform targeted observations to localize the source using interferometry techniques. The first known repeater, FRB 121102, was associated with a low-metallicity star-forming dwarf galaxy at redshift z = 0.19 [108], while a persistent and compact (<0.7 pc) radio source of unclear nature was discovered in association with the FRB direction [109]. High-resolution optical and infrared observations by the Hubble space telescope and the *Spitzer* telescope showed that the galaxy optical emission is dominated by an inner star-forming region whose position is consistent (within uncertainties) with the persistent radio source [110]. Such type of galaxy is also the typical host galaxy for extreme transient events such as GRBs or super-luminous supernovae. While the association of FRBs with cataclysmic events may sound natural and was originally proposed as counterpart of FRBs, the bursts of FRB 121102 have not revealed any signature of an afterglow emission and have been found to repeat at a rate short enough to rule any possible explosive mechanism to power them out. A second localized repeater, FRB 20180916B, shows an apparent ∼16-day (∼4 days active followed by 12 days of inactivity) periodicity [111]. It was found to be located at the edge of a star-forming region within a spiral galaxy, without any persistent counterpart associated. A possible periodicity of ∼150 days has been found also for FRB 121102 [112]. Some significant differences between the repeaters and the apparent one–off FRBs have been also reported in the literature [113]. In particular, repeater bursts seem to be intrinsically broader in width and narrower in bandwidth. The position of active repeating FRBs seems to be consistent with the one of young extreme objects such as magnetars. Magnetars are isolated NS with an extremely powerful magnetic field of the order of 10^{14}–10^{15} G, about 1000 times stronger than a normal NS. In these objects, the observed persistent electromagnetic radiation is likely powered by the decay in the intense magnetic field. On the other hand, magnetars can also undergo flaring episodes with outbursts on different timescales, detectable in X-rays and radio. These are probably caused by large-scale rearrangements of the surface and/or magnetospheric field. Interestingly, magnetars can additionally produce giant flares (GFs), which are among the most energetic (10^{44}–10^{47} erg s^{-1}) Galactic events.

On April 2020, the event FRB 200428 was detected by the Canadian Hydrogen Intensity Mapping Experiment (CHIME) telescope from a direction consistent with the Galactic magnetar (and soft gamma repeater) SGR 1935+2154 [114], located at a distance of 6.6–12.5 kpc and embedded in the supernova remnant SNR G57.2+0.8. This discovery represented the first detection of an FRB event from a known object, as well as the first FRB of Galactic origin. Contemporaneously to the FRB event, the detection of several X-ray flaring episodes was achieved by a wide range of instruments [115–117]. Figure 6 shows the X-ray light curve as measured by the INTEGRAL satellite where the radio emission is found to be in time coincidence with the X-ray flaring activity [115]. Furthermore, a long-lasting high-energy flaring activity in the form of a forest of intense X-ray bursts was detected by *Swift* [118] and *Fermi*-GBM [119] up to several hours after the initial episode. The discovery of the connection between hard X-ray bursts (HXRBs) of SGR 1935+2154 and FRBs significantly boosted the long-lasting idea of the theoretical interpretation of magnetars as progenitors of FRBs. However, deeper observations performed by the FAST radio telescope[2] showed that the majority of the X-ray bursts emitted by SGR 1935+2154 are actually *not* correlated with the FRBs [120]. Additionally, the further surprisiny detection of the repeater FRB 200120E in a position consistent with a globular cluster within the nearby galaxy M81 [121] challenges the young magnetar scenario as the only engine of FRB. Globular clusters are old enough to not have massive stars able to originate magnetars within. However, they do show high star densities and host short-period binaries which can lead to the production of magnetars via more exotic channels such as accretion-induced collapse (of a WD) or merger-induced collapse (of WDs and NSs). Thus, as a matter of fact, the progenitor of FRB remains a unanswered question. Nonetheless, in light of the so-far only robust hint of association with SGR 1935+2154, the magnetar paradigm is still considered the leading interpretative scenario and it has been discussed extensively in the literature (a non-exhaustive list in [122–125]).

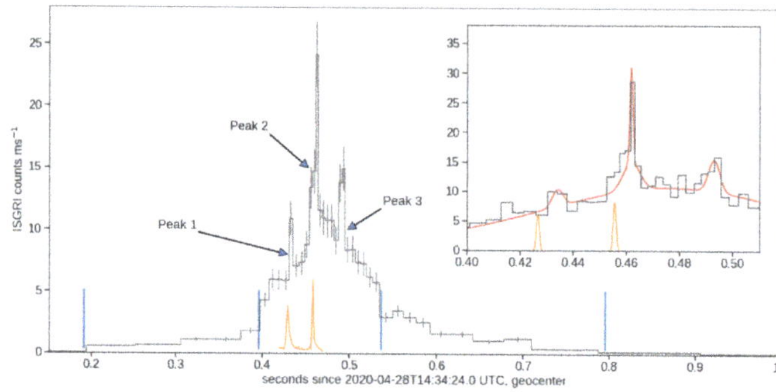

Figure 6. INTEGRAL (20–200 keV) light curve of one of the flare of SGR 1935+2154 referred to T_0 = 14:34:24 UTC of 28 April 2020. The vertical orange lines represent the time of the detected radio pulses. Reprinted with permission from [115].

Within this framework, the proposed emission processes involve coherent radiation mechanisms such as synchrotron maser radiation in magnetar internal [126] and external shock models [127] as well as magnetospheric pulsar-like models. The latter, however, do not foresee keV-MeV emission as observed in SGR 1935+2154. In [128], it is predicted that if FRBs are produced by magnetar flares, an afterglow emission peaking at the MeV-GeV band is expected with a total energy release in the X-ray/gamma-ray band at least a factor $\approx 10^4$ larger than the emitted radio energy. At the time of writing, the only magnetar flaring event detected in the gamma-ray regime was the GF from a magnetar in NGC 253, detected by *Fermi*-GBM [129] and *Fermi*-LAT with a photon of up to 1.7 GeV [130]. However, recent results published by *Fermi*-LAT on individual FRBs analysis reported no significant emission in the LAT energy band [131]. Nonetheless, the detection of hard X-ray bursts with a non-thermal spectrum in SGR 1935+2154 shows that at least some FRBs are able to accelerate particles and produce MeV non-thermal emission. Despite the puzzling scenario, FRB (and magnetars) are also an interesting target for IACTs. In fact, some theoretical models also predict VHE [128,132–134] emission correlated in time with FRBs. Not surprisingly, the flaring episodes of SGR 1935+2154 in April 2020 were also followed-up by current Cherenkov telescopes and monitoring campaigns on this magnetar have been active since then. The campaigns organized by MAGIC [135] and H.E.S.S. [136], coordinated within a larger multi-wavelength framework, did not reveal VHE emission to be neither persistent nor on shorter (minutes to milliseconds) time scales.

Current-generation IACTs have active follow-up programs on FRBs although no particularly stringent flux ULs in the VHE band were reported so far [137]. Some specific events have been the subjects of dedicated follow-up campaigns such as in the case of H.E.S.S. that obtained the first ULs on the potential VHE afterglow emission from FRB 20150418A [138] (Figure 7 right panel). This FRB was of particular interest as it showed hint of a $\approx$6 days-long-lasting radio afterglow detected by the Australia Telescope Compact Array (ATCA) [139]. The achieved limit on the VHE luminosity was of the order of 5×10^{47} erg s^{-1} at the energy of 1 TeV. The MAGIC and VERITAS collaborations reported VHE ULs on the repeater FRB 20121102A conducting coordinated observations with Arecibo [140,141]. In the case of MAGIC follow-up, five contemporaneous radio bursts were detected (at a central frequency of 1.38 GHz) although no millisecond timescale burst emission was detected in VHE or the optical band (Figure 7 left panel). Follow-up results on a sample of other repeaters (FRB 180814, FRB 180916, FRB 181030, and FRB 190116) were more recently reported by VERITAS [142], again with no detection achieved.

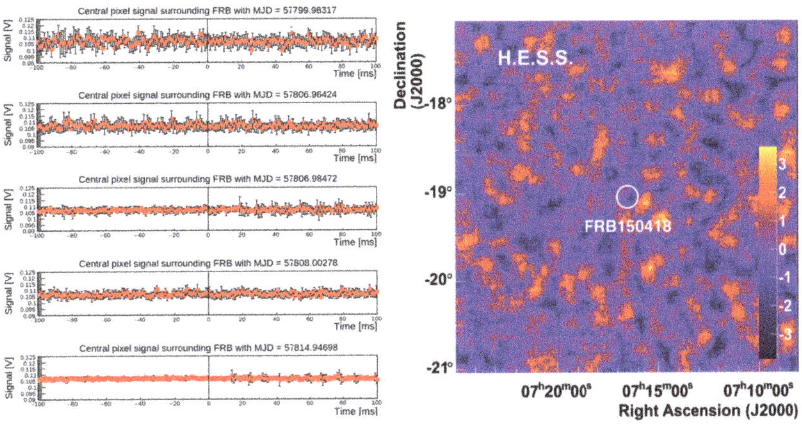

Figure 7. (**Left** plot): Optical light curves obtained by MAGIC and spanning 200 ms around the trigger times of 5 bursts from FRB 121102 detected simultaneously with MAGIC data. The vertical axis is proportional to the U-band flux. No significant excess is observed simultaneously with any of the 5 bursts. Reprinted with permission from [140]. (**Right** plot): Significance sky map from the H.E.S.S. follow-up observations of FRB 150418. Reprinted with permission from [138].

It is important to remark that IACTs are versatile instruments that, although designed to detect nanosecond pulses of Cherenkov light, are generally sensitive to millisecond timescale optical signals. Despite the modest quality of their mirrors when compared to standard optical telescopes, the typical large diameter of their primary mirror and their fast-response readout electronic make them effective high-time-resolution photometers. Current IACTs are indeed able to perform parallel VHE and optical observation on very short timescales up to a limiting magnitude significantly lower than standard optical telescopes [143–145]. This is a key feature that make IACTs excellent instruments for fast transient astronomy and with a relevant application in the case of FRBs. In fact, FRBs may be potentially accompanied by *fast optical bursts* (FOBs) via different mechanisms [146]. Optical counterparts have been detected in association with magnetars flaring episodes [147] and can therefore provide important insights into the physics of FRBs. The possibility to have parallel VHE and fast optical observations made IACTs key instruments for future follow-up. The improvement in VHE sensitivity as promised by next-generation instruments such as the CTA will finally allow for observations up to a gamma emission values comparable to the ones expected by magnetars.

7. Gravitational Waves

The possibility of performing astrophysical observations by means of non-electromagnetic signals such as gravitational waves (GWs) has become reality with the first scientific runs of the LIGO[3] and Virgo Scientific Collaborations[4] (LVC). The first LVC scientific observation run, named O1, opened the era of gravitational wave astronomy by means of the first direct detection of a GW signal [148] from a binary stellar-mass black hole merger (BBH). Not long after, during the O2 scientific run, the first GW signal (GW 170817) from the coalescence of a binary system composed of two NSs (BNS) was discovered [149] together with a new sample of BBH signals [150]. Approximately 2 s after the detection of GW 170817, the Fermi and INTEGRAL satellites detected a sub-threshold short GRB (sGRB), namely GRB 170817A [151,152]. The identification of a sGRB as electromagnetic counterpart of a GW signal triggered by a binary neutron star merger represented a groundbreaking observation that provided the first firm evidence on the nature of sGRB's progenitors. The potential link between GWs (from BNS mergers) and sGRBs has been widely explored and discussed in the literature in the past; see, e.g., [153], and references therein for a review. The discovery of GRB 170817A triggered an unprecedented follow-up campaign at all wavebands. It is important to remark that these observations are particularly challenging due to the very large localization uncertainties provided by GWs interferometers, up to tens of thousands of square degrees. Nonetheless, approximately 11 h after the GW trigger, an optical/IR counterpart, named AT 2017gfo (IAU naming) and hosted in the 40 Mpc-distant galaxy NGC 4993, was detected by the One-Meter Two-Hemisphere (1M2H) collaboration [154] and interpreted as a *kilonova*. Unlike BBH mergers, BNS mergers are expected to be source of optical/near-IR emission powered by the decay of radioactive nuclei generated by r-process nucleosynthesis in the outflow formed after the coalescence; see, e.g., [155,156]. The detection of AT 2017gfo represented the first confirmation of this theoretical prediction. In the days after the burst, an X-ray counterpart was detected and identified as the GRB afterglow non-thermal emission [157]. The late-time rising of an X-ray afterglow fits within the interpretation that the GRB is observed off-axis, with the jet-beamed ejecta pointing away from Earth. According to the hydrodynamic of a generic relativistic shock-wave model, the bulk Lorentz factor ($\Gamma(t)$) of the outflow is reduced by the deceleration of the jet, causing the relativistic beaming angle ($\propto 1/\Gamma(t)$) to increase [158]. The opening angle of the emission widens, eventually including the line of sight of the observer. From the observational point of view, a delayed emission, whose intensity and delay depends on the off-axis angle, may arise. The case of GW/GRB 170817A confirms this paradigm as successive radio observations did confirm the interpretation of the radio to X-ray emission as originated by an off-axis, structured jet (i.e., the energy and velocity of the ejected material scale with the angular distance from the jet axis) with a viewing angle of $\approx 30^\circ$ [157,159,160]. The radio and X-ray emission increased in the weeks following the initial trigger, peaking approximately 155 days after the merger.

The extensive multi-wavelength campaign triggered by the detection of GW 170817A also included follow-up at VHE by IACTs (see Figure 8). Less than two years after GW 170817A, the first detection of VHE gamma-ray emission from GRB 180720B [161], GRB 190114C [162], and GRB 190829A [163] was announced by the H.E.S.S. and the MAGIC collaborations, bringing an end to a quest lasting for more than twenty years. Although all of the GRBs detected so far[5] by current IACTs were long GRBs, sGRBs are also expected to emit VHE radiation. In this regard, a hint of VHE emission has been observed by MAGIC in the case of the short GRB 160821B [164], providing a compelling clue on the detectability of TeV emission from GW counterparts from compact object mergers. Few attempts by IACTs in following-up GW alerts were reported before the breakthrough of GW 170817A such as for GW 151226 [165,166], GW 170104 [167], and GW 170814 [168]. However, the VHE campaign organized for GW 170817A represented a step forward and a fundamental test-bench in exploring IACTs' capabilities in this challenging observations. The H.E.S.S. telescopes started a series of pointing over the uncertainty region of GW 170817A about 5 h after the first trigger [169], that made it the first ground telescope to point at the source location. Although the detection of AT 2017gfo was not yet announced [170], the pointing strategy proved to be efficient with the NGC 4993 location within the field of view of the H.E.S.S. first pointing. Nonetheless, no evidence of VHE emission was detected during this early monitoring campaign of SSS17a [169]. Starting from mid-December 2017, the sky position of the optical transient SSS17a became visible also to the MAGIC telescopes' site. This observation window roughly overlaps with the afterglow peaking time. Late time follow-up was then performed by MAGIC and, again H.E.S.S. covering the peak and the onset of the fading phase in the X-ray and radio lightcurves. Although no detection was achieved, the obtained ULs were used by the two collaborations to constrain physical emission models, although with a rather different prediction on the intensity of the TeV component (see Figure 9).

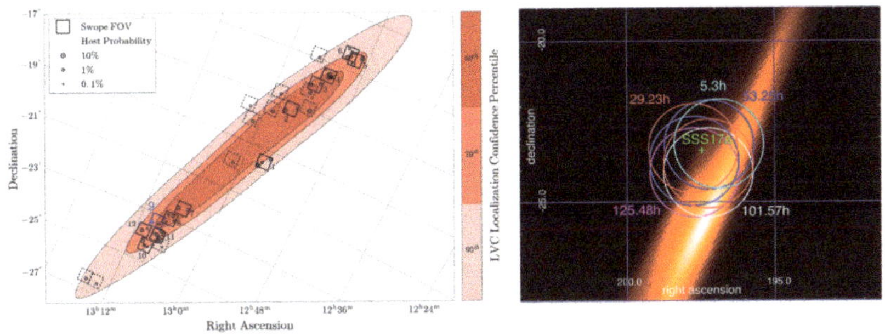

Figure 8. (**Left** plot): Sky map covering the 90% confidence-level region for the GW 170817 position. The positions of galaxies used in cross-correlating the large localization area and defining an optimized pointing strategy for the Swope telescope on 17–18 August 2017 are shown as gray circles. The size of the circle indicates the probability of a particular galaxy being the host galaxy for GW 170817. The square regions are individual Swope pointings labeled in the order that they were observed. Solid and dashed squares represent the square chosen to contain multiple and individual galaxies, respectively. The position of NGC 4993 and SSS17a are in the blue square. Reprinted with permission from [154]. (**Right** plot): H.E.S.S. pointing directions during the monitoring campaign of SSS17a. The circles denote an FoV with radius of 1.5° and the shown times are the start times of each observation with respect to GW 170817A. Colored background is the GW localization map. Reprinted with permission from [169].

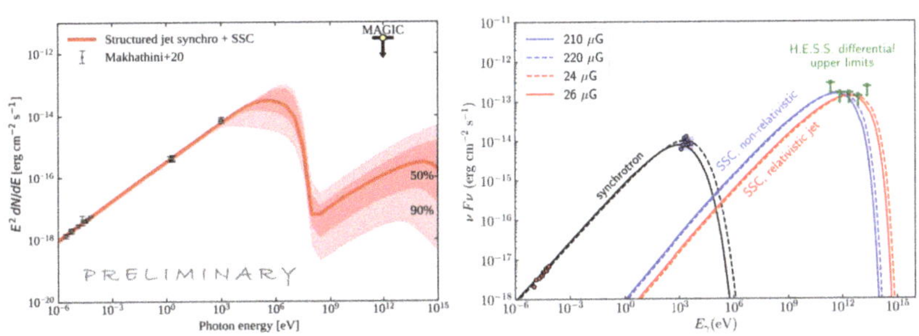

Figure 9. (**Left** plot): Expected SSC emission evaluated by MAGIC 155 days after the merger, using fit parameters constrained by the radio, optical, and X-ray data. Reprinted with permission from [171]. (**Right** plot): SSC spectra foreseen in [172] 110 days after the merger. The blue and red curves represent two possible geometry and expansion speed of the remnant: an isotropic, non-relativistic expansion (blue curves) and a relativistic jet (red SSC curves). The minimum magnetic field strength imposed by the H.E.S.S. ULs (green arrows) are also reported. Reprinted with permission from [171].

Both in [171,172], the broad-band SED is modeled by means of a synchrotron + synchrotron self-Compton (SSC) processes. However, the two proposed models are not directly comparable. In the structured jet approach used within the MAGIC interpretation and described in details in [173], the full time-evolution of the jet is taken into account in the evaluation of the expected emission.As the jet evolves, the observed radiation is the result of the convolutions of photons emitted at different times and different locations behind the shock. Such an evolution is not considered within the H.E.S.S. modeling wherein the emission is evaluated at specific single times. Within the uncertainties of assumed physical parameters for GRB 170817A, the structured jet model foresees a rather low TeV emission component, significantly lower than MAGIC upper limits, challenging the possibility of detection of such an event with current-generation IACTs. However, more favorable conditions in the emitting geometry and circumbusrt properties may mitigate these prospects as we will discuss in Section 9.

8. Other Transient Sources: Tidal Disruption Events and Gamma-Ray Bursts

The science topics of GRBs and tidal disruption events (TDEs) are discussed in more details in dedicated papers within this Special Issue [4]. Nevertheless, we briefly touch upon them in the following subsections for the completeness of this review.

8.1. Tidal Disruption Events and Neutrino Connection

Tidal disruptions events (TDEs) are powerful events that occur when a star is disrupted by tidal forces when approaching a massive BH. They are considered of extreme importance in particular in the framework of multi-messenger astrophysics. It is thought that the disruption of a stellar object may trigger the launch of a relativistic jet able to shock-accelerate particles from the star remnants. This material is naturally rich in light and heavy nuclei so that TDE may be a plausible acceleration site for ultra-high-energy cosmic rays (UHECR $> 10^{20}$ eV) and neutrino; see, e.g., [174,175]. While the cosmic neutrino flux has already been established through the measurements of the IceCube Neutrino Observatory [176], the association of this flux with specific astrophysical sources is still challenging. So far, few sources have been correlated with neutrinos: the flaring blazar TXS 0506+056, identified as the potential source of the IceCube neutrino alert IC 170922A [177]; the nearby star-forming galaxy NGC 1068, in spatial coincidence with a cluster of IceCube-detected neutrinos [178]; and the TDE AT 2019dsg, discovered in the optical band by the Zwicky Transient Facility (ZTF[6] [179]) and identified as the source of the event IC 191001A [180]. While NGC 1068 and TXS 0506+056 have already been the target of observations with IACTs [181,182], TDE is a relatively unexplored class of sources in the VHE band. In 2011, the remarkable TDE event Sw J1644+57, originally triggered as a GRB by *Swift*, was the subject of an extensive follow-up campaign by MAGIC [183] and VERITAS [184]. Although no significant VHE detection was found, these observations may potentially pave the road to future follow-ups with both current IACTs and the CTA Observatory.

In the near future, the Vera Rubin Observatory [185], will start operation and overlap with the CTA Observatory era. The Legacy Survey of Space and Time (LSST) will also dramatically enlarge the sample of detected TDEs, thereby providing an unprecedented number of possible triggers to CTA that may be able to detect VHE gamma-signature at least for nearby events ($\lesssim$20 Mpc) [186].

8.2. Gamma-Ray Bursts

GRBs are transient events last from milliseconds up to hundreds of seconds. They are the brightest electromagnetic events known and they are able to release an enormous amount of energy ($10^{52} \div 10^{54}$ erg). They show their phenomenology mainly in the 10 keV–1 MeV energy band. According to a relativistic shock model, described for example in [187,188], GRB emission is powered by the conversion of the kinetic energy of a relativistic outflow into electromagnetic emission. The details of this conversion remain poorly understood. A largely discussed possibility is that the observed photons are radiation from particles accelerated to ultra-relativistic energies by successive collisions within the magnetized medium. During the so-called *prompt phase*, GRB dynamic is thought to be driven by relativistic collisions between shells of plasma emitted by a central engine (internal shocks). Similarly, the emission during the *afterglow* phase seems to be connected to the shocks between these ejecta with the external medium (external shocks). The results of such internal/external shocks is the acceleration of particles through Fermi mechanisms. The accelerated particles can emit the observed high-energy photons through many possible non-thermal mechanisms. Within this framework, synchrotron emission has largely been considered as the most natural to explain the GRB sub-MeV emission [189–191]. Although alone it cannot fully explain the observed prompt spectrum for the majority of the events, synchrotron is believed to play an essential role in GRB dynamic. In particular, it has been suggested that the HE emission observed by *Fermi*-LAT extending after the end of the prompt emission is synchrotron radiation produced in the external shock that is driven by the jet into the circum-burst medium [192]. However, the recent detection of a VHE counterpart challenged the synchrotron-alone scenario, confirming the existence of a second emission component above the synchrotron burnoff limit. In the near future, the CTA will open the possibility of detecting ∼hundreds (or more) of photons from moderate-to-bright GRB, allowing for a significant improvement in the photon statistics and for the possibility to have good-quality time-resolved spectra. The first prototype of the 23 m class diameter LST-1, particularly suited for the follow-up of transient events due to the fast repositioning (∼30 s for 180°) and the relatively low energy threshold, is currently ending its commissioning phase at the CTA northern site. LST-1 already reported the follow-up of different GRBs and neutrino events, although with no reported significance yet [193].

Furthermore, many events have shown (somehow surprisingly) that long-lasting TeV signatures can also be detectable under favorable conditions. The close-by and very low luminosity burst GRB 190829A [163] was detected by the H.E.S.S. telescope up to a few TeV for three consecutive nights while the recent detection of GRB 221009A by the LHAASO experiment up to 13 TeV [194] has definitively proven that instruments operating in the energy range above few TeV band such as the ASTRI Mini-Array [195], although not specifically designed for transients and time-domain astrophysics, may also play a key role in the future follow-up programs of these events [196,197]. For more detailed review on GRBs, see [4,198].

9. Discussion and Prospects

The last two decades have proven to be the starting point of a golden era for multi-messenger time domain astrophysics. New facilities for non-electromagnetic astronomy such as neutrino and GW detectors have reached their nominal operational phase, joining a large network of telescopes and satellites covering an unprecedentedly wide energy band. New synergies and improving communication channels between these facilities have led to breakthrough discoveries such as the connection between sGRBs and GWs. The physics of extreme transient events both inside and outside our Galaxy has an intuitive connection with the highest energetic X- and gamma-ray radiation. Non-thermal emission processes, typical of the HE and VHE band, represent the signature of shock-powered radiation mechanisms, often invoked in explaining the dynamics of a wide range of extreme cosmic accelerators. Shock interactions may be at work as a particle acceleration mechanism in both a relativistic (like in GWs/GRBs) and non-relativistic (like in SNe) flavor. The corresponding radiation mechanisms at work may be shared among these sources, although showing a diverse phenomenology given the differences in shocks expanding velocity, external density, and surrounding environment. Hence, VHE observations provide a privileged channel to shed light into the physics of transient events in an energy range particularly important for the discrimination among different emitting scenarios. Although in operation since the beginning of the 2000s, current-generation IACTs can still lead to the discovery of new transient phenomena and to deepen our understanding of the TeV physics of these newly identified VHE sources. However, the IACT community is working toward the construction of the CTA Observatory, which is the next-generation ground-based observatory for Cherenkov astronomy. It will be composed by two arrays composed of telescopes of up to three different sizes, located in the northern (Roque de Los Muchachos Observatory, Spain) and southern (Paranal Observatory, Chile) hemispheres. It will cover the energy range between 20 GeV and 300 TeV and it will count with improved sensitivity with respect to current IACT experiments. It will have unprecedented sensitivity at short timescales (see Figure 10), making it a unique laboratory for VHE transient astrophysics. The *Transients Key Science Project* of the CTA Observatory [3] defines the core program for the follow-up of transient sources [199–201], including GRBs, GWs, neutrino counterparts or the large zoo of Galactic transient sources (novae, microquasars, magnetars, flaring PWNe, etc.), among other serendipitous transitory events.

The improved sensitivity of CTA, together with its better angular and energy resolution and large energy coverage, will allow for the discovery of new transient and multi-messenger phenomena, widening the population of current known sources.

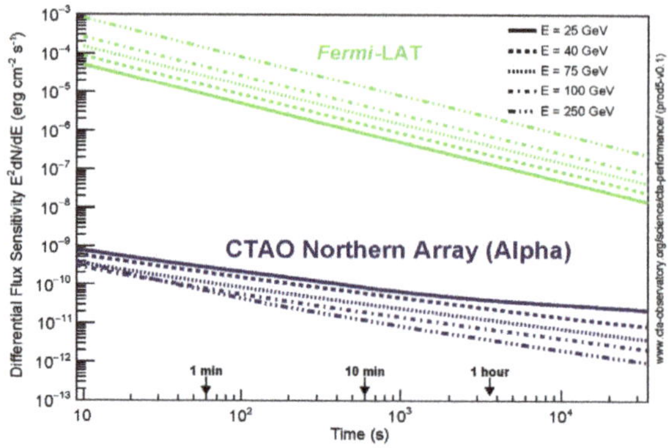

Figure 10. Differential flux sensitivity versus time of the CTA Observatory Northern Array in its Alpha configuration (blue), which accounts for the initial construction phase, compared to *Fermi*-LAT (green) at different energies. Figure taken from https://www.cta-observatory.org/science/ctao-performance/ (accessed on 14 March 2024).

9.1. Novae

We can expect to detect other recurrent symbiotic systems in the VHE regime. There are 10 confirmed recurrent novae in the Galaxy with recurrency times between 10 and 80 years approximately [8]. This number could, however, be larger since other systems with very massive WDs could also be recurrent, they have simply not yet been identified as so. The next imminent eruption is that of T CrB, a symbiotic nova which shows a recurrence of about 80 years [202] and for which the next explosion is calculated to happen on 2024.4 ± 0.3 [203]. The latest reports indicate that the source entered the so-called pre-eruption dip and its B and V magnitudes are slowly decreasing, as seen in Figure 11. The accretion disk reached a minimum in August–September 2023 and it is showing a fast rebrightening [204]. T CrB is closer than RS Oph (0.9 pc vs. 2.4 pc) and it is expected to reach a flux of about 10 times larger than RS Oph. Its peak optical magnitude is can reach magnitude 2.9 as in previous eruptions, being one of the brightest novae observed.

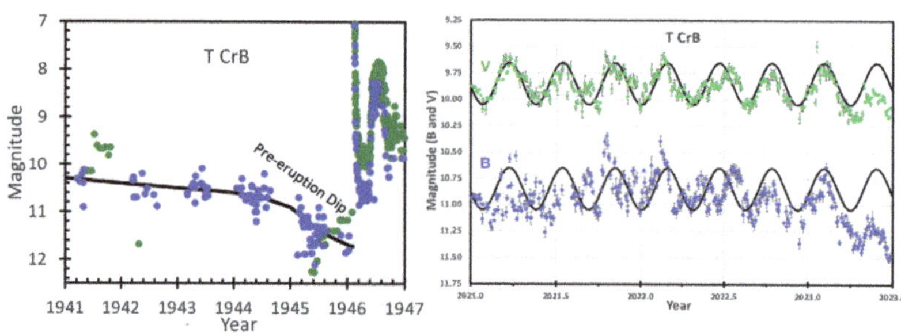

Figure 11. (**Left**) Optical lightcurve of T CrB during the 1946 eruption in B (blue dots) and V (green circles) magnitudes. A similar lightcurve is expected in the next eruptive event; (**Right**) Current fading of the B (blue circles) and V (green circles) of T CrB, revealing that the source has entered the pre-eruption dip. Figures from [202,203].

HE emission from classical novae was an unexpected discovery, due to the low density of the surrounding environment, but they are now the most frequently detected type of novae. Whether classical novae can emit at VHE is still an open question, although if shocks operate similarly in symbiotic and classical systems, then we could potentially expect VHE emission also from these systems.

9.2. *Microquasars*

The improved sensitivity of CTA will likely lead to the detection of TeV transient emission from flaring microquasars. When extrapolating the Cygnus X-1 hint observed by [55] in the VHE regime, we see that the CTA northern array will be able to detect a similar flare with high significance in only 30 min of observation (see Figure 12) [199]. We can expect the future CTA observatory to detect transient emission from other microquasars, probing particle acceleration in jets.

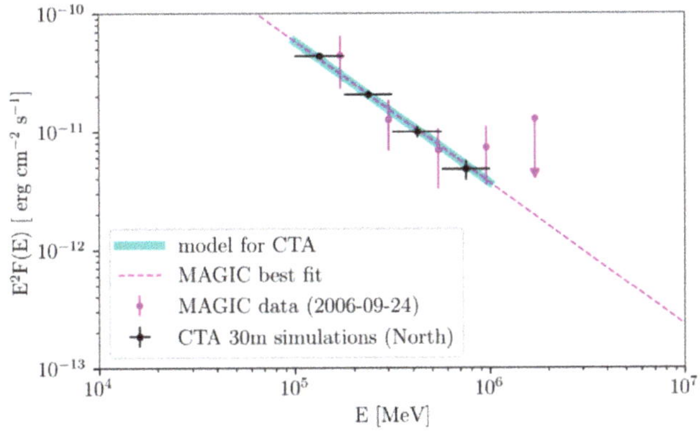

Figure 12. Simulated SED of Cyg X-1 as seen by CTA-North (black points) during a flaring episode similar to that reported in [55] (magenta points). Reprinted from [199] .

9.3. *Supernovae*

As discussed in Section 4, the CTA observatory is expected to detect CCSNe up to a few Mpc of distance. According to [205], an SN like the recent type II-P SN 2023ixf should be detectable by future experiments such as the CTA even at 7–10 Mpc. In the case Galactic CCSNe, there is the open possibility that current IACTs could detect a VHE counterpart. Since neutrino bursts take place during a core-collapse event, they are expected to precede the electromagnetic radiation from the SN when reaching the Earth, as it happened in SN 1987A. Hence, neutrino bursts are good alert trigger systems for a Galactic CCSNe event, which are rare events in our Galaxy. However, it is expected that the VHE emission is absorbed due to $\gamma\gamma$ annihilation during the first 7–10 days, approximately. It can be worth trying to catch the VHE counterpart during the first hours after explosion, since models do not manage to simulate the expected gamma-ray emission so early on and since an observation like this will definitely help constrain the theoretical scenarios for such unique events.

9.4. *Crab Nebula Flares*

Next-generation instrumentation such as the CTA observatory will count with an increased sensitivity[7] to short timescale transient events [201]. It has been explored how the northern array of the CTA observatory will be sensitive to flaring emission from the Crab Nebula [206,207]. The high sensitivity of the array will likely allow for the detection of both the synchrotron end at low energies in few ($\lesssim$5) hours in the case of hard synchrotron flares for magnetic fields with similar or larger intensity than that of the nebula. Even current facilities such as MAGIC could potentially detect bright flares (similar to that of 2011) or at least set strong constraints (see Figure 13). In the case of the IC component, TeV emission could be detectable if the energy of the electrons is boosted and under certain scenarios, such as soft spectra and mG magnetic fields (right panel of Figure 13).

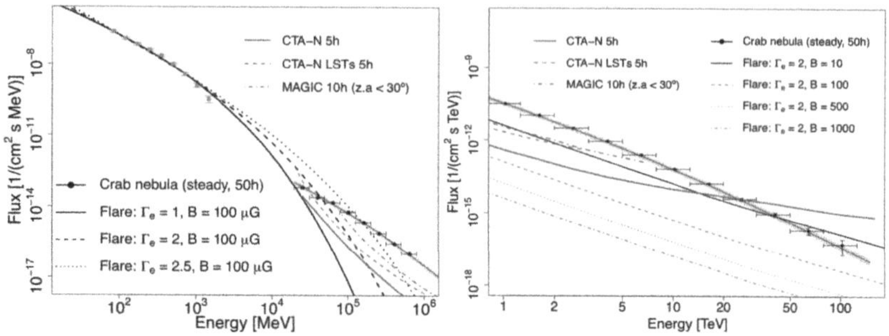

Figure 13. Simulated SEDs of the Crab Nebula during different flares compared to steady state. (**Left**) Synchrotron regime (**Right**) IC component. The 5 h sensitivity CTA-North array (bold line), the 5 h sensitivity of the 4 LSTs of CTA-North (dashed line), and the 10 h sensitivity of the MAGIC telescopes (dotted–dashed line) are represented. Figure rearranged and reprinted with permission from [206].

9.5. FRB and Magnetars

The nature of FRBs represents one of the most enigmatic (and recent) hot topics in time-domain astrophysics. The discovery of the association within an FRB-like emission and the Galactic magnetar SGR 1935+2154 provided possible evidence about the origin of these events. Magnetars already triggered the interest of IACT at VHE in the search for persistent emission; see, e.g., [208] and more recently as transient sources as they may undergo important GFs: rare and brief ($\sim$0.1 s) bursts of hard X-rays and soft gamma rays, recently detected up to the GeV range [209]. The energy release of a GF may be remarkable, reaching a total value of $10^{44} \div 10^{46}$ erg. Although many theoretical models do not envisage magnetars as VHE emitters during their quiescent state, the possibility of having VHE emission during flaring episodes cannot be ruled out. The April 2020 flaring activity of SGR 1935+2153 gathered an exceptional extended multi-wavelength coverage, mainly thanks to the above-mentioned FRB connection. The observed X-ray activity showed a harder spectra with respect to the typical bursts from SGR 1935+2154 (and other magnetars) although its intensity was relatively moderate and significantly too faint to be classified as a GF. Observations by H.E.S.S. and MAGIC ruled out possible extended emission up to the VHE band for this event [135,136]. Very recently, a candidate magnetar GF from the nearby galaxy M82 has been followed up by MAGIC with a (preliminary) non-detection at VHE [210].

The high sensitivity to short timescale signals foreseen for the CTA observatory will make it a perfect instrument to magnetars flaring activity follow-up. Furthermore, the new radio facilities that will operate at the time of the CTA will provide the detection of up to hundreds of FRBs per day. Many of these will have good localizations and will be inside the CTA field of view, making it possible to search for prompt and/or delayed VHE emission corresponding with radio activity, unveiling the still-puzzling connection between FRB and magnetars.

9.6. GRBs and GWs

Whether the sources of GWs are BNS merger or CCSNe, electromagnetic emission up to the VHE may be envisaged. Expectations for VHE emission from CCSNe likely pose these sources out of reach for current IACTs (see Section 4). On the other hand, although challenging, the VHE counterpart of BNS mergers stands in a better chance of detection for running facilities. The link between sGRB and GWs has indeed been proven by the detection of GW/GRB 170817A, while a hint for VHE also from sGRB (long GRBs are now known to be VHE emitters) has been achieved by MAGIC in the case of the short GRB 160821B [164]. Thus, it is justified to assume that each BNS merger may result in an sGRB launching a relativistic jet. However, GW 170817, the only event with a firmly detected electromagnetic counterpart, did not show any hint of GeV-TeV emission and detailed emission models for this event do not foresee a VHE component strong enough to be detectable with current IACTs [171]. Regardless, such negative prospects have to be considered as not conclusive. Within the framework of an off-axis GRB as the source of electromagnetic radiation in a BNS merger, geometry plays a key role in the expected emission at all wavebands. In the case of GW 170817, the relatively large viewing angle of $\theta \sim 15^\circ \div 25^\circ$ played a key role in suppressing the VHE emission component. Viewing angles closer to an on-axis geometry may certainly increase the flux expected at VHE although anticipating the peak time of the emission. Furthermore, the circumburst conditions may also have a significant impact on the expected spectrum. The low interstellar medium density for GW 170817 (10^{-4} cm^{-3}) stands as a disadvantage for a detectable VHE signal. In [171], an example of light curves at 1 TeV for a jet with the same parameters as that of GW 170817, but with a denser circumburst medium (5×10^{-2} cm^{-3}) (Figure 14 left plot).

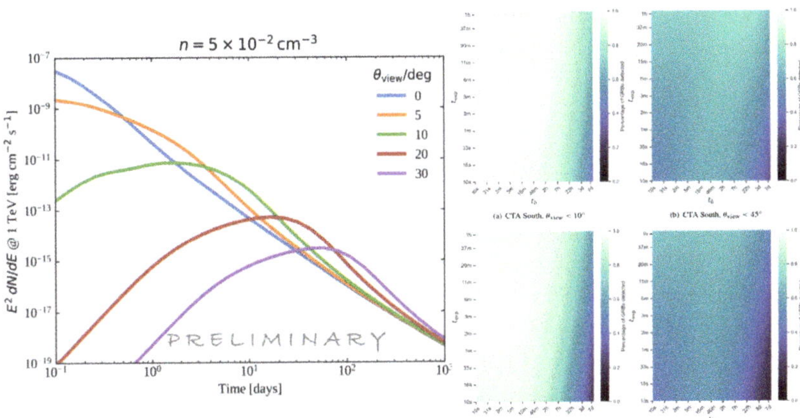

Figure 14. (**Left**) 1 TeV light curve expected for a GW 170817-like event under more favorable conditions. The circumburst medium density is fixed to (5×10^{-2} cm^{-3}), while different viewing angles are plotted. Reprinted with permission from [171]. (**Right**) Detectability of sGRBs with CTA-South (upper panels) and CTA-North (lower panels) array for the simulated events given latency and exposure time. The left panels show a subset of the sources with viewing angle <10°, while the right panels show all sources with view <45°. Reprinted with permission from [211].

It is important to remark that one of the keys for a successful GW follow-up lies in the synergies with other facilities and in the optimization of the observing strategy. As a matter of fact, GRB 180817A was a sub-threshold event, several orders of magnitudes less luminous than a standard GRB, although located much closer to us compared to the average GRB population. Hence, this event may not have been followed up if no gravitational wave was detected. The extensive multi-waveband follow-up has proved to be the key to identify the counterpart and its nature, representing a takeaway message for future observations. Within this framework, a large effort is currently taking place within the CTA consortium to optimize VHE follow-up strategy for near future observations (LVK run O5, planned for 2027),[8] where the parallel operation of more GW interferometers will allow for the localization of new merger events with much better precision.

In [211], a preliminary estimation of CTA detection capabilities on GWs/GRBs is reported. A set of simulated BNS mergers and their associated GW signals [212] are used taking into account realistic astrophysical distributions of masses, spins, distances, and sky locations of the neutron stars. Each merger is associated with a simulated sGRB. The authors use an empirical approach that does not need to assume any specific particle population or radiative process for the production of gamma-rays according to the empirical evidence collected by IACT observations of GRBs. Then, the luminosity in the TeV range is assumed to be comparable to the one at lower energies (in the soft X-ray range), and the spectra are assumed to have a photon index around −2.2. The synthetic spectra are then analyzed by means of CTA analysis tools and Instrument Response Functions (IRFs). The estimation of the integration time required to achieve a detection with CTA is reported in Figure 14 (right panel) as a function of the time needed by the telescopes to point at the region of interest.

It is clear that the CTA will represent a unique instrument to achieve a VHE detection of a GW counterpart, shedding light into the physics of GRB and BNS mergers dynamics and setting a key step for the future multi-messenger astronomy.

10. Conclusions

The past decade marked the beginning of the era of multi-messenger observations accompanied, in parallel, by the remarkable development of time-domain astronomy. In the GeV-TeV energy range, in particular, new sources of VHE gamma rays have been identified, opening new perspectives for transient astrophysics in this energy regime.

One of the newly identified class of Galactic VHE emitters are novae thanks to the recent discovery of VHE signal of hadronic origin in the recurrent symbiotic nova RS Oph. These novae create bubbles of enhanced cosmic-ray density in their close environment at ∼pc scales. Other (recurrent) symbiotic systems such as T CrB are expected to be detected by the current generation of IACTs in the very near future. The discussion on whether classical novae are TeV emitters is still open and will hopefully be addressed over the next years.

At extragalactic distances, the detection of the TeV counterpart of GRBs was finally achieved by current IACTs after a quest which lasted for more than 20 years. TeV GRBs were first detected in 2018–2019 and since then a total of four long GRBs located at redshifts between 0.0785 and 1.1 have been reported by IACTs during the afterglow phase. Furthermore, the brightest GRB of all times, GRB 221009A, has been recently detected by LHAASO above the 10 TeV, opening new possibilities for GRB study with instruments not originally thought for GRB follow-up due to their relatively large energy threshold such as particle array detectors or the ASTRI Mini-Array. The hint of detection at VHE from the short-GRB 160821 also proved the possible link between VHE and the GW emission from BNS mergers. In the near future, thanks to the improved sensitivity in the GW interferometers and the new-generation IACTs, follow-up observations of GW + VHE will connect the gamma-ray emission with the formation and evolution of the GW-central engine, shedding light into the physics of these extreme cosmic events.

New intriguing transients such as FRBs and their connection with magnetars represent a very recent development and a still marginally explored field for VHE transient astrophysics. Although magnetars *per se* are not found to be steady gamma-ray emitters, they have been detected in the GeV range by *Fermi*-LAT during giant flare episodes. Furthermore, magnetar-based models predict emission up to the VHE correlated in time with FRBs. The higher sensitivity to short IACTs compared to space-based instruments represents a unique feature for exploring the wide and complex range of transient phenomenology embedded in the magnetar–FRB scenario.

The current generation of IACTs is still on the catch of other transient events that are known HE emitters such as the enhanced flaring emission from PWNe, notably the Crab Nebula or flaring (massive) microquasars as, e.g., Cygnus X-1 or Cygnus X-3. Other transient phenomena are still elusive both in the HE and VHE regimes, such as core-collapse SNe (despite some candidate associations at HE), the VHE counterpart of kilonovae from GWs, or TDEs. The identification of any of these sources as TeV emitters will undoubtedly push the boundaries of our knowledge and open new research areas. In this regard, future instrumentation such as the CTA observatory, with enlarged energy range, improved (short timescale) sensitivity, and fast response capabilities will set new frontiers in time-domain TeV astrophysics.

Author Contributions: The authors contributed equally to this work. All authors have read and agreed to the published version of the manuscript.

Funding: This research is part of the project RYC2021-032991-I, funded by MICIN/AEI/10.13039/501100011033, and the European Union "NextGenerationEU"/PRTR.

Data Availability Statement: Not applicable.

Conflicts of Interest: The authors declare no conflicts of interest.

Abbreviations

The following abbreviations are used in this manuscript:

BH	Black hole
CCSNe	Core-collapse supernovae
FRB	Fast radio burst
GRB	Gamma-ray burst
GW	Gravitational wave
HE	High energy
IACT	Imaging Air Cherenkov Telescopes
IC	Inverse Compton
NS	Neutron star
RG	Red giant
SSC	Synchrotron self Compton
SED	Spectral energy distribution
sGRB	Short gamma-ray burst
SNe	Supernovae
TDE	Tidal disruption event
UL	Upper limit
VHE	Very-high-energy
WD	White dwarf

Notes

1. See https://asd.gsfc.nasa.gov/Koji.Mukai/novae/latnovae.html (accessed on 14 March 2024) for the list with LAT-detected novae and sub-significance hints.
2. https://fast.bao.ac.cn/ (accessed on 14 March 2024).
3. https://www.ligo.org/ (accessed on 14 March 2024).
4. https://www.virgo-gw.eu/ (accessed on 14 March 2024).
5. More GRBs have been detected at VHE since then such as GRB 201216C by MAGIC and the remarkable GRB 221009A, although not detected by IACTs, observed by LHAASO up to 13 TeV.
6. https://www.ztf.caltech.edu/ (accessed on 14 March 2024).
7. https://www.cta-observatory.org/science/ctao-performance/ (accessed on 14 March 2024).
8. https://observing.docs.ligo.org/plan/ (accessed on 14 March 2024).

References

1. Fioretti, V.; Ribeiro, D.; Humensky, T.B.; Bulgarelli, A.; Maier, G.; Moralejo, A.; Nigro, C. The Cherenkov Telescope Array sensitivity to the transient sky. In Proceedings of the 36th International Cosmic Ray Conference (ICRC2019), Madison, WI, USA, 24 July–1 August 2019; Volume 36, p. 673. [CrossRef]
2. Aleksić, J.; Ansoldi, S.; Antonelli, L.A.; Antoranz, P.; Babic, A.; Bangale, P.; Barceló, M.; Barrio, J.A.; Becerra González, J.; Bednarek, W.; et al. The major upgrade of the MAGIC telescopes, Part I: The hardware improvements and the commissioning of the system. *Astropart. Phys.* **2016**, *72*, 61–75. [CrossRef]
3. Cherenkov Telescope Array Consortium; Acharya, B.S.; Agudo, I.; Al Samarai, I.; Alfaro, R.; Alfaro, J.; Alispach, C.; Alves Batista, R.; Amans, J.P.; Amato, E.; et al. *Science with the Cherenkov Telescope Array*; World Scientific: Singapore, 2019. [CrossRef]
4. Vigliano, A.; Longo, F. Gamma-Ray Bursts: 50 Years and Counting! *Universe* **2024**, *10*, 57. [CrossRef]
5. Chomiuk, L.; Metzger, B.D.; Shen, K.J. New Insights into Classical Novae. *Annu. Rev. Astron. Astrophys.* **2021**, *59*, 391–444. [CrossRef]
6. Hernanz, M.; Tatischeff, V. High Energy Emission of Symbiotic Recurrent Novae: RS Oph and V407 Cyg. *Balt. Astron.* **2012**, *21*, 62–67. [CrossRef]
7. Mikolajewska, J. The Place of Recurrent Novae Among the Symbiotic Stars. In *Astronomical Society of the Pacific Conference Series, Proceedings of the RS Ophiuchi (2006) and the Recurrent Nova Phenomenon, Keele, UK, 12–14 June 2007*; Evans, A., Bode, M.F., O'Brien, T.J., Darnley, M.J., Eds.; Astronomical Society of the Pacific: San Francisco, CA, USA, 2008; Volume 401, p. 42. [CrossRef]

8. Schaefer, B.E. Comprehensive Photometric Histories of All Known Galactic Recurrent Novae. *Astrophys. J. Suppl. Ser.* **2010**, *187*, 275–373. [CrossRef]
9. Tatischeff, V.; Hernanz, M. Evidence for Nonlinear Diffusive Shock Acceleration of Cosmic Rays in the 2006 Outburst of the Recurrent Nova RS Ophiuchi. *Astrophys. J.* **2007**, *663*, L101–L104. [CrossRef]
10. Abdo, A.A.; Ackermann, M.; Ajello, M.; Atwood, W.B.; Baldini, L.; Ballet, J.; Barbiellini, G.; Bastieri, D.; Bechtol, K.; Bellazzini, R.; et al. Gamma-Ray Emission Concurrent with the Nova in the Symbiotic Binary V407 Cygni. *Science* **2010**, *329*, 817–821. [CrossRef]
11. Ackermann, M.; Ajello, M.; Albert, A.; Baldini, L.; Ballet, J.; Barbiellini, G.; Bastieri, D.; Bellazzini, R.; Bissaldi, E.; Blandford, R.D.; et al. Fermi establishes classical novae as a distinct class of gamma-ray sources. *Science* **2014**, *345*, 554–558. [CrossRef]
12. Cheung, C.C.; Jean, P.; Shore, S.N.; Stawarz, Ł.; Corbet, R.H.D.; Knödlseder, J.; Starrfield, S.; Wood, D.L.; Desiante, R.; Longo, F.; et al. Fermi-LAT Gamma-Ray Detections of Classical Novae V1369 Centauri 2013 and V5668 Sagittarii 2015. *Astrophys. J.* **2016**, *826*, 142. [CrossRef]
13. Sitarek, J.; Bednarek, W. GeV-TeV gamma rays and neutrinos from the Nova V407 Cygni. *Phys. Rev. D* **2012**, *86*, 063011. [CrossRef]
14. Aliu, E.; Archambault, S.; Arlen, T.; Aune, T.; Beilicke, M.; Benbow, W.; Bouvier, A.; Bradbury, S.M.; Buckley, J.H.; Bugaev, V.; et al. VERITAS Observations of the Nova in V407 Cygni. *Astrophys. J.* **2012**, *754*, 77. [CrossRef]
15. Ahnen, M.L.; Ansoldi, S.; Antonelli, L.A.; Antoranz, P.; Babic, A.; Banerjee, B.; Bangale, P.; Barres de Almeida, U.; Barrio, J.A.; Becerra González, J.; et al. Very high-energy γ-ray observations of novae and dwarf novae with the MAGIC telescopes. *Astron. Astrophys.* **2015**, *582*, A67. [CrossRef]
16. Mikołajewska, J.; Shara, M.M. The Massive CO White Dwarf in the Symbiotic Recurrent Nova RS Ophiuchi. *Astrophys. J.* **2017**, *847*, 99. [CrossRef]
17. Anupama, G.C.; Mikołajewska, J. Recurrent novae at quiescence: Systems with giant secondaries. *Astron. Astrophys.* **1999**, *344*, 177–187.
18. Brandi, E.; Quiroga, C.; Mikołajewska, J.; Ferrer, O.E.; García, L.G. Spectroscopic orbits and variations of RS Ophiuchi. *Astron. Astrophys.* **2009**, *497*, 815–825. [CrossRef]
19. Hachisu, I.; Kato, M. A Theoretical Light-Curve Model for the 1985 Outburst of RS Ophiuchi. *Astrophys. J.* **2000**, *536*, L93–L96. [CrossRef]
20. Hernanz, M.; José, J. The recurrent nova RS Oph: A possible scenario for type Ia supernovae. *New Astron. Rev.* **2008**, *52*, 386–389. [CrossRef]
21. Patat, F.; Chugai, N.N.; Podsiadlowski, P.; Mason, E.; Melo, C.; Pasquini, L. Connecting RS Ophiuchi to [some] type Ia supernovae. *Astron. Astrophys.* **2011**, *530*, A63. [CrossRef]
22. Geary, K. Outburst of RS Ophiuchi. *Vsnet-Alert* **2021**, *26131*. http://ooruri.kusastro.kyoto-u.ac.jp/mailarchive/vsnet-alert/26131
23. Cheung, C.C.; Ciprini, S.; Johnson, T.J. Fermi-LAT Gamma-ray Detection of the Recurrent Nova RS Oph. *Astron. Telegr.* **2021**, *14834*, 1.
24. H. E. S. S. Collaboration; Aharonian, F.; Ait Benkhali, F.; Angüner, E.O.; Ashkar, H.; Backes, M.; Baghmanyan, V.; Martins, V.B.; Batzofin, R.; Becherini, Y.; et al. Time-resolved hadronic particle acceleration in the recurrent nova RS Ophiuchi. *Science* **2022**, *376*, 77–80. [CrossRef]
25. Acciari, V.A.; Ansoldi, S.; Antonelli, L.A.; Arbet Engels, A.; Artero, M.; Asano, K.; Baack, D.; Babić, A.; Baquero, A.; Barres de Almeida, U.; et al. Proton acceleration in thermonuclear nova explosions revealed by gamma rays. *Nat. Astron.* **2022**, *6*, 689–697. [CrossRef]
26. Abe, H.; Abe, K.; Abe, S.; Aguasca-Cabot, A.; Agudo, I.; Alvarez Crespo, N.; Antonelli, L.A.; Aramo, C.; Arbet-Engels, A.; Artero, M.; et al. RS Ophiuchi nova outburst detection by the LST-1. *PoS* **2023**, *Gamma2022*, 055. [CrossRef]
27. Kobayashi, Y.; Aguasca-Cabot, A.; Bernardos Martín, M.I.; Green, D.; López-Coto, R. Detection of the 2021 Outburst of RS Ophiuchi with the LST-1. *arXiv* **2023**, arXiv:2310.09683. https://doi.org/10.48550/arXiv.2310.09683.
28. De Sarkar, A.; Nayana, A.J.; Roy, N.; Razzaque, S.; Anupama, G.C. Lepto-hadronic Interpretation of 2021 RS Ophiuchi Nova Outburst. *Astrophys. J.* **2023**, *951*, 62. [CrossRef]
29. Chomiuk, L.; Linford, J.D.; Yang, J.; O'Brien, T.J.; Paragi, Z.; Mioduszewski, A.J.; Beswick, R.J.; Cheung, C.C.; Mukai, K.; Nelson, T.; et al. Binary orbits as the driver of γ-ray emission and mass ejection in classical novae. *Nature* **2014**, *514*, 339–342. [CrossRef]
30. Chesneau, O.; Nardetto, N.; Millour, F.; Hummel, C.; Domiciano de Souza, A.; Bonneau, D.; Vannier, M.; Rantakyrö, F.; Spang, A.; Malbet, F.; et al. AMBER/VLTI interferometric observations of the recurrent Nova RS Ophiuchii 5.5 days after outburst. *Astron. Astrophys.* **2007**, *464*, 119–126. [CrossRef]
31. Bode, M.F.; Harman, D.J.; O'Brien, T.J.; Bond, H.E.; Starrfield, S.; Darnley, M.J.; Evans, A.; Eyres, S.P.S. Hubble Space Telescope Imaging of the Expanding Nebular Remnant of the 2006 Outburst of the Recurrent Nova RS Ophiuchi. *Astrophys. J.* **2007**, *665*, L63–L66. [CrossRef]
32. Ribeiro, V.A.R.M.; Bode, M.F.; Darnley, M.J.; Harman, D.J.; Newsam, A.M.; O'Brien, T.J.; Bohigas, J.; Echevarría, J.M.; Bond, H.E.; Chavushyan, V.H.; et al. The Expanding Nebular Remnant of the Recurrent Nova RS Ophiuchi (2006). II. Modeling of Combined Hubble Space Telescope Imaging and Ground-based Spectroscopy. *Astrophys. J.* **2009**, *703*, 1955–1963. [CrossRef]
33. Munari, U.; Giroletti, M.; Marcote, B.; O'Brien, T.J.; Veres, P.; Yang, J.; Williams, D.R.A.; Woudt, P. Radio interferometric imaging of RS Oph bipolar ejecta for the 2021 nova outburst. *Astron. Astrophys.* **2022**, *666*, L6. [CrossRef]

34. Nikolov, Y.; Luna, G.J.M.; Stoyanov, K.A.; Borisov, G.; Mukai, K.; Sokoloski, J.L.; Avramova-Boncheva, A. Transient and asymmetric dust structures in the TeV-bright nova RS Oph revealed by spectropolarimetry. *arXiv* **2023**, arXiv:2309.11288. https://doi.org/10.48550/arXiv.2309.11288.
35. Montez, R.; Luna, G.J.M.; Mukai, K.; Sokoloski, J.L.; Kastner, J.H. Expanding Bipolar X-Ray Structure After the 2006 Eruption of RS Oph. *Astrophys. J.* **2022**, *926*, 100. [CrossRef]
36. Diesing, R.; Metzger, B.D.; Aydi, E.; Chomiuk, L.; Vurm, I.; Gupta, S.; Caprioli, D. Evidence for Multiple Shocks from the γ-Ray Emission of RS Ophiuchi. *Astrophys. J.* **2023**, *947*, 70. [CrossRef]
37. Paredes, J.M.; Bordas, P. Phenomenology of gamma-ray emitting binaries. *Rend. Lincei Sci. Fis. Nat.* **2019**, *30*, 107–113. [CrossRef]
38. Ahnen, M.L.; Ansoldi, S.; Antonelli, L.A.; Antoranz, P.; Babic, A.; Banerjee, B.; Bangale, P.; Barres de Almeida, U.; Barrio, J.A.; Becerra González, J.; et al. Super-orbital variability of LS I +61°303 at TeV energies. *Astron. Astrophys.* **2016**, *591*, A76. [CrossRef]
39. Archambault, S.; Archer, A.; Aune, T.; Barnacka, A.; Benbow, W.; Bird, R.; Buchovecky, M.; Buckley, J.H.; Bugaev, V.; Byrum, K.; et al. Exceptionally Bright TeV Flares from the Binary LS I +61 303. *Astrophys. J.* **2016**, *817*, L7. [CrossRef]
40. Adams, C.B.; Benbow, W.; Brill, A.; Buckley, J.H.; Capasso, M.; Chromey, A.J.; Errando, M.; Falcone, A.; A Farrell, K.; Feng, Q.; et al. Observation of the Gamma-Ray Binary HESS J0632+057 with the H.E.S.S., MAGIC, and VERITAS Telescopes. *Astrophys. J.* **2021**, *923*, 241. [CrossRef]
41. Abeysekara, A.U.; Benbow, W.; Bird, R.; Brill, A.; Brose, R.; Buckley, J.H.; Chromey, A.J.; Daniel, M.K.; Falcone, A.; Finley, J.P.; et al. Periastron Observations of TeV Gamma-Ray Emission from a Binary System with a 50-year Period. *Astrophys. J.* **2018**, *867*, L19. [CrossRef]
42. H. E. S. S. Collaboration; Abdalla, H.; Adam, R.; Aharonian, F.; Ait Benkhali, F.; Angüner, E.O.; Arakawa, M.; Arcaro, C.; Armand, C.; Ashkar, H.; et al. H.E.S.S. and Fermi-LAT observations of PSR B1259-63/LS 2883 during its 2014 and 2017 periastron passages. *Astron. Astrophys.* **2020**, *633*, A102. [CrossRef]
43. Bulgarelli, A.; Pittori, C.; Lucarelli, F.; Striani, E.; Gianotti, F.; Trifoglio, M.; Sabatini, S.; Tavani, M.; Verrecchia, F.; Trois, A.; et al. AGILE detection of a gamma ray flare from the Cygnus X-1 region. *Astron. Telegr.* **2010**, *2512*, 1.
44. Sabatini, S.; Tavani, M.; Striani, E.; Bulgarelli, A.; Vittorini, V.; Piano, G.; Del Monte, E.; Feroci, M.; de Pasquale, F.; Trifoglio, M.; et al. Episodic Transient Gamma-ray Emission from the Microquasar Cygnus X-1. *Astrophys. J.* **2010**, *712*, L10–L15. [CrossRef]
45. Sabatini, S.; Tavani, M.; Coppi, P.; Pooley, G.; Del Santo, M.; Campana, R.; Chen, A.; Evangelista, Y.; Piano, G.; Bulgarelli, A.; et al. Gamma-Ray Observations of Cygnus X-1 above 100 MeV in the Hard and Soft States. *Astrophys. J.* **2013**, *766*, 83. [CrossRef]
46. Tavani, M.; Bulgarelli, A.; Piano, G.; Sabatini, S.; Striani, E.; Evangelista, Y.; Trois, A.; Pooley, G.; Trushkin, S.; Nizhelskij, N.A.; et al. Extreme particle acceleration in the microquasar CygnusX-3. *Nature* **2009**, *462*, 620–623. [CrossRef]
47. Fermi LAT Collaboration; Abdo, A.A.; Ackermann, M.; Ajello, M.; Axelsson, M.; Baldini, L.; Ballet, J.; Barbiellini, G.; Bastieri, D.; Baughman, B.M.; et al. Modulated High-Energy Gamma-Ray Emission from the Microquasar Cygnus X-3. *Science* **2009**, *326*, 1512. [CrossRef] [PubMed]
48. Loh, A.; Corbel, S.; Dubus, G.; Rodriguez, J.; Grenier, I.; Hovatta, T.; Pearson, T.; Readhead, A.; Fender, R.; Mooley, K. High-energy gamma-ray observations of the accreting black hole V404 Cygni during its 2015 June outburst. *Mon. Not. R. Astron. Soc. Lett.* **2016**, *462*, L111–L115. [CrossRef]
49. Piano, G.; Munar-Adrover, P.; Verrecchia, F.; Tavani, M.; Trushkin, S.A. High-energy Gamma-Ray Activity from V404 Cygni Detected by AGILE during the 2015 June Outburst. *Astrophys. J.* **2017**, *839*, 84. [CrossRef]
50. Aleksić, J.; Alvarez, E.A.; Antonelli, L.A.; Antoranz, P.; Asensio, M.; Backes, M.; Barrio, J.A.; Bastieri, D.; Becerra González, J.; Bednarek, W.; et al. A Search for Very High Energy Gamma-Ray Emission from Scorpius X-1 with the Magic Telescopes. *Astrophys. J.* **2011**, *735*, L5. [CrossRef]
51. Aleksić, J.; Ansoldi, S.; Antonelli, L.A.; Antoranz, P.; Babic, A.; Bangale, P.; Barrio, J.A.; Becerra González, J.; Bednarek, W.; Bernardini, E.; et al. MAGIC observations of MWC 656, the only known Be/BH system. *Astron. Astrophys.* **2015**, *576*, A36. [CrossRef]
52. Ahnen, M.L.; Ansoldi, S.; Antonelli, L.A.; Arcaro, C.; Babić, A.; Banerjee, B.; Bangale, P.; Barres de Almeida, U.; Barrio, J.A.; Becerra González, J.; et al. MAGIC observations of the microquasar V404 Cygni during the 2015 outburst. *Mon. Not. R. Astron. Soc. Lett.* **2017**, *471*, 1688–1693. [CrossRef]
53. H. E. S. S. Collaboration; Abdalla, H.; Abramowski, A.; Aharonian, F.; Ait Benkhali, F.; Akhperjanian, A.G.; Angüner, E.O.; Arrieta, M.; Aubert, P.; Backes, M.; et al. A search for very high-energy flares from the microquasars GRS 1915+105, Circinus X-1, and V4641 Sgr using contemporaneous H.E.S.S. and RXTE observations. *Astron. Astrophys.* **2018**, *612*, A10. [CrossRef]
54. Abe, H.; Abe, S.; Acciari, V.A.; Aniello, T.; Ansoldi, S.; Antonelli, L.A.; Arbet Engels, A.; Arcaro, C.; Artero, M.; Asano, K.; et al. Gamma-ray observations of MAXI J1820+070 during the 2018 outburst. *Mon. Not. R. Astron. Soc. Lett.* **2022**, *517*, 4736–4751. [CrossRef]
55. Albert, J.; Aliu, E.; Anderhub, H.; Antoranz, P.; Armada, A.; Baixeras, C.; Barrio, J.A.; Bartko, H.; Bastieri, D.; Becker, J.K.; et al. Very High Energy Gamma-Ray Radiation from the Stellar Mass Black Hole Binary Cygnus X-1. *Astrophys. J.* **2007**, *665*, L51–L54. [CrossRef]
56. Abeysekara, A.U.; Albert, A.; Alfaro, R.; Alvarez, C.; Álvarez, J.D.; Arceo, R.; Arteaga-Velázquez, J.C.; Avila Rojas, D.; Ayala Solares, H.A.; Belmont-Moreno, E.; et al. Very-high-energy particle acceleration powered by the jets of the microquasar SS 433. *Nature* **2018**, *562*, 82–85. [CrossRef]
57. Tibolla, O. Recent results from the HAWC experiment. *J. Phys. Conf. Ser.* **2023**, *2429*, 012017. [CrossRef]

58. H. E. S. S. Collaboration; Olivera-Nieto, L.; Reville, B.; Hinton, J.; Tsirou, M. Acceleration and transport of relativistic electrons in the jets of the microquasar SS 433. *Science* **2024**, *383*, 402–406. [CrossRef]
59. Filippenko, A.V. Optical Spectra of Supernovae. *Annu. Rev. Astron. Astrophys.* **1997**, *35*, 309–355. [CrossRef]
60. Georgy, C.; Meynet, G.; Walder, R.; Folini, D.; Maeder, A. The different progenitors of type Ib, Ic SNe, and of GRB. *Astron. Astrophys.* **2009**, *502*, 611–622. [CrossRef]
61. Cao, Z.; Aharonian, F.A.; An, Q.; Axikegu, Bai, L.X.; Bai, Y.X.; Bao, Y.W.; Bastieri, D.; Bi, X.J.; Bi, Y.J.; et al. Ultrahigh-energy photons up to 1.4 petaelectronvolts from 12 γ-ray Galactic sources. *Nature* **2021**, *594*, 33–36. [CrossRef]
62. Cao, Z.; Aharonian, F.; An, Q.; Axikegu; Bai, Y.X.; Bao, Y.W.; Bastieri, D.; Bi, X.J.; Bi, Y.J.; Cai, J.T.; et al. The First LHAASO Catalog of Gamma-Ray Sources. *arXiv* **2023**, arXiv:2305.17030. https://doi.org/10.48550/arXiv.2305.17030.
63. Smith, N.; Li, W.; Filippenko, A.V.; Chornock, R. Observed fractions of core-collapse supernova types and initial masses of their single and binary progenitor stars. *Mon. Not. R. Astron. Soc. Lett.* **2011**, *412*, 1522–1538. [CrossRef]
64. Murase, K.; Thompson, T.A.; Lacki, B.C.; Beacom, J.F. New class of high-energy transients from crashes of supernova ejecta with massive circumstellar material shells. *Phys. Rev. D* **2011**, *84*, 043003. [CrossRef]
65. Cristofari, P.; Marcowith, A.; Renaud, M.; Dwarkadas, V.V.; Tatischeff, V.; Giacinti, G.; Peretti, E.; Sol, H. The first days of Type II-P core collapse supernovae in the gamma-ray range. *Mon. Not. R. Astron. Soc. Lett.* **2022**, *511*, 3321–3329. [CrossRef]
66. Brose, R.; Sushch, I.; Mackey, J. Core-collapse supernovae in dense environments—Particle acceleration and non-thermal emission. *Mon. Not. R. Astron. Soc. Lett.* **2022**, *516*, 492–505. [CrossRef]
67. Ahnen, M.L.; Ansoldi, S.; Antonelli, L.A.; Antoranz, P.; Arcaro, C.; Babic, A.; Banerjee, B.; Bangale, P.; Barres de Almeida, U.; Barrio, J.A.; et al. Very-high-energy gamma-ray observations of the Type Ia Supernova SN 2014J with the MAGIC telescopes. *Astron. Astrophys.* **2017**, *602*, A98. [CrossRef]
68. Dwarkadas, V.V. Exploring the γ-ray emissivity of young supernova remnants—I. Hadronic emission. *Mon. Not. R. Astron. Soc. Lett.* **2013**, *434*, 3368–3377. [CrossRef]
69. Yuan, Q.; Liao, N.H.; Xin, Y.L.; Li, Y.; Fan, Y.Z.; Zhang, B.; Hu, H.B.; Bi, X.J. Fermi Large Area Telescope Detection of Gamma-Ray Emission from the Direction of Supernova iPTF14hls. *Astrophys. J.* **2018**, *854*, L18. [CrossRef]
70. Xi, S.Q.; Liu, R.Y.; Wang, X.Y.; Yang, R.Z.; Yuan, Q.; Zhang, B. A Serendipitous Discovery of GeV Gamma-Ray Emission from Supernova 2004dj in a Survey of Nearby Star-forming Galaxies with Fermi-LAT. *Astrophys. J.* **2020**, *896*, L33. [CrossRef]
71. Prokhorov, D.A.; Moraghan, A.; Vink, J. Search for gamma rays from SNe with a variable-size sliding-time-window analysis of the Fermi-LAT data. *Mon. Not. R. Astron. Soc. Lett.* **2021**, *505*, 1413–1421. [CrossRef]
72. H. E. S. S. Collaboration; Abdalla, H.; Aharonian, F.; Ait Benkhali, F.; Angüner, E.O.; Arakawa, M.; Arcaro, C.; Armand, C.; Ashkar, H.; Backes, M.; et al. Upper limits on very-high-energy gamma-ray emission from core-collapse supernovae observed with H.E.S.S. *Astron. Astrophys.* **2019**, *626*, A57. [CrossRef]
73. Acharyya, A.; Adams, C.B.; Bangale, P.; Benbow, W.; Buckley, J.H.; Capasso, M.; Dwarkadas, V.V.; Errando, M.; Falcone, A.; Feng, Q.; et al. VERITAS and Fermi-LAT Constraints on the Gamma-Ray Emission from Superluminous Supernovae SN2015bn and SN2017egm. *Astrophys. J.* **2023**, *945*, 30. [CrossRef]
74. Hirata, K.; Kajita, T.; Koshiba, M.; Nakahata, M.; Oyama, Y.; Sato, N.; Suzuki, A.; Takita, M.; Totsuka, Y.; Kifune, T.; et al. Observation of a neutrino burst from the supernova SN1987A. *Phys. Rev. Lett.* **1987**, *58*, 1490–1493. [CrossRef]
75. Bionta, R.M.; Blewitt, G.; Bratton, C.B.; Casper, D.; Ciocio, A.; Claus, R.; Cortez, B.; Crouch, M.; Dye, S.T.; Errede, S.; et al. Observation of a neutrino burst in coincidence with supernova 1987A in the Large Magellanic Cloud. *Phys. Rev. Lett.* **1987**, *58*, 1494–1496. [CrossRef]
76. Cigan, P.; Matsuura, M.; Gomez, H.L.; Indebetouw, R.; Abellán, F.; Gabler, M.; Richards, A.; Alp, D.; Davis, T.A.; Janka, H.T.; et al. High Angular Resolution ALMA Images of Dust and Molecules in the SN 1987A Ejecta. *Astrophys. J.* **2019**, *886*, 51. [CrossRef]
77. Greco, E.; Miceli, M.; Orlando, S.; Olmi, B.; Bocchino, F.; Nagataki, S.; Ono, M.; Dohi, A.; Peres, G. Indication of a Pulsar Wind Nebula in the Hard X-Ray Emission from SN 1987A. *Astrophys. J.* **2021**, *908*, L45. [CrossRef]
78. McCray, R.; Fransson, C. The Remnant of Supernova 1987A. *Annu. Rev. Astron. Astrophys.* **2016**, *54*, 19–52. [CrossRef]
79. H. E. S. S. Collaboration; Abramowski, A.; Aharonian, F.; Ait Benkhali, F.; Akhperjanian, A.G.; Angüner, E.O.; Backes, M.; Balenderan, S.; Balzer, A.; Barnacka, A.; et al. The exceptionally powerful TeV γ-ray emitters in the Large Magellanic Cloud. *Science* **2015**, *347*, 406–412. [CrossRef]
80. Marti-Devesa, G. Fermi-LAT gamma-ray observations of SN 2023ixf. *Astron. Telegr.* **2023**, *16075*, 1.
81. Gaensler, B.M.; Slane, P.O. The Evolution and Structure of Pulsar Wind Nebulae. *Annu. Rev. Astron. Astrophys.* **2006**, *44*, 17–47. [CrossRef]
82. Giacinti, G.; Mitchell, A.M.W.; López-Coto, R.; Joshi, V.; Parsons, R.D.; Hinton, J.A. Halo fraction in TeV-bright pulsar wind nebulae. *Astron. Astrophys.* **2020**, *636*, A113. [CrossRef]
83. Olmi, B. Evolved Pulsar Wind Nebulae. *Universe* **2023**, *9*, 402. [CrossRef]
84. Abeysekara, A.U.; Albert, A.; Alfaro, R.; Alvarez, C.; Álvarez, J.D.; Arceo, R.; Arteaga-Velázquez, J.C.; Avila Rojas, D.; Ayala Solares, H.A.; Barber, A.S.; et al. Extended gamma-ray sources around pulsars constrain the origin of the positron flux at Earth. *Science* **2017**, *358*, 911–914. [CrossRef]
85. Weekes, T.C.; Cawley, M.F.; Fegan, D.J.; Gibbs, K.G.; Hillas, A.M.; Kowk, P.W.; Lamb, R.C.; Lewis, D.A.; Macomb, D.; Porter, N.A.; et al. Observation of TeV Gamma Rays from the Crab Nebula Using the Atmospheric Cerenkov Imaging Technique. *Astrophys. J.* **1989**, *342*, 379. [CrossRef]

86. Aliu, E.; Anderhub, H.; Antonelli, L.A.; Antoranz, P.; Backes, M.; Baixeras, C.; Barrio, J.A.; Bartko, H.; Bastieri, D.; Becker, J.K.; et al. Observation of Pulsed γ-Rays Above 25 GeV from the Crab Pulsar with MAGIC. *Science* **2008**, *322*, 1221. [CrossRef]
87. Ansoldi, S.; Antonelli, L.A.; Antoranz, P.; Babic, A.; Bangale, P.; Barres de Almeida, U.; Barrio, J.A.; Becerra González, J.; Bednarek, W.; Bernardini, E.; et al. Teraelectronvolt pulsed emission from the Crab Pulsar detected by MAGIC. *Astron. Astrophys.* **2016**, *585*, A133. [CrossRef]
88. H. E. S. S. Collaboration. Resolving the Crab pulsar wind nebula at teraelectronvolt energies. *Nat. Astron.* **2020**, *4*, 167–173. [CrossRef]
89. Amenomori, M.; Bao, Y.W.; Bi, X.J.; Chen, D.; Chen, T.L.; Chen, W.Y.; Chen, X.; Chen, Y.; Cirennima.; Cui, S.W.; et al. First Detection of Photons with Energy beyond 100 TeV from an Astrophysical Source. *Phys. Rev. Lett.* **2019**, *123*, 051101. [CrossRef]
90. Abeysekara, A.U.; Albert, A.; Alfaro, R.; Alvarez, C.; Álvarez, J.D.; Camacho, J.R.A.; Arceo, R.; Arteaga-Velázquez, J.C.; Arunbabu, K.P.; Avila Rojas, D.; et al. Measurement of the Crab Nebula Spectrum Past 100 TeV with HAWC. *Astrophys. J.* **2019**, *881*, 134. [CrossRef]
91. MAGIC Collaboration; Acciari, V.A.; Ansoldi, S.; Antonelli, L.A.; Arbet Engels, A.; Baack, D.; Babić, A.; Banerjee, B.; Barres de Almeida, U.; Barrio, J.A.; et al. MAGIC very large zenith angle observations of the Crab Nebula up to 100 TeV. *Astron. Astrophys.* **2020**, *635*, A158. [CrossRef]
92. Tavani, M.; Bulgarelli, A.; Vittorini, V.; Pellizzoni, A.; Striani, E.; Caraveo, P.; Weisskopf, M.C.; Tennant, A.; Pucella, G.; Trois, A.; et al. Discovery of Powerful Gamma-Ray Flares from the Crab Nebula. *Science* **2011**, *331*, 736. [CrossRef]
93. Abdo, A.A.; Ackermann, M.; Ajello, M.; Allafort, A.; Baldini, L.; Ballet, J.; Barbiellini, G.; Bastieri, D.; Bechtol, K.; Bellazzini, R.; et al. Gamma-Ray Flares from the Crab Nebula. *Science* **2011**, *331*, 739. [CrossRef]
94. Bühler, R.; Blandford, R. The surprising Crab pulsar and its nebula: A review. *Rep. Prog. Phys.* **2014**, *77*, 066901. [CrossRef]
95. Mariotti, M. No significant enhancement in the VHE gamma-ray flux of the Crab Nebula measured by MAGIC in September 2010. *Astron. Telegr.* **2010**, *2967*, 1.
96. Ong, R.A. Search for an Enhanced TeV Gamma-Ray Flux from the Crab Nebula with VERITAS. *Astron. Telegr.* **2010**, *2968*, 1.
97. Ojha, R.; Hays, E.; Buehler, R.; Dutka, M. Fermi LAT detection of a new gamma-ray flare from the Crab Nebula region. *Astron. Telegr.* **2013**, *4855*, 1.
98. H. E. S. S. Collaboration; Abramowski, A.; Aharonian, F.; Ait Benkhali, F.; Akhperjanian, A.G.; Angüner, E.; Anton, G.; Balenderan, S.; Balzer, A.; Barnacka, A.; et al. H.E.S.S. observations of the Crab during its March 2013 GeV gamma-ray flare. *Astron. Astrophys.* **2014**, *562*, L4. [CrossRef]
99. Aliu, E.; Archambault, S.; Aune, T.; Benbow, W.; Berger, K.; Bird, R.; Bouvier, A.; Buckley, J.H.; Bugaev, V.; Byrum, K.; et al. A Search for Enhanced Very High Energy Gamma-Ray Emission from the 2013 March Crab Nebula Flare. *Astrophys. J.* **2014**, *781*, L11. [CrossRef]
100. van Scherpenberg, J.; Mirzoyan, R.; Vovk, I.; Peresano, M.; Zaric, D.; Temnikov, P.; Godinović, N.; Besenrieder, J. Searching for Variability of the Crab Nebula Flux at TeV Energies using MAGIC Very Large Zenith Angle Observations. In Proceedings of the 36th International Cosmic Ray Conference (ICRC2019), Madison, WI, USA, 24 July–1 August 2019; Volume 36, p. 812.
101. Lyutikov, M.; Komissarov, S.; Sironi, L.; Porth, O. Particle acceleration in explosive relativistic reconnection events and Crab Nebula gamma-ray flares. *J. Plasma Phys.* **2018**, *84*, 635840201. [CrossRef]
102. Kirk, J.G.; Giacinti, G. Inductive Spikes in the Crab Nebula: A Theory of γ -Ray Flares. *Phys. Rev. Lett.* **2017**, *119*, 211101. [CrossRef]
103. Nizamov, B.A.; Pshirkov, M.S. Gamma-ray flares from pulsar wind nebulae in the Large Magellanic Cloud. *Mon. Not. R. Astron. Soc. Lett.* **2023**, *520*, 4456–4462. [CrossRef]
104. Lorimer, D.R.; Bailes, M.; McLaughlin, M.A.; Narkevic, D.J.; Crawford, F. A Bright Millisecond Radio Burst of Extragalactic Origin. *Science* **2007**, *318*, 777. [CrossRef]
105. Cordes, J.M.; Chatterjee, S. Fast Radio Bursts: An Extragalactic Enigma. *Annu. Rev. Astron. Astrophys.* **2019**, *57*, 417–465. [CrossRef]
106. Petroff, E.; Hessels, J.W.T.; Lorimer, D.R. Fast radio bursts. *Astron. Astrophys. Rev.* **2019**, *27*, 4. [CrossRef]
107. Chime/Frb Collaboration; Andersen, B.C.; Bandura, K.; Bhardwaj, M.; Boyle, P.J.; Brar, C.; Cassanelli, T.; Chatterjee, S.; Chawla, P.; Cook, A.M.; et al. CHIME/FRB Discovery of 25 Repeating Fast Radio Burst Sources. *Astrophys. J.* **2023**, *947*, 83. [CrossRef]
108. Tendulkar, S.P.; Bassa, C.G.; Cordes, J.M.; Bower, G.C.; Law, C.J.; Chatterjee, S.; Adams, E.A.K.; Bogdanov, S.; Burke-Spolaor, S.; Butler, B.J.; et al. The Host Galaxy and Redshift of the Repeating Fast Radio Burst FRB 121102. *Astrophys. J.* **2017**, *834*, L7. [CrossRef]
109. Chatterjee, S.; Law, C.J.; Wharton, R.S.; Burke-Spolaor, S.; Hessels, J.W.T.; Bower, G.C.; Cordes, J.M.; Tendulkar, S.P.; Bassa, C.G.; Demorest, P.; et al. A direct localization of a fast radio burst and its host. *Nature* **2017**, *541*, 58–61. [CrossRef]
110. Bassa, C.G.; Tendulkar, S.P.; Adams, E.A.K.; Maddox, N.; Bogdanov, S.; Bower, G.C.; Burke-Spolaor, S.; Butler, B.J.; Chatterjee, S.; Cordes, J.M.; et al. FRB 121102 Is Coincident with a Star-forming Region in Its Host Galaxy. *Astrophys. J.* **2017**, *843*, L8. [CrossRef]
111. Chime/Frb Collaboration; Amiri, M.; Andersen, B.C.; Bandura, K.M.; Bhardwaj, M.; Boyle, P.J.; Brar, C.; Chawla, P.; Chen, T.; Cliche, J.F.; et al. Periodic activity from a fast radio burst source. *Nature* **2020**, *582*, 351–355. [CrossRef]
112. Rajwade, K.M.; Mickaliger, M.B.; Stappers, B.W.; Morello, V.; Agarwal, D.; Bassa, C.G.; Breton, R.P.; Caleb, M.; Karastergiou, A.; Keane, E.F.; et al. Possible periodic activity in the repeating FRB 121102. *Mon. Not. R. Astron. Soc. Lett.* **2020**, *495*, 3551–3558. [CrossRef]

113. Pleunis, Z.; Good, D.C.; Kaspi, V.M.; Mckinven, R.; Ransom, S.M.; Scholz, P.; Bandura, K.; Bhardwaj, M.; Boyle, P.J.; Brar, C.; et al. Fast Radio Burst Morphology in the First CHIME/FRB Catalog. *Astrophys. J.* **2021**, *923*, 1. [CrossRef]
114. CHIME/FRB Collaboration; Andersen, B.C.; Bandura, K.M.; Bhardwaj, M.; Bij, A.; Boyce, M.M.; Boyle, P.J.; Brar, C.; Cassanelli, T.; Chawla, P.; et al. A bright millisecond-duration radio burst from a Galactic magnetar. *Nature* **2020**, *587*, 54–58. [CrossRef]
115. Mereghetti, S.; Savchenko, V.; Ferrigno, C.; Götz, D.; Rigoselli, M.; Tiengo, A.; Bazzano, A.; Bozzo, E.; Coleiro, A.; Courvoisier, T.J.L.; et al. INTEGRAL Discovery of a Burst with Associated Radio Emission from the Magnetar SGR 1935+2154. *Astrophys. J.* **2020**, *898*, L29. [CrossRef]
116. Tavani, M.; Casentini, C.; Ursi, A.; Verrecchia, F.; Addis, A.; Antonelli, L.A.; Argan, A.; Barbiellini, G.; Baroncelli, L.; Bernardi, G.; et al. An X-ray burst from a magnetar enlightening the mechanism of fast radio bursts. *Nat. Astron.* **2021**, *5*, 401–407. [CrossRef]
117. Ridnaia, A.; Svinkin, D.; Frederiks, D.; Bykov, A.; Popov, S.; Aptekar, R.; Golenetskii, S.; Lysenko, A.; Tsvetkova, A.; Ulanov, M.; et al. A peculiar hard X-ray counterpart of a Galactic fast radio burst. *Nat. Astron.* **2021**, *5*, 372–377. [CrossRef]
118. Palmer, D.M. A Forest of Bursts from SGR 1935+2154. *Astron. Telegr.* **2020**, *13675*, 1.
119. Fletcher, C.; Fermi GBM Team. Fermi GBM observation of a bright flare from magnetar SGR 1935+2154. *GRB Coord. Netw.* **2020**, *27659*, 1.
120. Lin, L.; Zhang, C.F.; Wang, P.; Gao, H.; Guan, X.; Han, J.L.; Jiang, J.C.; Jiang, P.; Lee, K.J.; Li, D.; et al. No pulsed radio emission during a bursting phase of a Galactic magnetar. *Nature* **2020**, *587*, 63–65. [CrossRef]
121. Kirsten, F.; Marcote, B.; Nimmo, K.; Hessels, J.W.T.; Bhardwaj, M.; Tendulkar, S.P.; Keimpema, A.; Yang, J.; Snelders, M.P.; Scholz, P.; et al. A repeating fast radio burst source in a globular cluster. *Nature* **2022**, *602*, 585–589. [CrossRef] [PubMed]
122. Popov, S.B.; Postnov, K.A. Hyperflares of SGRs as an engine for millisecond extragalactic radio bursts. In *Evolution of Cosmic Objects through Their Physical Activity; Proceedings of the Conference Dedicated to Viktor Ambartsumian's 100th Anniversary, Yerevan and Byurakan, Armenia, 15–18 September 2008*; Harutyunian, H.A., Mickaelian, A.M., Terzian, Y., Eds.; Cornell University: Ithaca, NY, USA, 2010; pp. 129–132. [CrossRef]
123. Kulkarni, S.R.; Ofek, E.O.; Neill, J.D.; Zheng, Z.; Juric, M. Giant Sparks at Cosmological Distances? *Astrophys. J.* **2014**, *797*, 70. [CrossRef]
124. Katz, J.I. How Soft Gamma Repeaters Might Make Fast Radio Bursts. *Astrophys. J.* **2016**, *826*, 226. [CrossRef]
125. Zhang, B. The physical mechanisms of fast radio bursts. *Nature* **2020**, *587*, 45–53. [CrossRef] [PubMed]
126. Beloborodov, A.M. Blast Waves from Magnetar Flares and Fast Radio Bursts. *Astrophys. J.* **2020**, *896*, 142. [CrossRef]
127. Metzger, B.D.; Margalit, B.; Sironi, L. Fast radio bursts as synchrotron maser emission from decelerating relativistic blast waves. *Mon. Not. R. Astron. Soc. Lett.* **2019**, *485*, 4091–4106. [CrossRef]
128. Metzger, B.D.; Fang, K.; Margalit, B. Neutrino Counterparts of Fast Radio Bursts. *Astrophys. J.* **2020**, *902*, L22. [CrossRef]
129. Bissaldi, E.; Briggs, M.; Burns, E.; Roberts, O.J.; Veres, P.; Fermi GBM Team. GRB 200415A: Fermi GBM observation. *GRB Coord. Netw.* **2020**, *27587*, 1.
130. Omodei, N.; Axelsson, M.; Piron, F.; Longo, F.; Kocevski, D.; Bissaldi, E.; Berretta, A.; Fermi-LAT Collaboration. GRB 200415A: Fermi-LAT detection. *GRB Coord. Netw.* **2020**, *27586*, 1.
131. Principe, G.; Di Venere, L.; Negro, M.; Di Lalla, N.; Omodei, N.; Di Tria, R.; Mazziotta, M.N.; Longo, F. Hunting for gamma-ray emission from fast radio bursts. *Astron. Astrophys.* **2023**, *675*, A99. [CrossRef]
132. Lyubarsky, Y. A model for fast extragalactic radio bursts. *Mon. Not. R. Astron. Soc. Lett.* **2014**, *442*, L9–L13. [CrossRef]
133. Murase, K.; Kashiyama, K.; Mészáros, P. A burst in a wind bubble and the impact on baryonic ejecta: High-energy gamma-ray flashes and afterglows from fast radio bursts and pulsar-driven supernova remnants. *Mon. Not. R. Astron. Soc. Lett.* **2016**, *461*, 1498–1511. [CrossRef]
134. Margalit, B.; Beniamini, P.; Sridhar, N.; Metzger, B.D. Implications of a Fast Radio Burst from a Galactic Magnetar. *Astrophys. J.* **2020**, *899*, L27. [CrossRef]
135. López-Oramas, A.; Jiménez Martínez, I.; Hassan, T.; Hoang, J.; Inoue, S.; Acciari, V.A.; Ansoldi, S.; Antonelli, L.A.; Arbet Engels, A.; Artero, M.; et al. Monitoring the magnetar SGR 1935+2154 with the MAGIC telescopes. *PoS* **2021**,*ICRC2021*, 783. [CrossRef]
136. Abdalla, H.; Aharonian, F.; Ait Benkhali, F.; Angüner, E.O.; Arcaro, C.; Armand, C.; Armstrong, T.; Ashkar, H.; Backes, M.; Baghmanyan, V.; et al. Searching for TeV Gamma-Ray Emission from SGR 1935+2154 during Its 2020 X-Ray and Radio Bursting Phase. *Astrophys. J.* **2021**, *919*, 106. [CrossRef]
137. Holder, J.; VERITAS Collaboration; Lynch, R.S. VERITAS Observations of Fast Radio Bursts. In Proceedings of the 36th International Cosmic Ray Conference (ICRC2019), Madison, WI, USA, 24 July–1 August 2019; Volume 36, p. 698. [CrossRef]
138. H. E. S. S. Collaboration; Abdalla, H.; Abramowski, A.; Aharonian, F.; Ait Benkhali, F.; Akhperjanian, A.G.; Andersson, T.; Angüner, E.O.; Arakawa, M.; Arrieta, M.; et al. First limits on the very-high energy gamma-ray afterglow emission of a fast radio burst. H.E.S.S. observations of FRB 150418. *Astron. Astrophys.* **2017**, *597*, A115. [CrossRef]
139. Keane, E.F.; Johnston, S.; Bhandari, S.; Barr, E.; Bhat, N.D.R.; Burgay, M.; Caleb, M.; Flynn, C.; Jameson, A.; Kramer, M.; et al. The host galaxy of a fast radio burst. *Nature* **2016**, *530*, 453–456. [CrossRef]
140. MAGIC Collaboration; Acciari, V.A.; Ansoldi, S.; Antonelli, L.A.; Arbet Engels, A.; Arcaro, C.; Baack, D.; Babić, A.; Banerjee, B.; et al. Constraining very-high-energy and optical emission from FRB 121102 with the MAGIC telescopes. *Mon. Not. R. Astron. Soc. Lett.* **2018**, *481*, 2479–2486. [CrossRef]

141. Bird, R.; VERITAS Collaboration. Observing FRB 121102 with VERITAS; Searching for Associated TeV Emission. In Proceedings of the 35th International Cosmic Ray Conference (ICRC2017), Busan, Republic of Korea, 12–20 July 2017; Volume 301, p. 621. [CrossRef]
142. Veritas; Capasso, M.; Ong, R.; Sadeh, I.; Kaaret, P.; Jin, W.; Benbow, W.; Mukherjee, R.; Prado, R.; Lundy, M.; et al. Gamma-ray and Optical Observations of Repeating Fast Radio Bursts with VERITAS. In Proceedings of the 37th International Cosmic Ray Conference, Berlin, Germany, 12–23 July 2022; p. 857. [CrossRef]
143. Hoang, J.; Hassan, T.; Tejedor, L.A.; Barrio, J.A.; López, M.; Fink, D.; Will, M.; Contreras, J.L. MAGIC-II's central pixel system for simultaneous optical and gamma-ray observation. *J. Astron. Telesc. Instruments Syst.* **2020**, *6*, 036002. [CrossRef]
144. Griffin, S.; Hanna, D.; Gilbert, A. Searching for Fast Optical Transients using VERITAS Cherenkov Telescopes. In Proceedings of the International Cosmic Ray Conference, Beijing, China, 11–18 August 2011; Volume 9, p. 38. [CrossRef]
145. Franzen, A.; Gillessen, S.; Hermann, G.; Hinton, J.; H. E. S. S. Collaboration. Optical Observations of the Crab Pulsar Using the First H.E.S.S. Cherenkov Telescope. In Proceedings of the International Cosmic Ray Conference, Tsukuba, Japan, 31 July–7 August 2003; Volume 5, p. 2987.
146. Yang, Y.P.; Zhang, B.; Wei, J.Y. How Bright Are Fast Optical Bursts Associated With Fast Radio Bursts? *Astrophys. J.* **2019**, *878*, 89. [CrossRef]
147. Stefanescu, A.; Kanbach, G.; Słowikowska, A.; Greiner, J.; McBreen, S.; Sala, G. Very fast optical flaring from a possible new Galactic magnetar. *Nature* **2008**, *455*, 503–505. [CrossRef]
148. Abbott, B.P.; Abbott, R.; Abbott, T.D.; Abernathy, M.R.; Acernese, F.; Ackley, K.; Adams, C.; Adams, T.; Addesso, P.; Adhikari, R.X.; et al. Observation of Gravitational Waves from a Binary Black Hole Merger. *Phys. Rev. Lett.* **2016**, *116*, 061102. [CrossRef]
149. Abbott, B.P.; Abbott, R.; Abbott, T.D.; Acernese, F.; Ackley, K.; Adams, C.; Adams, T.; Addesso, P.; Adhikari, R.X.; Adya, V.B.; et al. GW170817: Observation of Gravitational Waves from a Binary Neutron Star Inspiral. *Phys. Rev. Lett.* **2017**, *119*, 161101. [CrossRef]
150. Abbott, B.P.; Abbott, R.; Abbott, T.D.; Abraham, S.; Acernese, F.; Ackley, K.; Adams, C.; Adhikari, R.X.; Adya, V.B.; Affeldt, C.; et al. GWTC-1: A Gravitational-Wave Transient Catalog of Compact Binary Mergers Observed by LIGO and Virgo during the First and Second Observing Runs. *Phys. Rev. X* **2019**, *9*, 031040. [CrossRef]
151. Goldstein, A.; Veres, P.; Burns, E.; Briggs, M.S.; Hamburg, R.; Kocevski, D.; Wilson-Hodge, C.A.; Preece, R.D.; Poolakkil, S.; Roberts, O.J.; et al. An Ordinary Short Gamma-Ray Burst with Extraordinary Implications: Fermi-GBM Detection of GRB 170817A. *Astrophys. J.* **2017**, *848*, L14. [CrossRef]
152. Savchenko, V.; Ferrigno, C.; Kuulkers, E.; Bazzano, A.; Bozzo, E.; Brandt, S.; Chenevez, J.; Courvoisier, T.J.L.; Diehl, R.; Domingo, A.; et al. INTEGRAL Detection of the First Prompt Gamma-Ray Signal Coincident with the Gravitational-wave Event GW170817. *Astrophys. J.* **2017**, *848*, L15. [CrossRef]
153. Levan, A.; Crowther, P.; de Grijs, R.; Langer, N.; Xu, D.; Yoon, S.C. Gamma-Ray Burst Progenitors. *Space Sci. Rev.* **2016**, *202*, 33–78. [CrossRef]
154. Coulter, D.A.; Foley, R.J.; Kilpatrick, C.D.; Drout, M.R.; Piro, A.L.; Shappee, B.J.; Siebert, M.R.; Simon, J.D.; Ulloa, N.; Kasen, D.; et al. Swope Supernova Survey 2017a (SSS17a), the optical counterpart to a gravitational wave source. *Science* **2017**, *358*, 1556–1558. [CrossRef]
155. Li, L.X.; Paczyński, B. Transient Events from Neutron Star Mergers. *Astrophys. J.* **1998**, *507*, L59–L62. [CrossRef]
156. Metzger, B.D.; Martínez-Pinedo, G.; Darbha, S.; Quataert, E.; Arcones, A.; Kasen, D.; Thomas, R.; Nugent, P.; Panov, I.V.; Zinner, N.T. Electromagnetic counterparts of compact object mergers powered by the radioactive decay of r-process nuclei. *Mon. Not. R. Astron. Soc. Lett.* **2010**, *406*, 2650–2662. [CrossRef]
157. Troja, E.; Piro, L.; van Eerten, H.; Wollaeger, R.T.; Im, M.; Fox, O.D.; Butler, N.R.; Cenko, S.B.; Sakamoto, T.; Fryer, C.L.; et al. The X-ray counterpart to the gravitational-wave event GW170817. *Nature* **2017**, *551*, 71–74. [CrossRef]
158. Blandford, R.D.; McKee, C.F. Fluid dynamics of relativistic blast waves. *Phys. Fluids* **1976**, *19*, 1130–1138. [CrossRef]
159. Pian, E.; D'Avanzo, P.; Benetti, S.; Branchesi, M.; Brocato, E.; Campana, S.; Cappellaro, E.; Covino, S.; D'Elia, V.; Fynbo, J.P.U.; et al. Spectroscopic identification of r-process nucleosynthesis in a double neutron-star merger. *Nature* **2017**, *551*, 67–70. [CrossRef]
160. Ghirlanda, G.; Salafia, O.S.; Paragi, Z.; Giroletti, M.; Yang, J.; Marcote, B.; Blanchard, J.; Agudo, I.; An, T.; Bernardini, M.G.; et al. Compact radio emission indicates a structured jet was produced by a binary neutron star merger. *Science* **2019**, *363*, 968–971. [CrossRef]
161. Abdalla, H.; Adam, R.; Aharonian, F.; Ait Benkhali, F.; Angüner, E.O.; Arakawa, M.; Arcaro, C.; Armand, C.; Ashkar, H.; Backes, M.; et al. A very-high-energy component deep in the γ-ray burst afterglow. *Nature* **2019**, *575*, 464–467. [CrossRef]
162. MAGIC Collaboration; Acciari, V.A.; Ansoldi, S.; Antonelli, L.A.; Arbet Engels, A.; Baack, D.; Babić, A.; Banerjee, B.; Barres de Almeida, U.; Barrio, J.A.; et al. Teraelectronvolt emission from the γ-ray burst GRB 190114C. *Nature* **2019**, *575*, 455–458. [CrossRef]
163. H. E. S. S. Collaboration; Abdalla, H.; Aharonian, F.; Ait Benkhali, F.; Angüner, E.O.; Arcaro, C.; Armand, C.; Armstrong, T.; Ashkar, H.; Backes, M.; et al. Revealing X-ray and gamma ray temporal and spectral similarities in the GRB 190829A afterglow. *Science* **2021**, *372*, 1081–1085. [CrossRef] [PubMed]
164. Acciari, V.A.; Ansoldi, S.; Antonelli, L.A.; Arbet Engels, A.; Asano, K.; Baack, D.; Babić, A.; Baquero, A.; Barres de Almeida, U.; Barrio, J.A.; et al. MAGIC Observations of the Nearby Short Gamma-Ray Burst GRB 160821B. *Astrophys. J.* **2021**, *908*, 90. [CrossRef]
165. Antonelli, A.; Carosi, A.; de Lotto, B.; Mirzoyan, R.; Stamerra, A. LIGO/Virgo G211117: MAGIC very-high energy gamma-ray observations. *GRB Coord. Netw.* **2015**, *18776*, 1.

166. de Lotto, B.; Ansoldi, S.; Antonelli, A.; Berti, A.; Carosi, A.; Longo, F.; Stamerra, A. MAGIC electromagnetic follow-up of gravitational wave alerts. In *New Frontiers in Black Hole Astrophysics, Ljubljana, Slovenia, Proceedings of the International Astronomical Union, IAU Symposium*; Cambridge University Press: Cambridge, UK, 2017; Volume 324, pp. 287–290. [CrossRef]
167. Mukherjee, R.; VERITAS Collaboration. LIGO/Virgo G268556: VERITAS Very-High-Energy Gamma-Ray Observations. *GRB Coord. Netw.* **2017**, *21153*, 1.
168. Ashka, H.; Schüssler, F.; Seglar-Arroyo, M.; H. E. S. S. Collaboration. Searches for TeV gamma-ray counterparts to gravitational wave events with H.E.S.S. *Mem. Soc. Astron. Ital.* **2019**, *90*, 49. [CrossRef]
169. Abdalla, H.; Abramowski, A.; Aharonian, F.; Ait Benkhali, F.; Angüner, E.O.; Arakawa, M.; Arrieta, M.; Aubert, P.; Backes, M.; Balzer, A.; et al. TeV Gamma-Ray Observations of the Binary Neutron Star Merger GW170817 with H.E.S.S. *Astrophys. J.* **2017**, *850*, L22. [CrossRef]
170. Coulter, D.A.; Kilpatrick, C.D.; Siebert, M.R.; Foley, R.J.; Shappee, B.J.; Drout, M.R.; Simon, J.S.; Piro, A.L.; Rest, A.; One-Meter Two-Hemisphere (1M2H) Collaboration. LIGO/Virgo G298048: Potential optical counterpart discovered by Swope telescope. *GRB Coord. Netw.* **2017**, *21529*, 1.
171. Salafia, O.S.; Berti, A.; Covino, S.; D'Elia, V.; Miceli, D.; Nava, L.; Patricelli, B.; Righi, C.; Inoue, S.; Antonelli, L.A.; et al. Follow-up observations of GW170817 with the MAGIC telescopes. In Proceedings of the 37th International Cosmic Ray Conference, (ICRC2021), Berlin, Germany, 12–23 July 2021; p. 944. [CrossRef]
172. Abdalla, H.; Adam, R.; Aharonian, F.; Ait Benkhali, F.; Angüner, E.O.; Arakawa, M.; Arcaro, C.; Armand, C.; Armstrong, T.; Ashkar, H.; et al. Probing the Magnetic Field in the GW170817 Outflow Using H.E.S.S. Observations. *Astrophys. J.* **2020**, *894*, L16. [CrossRef]
173. Salafia, O.S.; Ghirlanda, G.; Ascenzi, S.; Ghisellini, G. On-axis view of GRB 170817A. *Astron. Astrophys.* **2019**, *628*, A18. [CrossRef]
174. Farrar, G.R.; Piran, T. Tidal disruption jets as the source of Ultra-High Energy Cosmic Rays. *arXiv* **2014**, arXiv:1411.0704. https://doi.org/10.48550/arXiv.1411.0704.
175. Guépin, C.; Kotera, K.; Barausse, E.; Fang, K.; Murase, K. Ultra-high-energy cosmic rays and neutrinos from tidal disruptions by massive black holes (Corrigendum). *Astron. Astrophys.* **2020**, *636*, C3. [CrossRef]
176. Aartsen, M.G.; Abraham, K.; Ackermann, M.; Adams, J.; Aguilar, J.A.; Ahlers, M.; Ahrens, M.; Altmann, D.; Anderson, T.; Archinger, M.; et al. Evidence for Astrophysical Muon Neutrinos from the Northern Sky with IceCube. *Phys. Rev. Lett.* **2015**, *115*, 081102. [CrossRef]
177. IceCube Collaboration; Aartsen, M.G.; Ackermann, M.; Adams, J.; Aguilar, J.A.; Ahlers, M.; Ahrens, M.; Al Samarai, I.; Altmann, D.; Andeen, K.; et al. Multimessenger observations of a flaring blazar coincident with high-energy neutrino IceCube-170922A. *Science* **2018**, *361*, eaat1378. [CrossRef]
178. IceCube Collaboration; Abbasi, R.; Ackermann, M.; Adams, J.; Aguilar, J.A.; Ahlers, M.; Ahrens, M.; Alameddine, J.M.; Alispach, C.; Alves, A. A., J.; et al. Evidence for neutrino emission from the nearby active galaxy NGC 1068. *Science* **2022**, *378*, 538–543. [CrossRef] [PubMed]
179. Bellm, E.C.; Kulkarni, S.R.; Graham, M.J.; Dekany, R.; Smith, R.M.; Riddle, R.; Masci, F.J.; Helou, G.; Prince, T.A.; Adams, S.M.; et al. The Zwicky Transient Facility: System Overview, Performance, and First Results. *Publ. Astron. Soc. Pac.* **2019**, *131*, 018002. [CrossRef]
180. Stein, R.; van Velzen, S.; Kowalski, M.; Franckowiak, A.; Gezari, S.; Miller-Jones, J.C.A.; Frederick, S.; Sfaradi, I.; Bietenholz, M.F.; Horesh, A.; et al. A tidal disruption event coincident with a high-energy neutrino. *Nat. Astron.* **2021**, *5*, 510–518. [CrossRef]
181. Acciari, V.A.; Aniello, T.; Ansoldi, S.; Antonelli, L.A.; Arbet Engels, A.; Artero, M.; Asano, K.; Baack, D.; Babić, A.; Baquero, A.; et al. Investigating the Blazar TXS 0506+056 through Sharp Multiwavelength Eyes During 2017-2019. *Astrophys. J.* **2022**, *927*, 197. [CrossRef]
182. Acciari, V.A.; Ansoldi, S.; Antonelli, L.A.; Arbet Engels, A.; Baack, D.; Babić, A.; Banerjee, B.; Barres de Almeida, U.; Barrio, J.A.; Becerra González, J.; et al. Constraints on Gamma-Ray and Neutrino Emission from NGC 1068 with the MAGIC Telescopes. *Astrophys. J.* **2019**, *883*, 135. [CrossRef]
183. Aleksić, J.; Antonelli, L.A.; Antoranz, P.; Asensio, M.; Backes, M.; Barres de Almeida, U.; Barrio, J.A.; Becerra González, J.; Bednarek, W.; Berger, K.; et al. Very high energy gamma-ray observation of the peculiar transient event Swift J1644+57 with the MAGIC telescopes and AGILE. *Astron. Astrophys.* **2013**, *552*, A112. [CrossRef]
184. Aliu, E.; Arlen, T.; Aune, T.; Beilicke, M.; Benbow, W.; Böttcher, M.; Bouvier, A.; Bradbury, S.M.; Buckley, J.H.; Bugaev, V.; et al. VERITAS Observations of the Unusual Extragalactic Transient Swift J164449.3+573451. *Astrophys. J.* **2011**, *738*, L30. [CrossRef]
185. Ivezić, Ž.; Kahn, S.M.; Tyson, J.A.; Abel, B.; Acosta, E.; Allsman, R.; Alonso, D.; AlSayyad, Y.; Anderson, S.F.; Andrew, J.; et al. LSST: From Science Drivers to Reference Design and Anticipated Data Products. *Astrophys. J.* **2019**, *873*, 111. [CrossRef]
186. Chen, X.; Gómez-Vargas, G.A.; Guillochon, J. The γ-ray afterglows of tidal disruption events. *Mon. Not. R. Astron. Soc. Lett.* **2016**, *458*, 3314–3323. [CrossRef] [PubMed]
187. Paczynski, B. Gamma-ray bursters at cosmological distances. *Astrophys. J.* **1986**, *308*, L43–L46. [CrossRef]
188. Piran, T. Gamma-ray bursts and the fireball model. *Phys. Rep.* **1999**, *314*, 575–667. [CrossRef]
189. Sari, R.; Esin, A.A. On the Synchrotron Self-Compton Emission from Relativistic Shocks and Its Implications for Gamma-Ray Burst Afterglows. *Astrophys. J.* **2001**, *548*, 787–799. [CrossRef]
190. Zhang, B.; Mészáros, P. High-Energy Spectral Components in Gamma-Ray Burst Afterglows. *Astrophys. J.* **2001**, *559*, 110–122. [CrossRef]

191. Gupta, N.; Zhang, B. Prompt emission of high-energy photons from gamma ray bursts. *Mon. Not. R. Astron. Soc. Lett.* **2007**, *380*, 78–92. [CrossRef]
192. Ghisellini, G.; Ghirlanda, G.; Nava, L.; Celotti, A. GeV emission from gamma-ray bursts: A radiative fireball? *Mon. Not. R. Astron. Soc. Lett.* **2010**, *403*, 926–937. [CrossRef]
193. Carosi, A.; Ashkar, H.; Berti, A.; Bordas, P.; de Bony Lavergne, M.; Donini, A.; Dalchenko, M.; Fiasson, A.; Foffano, L.; Fukami, S.; et al. First follow-up of transient events with the CTA Large Size Telescope prototype. *arXiv* **2021**, arXiv:2108.04309. https://doi.org/10.48550/arXiv.2108.04309.
194. LHAASO Collaboration; Cao, Z.; Aharonian, F.; An, Q.; Axikegu, A.; Bai, L.X.; Bai, Y.X.; Bao, Y.W.; Bastieri, D.; Bi, X.J.; et al. A tera-electron volt afterglow from a narrow jet in an extremely bright gamma-ray burst. *Science* **2023**, *380*, 1390–1396. [CrossRef]
195. Scuderi, S.; Giuliani, A.; Pareschi, G.; Tosti, G.; Catalano, O.; Amato, E.; Antonelli, L.A.; Becerra Gonzàles, J.; Bellassai, G.; Bigongiari, C.; et al. The ASTRI Mini-Array of Cherenkov telescopes at the Observatorio del Teide. *J. High Energy Astrophys.* **2022**, *35*, 52–68. [CrossRef]
196. Stamerra, A.; Saturni, F.G.; Green, J.G.; Nava, L.; Lucarelli, F.; Antonelli, L.A. TeV Transients with the ASTRI Mini-Array: A case study with GRB 190114C. In Proceedings of the ICRC2021, Berlin, Germany, 12–23 July 2021; p. 890. [CrossRef]
197. Carosi, A.; Stamerra, A.; Nava, L.; Pintore, F.; Antonelli, L.A.; D'Aì, A.; Lombardi, S.; Lucarelli, F.; Saturni, F.G.; Ghirlanda, G.; et al. The ASTRI Mini-Array observations of TeV transient events. In Proceedings of the 38th International Cosmic Ray Conference—PoS(ICRC2023), Nagoya, Japan, 26 July–3 August 2023; p. 785. [CrossRef]
198. Berti, A.; Carosi, A. The Detection of GRBs at VHE: A Challenge Lasting for More than Two Decades, What Is Next? *Galaxies* **2022**, *10*, 67. [CrossRef]
199. Bošnjak, Ž.; Brown, A.M.; Carosi, A.; Chernyakova, M.; Cristofari, P.; Longo, F.; López-Oramas, A.; Santander, M.; Satalecka, K.; Schüssler, F.; et al. Multi-messenger and transient astrophysics with the Cherenkov Telescope Array. *arXiv* **2021**, arXiv:2106.03621. https://doi.org/10.48550/arXiv.2106.03621.
200. Carosi, A.; López-Oramas, A.; Longo, F.; CTA Collaboration. The Cherenkov Telescope Array transient and multi-messenger program. In Proceedings of the 37th International Cosmic Ray Conference, (ICRC2021), Berlin, Germany, 12–23 July 2021; p. 736. [CrossRef]
201. López-Oramas, A. Transient and multi-messenger astrophysics with the Cherenkov Telescope Array. In Proceedings of the Highlights on Spanish Astrophysics XI, La Laguna, Spain, 5–9 September 2022; p. 159.
202. Schaefer, B.E. The B & V light curves for recurrent nova T CrB from 1842-2022, the unique pre- and post-eruption high-states, the complex period changes, and the upcoming eruption in 2025.5 $\pm$ 1.3. *Mon. Not. R. Astron. Soc. Lett.* **2023**, *524*, 3146–3165. [CrossRef]
203. Schaefer, B.E.; Kloppenborg, B.; Waagen, E.O.; Observers, T.A. Recurrent nova T CrB has just started its Pre-eruption Dip in March/April 2023, so the eruption should occur around 2024.4 +- 0.3. *Astron. Telegr.* **2023**, *16107*, 1.
204. Munari, U.; Ochner, P.; Dallaporta, S.; Valisa, P.; Vagnozzi, A.; Moretti, S.; Bergamini, A.; Cherini, G. Fast and steady re-brightening of the accretion disk of T CrB past the deep minimum of August-September 2023. *Astron. Telegr.* **2024**, *16404*, 1.
205. Sarmah, P. New constraints on the gamma-ray and high energy neutrino fluxes from the circumstellar interaction of SN 2023ixf. *arXiv* **2023**, arXiv:2307.08744. https://doi.org/10.48550/arXiv.2307.08744.
206. Mestre, E.; de Oña Wilhelmi, E.; Khangulyan, D.; Zanin, R.; Acero, F.; Torres, D.F. The Crab nebula variability at short time-scales with the Cherenkov telescope array. *Mon. Not. R. Astron. Soc. Lett.* **2021**, *501*, 337–346. [CrossRef]
207. López-Oramas, A.; Bulgarelli, A.; Chaty, S.; Chernyakova, M.; Gnatyk, R.; Hnatyk, B.; Kantzas, D.; Markoff, S.; McKeague, S.; Mereghetti, S.; et al. Prospects for Galactic transient sources detection with the Cherenkov Telescope Array. In Proceedings of the 37th International Cosmic Ray Conference, (ICRC2021), Berlin, Germany, 12–23 July 2021; p. 784. [CrossRef]
208. Aleksić, J.; Antonelli, L.A.; Antoranz, P.; Asensio, M.; Barres de Almeida, U.; Barrio, J.A.; Becerra González, J.; Bednarek, W.; Berger, K.; Bernardini, E.; et al. Observations of the magnetars 4U 0142+61 and 1E 2259+586 with the MAGIC telescopes. *Astron. Astrophys.* **2013**, *549*, A23. [CrossRef]
209. Fermi-LAT Collaboration; Ajello, M.; Atwood, W.B.; Axelsson, M.; Baldini, L.; Barbiellini, G.; Baring, M.G.; Bastieri, D.; Bellazzini, R.; Berretta, A.; et al. High-energy emission from a magnetar giant flare in the Sculptor galaxy. *Nat. Astron.* **2021**, *5*, 385–391. [CrossRef]
210. Paneque, D. MAGIC observation of GRB 231115A, a possible magnetar giant flare. *GRB Coord. Netw.* **2023**, *35068*, 1.
211. Green, J.G.; Seglar-Arroyo, M.; Consortium, C.; Abe, K.; Abe, S.; Acharyya, A.; Adam, R.; Aguasca-Cabot, A.; Agudo, I.; et al. Chasing Gravitational Waves with the Cherenkov Telescope Array. *arXiv* **2023**, arXiv:2310.07413. https://doi.org/10.48550/arXiv.2310.07413.
212. Petrov, P.; Singer, L.P.; Coughlin, M.W.; Kumar, V.; Almualla, M.; Anand, S.; Bulla, M.; Dietrich, T.; Foucart, F.; Guessoum, N. Data-driven Expectations for Electromagnetic Counterpart Searches Based on LIGO/Virgo Public Alerts. *Astrophys. J.* **2022**, *924*, 54. [CrossRef]

universe MDPI

Review

The ASTRI Mini-Array: A New Pathfinder for Imaging Cherenkov Telescope Arrays

Salvatore Scuderi † on behalf of the ASTRI Project

INAF IASF-Milano, Via A. Corti, 12, 20133 Milano, Italy; salvatore.scuderi@inaf.it
† http://www.astri.inaf.it/en/library/ (accessed on 5 March 2024).

Abstract: The ASTRI Mini-Array is an Istituto Nazionale di Astrofisica (INAF) project to build and operate an array of nine Imaging Atmospheric Cherenkov Telescopes (IACTs) at the Teide Astronomical Observatory of the Instituto de Astrofisica de Canarias in Tenerife (Spain) based on a host agreement with INAF and, as such, it will be the largest IACT array until the Cherenkov Telescope Array Observatory starts operations. Implementing the ASTRI Mini-Array poses several challenges from technical, logistic, and management points of view. Starting from the description of the innovative technologies adopted to build the telescopes, we will discuss the solutions adopted to overcome these challenges, making the ASTRI Mini-Array a great instrument to perform deep observations of the galactic and extra-galactic sky at very high energies.

Keywords: γ-ray astrophysics; IACTs; telescope array

Citation: Scuderi, S., on behalf of the ASTRI Project. The ASTRI Mini-Array: A New Pathfinder for Imaging Cherenkov Telescope Arrays. *Universe* **2024**, *10*, 146. https://doi.org/10.3390/universe10030146

Academic Editor: Binbin Zhang

Received: 14 February 2024
Revised: 5 March 2024
Accepted: 12 March 2024
Published: 16 March 2024

1. Introduction

The ASTRI Mini-Array is an INAF project to build and operate a facility to study astronomical sources emitting very high energy in the TeV spectral band. It consists of a group of nine innovative aplanatic dual-mirror Imaging Atmospheric Cherenkov Telescopes (IACTs) that are in the process of being installed at the Teide Astronomical Observatory in collaboration with the Instituto de Astrofisica de Canarias in Tenerife (Canary Islands, Spain). The project involves researchers from several INAF institutes and Italian universities. The Istituto Nazionale di Fisica Nucleare (INFN) sections of Roma Tor Vergata and Perugia are also participating in the project. International partners, specifically, the University of Sao Paulo in Brazil, the North Western University in South Africa, the University of Geneve in Switzerland, and the Instituto de Astrofisica de Canarias (IAC) in Spain, are contributing to the project in different forms. The ASTRI Mini-Array project is also supported by the "Fundación Galileo Galilei-INAF, Fundación Canaria" (FGG). The FGG is a Spanish non-profit institution, constituted by INAF, whose aim is to manage and run the Telescopio Nazionale Galileo[1] (TNG) and to promote astrophysical research in the Canary Islands on behalf of INAF.

The project is part of the ASTRI program, whose initial aim was to design, produce, and validate a prototype of Imaging Atmospheric Cherenkov Telescopes (IACTs) of 4-meter class in the framework of the development of the Cherenkov Telescope Array Observatory CTAO [1]. The prototype telescope, named ASTRI-Horn, was installed in 2014 at the M.G. Fracastoro station of the INAF–Catania Astrophysical Observatory [2] and was fully tested and scientifically validated with the detection of Crab nebula [3].

The ASTRI Mini-Array in a Nutshell

The ASTRI Mini-Array observational site is at the Observatorio del Teide. The area occupied by the nine telescopes is a strip approximately $300 \times 700\ \mathrm{m}^2$ in size. The telescopes are dual mirrors of the 4-meter class, with an alt-azimuth mount, and are equipped with SiPM-based cameras. The cameras will implement a field of view (FoV) of more than 10°

in diameter. Apart from the telescopes, also located at the site are the onsite data center, the local control room, and several auxiliary instruments to monitor the environment, to characterize the atmosphere above the site, and to calibrate the array. Remote operations centers are nearby in Tenerife and in Italy, as well as the offsite data center. Figure 1 shows a view of the site taken from the Themis solar observatory. A detailed technical description of the ASTRI Mini-Array can be found in Ref. [4].

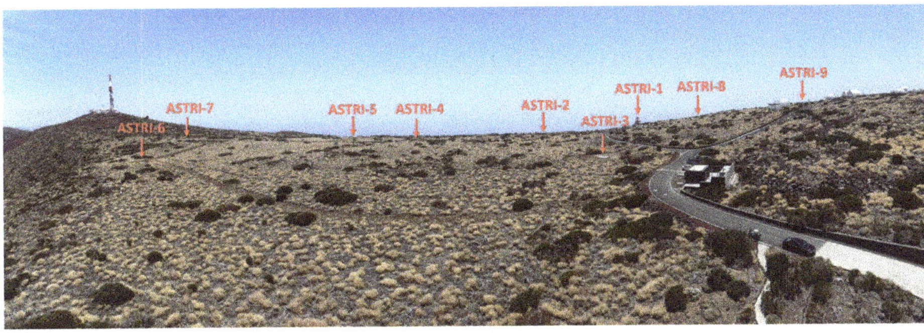

Figure 1. View of the ASTRI Mini-Array site from the terrace of the Themis telescope. Arrows indicate the positions of the nine telescopes. Only one telescope, ASTRI-1, is currently installed at the site. Adapted from Ref. [4].

Compared to currently operating IACT systems the ASTRI Mini-Array will be more sensitive at energies larger than a few TeV (see Figure 9 in Ref. [4]) and will extend the sensitivity up to 100 TeV and beyond, an almost unexplored energy range by IACTs. The large FoV will allow simultaneous monitoring of several sources during the same pointing. The combination of the sensitivity and the homogeneous performance across the FoV will allow us to study the emissions from extended sources such as SNRs and PWNs at $E > 10$ TeV, and to investigate the presence of spectral cut-offs. Coordinated observations with the current (MAGIC, VERITAS) and next generation (CTAO-N) IACT arrays in the northern hemisphere are clearly foreseen. Synergies will also be explored with wide-field particle shower arrays like HAWC [5] and with LHAASO [6]. In fact, these facilities survey a very large area of the northern sky and the ASTRI Mini-Array will be complementary, with pointed observations to characterize the morphology of extended sources detected at the extremely high energies by them.

During the first 4 years of operations the ASTRI Mini-Array will be run as an experiment and not as an observatory. After this initial period, the ASTRI Mini-Array will gradually move towards an observatory model. The scientific program during the first observing years will be devoted to the following core science topics: the origin of cosmic rays, the extra-galactic background light, and the study of fundamental physics, gamma-ray bursts and multi-messenger transients. A detailed description of the science that the ASTRI Mini-Array will perform can be found in Ref. [7].

Even if γ-ray astrophysics is undoubtedly its core science, the ASTRI Mini-Array will also be capable of exploring other scientific topics. In particular:

Stellar Hanbury Brown intensity interferometry: each telescope of the ASTRI Mini-Array will be equipped with an intensity interferometry module. With an expected angular resolution of 50 μ-arcsec, it will be possible, for example, to reveal details on the surface of bright stars and of the environment surrounding them.

Direct measurements of cosmic rays: in total, 99 % of the observable component of the Cherenkov light is hadronic in nature. This background, recorded during normal γ-ray observations, will be used to perform direct measurements and detailed studies on the cosmic rays.

The ASTRI Mini-Array will then be a remarkable instrument that is able perform seminal studies on both galactic and extra-galactic astrophysics, and also tackling frontier issues at the intersection of the fields of astrophysics, cosmology, particle physics, and fundamental physics [8,9]. Furthermore, for some time, the ASTRI Mini-Array will be the largest IACT facility operating in the world, until CTAO starts operations.

The ASTRI Mini-Array is, however, a complex instrument. Building it and then operating it and maintaining it present us with several challenges (technical, logistic, and managerial). The aim of this paper is, thus, to present the solutions, based on best practices and technological innovations, that allow us to mitigate, or better, overcome these challenges.

2. Challenges and Innovations

The construction of nine identical telescopes is something that has more to do with mass production than with what is common in the world of astrophysical research where, most of the time, each instrument is unique. However, in order to guarantee the required scientific performance, this kind of mass production shall have to satisfy strict technical requirements and undergo extremely thorough quality control processes. Furthermore, the ASTRI Mini-Array can be considered as a forerunner for CTAO Small-Sized Telescopes (SST, see for example [10]). In the case of CTAO, the SST Consortium will have to deliver up to 42 complete Cherenkov telescopes. As the optics and the electro-mechanical structures will be very similar if not, in some cases, identical to the ASTRI Mini-Array ones, the project represents an essential training ground to optimize methods and approaches to be applied to production and quality assurance processes for the SST telescopes. In practice, this sometimes simply translates to verifying that the manufacturer applies the best practices in terms of norms and standards (included in the technical requirements) in its production activities when, for example, it comes to the characteristics, verification, certification, and traceability of materials, and the welding, thermal, or anticorrosion treatments. To face other instances, specific procedures have been developed. An example of how the project has dealt with such complex tasks can be found in Ref. [11] where the large-scale production of mirrors for the ASTRI Mini-Array (see also Section 2.1.1) as well as the adopted testing methods and approaches, first to qualify the production process and then to verify the mirrors' performance, are described. At the end of the production and after the verification process, each mirror is delivered with a unique identity card where its characteristics are reported and traced.

As we will see in the next Sections, technological innovations have also played an important role in simplifying manufacturing activities, operations, and maintenance.

2.1. The Optical Design

Most of the technological innovations of the ASTRI Mini-Array telescopes derive from the selected optical design. Some have an impact only on the scientific performance, while others simplify the complexity of the system (building, operations, and maintenance).

The capability to properly image the signal produced by γ-rays emitted by astronomical sources when they interact with the Earth's atmosphere and to distinguish it from the background photons induced by hadron showers is the fundamental requirement to consider in the optical design of a Cherenkov telescope.

The ASTRI design was developed starting from the idea of an aplanatic two-mirror telescope proposed by Ref. [12] for application to Cherenkov telescopes. The design is based and further elaborates on a Schwarzschild–Couder configuration, as described in Ref. [13] where a polynomial optimization leads to a two-mirror design free of aberrations and characterized by a large FoV, small plate scale, low vignetting, and isochrony. Table 1 shows the optical parameters related to the design. The primary (M1) and the secondary (M2) mirrors are both aspherical and have diameters of 4.3 m and 1.8 m, respectively, while the focal surface is curved but spherical. The design is very compact as can be derived by the distances between the optical elements. Finally, the FoV is 10.5 degrees, which, given the plate scale, corresponds to a linear size of about 400 mm.

Table 1. ASTRI Mini-Array optical system parameters.

Parameter	Value
M1 (diameter)	4300 mm
M2 (diameter)	1800 mm
Distance M1–M2	3108.4 mm
Distance M2–focal surface	519.6 mm
Effective focal length	2154 mm
F-number	0.5
Field of view	10.5 degree
Plate scale	37.64 mm/degree

Figure 2 is the spot diagram resulting from ray tracing analysis that shows how the shape of the point spread function (PSF) changes going from on axis to 5 degrees off axis. Figure 3, instead, shows the behavior across the FoV of D80, a parameter which is obtained by integrating the PSF in radial direction with respect to its barycenter until 80% of the total number of photons used in the ray tracing analysis is obtained. As shown, the D80 is contained in a Cherenkov pixel of 7 mm. The two figures show that the residual aberrations from the optical design produce uniform behavior of the PSF across an FoV of 10° of the PSF, guaranteeing, at the same time, the necessary optical resolution.

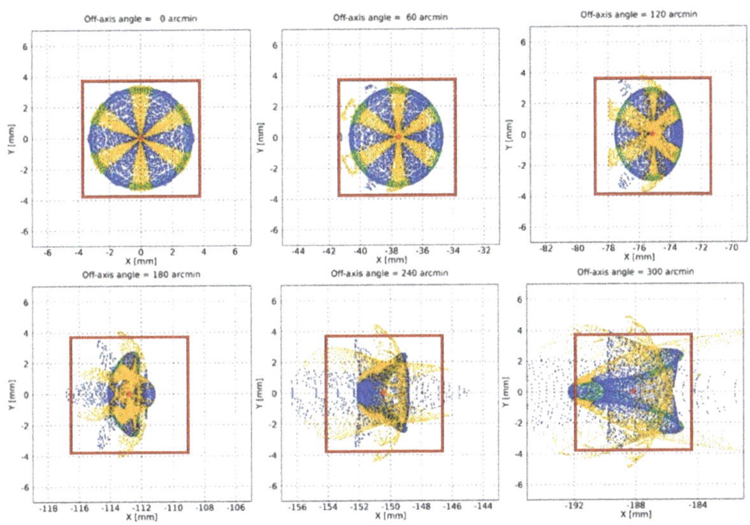

Figure 2. PSF versus position on the FoV. X and Y axes are in mm. The red square corresponds to the dimensions of a pixel of the ASTRI Mini-Array Cherenkov camera. The red star is the barycenter of the rays' distribution. Different colors correspond to the contribution to the PSF of the various panels forming M1 (see Section 2.1.1).

Figure 4 shows the various contributions to vignetting introduced by the secondary mirror and by some parts of the telescope structure. The vignetting increases going from the center to the border of the FoV reaching about 40% of the geometrical area of the primary mirror (slightly more than 11 m^2).

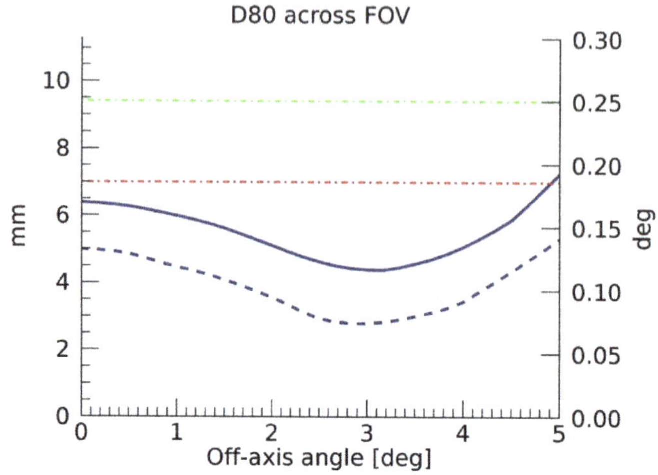

Figure 3. D80 across the FoV. D80 is the standard parameter for characterizing the optical PSF of a telescope. It corresponds to the diameter of the circle within which 80% of the photons fall. Green dashed–dotted line is the requirement, the red dashed–dotted line represents the dimensions of the pixel of the ASTRI Cherenkov camera, the blue line is the D80 across the FoV, and the blue dashed line is the D50.

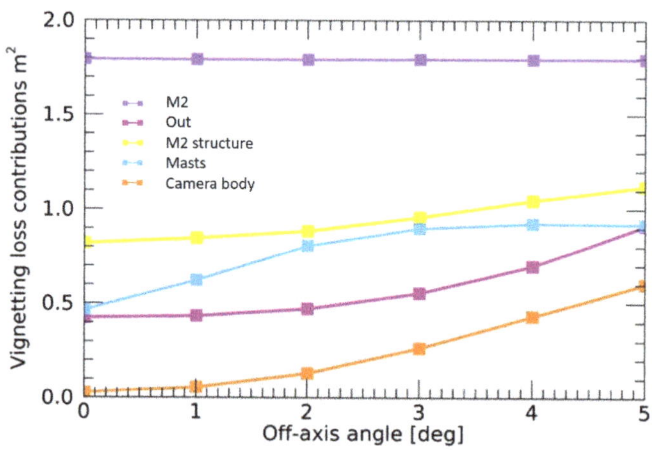

Figure 4. Contribution of various components to the vignetting. All the lines are self-explanatory apart from the magenta line, which represents the fraction of photons reflected by M1 falling outside M2.

Finally, an important characteristic of the Schwarzschild–Couder configuration is the isochronicity behavior. The time spread of the photons impinging at various angles on the focal surface, introduced by the optical system, should be smaller compared to the intrinsic time dispersion of a Cherenkov signal (a few nanoseconds). For on-axis rays, this optical configuration is isochronous and only a small time dispersion is introduced when photons enter at large field angles. In the case of the ASTRI telescopes, it is in the range 0.65–1.55 ns, as Figure 5 shows.

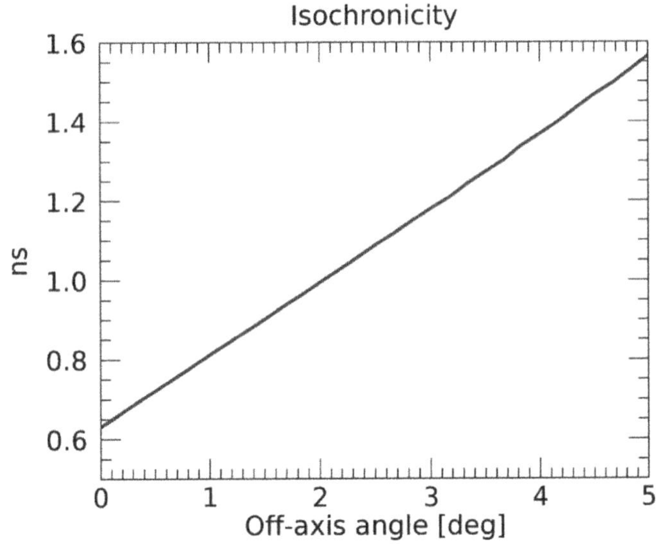

Figure 5. Isochronicity of the optical system across the FoV.

The validation of the Schwarzschild–Couder optical concept developed for the ASTRI Mini-Array was obtained in Ref. [14] during the commissioning of the ASTRI-Horn telescope prototype. A comprehensive article on the ASTRI Mini-Array optical design is under submission by G. Sironi.

The ASTRI telescopes are not the only IACTs implementing a Schwarzschild–Couder configuration. In the framework of CTAO development, two more prototype telescopes were built and tested: the small-size Gamma-ray Cherenkov Telescope (GCT, Ref. [15]) and the medium-size Schwarzschild–Couder Telescope (SCT, Ref. [16]).

2.1.1. Mirrors' Production

The implementation of the optical design translates to a strong aspherical surface for both M1 and M2, while the focal plane is spherical. The M1 mirror is segmented with 18 hexagonal panels placed in three concentric rings of six panels each, whose centers have different distances from the optical axis. All panels of each ring have the same radius of curvature but different from that of the other two. The M2 mirror is monolithic. Details can be found in Ref. [14].

The challenge posed by the characteristics of the M1 panels and M2 were faced using, as a manufacturing technique, the slumping method, which consists of the production of a metallic mold, machined to reproduce the aspheric surface of the mirror, on which a slab of glass is placed to make, by replica, an optical surface with the requested form. The slumping can be obtained using two different techniques. The first technique, called hot slumping, consists of heating up, inside a special oven, the glass slab placed on the mold that then "adapts" to the mold itself (see Ref. [17] for details). Alternatively, the slab is placed on the mold and, through suction, is bent to the required form without any heating. This technique, called cold slumping, is used on thin slabs that, once bent, need to be reinforced through a honeycomb structure that is finished with a second glass slab, thus giving the final product the shape of a sandwich (see Ref. [11] for details). The technique was developed by Media Lario Technologies Company under the scientific supervision of INAF, starting from the development for the ALMA panels, and subsequently applied to the MAGIC telescopes [18].

The selection of the method to be used depends on the characteristics of the mirror to be produced. The M2 mirror was manufactured using the hot slumping technique due to its dimensions. The mirror is 180 cm in diameter; therefore, too large for slabs as thin as those used in cold slumping. Being 19 mm in thickness, even the M2 mirror produced with the hot slumping technique has some criticality: we found that these critical issues are not structural but regarding the handling procedures since the mirror is large and heavy (about 150 kg in weight).

For the panels of the M1 mirror, the cold slumping technique was by far the most advantageous. Dimensions and characteristics (radius of curvature) were proven not to be a problem and the sandwich structure made them much more lightweight (density of 15 kg/m^2) compared to a slab of solid glass with the same thickness (25 mm) of the sandwich. Figure 6 shows one of the M1 panels placed on the mold during the slumping process in Media Lario.

Figure 6. The phase of bending the glass slab in the cold slumping technique. Photo courtesy of Media Lario SrL.

Finally, the production by replica makes both techniques very suitable for mass production, especially in the case of M1 panels of which 198 units were produced.

2.2. The Cherenkov Camera

One of the major advantages of the Schwarzschild–Couder configuration is the plate scale at the focal surface that allows us to have a compact camera. The small plate scale of 37.64 mm degree^{-1} and the 0.19 degree angular resolution of the ASTRI optical design led to the development of a camera with a focal plane of less than 400 mm linear dimensions that covers an FoV of about 10.5 degrees. A detailed description of the ASTRI Mini-Array camera can be found in Ref. [19]. In this paper, we focus only on the technological novelties.

2.2.1. Silicon Photomultipliers

The linear dimensions of the PSF (D80) are about 7 mm, which fit the size of the SiPM detectors well. SiPMs are basically an array of Avalanche Photodiodes working in Geiger-mode that have several advantages compared to photomultiplier detectors traditionally used in Cherenkov cameras. In particular, aside from the size, they have a Photon Detection Efficiency up to 50%, bias voltage down to 30 V, excellent single photon resolution, are not sensitive to magnetic fields, and are not damaged by a high level of light exposure. The use of an SiPM-based camera will improve the duty cycle of the system allowing safe and effective operation with any level of Moon condition as already demonstrated by the FACT telescope [20] and very recently by LHAASO [21]. On the other hand, they are affected by high dark counts, after pulses, and optical crosstalk and have a gain that is temperature dependent.

The SiPM detectors chosen for the ASTRI Mini-Array cameras have been developed by Hamamatsu Photonics specifically for the ASTRI project. The main characteristics of the detectors are summarized in Table 2. Figure 7 shows an 8 × 8 SiPM matrix. In the figure, we show a single pixel with the effective photosensitive area of 6.975 × 6.975 mm with a 0.2 mm interspace between pixels and 0.2 mm tile edge. This yields a physical dimension of 57.6 mm with a geometrical filling factor of 93.18%.

Table 2. ASTRI Mini-Array SiPM detectors' characteristics.

Parameter	Value
Photosensitive area (pixel size)	6.975 × 6.975 mm
Number of channels	64 (8 × 8 matrix)
Micro-cells' size	75 × 75 µm
Optical Crosstalk (OCT) [1]	5%
Dark Count Rate (DCR) [1]	4000 kHz
Photon Detection Efficiency (PDE) @ 400 nm [2]	51%
Coating	None

[1] These values have been measured by Hamamatsu Photonics at T = 25 °C and Overvoltage = 3 V; [2] PDE has been measured in Ref. [22] at T = 25 C and Overvoltage = 3 V.

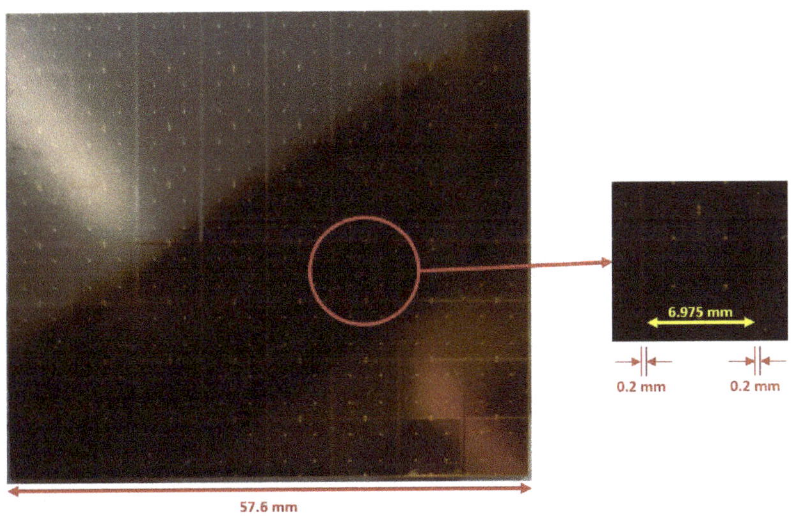

Figure 7. Image of an 8 × 8 SiPM array with the indication of the size. The zoomed image shows a single pixel. The dimension of the active area of a pixel is 6.975 mm with an interspace between pixels of 0.2 mm.

The 7 × 7 mm pixel size, in addition to the choice not to apply any coating, allowed us to have an enhanced PDE while keeping the Dark Count Rate and Optical Crosstalk within the requirements.

2.2.2. Front End Electronics

The Front End Electronics (FEE) has the fundamental function to process the output signals of the SiPM detectors and, for this reason, it is the heart of the acquisition electronics. To fulfill this task, an ASIC, specifically designed for the ASTRI project, is used. The ASIC is the CITIROC-1A produced by the Weeroc company[2]. The two main innovations are the peak detection technique and the variance technique.

The peak detection technique is an alternative method to measure the signal generated by an SiPM pixel. The traditional method called waveform sampling consists of following the temporal evolution of the SiPM output signal and sampling it at different times to recover information on the amplitude of the signal and on the time gradient. With the peak detection technique, only a single sampling point is necessary that allows a value to be identified that is proportional to the charge injected by an SiPM pixel. This allows the time gradient issues associated with the signal detection to be resolved while reducing the data flux.

Figure 8 shows a simplified block diagram of the read-out scheme of a single pixel of the ASTRI Cherenkov camera. The figure shows the processing chain of the analog signal produced by an SiPM pixel. The CITIROC inputs have a DAC converter that allows adjustment of the SiPM operating voltage to compensate for pixel-to-pixel gain variation but also those induced by temperature. The signal then enters in two separate chains that feature two pre-amplifiers, working in parallel, with different gains to maximize the dynamic range of the ASIC, which, in photoelectrons, spans from 1 to 2000. The High Gain (HG) channel, with higher resolution, samples the dynamic range up to 60 photoelectrons, while the Low Gain (LG) channel samples the entire dynamic range. The signal going to the HG channel enters a part of the ASIC, called a Fast Shaper (FSH), that is able to detect and analyze fast varying signals. The main function of the FSH is to produce a digital trigger, called a first-level trigger, every time the input signal exceeds a preset threshold. The search is conducted by integrating the signal in a 12.5 ns time interval so that, once a fast varying signal is detected, the FSH follows it for 15 ns, after which, if the value is below the threshold, the shaper goes back to its steady-state value until a new signal is detected. Once the trigger is generated, the peak detection mode is activated. As can be seen from the figure, the signal goes into two Slow Shapers (SSHs), which behave exactly like the fast shaper but with the difference being that the time window to follow the signal is programmable in 12.5 ns steps from 12.5 ns to 87.5 ns. The two shapers follow the varying signal and store its maximum value within the preset integration window. Once the integration time is over, the peak detector is disconnected from the shaper and the signal is held until the ASIC is read-out and the signals digitized and passed to the FPGA board. This board implements an algorithm that looks for adjacent patterns of pixels that yield a signal amplitude above a certain threshold level (topological trigger). If that happens, the entire focal plane is read-out (camera trigger) and the signal together with further information (time stamp, temperatures) are formatted and sent to the camera server. At the end of the pixels' read-out, the peak detectors are reset. Figure 9 shows a schematic representation of how the peak detector technique works.

The Variance technique allows the retrieval of a signal that is proportional to the photon flux impinging on a pixel, allowing, for example, the sky background signal to be measured. This technique is based on the statistical analysis of the variability in the signal detected by the camera front end electronics. It considers the asynchronous randomly repeated sampling of the electric signal (pulse amplitude) generated by each pixel when not triggered by the first-level trigger. The net result is, for each pixel, a sequence of ADC values whose average is constant with time but whose variance is proportional to the photon flux impinging on the pixel. Variance data are transmitted by default to the camera server every second.

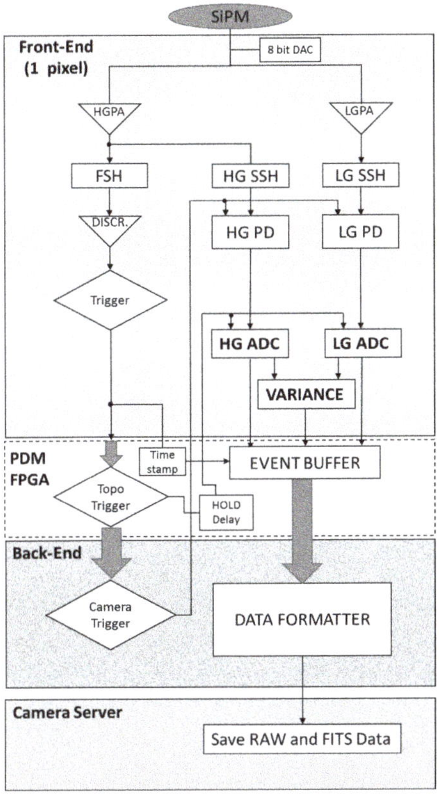

Figure 8. Simplified read-out scheme of the ASTRI camera for a single channel. Each front end electronics module consists of an SiPM tile (64 pixels), an ASIC board offering 64-channel read-out capabilities (two CITIROCs and two dual-channel ADCs), and an FPGA board for digital processing. Digital data from the 37 FPGAs are transmitted to a common BEE, which provides suitably formatted data packets to the camera server.

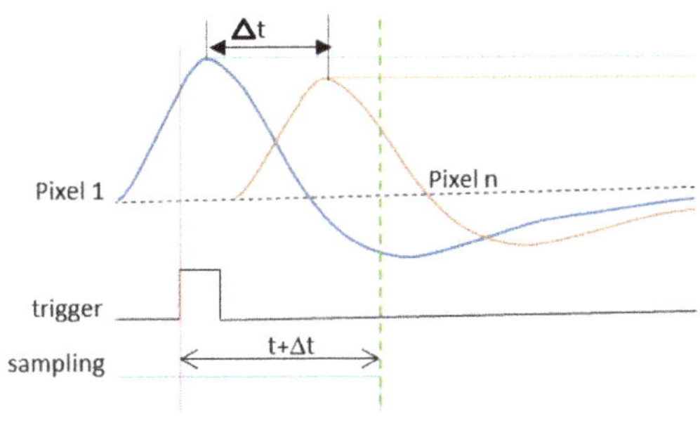

Figure 9. Schematic representation of the peak detector technique. The n pixel signals (blue and orange lines) arrive within a time interval of Δt ns. At the occurrence of a camera trigger, the peak detector is armed for all the pixels (channels). Peak detection is activated if the camera trigger signal occurs before the set peaking time. Peak values are then kept constant (blue and orange straight lines) for those pixels and the reading of these values can be achieved at the desired time t + Δt. Vertical red line is the time when the trigger is received and the peak detector activated. Green dashed vertical line is the time when the peak detector is disarmed. Horizontal black line is the threshold for detection of the sinlge pixel.

2.2.3. Stereo Event Builder

As explained in Section 2.2.2, the use of the peak detector technique allows us to reduce the data flux related to Cherenkov events. Even assuming a rate of events of 600 Hz, the amount of data produced by a single ASTRI Mini-Array telescope in one hour is of the order of 50 GByte. At the site, there is only a limited storage capacity, basically a buffer memory to store a week of data, so the normal procedure will be the transfer of the data directly to Italy to the offsite data center in Rome. The nominal bandwidth of this link is 10 Gbit s^{-1} so, in this scenario, the data produced in one hour by the entire set of telescopes will be transferred in about six minutes. Preliminary tests on the actual bandwidth give a maximum of 8 Gbit s^{-1} with a margin to improve it. However, even degrading this bandwidth further to 5 Gbit s^{-1}, the transfer time increases just to 12 min. This implies that we can easily manage the data transfer without any need for data storage or preprocessing. In particular, there will be no need to identify stereoscopic events, a procedure that is used to reduce the amount of data to transfer for Cherenkov telescopes using waveform sampling and/or facing a higher rate of events that are sensitive to lower energy γ-rays.

In the case of the ASTRI Mini-Array, no analog array stereo trigger (see, for example, [23]) will be implemented at the site. All the events generated by a single telescope will be transferred to and stored at the ASTRI Mini-Array data center in Rome, and the search for Cherenkov events detected in coincidence by more than one telescope will be performed offline as the initial step of the data processing chain. The software we have set up to perform this task is called the Stereo Event Builder software system. A complete description of the Stereo Event Builder algorithm is given in Ref. [24]. The software analyzes all the recorded events for their multiplicity and those that have a multiplicity greater than 1 are considered stereo events, while the others are still kept to be used, for example, for calibration purposes (muon events). Preliminary tests carried out using Monte Carlo simulations show an efficiency above 99%.

2.2.4. Thermal Control System

Another advantage of the use of the CITIROC-1A ASIC is that the power necessary to work, i.e. to process the signal, is of the order of 300 mW [19]; hence, quite low. As a consequence, the ASIC can be placed very close to the SiPM to reduce noise issues, simplifying the mechanical design but also that of the thermal control system. This subsystem is very important in an SiPM-based camera as gain and dark noise are temperature dependent so keeping the temperature of the focal plane low and stable is essential. In the case of the ASTRI Mini-Array camera, the temperature of the focal plane has to kept at 15 °C with a temperature gradient along it below ±1 °C. Thanks to the characteristics of the electronics, this can be achieved using a thermal control system based on ThermoElectric Coolers (TECs) deployed uniformly below the focal plane and heat pipes embedded in it. The heat produced by the TECs to keep the SiPM cold is dissipated by air circulation through a number of fans. The resulting system is compact and needs little power. Again, we have a difference with respect to other SiPM-based cameras used in Cherenkov telescopes as the latter are cooled through an external chiller, and this makes the system more complex and require more electrical power to work and also more maintenance activities.

2.2.5. The NSB Filter

Among the few disadvantages related to the use of the SiPM detectors in Cherenkov astronomy, their high response in the red part ($\lambda > 600$ nm) of the optical spectrum is one of the most troublesome. Being basically silicon-based detectors, they are sensitive to the electromagnetic radiation with wavelengths up to 1 µm. The Cherenkov radiation produced by the atmospheric showers span from about 300 nm to 600 nm, peaking around 350 nm. Everything outside this range has to be considered background noise that affects the signal to noise ratio of the Cherenkov signal. In particular, the night sky background, that is the brightness of the sky in a moonless night, increases rapidly above 600 nm due to OH airglow lines (see Ref. [25] and Figure 10). Classical PMTs used in Cherenkov cameras are not affected by this problem because a specific coating, cutting radiation above the range of interest, can be deposited on the photomultiplier cathode.

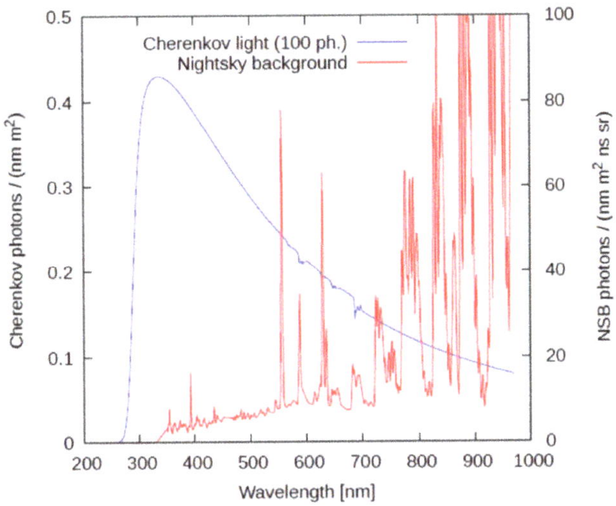

Figure 10. The spectrum of night sky background from Ref. [25] with the spectrum of Cherenkov radiation superimposed.

To face this problem, we studied the possibility of using the window that covers the focal surface of the Cherenkov camera as a filter. This window is made from a stack of three circular Spectrosil glass foils. Both faces of the foils are coated with a dielectric multilayer. The number of layers and composition of the coating were optimized to cut the signal at wavelengths below 300 nm and above 550 nm. For more details, see Ref. [26] from which Figure 11, showing the transmittance of the filter, is reproduced.

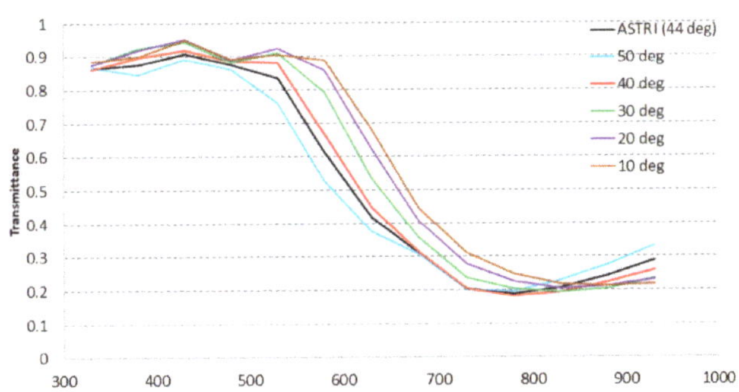

Figure 11. Transmittance of the Cherenkov camera filter as function of wavelength at various angles of incidence. Reproduced from Ref. [26].

2.3. The Mechanical Structure

The ASTRI telescope has an alt-azimuthal mount but, with respect to optical astronomical telescopes, uses a preloaded ball screw jack for the motion along the elevation axis, a configuration common among radio antennas. A detailed description of the mechanical structure can be found in Ref. [27]. Here, we will focus on those aspects that were useful to simplify the project.

One peculiar aspect of the telescope is the absence of permanent mirror actuators on the panels of the primary mirrors. The telescopes of the ASTRI Mini-Array are provided with a set of removable actuators that are used during the AIT/V phase to align the mirrors for the first time and then are dismounted. The actuators (or a subset of them) are remounted only if one or more mirrors become misaligned. This is due to the fact that the telescope stiffness is such that the mirrors' position is not sensitive to gravity, wind, or seasonal changes in average temperatures [28], making the presence of permanent actuators useless, which, instead, are always present in all of the other Cherenkov telescopes. This characteristic of the telescope simplifies:

- Operations as the software does not need to control at least 36 actuators per telescope during every night throughout the year.
- Maintenance as there will be no permanent mechanisms and corresponding control electronics that will need maintenance, either preventive or corrective. In case of mirror misalignment, it is true that one needs to remount the actuators (or more likely a subset of them), but this will not increase the technical downtime of the telescope because the realignment operation will need an optical camera to replace the Cherenkov camera and that operation will happen simultaneously with mounting of the actuators.

Another aspect that has an impact on the logistics and the integration activities is that the telescope is generally integrated at the production/integration site in Italy to be tested, but before being shipped, it is not completely disassembled, thus traveling in a small number of pieces. In particular, the M2 support structure is shipped completely assembled, meaning with the mirror, actuators, auxiliaries, and corresponding control electronics mounted. Furthermore, everything below the M1 dish (base, AZ platform, electrical cabinets, motors, etc.) is shipped completely mounted and equipped (encoder and limit switches mounted, cables routed, etc.). This will allow us to optimize the shipping of the telescopes in terms of space and then of costs. Additionally, the integration operations at the site will be shortened, simplified, and made safer, with all these aspects being extremely important when operating at high altitudes.

3. Operations: An Array of Nine

Operating an array of telescopes has been a long ongoing business in astronomy, running back to 1980 when the Very Large Array[3] (VLA) was completed. The ASTRI Mini-Array will have nine Cherenkov telescopes, being the largest IACT array before CTAO commences operations, but not so large when compared to ALMA[4], which, with 66 antennas, is the largest array of telescopes in the world. This does not imply that operating such an array is a simple matter. The philosophy behind the operation concept of the ASTRI Mini-Array is to minimize local operations, ideally restricting them only to maintenance activities, and to automatize as much as possible all the operations necessary to manage the array. Of course, this approach will impact the software that has to handle the life-cycle of the array, that goes from the creation of an observing project to the production of the final results to be used then by science, but also the infrastructures to support it. The successful application of this philosophy will result in a reduction in costs and manpower to manage the array.

3.1. ASTRI Mini-Array Software

Ref. [29] describes the architecture and the development approach of the ASTRI Mini-Array software. Figure 12 is the context view of the ASTRI Mini-Array software showing all the software systems that compose it and the relationship/interface between them and with the external world (systems and actors). In the figure, we also show those software subsystems that work onsite and those that instead work offsite. The archive is central to the software architecture and it is located offsite, but parts of it, necessary for site activities, are also replicated onsite. As is clear from the figure, onsite and offsite software have different tasks that are complementary to each other. The onsite software manages all the activities necessary to produce and transfer scientific, calibration, and engineering data. Basically, no data analysis/reduction needs to be performed at the site, also thanks to the characteristics of the camera we described in Sections 2.2.2 and 2.2.3. Real-time analysis is performed on some instrumental parameters that allows us to have a quality check of the data, giving the operator immediate feedback on abnormal conditions affecting the instruments and then the observations (see Ref. [30]). The offsite software, on the other hand, is responsible for storing and reducing those data produced onsite [31].

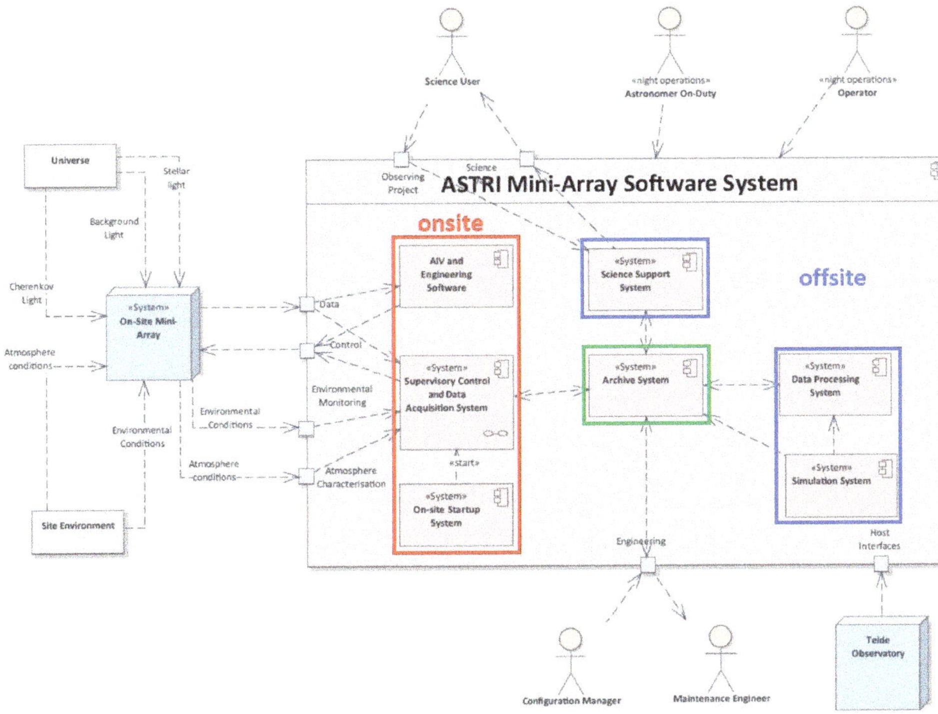

Figure 12. Context diagram of the ASTRI Mini-Array software adapted from Ref. [29]. The diagram shows the main software subsystems and their internal and external relationships. Onsite deployed software is shown inside the red square, offsite one inside the blue one. Archive software is shown in a different color because it is central to the software architecture.

The core of the online software is the Supervisor Control and Data Acquisition System (SCADA), and its function is to control all the operations carried out at the Mini-Array site, including the startup of the Mini-Array system, interfacing and communicating with all the equipment and dedicated software installed onsite. The SCADA software is being developed to minimize the interaction between human actors and the system. First of all, no human presence is foreseen at the observing site during observations so all operations will be managed remotely by an operator and an astronomer on duty. The observation sequence for the entire night, initiated by the operator, will be automatically executed by SCADA. The software is also in charge of verifying that the required conditions for those specific observations are met, before proceeding or moving to a more appropriate set of observations. Furthermore, SCADA will react to critical environmental conditions by automatically sending the array system to a safe state. Basically, all the operations are performed in an automated way with the supervision of the operator that intervenes only to react to external science alerts or to an alarm not directly managed by SCADA itself. In the first case, when the alert is produced, it has already gone through a selection process that classified it as interesting and flagged it as observable, so a new observation plan is created and provided to be executed by SCADA; then, the operator, after stopping the ongoing observation, starts the new plan. The second case occurs when the online quality check notifies workers of a problem with one of the telescopes that cannot be solved remotely. In this case, the operator, in addition to notifying the maintenance team of the problem, will have different options including to exclude the telescope from the operations. Once an observation is completed, SCADA transfers the data to the offsite data center where they are stored in the archive. No real-time data reduction is foreseen, but the data analysis policy adopted will then be the next-day processing.

The management of the ASTRI Mini-Array operations just described has several positive implications for the infrastructure supporting them; in particular, the operation centers and the onsite and offsite data centers.

As explained in Ref. [32], the ASTRI Mini-Array will have several operation centers: a local operation center and several remote ones. The local control room is located in the Themis observatory[5] and will be used during the AIT/V and commissioning phases and then for maintenance activities. One remote operation center will be located at the La Laguna IAC premises on Tenerife island, while all the others will be in Italy at some of the institutes participating in the project. Remote operations translate to there being no need for a complex local (at the site) control room and in cost reduction due to less manpower being needed at the site and less travel expenses for observers.

Two data centers, one onsite and one offsite in Rome, will support the operation centers. Thanks to the software architecture previously described, the hardware architectures of the two data centers have been specialized for their specific tasks, avoiding unnecessary duplication and so reducing the costs for their procurement. Less or more specialized hardware, especially for the onsite data center, also means lower electric power requirements and fewer maintenance activities, which, again, translates to a reduction in operation management costs.

3.2. Maintenance Activities

According to the hosting agreement between INAF and IAC, the operational lifetime of the ASTRI Mini-Array will be at least 8 years. The use of SiPM detectors, which allow work to be conducted under moderate moonlight conditions, but also the possibility to perform, alternatively, stellar intensity interferometry measurements under bright sky conditions, ensures that most nights, not affected by bad weather, will be available for observations with the ASTRI Mini-Array. In principle, to make the most efficient use of this time, the ASTRI Mini-Array system must always be available. This is clearly not possible and failures will happen. Subsystem availability requirements range from 98.5% of the mechanical structure and Cherenkov camera to 99.5% of the onsite data center and of some software subsystems. Preventing equipment failure and degradation in performance in the first instance, and then reducing the frequency of their occurrence and the time to resolve them, to keep the operational availability within the requests, is the challenge of the maintenance of any system and therefore of the ASTRI Mini-Array.

Maintenance of a complex system like the ASTRI Mini-Array starts already during its design phase. In particular, starting from lifetime, reliability, availability, and maintainability (RAM) requirements, and also taking into consideration the environmental conditions during operations, a RAM analysis on the various subsystems of the ASTRI Mini-Array and then on the integrated system has been performed. This analysis yields information, for example, on the failure rates of the most critical components, driving the choice of the number and type of spare parts necessary. Thus, following best practices (RAM analysis, management plans, etc.) is the first step in having efficient maintenance. Then, again, technological innovations/solutions can simplify some procedures, thus reducing time, manpower, and costs.

The maintenance activities for the ASTRI Mini-Array will be of the three usual types: preventive, predictive, and corrective.

3.2.1. Preventive Maintenance

Preventive maintenance consists of scheduled activities on equipment as recommended by suppliers or based on experience. Table 3 shows the preventive maintenance tasks foreseen for a single telescope focusing on the mechanical structure. All these activities require at least two people to be performed. One of the most frequent and time-consuming tasks is the greasing of some parts of the mechanical structure. In the ASTRI-Horn prototype telescope, this is performed manually. For the ASTRI Mini-Array telescopes, some technical solutions to simplify the task are under development. In particular, we are implementing a centralized lubrication system, possibly automated. In this case, the frequency and the duration of the task will be reduced as it will consist of checking and eventually refilling the main grease tank.

Table 3. ASTRI Mini-Array telescope preventive maintenance tasks.

Item	Operation	Frequency	Time
AZ bearing spur gear	Greasing	3 M	90 min
M2 load spreader	Lubrication	3 M	10 min
Electrical cabinets filters	Change	3 M	10 min
AZ bearing	Greasing	6 M	70 min
AZ bearing seals	Inspection	6 M	55 min
AZ motor	Inspection	6 M	20 min
EL actuator	Inspection and greasing	6 M	60 min
AZ/EL limit switches	Test	6 M	30 min
AZ bearing screws	Inspection	1 Y	55 min
AZ encoder	Inspection	1 Y	20 min
EL axis bearing	Inspection	1 Y	25 min
AZ stow pin	Inspection	1 Y	30 min
EL stow pin	Inspection	1 Y	40 min
M2 load spreader assembly	Inspection	1 Y	45 min
Electrical cabinets	Inspection	1 Y	45 min
LPS and grounding	Inspection	1 Y	60 min
Base structure	Inspection	3 Y	55 min
AZ fork structure	Inspection	3 Y	30 min
EL axis bearing	Greasing	3 Y	15 min
EL hinges	Greasing	3 Y	20 min
AZ stow pin	Greasing	3 Y	30 min
EL stow pin	Greasing	3 Y	35 min
External electrical conduits	Inspection	3 Y	120 min
AZ motor	Oil change	5 Y	90 min
M2 load spreader	Greasing	10 Y	120 min

3.2.2. Predictive Maintenance

Predictive maintenance is based on monitoring, collecting data, and analyzing trends, raising a flag when certain limits, indicating the equipment will fail, are reached, so triggering the replacement of the equipment before its actual failure. To this aim, each telescope is equipped with a condition monitoring system made of several sensors that monitor the behavior of critical elements (e.g., motors). Furthermore, we are developing specific software tools/models to analyze data produced by the condition monitoring system, using the ASTRI-Horn prototype telescope, that has been in operation since 2015, as a test bench [33].

3.2.3. Corrective Maintenance

Corrective maintenance consists of repairing or replacing a defective element. We foresee three procedures:

- Removal and Replacement of the defective item. Depending on the subsystem, the item can be the subsystem itself or an element of it. This will be the normal corrective maintenance procedure. In the case where the item is repairable, then it will be

repaired either at the La Laguna site or sent to the manufacturer. The restored item will remain as a spare part.
- Removal and Repair of the defective item. In this case, once repaired, the item is restored. The repair will happen at the site if possible.
- Repair of the defective item without removal. An example of this is the misalignment of the panels of the primary mirrors, considered a maintenance activity.

For the ASTRI Mini-Array, corrective maintenance is therefore crucial to identify those items, called Lowest Line Replaceable Units (LLRUs), which are the ones that have to be removed and replaced in case of failure and for which an appropriate number of spares must be available. Some complex but still manageable subsystems, such as the Cherenkov camera or the M2 subsystem, were identified as LLRUs. The rationale behind this choice is to minimize the downtime of the ASTRI Mini-Array, since repairing the subsystem would require longer than simply replacing it.

Another important feature of the management of maintenance activities is the use of a Computerized Maintenance Management System (CMMS) application developed for the management of mobile assets (buildings, infrastructure, etc.) and technical devices (https://www.openmaint.org/en/ accessed 15 March 2024). The software has been specifically configured to satisfy the ASTRI Mini-Array necessities. The application has complete knowledge of the ASTRI Mini-Array inventory and of its status, with it being linked to the monitoring, logging, and alarm system [34], that is part of SCADA. From the point of view of preventive maintenance, this leads to the accurate scheduling and assignment of activities and then to an efficient response, while in the case of corrective maintenance, this allows for the prompt identification of problems and, consequently, a fast response. A database of spare parts and consumables used in maintenance activities is fully integrated in the system.

4. Conclusions

The ASTRI Mini-Array is an INAF project to build, install, and operate nine innovative Imaging Atmospheric Cherenkov telescopes at the Teide Astronomical Observatory in collaboration with FGG and IAC. We are currently installing the telescopes at the site. The facility will operate for at least 8 years and will be the largest facility of IACT arrays until the CTAO starts operations. Based on the currently available information, operations should start in the second half of 2025. In this paper, we have reviewed technological solutions and innovations, and we have adopted or developed these to face the challenges of building, installing, operating and maintaining this facility. We have shown how, starting from the selected optical design, in itself a novelty, a number of technological innovations have been derived. Some of them had an impact just on the performance, while others also led also to simplification of the system in terms of production, maintainability, and requests on the infrastructure (for example, power and data management). When it comes to operations management, a central role is played by the software. In the case of the ASTRI Mini-Array, automation has been the goal. Automatic procedures remotely controlled and supervised will run the facility and also analyze the scientific data. Last but not least, maintenance activities will be in place whose basic philosophy is to have a system aimed at preventing equipment failure and degradation in performance, in order to guarantee the high degree of availability necessary to guarantee the scientific results, but at the same time, consistent with safety and the low cost of operation.

Funding: This work was conducted in the context of the ASTRI Project thanks to the support of the Italian Ministry of University and Research (MUR) as well as the Ministry for Economic Development (MISE), with funds explicitly assigned to the Italian National Institute of Astrophysics (INAF). We acknowledge the support of the Brazilian Funding Agency FAPESP (Grant No. 2013/10559-5), the South African Department of Science and Technology through Funding Agreement 0227/2014 for the South African Gamma-Ray Astronomy Program, and the ANID-Basal Fund, Project FB0008 (AC3E). IAC is supported by the Spanish Ministry of Science and Innovation (MICIU). They are partially supported by H2020-ASTERICS, a project funded by the European Commission Framework Programme Horizon 2020 Research and Innovation action under Grant Agreement No. 653477.

Data Availability Statement: No new data were created or analyzed in this study. Data sharing is not applicable to this article.

Acknowledgments: The ASTRI project is becoming a reality thanks to Giovanni "Nanni" Bignami and Nicolò "Nichi" D'Amico, two outstanding scientists who, in their capability as INAF Presidents, provided continuous support and invaluable guidance. While Nanni was instrumental in starting the ASTRI telescope, Nichi transformed it into the Mini Array in Tenerife. Now, the project is being built owing to the unfaltering support of Marco Tavani, the current INAF President. Paolo Vettolani and Filippo Zerbi, the past and current INAF Science Directors, and Massimo Cappi, the Coordinator of the High Energy branch of INAF, have also been very supportive of our work. We are very grateful to all of them. Unfortunately, Nanni and Nichi passed away, but their vision still guides us. Figures 2–5 are part of an internal report of the ASTRI Mini-Array and are reproduced with the permission of the author Giorgia Sironi. Figures 8 and 9 are part of an internal report of the ASTRI Mini-Array and are reproduced with the permission of the author Osvaldo Catalano. This article went through the internal ASTRI review process.

Conflicts of Interest: The author declares no conflict of interest.

Abbreviations

The following abbreviations are used in this manuscript:

ALMA	Atacama Large Millimeter/Submillimeter Array
ADC	Analog to Digital Converter
AIT/V	Assembly, Integration, Test/Verification
ASTRI	Astrofisica con Specchi a Tecnologia Replicante Italiana
AZ	Azimuth
CMMS	Computerized Maintenance Management System
CTAO	Cherenkov Telescope Array Observatory
D80	The diameter of the circle within which 80% of the photons in the ray tracing analysis fall.
EL	Elevation
FPGA	Field Programmable Gate Array
FGG	Fundacion Galileo Galilei
FoV	Field of View
GCT	Gamma-ray Cherenkov Telescope
HAWC	High-Altitude Water Cherenkov
IAC	Instituto de Astrofisica de Canarias
IACT	Imaging Atmospheric Cherenkov Telescope
INAF	Istituto Nazionale di Astrofisica
LHAASO	Large High Altitude Air Shower Observatory
LLRU	Lowest Line Replaceable Unit
NSB	Night Sky Background
PDE	Photon Detection Efficiency
PSF	Point Spread Function
PWN	Planetary Wind Nebulae
RAM	Reliability, Availability, Maintainability
SCADA	Supervisor Control and Data Acquisition System
SCT	Schwarzschild–Couder Telescope
SiPM	Silicon Photomultiplier
SNR	Supernova Remnant
SST	Small-Sized Telescope
TNG	Telescopio Nazionale Galileo
VLA	Very Large Array

Notes

1 www.tng.iac.es (accessed on 5 March 2024)

2 https://www.weeroc.com (accessed on 5 March 2024)

3 https://science.nrao.edu/facilities/vla (accessed on 5 March 2024)

4 https://www.almaobservatory.org/en/home/ (accessed on 5 March 2024)

5 http://themis.iac.es/ (accessed on 5 March 2024)

References

1. Cherenkov Telescope Array Consortium; Acharya, B.S.; Agudo, I.; Al Samarai, I.; Alfaro, R.; Alfaro, J.; Alispach, C.; Alves Batista, R.; Amans, J.P.; Amato, E.; et al. *Science with the Cherenkov Telescope Array*; World Scientific: Singapore, 2019. [CrossRef]
2. Pareschi, G. The ASTRI SST-2M prototype and mini-array for the Cherenkov Telescope Array (CTA). In Proceedings of the Ground-Based and Airborne Telescopes VI, Edinburgh, UK, 26 June–1 July 2016; p. 99065T. [CrossRef]
3. Lombardi, S.; Catalano, O.; Scuderi, S.; Antonelli, L.A.; Pareschi, G.; Antolini, E.; Arrabito, L.; Bellassai, G.; Bernlöhr, K.; Bigongiari, C.; et al. First detection of the Crab Nebula at TeV energies with a Cherenkov telescope in a dual-mirror Schwarzschild-Couder configuration: The ASTRI-Horn telescope. *Astron. Astrophys.* **2020**, *634*, A22. [CrossRef]
4. Vercellone, S.; Bigongiari, C.; Burtovoi, A.; Cardillo, M.; Catalano, O.; Franceschini, A.; Lombardi, S.; Nava, L.; Pintore, F.; Stamerra, A.; et al. ASTRI Mini-Array core science at the Observatorio del Teide. *J. High Energy Astrophys.* **2022**, *35*, 1–42. [CrossRef]
5. Abeysekara, A.U.; Albert, A.; Alfaro, R.; Alvarez, C.; Álvarez, J.D.; Arceo, R.; Arteaga-Velázquez, J.C.; Ayala Solares, H.A.; Barber, A.S.; Bautista-Elivar, N.; et al. Observation of the Crab Nebula with the HAWC Gamma-Ray Observatory. *Astrophys. J.* **2017**, *843*, 39. [CrossRef]
6. Cao, Z. A future project at tibet: the large high altitude air shower observatory (LHAASO). *Chin. Phys. C* **2010**, *34*, 249–252. [CrossRef]
7. Vercellone, S. Science with the ASTRI Mini-Array: From Experiment to Open Observatory. *Universe* **2024**, *10*, 94. [CrossRef]
8. Saturni, F.G.; Arcaro, C.H.E.; Balmaverde, B.; Becerra González, J.; Caccianiga, A.; Capalbi, M.; Lamastra, A.; Lombardi, S.; Lucarelli, F.; Alves Batista, R.; et al. Extragalactic observatory science with the ASTRI mini-array at the Observatorio del Teide. *J. High Energy Astrophys.* **2022**, *35*, 91–111. [CrossRef]
9. D'Aì, A.; Amato, E.; Burtovoi, A.; Compagnino, A.A.; Fiori, M.; Giuliani, A.; La Palombara, N.; Paizis, A.; Piano, G.; Saturni, F.G.; et al. Galactic observatory science with the ASTRI Mini-Array at the Observatorio del Teide. *J. High Energy Astrophys.* **2022**, *35*, 139–175. [CrossRef]
10. White, R.; Amans, J.P.; Berge, D.; Bonanno, G.; Bose, R.B.; Brown, A.M.; Buckley, J.H.; Chadwick, P.M.; Conte, F.; Cotter, G.; et al. The Small-Sized Telescopes for the Southern Site of the Cherenkov Telescope Array. In Proceedings of the 37th International Cosmic Ray Conference, Berlin, Germany, 12–23 July 2021; p. 728, [CrossRef]
11. La Palombara, N.; Sironi, G.; Giro, E.; Scuderi, S.; Canestrari, R.; Iovenitti, S.; Garczarczyk, M.; Krause, M.; Diebold, S.; Millul, R.; et al. Mirror production for the Cherenkov telescopes of the ASTRI mini-array and the MST project for the Cherenkov Telescope Array. *J. Astron. Telesc. Instrum. Syst.* **2022**, *8*, 014005. [CrossRef]
12. Vassiliev, V.; Fegan, S.; Brousseau, P. Wide field aplanatic two-mirror telescopes for ground-based γ-ray astronomy. *Astroparticle Physics* **2007**, *28*, 10–27. [CrossRef]
13. Sironi, G. Aplanatic telescopes based on Schwarzschild optical configuration: from grazing incidence Wolter-like X-ray optics to Cherenkov two-mirror normal incidence telescopes. In Proceedings of the 2017 Society of Photo-Optical Instrumentation Engineers (SPIE) Conference Series, San Diego, CA, USA, 6–10 August 2017; Volume 10399, p. 1039903. [CrossRef]
14. Giro, E.; Canestrari, R.; Sironi, G.; Antolini, E.; Conconi, P.; Fermino, C.E.; Gargano, C.; Rodeghiero, G.; Russo, F.; Scuderi, S.; et al. First optical validation of a Schwarzschild Couder telescope: the ASTRI SST-2M Cherenkov telescope. *Astron. Astrophys.* **2017**, *608*, A86. [CrossRef]
15. Sol, H.; Greenshaw, T.; Le Blanc, O.; White, R.; GCT project, C. Observing the sky at extremely high energies with CTA: Status of the GCT project. In Proceedings of the 2017 35th International Cosmic Ray Conference (ICRC2017), Busan, Republic of Korea, 10–20 July 2017; Volume 301, p. 822. [CrossRef]
16. Adams, C.B.; Alfaro, R.; Ambrosi, G.; Ambrosio, M.; Aramo, C.; Arlen, T.; Batista, P.I.; Benbow, W.; Bertucci, B.; Bissaldi, E.; et al. Detection of the Crab Nebula with the 9.7 m prototype Schwarzschild-Couder telescope. *Astropart. Phys.* **2021**, *128*, 102562, [CrossRef]
17. Ghigo, M.; Basso, S.; Canestrari, R.; Proserpio, L. Development of hot slumping technique and last optical performances obtained on a 500mm diameter slumped segment prototype for adaptive optics. In Proceedings of the Astronomical and Space Optical Systems, San Diego, CA, USA, 2–6 August 2009; p. 74390M. [CrossRef]
18. Canestrari, R.; Pareschi, G.; Parodi, G.; Martelli, F.; Missaglia, N.; Banham, R. Cold-shaping of thin glass foils as a method for mirror processing: from basic concepts to mass production of mirrors. *Opt. Eng.* **2013**, *52*, 051204. [CrossRef]
19. Sottile, G.; Sangiorgi, P.; Gargano, C.; Lo Gerfo, F.; Corpora, M.; Catalano, O.; Impiombato, D.; Mollica, D.; Capalbi, M.; Mineo, T.; et al. The ASTRI Cherenkov Camera: from the prototype to the industrial version for the Mini-Array. *arXiv* **2023**, arXiv:2301.09915. [CrossRef]
20. Dorner, D.; Adam, J.; Ahnen, L.M.; Baack, D.; Balbo, M.; Biland, A.; Blank, M.; Bretz, T.; Bruegge, K.; Bulinski, M.; et al. FACT—Highlights from more than Five Years of Unbiased Monitoring at TeV Energies. In Proceedings of the 35th International Cosmic Ray Conference (ICRC2017), Busan, Republic of Korea, 10–20 July 2017; Volume 301, p. 609. [CrossRef]

21. Aharonian, F.; An, Q.; Axikegu; Bai, L.X.; Bai, Y.X.; Bao, Y.W.; Bastieri, D.; Bi, X.J.; Bi, Y.J.; Cai, H.; et al. Construction and on-site performance of the LHAASO WFCTA camera. *Eur. Phys. J. C* **2021**, *81*, 657. [CrossRef]
22. Bonanno, G.; Romeo, G.; Occhipinti, G.; Timpanaro, M.C.; Grillo, A. Characterization method to achieve simultaneous absolute PDE measurements of all pixels of an ASTRI Mini-Array camera tile. *Nucl. Instrum. Methods Phys. Res. A* **2020**, *980*, 164489. [CrossRef]
23. Tejedor, L.A.; Barrio, J.A.; Peñil, P.; Pérez, A.; Herranz, D.; Martín, J. A Trigger Interface Board for the Large and Medium Sized telescopes of the Cherenkov Telescope Array. *Nucl. Instrum. Methods Phys. Res. A* **2022**, *1027*, 166058. [CrossRef]
24. Germani, S.; Lombardi, S.; La Parola, V.; Lucarelli, F.; Saturni, F.G.; Bigongiari, C.; Cardillo, M.; Mineo, T. The Stereo Event Builder software system of the ASTRI Mini-Array project. In Proceedings of the Society of Photo-Optical Instrumentation Engineers (SPIE) Conference Series, Montréal, QB, Canada, 17–23 July 2022; Volume 12189, p. 121891R. [CrossRef]
25. Benn, C.R.; Ellison, S.L. Brightness of the night sky over La Palma. *New Astron. Rev.* **1998**, *42*, 503–507. [CrossRef]
26. Romeo, G.; Bonanno, G.; Sironi, G.; Timpanaro, M.C. Novel silicon photomultipliers suitable for dual-mirror small-sized telescopes of the Cherenkov telescope array. *Nucl. Instrum. Methods Phys. Res. A* **2018**, *908*, 117–127. [CrossRef]
27. Marchiori, G.; Busatta, A.; Marcuzzi, E.; Manfrin, C.; Folla, I.; Pareschi, G.; Scuderi, S.; Giro, E.; Canestrari, R.; Tosti, G.; et al. ASTRI SST-2M: the design evolution from the prototype to the array telescope. In Proceedings of the Ground-Based and Airborne Telescopes VII, Austin, TX, USA, 10–15 June 2018; p. 107005W. [CrossRef]
28. Canestrari, R.; Giro, E.; Sironi, G.; Antolini, E.; Fermino, C.E.; Fugazza, D.; Gargano, C.; Russo, F.; Scuderi, S.; Tosti, G.; et al. The ASTRI SST-2M prototype for the Cherenkov Telescope Array: status after the commissioning phase of the telescope. In Proceedings of the Society of Photo-Optical Instrumentation Engineers (SPIE) Conference Series, San Diego, CA, USA, 6–10 August 2017; Volume 10399, p. 1039904. [CrossRef]
29. Bulgarelli, A.; Lucarelli, F.; Tosti, G.; Conforti, V.; Parmiggiani, N.; Schwarz, J.H.; Gallardo, J.G.A.; Antonelli, L.A.; Araya, M.; Balbo, M.; et al. Software architecture and development approach for the ASTRI Mini-Array project at the Teide Observatory. *J. Astron. Telesc. Instrum. Syst.* **2024**, *10*, 017001. [CrossRef]
30. Parmiggiani, N.; Bulgarelli, A.; Baroncelli, L.; Addis, A.; Fioretti, V.; Di Piano, A.; Capalbi, M.; Catalano, O.; Conforti, V.; Fiori, M.; et al. The online observation quality system software architecture for the ASTRI Mini-Array project. In Proceedings of the 2022 Software and Cyberinfrastructure for Astronomy VII, Montréal, QB, Canada, 17–23 July 2022; Volume 12189, p. 121892H. [CrossRef]
31. Lombardi, S.; Lucarelli, F.; Bigongiari, C.; Gallozzi, S.; Cardillo, M.; Mastropietro, M.; Saturni, F.G.; Visconti, F.; Antonelli, L.A.; Bulgarelli, A.; et al. The data processing, simulation, and archive systems of the ASTRI Mini-Array project. In Proceedings of the 2022 Society of Photo-Optical Instrumentation Engineers (SPIE) Conference Series, Montréal, QB, Canada, 17–23 July 2022; Volume 12189, p. 121890P. [CrossRef]
32. Scuderi, S.; Giuliani, A.; Pareschi, G.; Tosti, G.; Catalano, O.; Amato, E.; Antonelli, L.A.; Becerra Gonzàles, J.; Bellassai, G.; Bigongiari, C.; et al. The ASTRI Mini-Array of Cherenkov telescopes at the Observatorio del Teide. *J. High Energy Astrophys.* **2022**, *35*, 52–68. [CrossRef]
33. Incardona, F.; Costa, A.; Munari, K. Failure type detection and predictive maintenance for the next generation of imaging atmospheric Cherenkov telescopes. *arXiv* **2022**, arXiv:2212.12381. [CrossRef]
34. Incardona, F.; Costa, A.; Munari, K.; Gambadoro, S.; Germani, S.; Bruno, P.; Bulgarelli, A.; Conforti, V.; Gianotti, F.; Grillo, A.; et al. The monitoring, logging, and alarm system of the ASTRI mini-array gamma-ray air-Cherenkov experiment at the Observatorio del Teide. In Proceedings of the 2022 Software and Cyberinfrastructure for Astronomy VII, Montréal, QB, Canada, 17–23 July 2022; Volume 12189, p. 121891E. [CrossRef]

universe MDPI

Review

Science with the ASTRI Mini-Array: From Experiment to Open Observatory

Stefano Vercellone † on behalf of the ASTRI Project

INAF Osservatorio Astronomico di Brera, Via E. Bianchi 46, 23807 Merate, LC, Italy; stefano.vercellone@inaf.it; Tel.: +39-02-72320-509
† http://www.astri.inaf.it/en/library/ (accessed on 21 January 2024).

Abstract: Although celestial sources emitting in the few tens of GeV up to a few TeV are being investigated by imaging atmospheric Čerenkov telescope arrays such as H.E.S.S., MAGIC, and VERITAS, at higher energies, up to PeV, more suitable instrumentation is required to detect ultra-high-energy photons, such as extensive air shower arrays, as HAWC, LHAASO, Tibet AS-γ. The Italian National Institute for Astrophysics has recently become the leader of an international project, the ASTRI Mini-Array, with the aim of installing and operating an array of nine dual-mirror Čerenkov telescopes at the Observatorio del Teide in Spain starting in 2025. The ASTRI Mini-Array is expected to span a wide range of energies (1–200 TeV), with a large field of view (about 10 degrees) and an angular and energy resolution of $\sim$3 arcmin and $\sim$10%, respectively. The first four years of operations will be dedicated to the exploitation of Core Science, with a small and selected number of pointings with the goal of addressing some of the fundamental questions on the origin of cosmic rays, cosmology, and fundamental physics, the time-domain astrophysics and non γ-ray studies (e.g., stellar intensity interferometry and direct measurements of cosmic rays). Subsequently, four more years will be dedicated to Observatory Science, open to the scientific community through the submission of observational proposals selected on a competitive basis. In this paper, I will review the Core Science topics and provide examples of possible Observatory Science cases, taking into account the synergies with current and upcoming observational facilities.

Keywords: ASTRI; imaging atmospheric Čerenkov arrays; very high-energy; γ-ray astrophysics; astro-particle

Citation: Vercellone, S., on behalf of the ASTRI Project. Science with the ASTRI Mini-Array: From Experiment to Open Observatory. *Universe* **2024**, *10*, 94. https://doi.org/10.3390/universe10020094

Academic Editor: Fridolin Weber

Received: 22 January 2024
Revised: 8 February 2024
Accepted: 13 February 2024
Published: 16 February 2024

1. Introduction

About 300 celestial sources are currently known to emit in the $0.1 < E < 30$ TeV energy range (see the TeVCat Webpage[1] [1]) based on their detection by the major imaging atmospheric Čerenkov telescope arrays (IACTs), such as H.E.S.S. [2], MAGIC [3], and VERITAS [4], whose energy range extends up to a few tens of TeV. Alternatively, extended air shower arrays (EAS) such as HAWC [5], LHAASO [6] and Tibet AS-γ [7], adopt a different detection technique that allows us to investigate energies up to several hundreds of TeV, reaching the PeV limit. The sources detected by the current generation of EAS at energies $E > 100$ TeV, and up to a few PeVs are a factor of ten fewer.

The Čerenkov Telescope Array Observatory (CTAO [8]) will be the next large scale Čerenkov array and will cover an energy range from a few tens of GeV up to a few hundreds of TeV by means of telescopes of different sizes (see, e.g., [9]). It will be deployed in both hemispheres to observe the full sky. A few telescope prototypes were developed in recent years, among them the ASTRI-Horn dual-mirror, Schwarzschild–Couder (SC) telescope [10], currently operating on Mount Etna in Sicily (Italy), which obtained the first-light optical qualification by means of observation of Polaris, using a dedicated optical camera [11], and the first detection of very high-energy γ-ray emission from the Crab Nebula by a Čerenkov telescope in dual-mirror SC configuration [12].

In this review, I will first describe the ASTRI Mini-Array characteristics and performance in the context of currently available very high- and ultra-high energy (VHE and UHE, respectively) instrumentation (Section 2), then briefly highlight the ASTRI Mini-Array science topics that will be pursued during the first four years of operation (Section 3). In Section 4, I will describe in detail the subsequent four years of operation and the Open Observatory Phase, when the scientific investigation is mainly driven by the community.

2. The ASTRI Mini-Array

The ASTRI Mini-Array [13,14] consists of nine ASTRI dual-mirror small-sized (SSTs) Čerenkov telescopes, currently being deployed at the Observatorio del Teide (Spain), which will commence its scientific operations in late 2025. The ASTRI Mini-Array will provide a large field of view (FoV) of about 10°, a wide energy range from 1 TeV to 200 TeV, an angular resolution of $\sim 3'$, and an energy resolution of $\sim 10\,\%$. Table 1 compares the ASTRI Mini-Array performance with that of the current IACTs.

Table 1. Performance of the ASTRI Mini-Array compared with the main current IACT arrays. References: ASTRI Mini-array [15], MAGIC [16], VERITAS [17] and https://veritas.sao.arizona.edu (accessed on 14 February 2024), H.E.S.S. [2].

Quantity	ASTRI Mini-Array	MAGIC	VERITAS	H.E.S.S.
Location	28° 18′ 04″ N 16° 30′ 38″ W	28° 45′ 22″ N 17° 53′ 30″ W	31° 40′ 30″ N 110° 57′ 7.8″ W	23° 16′ 18″ S 16° 30′ 00″ E
Altitude [m]	2390	2396	1268	1800
FoV	$\sim 10°$	$\sim 3.5°$	$\sim 3.5°$	$\sim 5°$
Angular Res.	0.05° (10 TeV)	0.07° (1 TeV)	0.07° (1 TeV)	0.06° (1 TeV)
Energy Res.	10% (10 TeV)	16% (1 TeV)	17% (1 TeV)	15% (1 TeV)
Energy Range	(0.5–200) TeV	(0.05–20) TeV	(0.08–30) TeV	(0.02–30) TeV [(a)]

Notes: [(a)]: considering the contribution of H.E.S.S.— —II telescope unit [18].

A detailed description of the ASTRI Mini-Array performance is reported in [19]. Figure 1 shows the ASTRI Mini-Array differential sensitivity (turquoise points, 50 h integration time, 5σ confidence level, C.L.) compared with those of the current major IACTs (H.E.S.S., MAGIC, and VERITAS) and of the planned CTAO. The ASTRI Mini-Array will improve the current IACT sensitivity at energies greater than a few TeVs and will be of the same order as that of CTAO North in the "alpha" (4 LSTs[2] + 9 MSTs[3] [9] configuration, and slightly better in the "science verification" (4 LSTs + 5 MSTs [20]) configuration, respectively, at energies greater than a few tens of TeVs.

The ASTRI Mini-Array will reserve, as we shall describe in Section 3, its first four years of operation for the investigation of a few specific science topics. This implies that most of the science operations will be performed as deep pointings, with exposures in the order of 200 h or even 500 h, towards specific sky regions. Figure 2 shows the ASTRI Mini-Array differential sensitivity curves for 200 h (turquoise squares) and 500 h (turquoise triangles) integration time, respectively. For such long integration times, the most appropriate comparison is with current EAS differential sensitivity curves, HAWC [21], Tibet AS-γ (Takita M., priv. comm. based on [22]), and LHAASO [23].

The main advantage of EAS with respect to IACTs is the former's 2 sr FoV and their larger duty cycle. On the other hand, as reported in Table 2, their energy and angular resolution in the same energy range as the ASTRI Mini-Array (about 10 TeV) are at least a factor of 3 to 4 times worse. Clearly, this makes the ASTRI Mini-Array extremely competitive in studying the morphology of extended sources and crowded fields and accurately monitoring multiple targets in the same pointing.

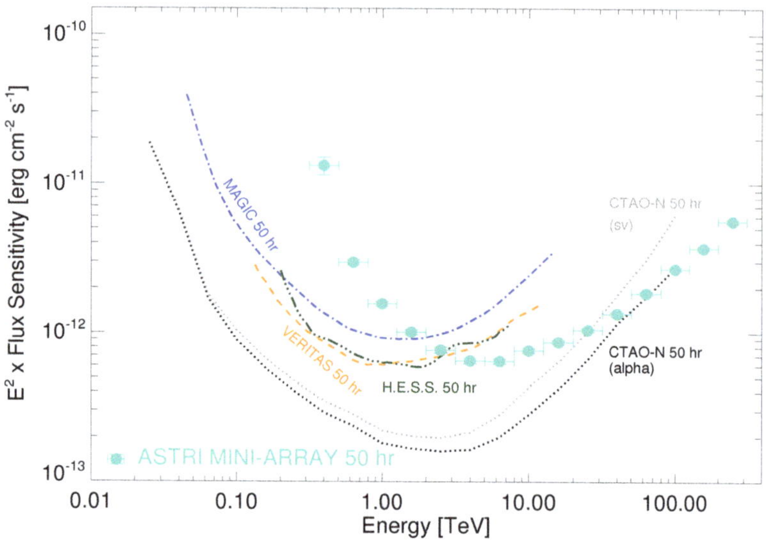

Figure 1. ASTRI Mini-Array differential sensitivity for 50 h integration compared with those of MAGIC, H.E.S.S., VERITAS, and CTAO North. The differential sensitivity curves are drawn from [19] (ASTRI Mini-Array), [16] (MAGIC), the VERITAS official website https://veritas.sao.arizona.edu (accessed on 14 February 2024), and [24] (sensitivity curve for H.E.S.S.−I, stereo reconstruction). CTAO−N "alpha configuration" (alpha) sensitivity curve comes from [9]. The CTAO−N "science verification" (sv) sensitivity is drawn from [20].

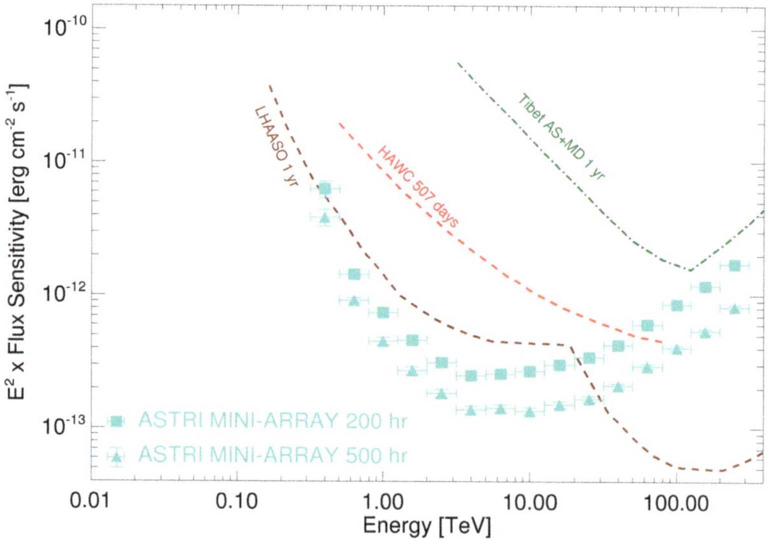

Figure 2. ASTRI Mini-Array differential sensitivity for 200 h (turquoise squares) and 500 h (turquoise triangles) integration times compared with those of HAWC (507 d), Tibet AS-γ (1 yr), and LHAASO (1 yr). The differential sensitivity curves are drawn from ASTRI Mini-Array [19], HAWC [21], LHAASO [23], and Takita M. (priv. comm.) based on [22] (Tibet AS + MD). We note that the 507-day HAWC differential sensitivity curve corresponds to about 3000 h of acquisition on a source at a declination of 22° within its field of view [21].

Table 2. Summary of the performance of the current main particle sampling arrays compared with those of the ASTRI Mini-Array. References: ASTRI Mini-array [15], HAWC [5,25], LHAASO [6], Tibet AS-γ [22,26].

Quantity	ASTRI Mini-Array	HAWC	LHAASO	Tibet AS-γ
Location	28° 18′ 04″ N 16° 30′ 38″ W	18° 59′ 41″ N 97° 18′ 27″ W	29° 21′ 31″ N 100° 08′ 15″ E	30° 05′ 00″ N 90° 33′ 00″ E
Altitude [m]	2390	4100	4410	4300
FoV	∼0.024 sr	2 sr	2 sr	2 sr
Angular Res.	0.05° (10 TeV)	0.15° [(a)] (10 TeV)	(0.24–0.32)° [(b)] (100 TeV)	0.2° [(c)] (100 TeV)
Energy Res.	10% (10 TeV)	30% (10 TeV)	(13–36)% (100 TeV) [(b)]	20% [(c)] (100 TeV)
Energy Range	(0.5–200) TeV	(0.1–1000) TeV	(0.1–1000) TeV	(0.1–1000) TeV

Notes: [(a)]: (0.15–1)° as a function of the event size. [(b)]: angular resolution is (0.70–0.94)° at 10 TeV; (0.24–0.32)° at 100 TeV; 0.15° at 1000 TeV. Energy resolution is (30–45)% at 10 TeV; (13–36)% at 100 TeV; (8–20)% at 1000 TeV [27]. [(c)]: angular resolution is ∼ 0.5°at 10 TeV and ∼0.2°at 10 TeV at 50% containment radius [22]. Energy resolution is ∼40% at 10 TeV and ∼20% at 100 TeV [26]. The different values of the LHAASO angular and energy resolution performance at a given energy have been computed at different Zenith angles, $0 < \theta < 20$, $20 < \theta < 35$, and $35 < \theta < 50$ degrees, respectively. At lower Zenith angles, the performance is better.

Furthermore, the ASTRI Mini-Array angular resolution will allow us to investigate the LHAASO uncertainty error box of Galactic sources, which is of the order of one degree [28] and study the different sources possibly associated with the PeV emission, in order to unambiguously identify them, when in synergy with GeV and X-ray facilities.

Figure 3 shows the ASTRI Mini-Array angular (left panel) and energy (right panel) resolution as a function of the energy as reported in [19].

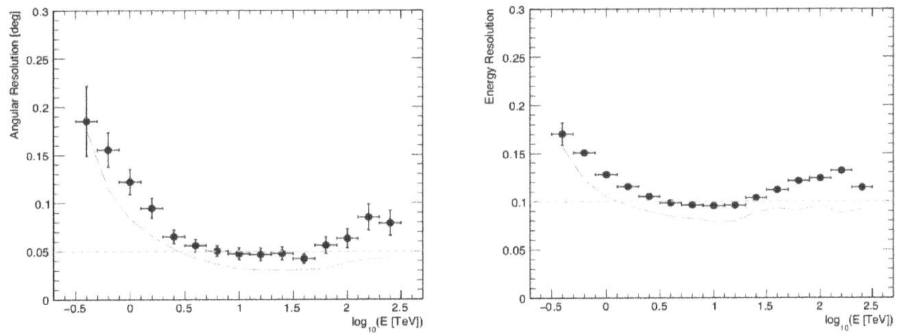

Figure 3. ASTRI Mini-Array angular (**left panel**) and energy (**right panel**) resolution as a function of the energy. The angular resolution is defined as the 68% γ-ray event containment radius (in degrees). The black points were computed with analysis cuts optimizing the differential sensitivity in 50 h; the long-dashed, dark-green lines were instead derived with analysis cuts taking into account also the angular/energy resolution in the optimization; the short-dashed, light-green lines mark the 0.05° and the 0.1% threshold for the angular and energy resolution, respectively. Adapted from [19].

3. Core Science Topics

The ASTRI Mini-Array science program will develop in two phases. During the first four years of operations, the ASTRI Mini-Array will be run as an *experiment*, while in the subsequent four years, it will gradually evolve into an *observatory* open to the scientific community.

A graphical description of the main Core Science topics that we plan to investigate during the first four years of operations is shown in Figure 4. Our Core Science Program is based on "Main Pillars". They are science fields in which the ASTRI Mini-Array will contribute breakthrough pieces of evidence to improve our understanding of a few key science questions.

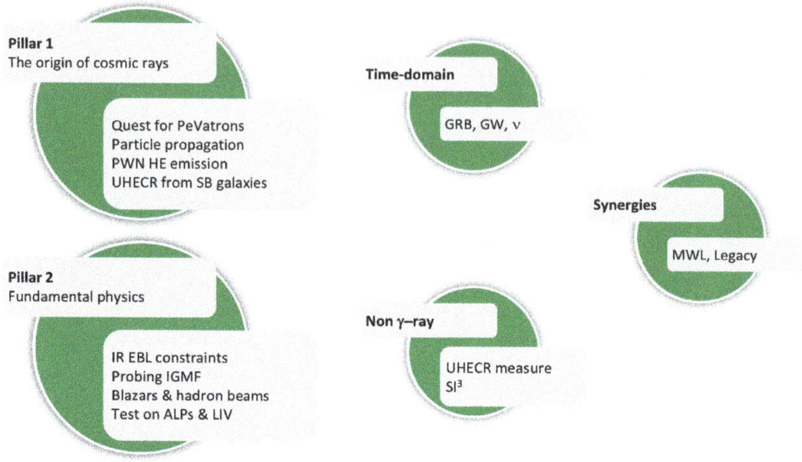

Figure 4. Graphical description of the ASTRI Mini-Array Core Science program.

Recently, [15] discussed the ASTRI Mini-Array Core Science, which includes the study of the: (1) origin of cosmic rays; (2) cosmology and fundamental physics; (3) GRBs and time-domain astrophysics; (4) direct measurements of cosmic rays; (5) stellar-intensity interferometry. Here, I will review some of the results on all science topics, while Section 4.1 will focus on the Observatory Science ones, presented in [29,30].

3.1. The Origin of Cosmic Rays

The LHAASO Collaboration [28] reported the discovery of twelve Galactic sources emitting γ-rays at several hundreds TeV up to 1.4 PeV. These sources are able to accelerate particles up to $\sim 10^{15}$ eV, making them "PeVatron candidates". We note that the majority of these sources are diffuse γ-ray structures with angular extensions up to 1°, which, together with the LHAASO limited angular resolution, make the identification of the actual sources responsible for the ultra high-energy γ-ray emission not univocal (except for the Crab Nebula). The recent publication of the First LHAASO Catalog of γ-ray Sources (1LHHASO [31]) containing 90 sources, 43 of them with emissions at energies $E > 0.1$ PeV, marks a fundamental step for the astrophysics at very high- and ultra high-energies. This discovery is extremely important for the ASTRI Mini-Array science, especially because of its angular resolution, which, at energies of about 100 TeV, is a factor of 3 to 4 times better in radius than the LHAASO one: 0.08° vs. 0.24–0.32°. We should also mention that both the angular resolution and the energy resolution can be improved by means of specific analysis cuts, as shown in Figure 3. The ASTRI Mini-Array will investigate these and future PeVatron sources, providing important information on their morphology above 10 TeV. The ASTRI Mini-Array wide FoV will be extremely important in the investigation of extended regions and point-like sources. A single pointing will allow us to investigate the Galactic Center or the Cygnus regions. In these regions, we can accumulate several hundreds of hours by also including the epochs of moderate Moon condition [32]. This is crucial to investigate both the (energy-dependent) morphology of the sources in these regions and their possible variability on a long time scale. The ASTRI Mini-Array will investigate the Galactic Center at a high Zenith angle (maximum culmination angle of $\sim$57°). We expect to be able to study this region up to $E \sim 200$ TeV for an exposure time of 260 h, significantly improving the current results of other IACTs (see for further details [15]). The high-energy boundaries of the ASTRI Mini-Array will also be important to study the

Crab Nebula, the only Galactic PeVatron[4] currently known [33,34]. The origin of the Crab Nebula γ-ray emission detected by LHAASO does not require a hadronic contribution but cannot exclude it either. A deep ASTRI Mini-Array observation lasting about 500 h in the $E > 100$ TeV energy range should definitely be able to provide constraints on the proton component in this source.

3.2. Cosmology and Fundamental Physics

IACT arrays detected extra-galactic sources since the early nineties [35]. Since then, among the 280 sources listed in TeVCat, 93 are found to be extra-galactic: 55 high-peaked BL Lacs (HBLs), 10 intermediate-peaked BL Lacs (IBLs), 9 flat-spectrum radio quasars (FSRQs), 4 Blazars, 4 Fanaroff-Riley Type (FR-I) galaxies, 2 star-bursting galaxies (SBGs), 2 BL Lacertae objects with class unclear (BL Lacs), 2 unknown type AGNs, and 5 γ-ray bursts (GRBs). Extra-galactic jetted sources are excellent probes for several science cases. They can be used to investigate the extra-galactic background light, as well as to probe, by means of variability studies, the properties of the γ-ray emitting region. They can also be useful to investigate peculiar physical phenomena, such as the existence of the axion-like particle, to test the Lorentz invariance violation, and to study the intergalactic magnetic fields.

THE EXTRA-GALACTIC BACKGROUND LIGHT (EBL)—The EBL significantly affects the spectra of jetted sources at energies that can be explored by the ASTRI Mini-Array. Moreover, the EBL direct measurement in the infra-red (IR) portion of the spectrum is particularly challenging because of the dominant contribution of our Galaxy at these wavelengths. Nevertheless, the ASTRI Mini-Array can contribute to the study of the EBL IR component given the well-known relation $\lambda_{\rm max} \simeq 1.24 \times E_{\rm TeV}$ µm, between the wavelength of the target EBL photon, $\lambda_{\rm max}$ and the energy of the γ-ray, $E_{\rm TeV}$. The IR component in the $(10 < \lambda < 100)$ µm regime represents a challenge because of the dominance of local emission from both the Galaxy and our Solar system. The preferred candidates for observations with the ASTRI Mini-Array are TeV-emitting low-redshift radio-galaxies and local star-bursting galaxies. Among the sources fulfilling these criteria, we investigated the low-redshift radio-galaxies IC 310 (z$\sim$0.0189) and M 87 (z$\sim$0.00428). For IC 310 we assumed three different spectral states: flare [36], high and low [37]. Short (5 h) and deep (200 h) observations will allow us to detect these sources in different spectral states at energies $E > 10$ TeV, thus probing the IR EBL component. Similar results can be obtained for M 87 in different spectral states, as reported by [38] (low state), [39] (high state), and [40] (flaring state).

FUNDAMENTAL AND EXOTIC PHYSICS—Blazar spectra above a few TeV are excellent probes of non-standard γ-ray propagation effects such as the presence of hadron beams (HB) in the jet of extreme BL Lac objects (E-HBL and references therein [41]), the existence of axion-like particles [42] (ALP), or the effects of the Lorentz invariance violation (LIV and references therein [41]) and the properties of inter-galactic magnetic fields (IGMF and references therein [43]). The most promising sources to detect spectral signatures induced by these effects are 1ES 0229+200 (E-HBL, z$\sim$0.139) and Mrk 501 (HBL, z$\sim$0.03298). The presence of HB implies that the spectrum of blazars extends at energies above those allowed by the standard EBL model [44]. A detection of a γ-ray photons at energies of a few tens of TeV, when the standard EBL model would imply a roll-off at a few TeV, could be the signature for the presence of the HB scenario. On the other hand, this excess at energies above a few tens of TeV could also be induced by both the ALPs and LIV effects. ALPs produce a distinctive oscillation pattern in the blazar spectrum, that could represent the unique marker for this process, but it would require a much finer energy resolution than that of the ASTRI Mini-Array to be revealed. Also, in the context of the widely studied dark matter (DM) weakly-interacting massive particles scenario, the ASTRI Mini-Array may provide interesting DM-related insights from dwarf spheroidal galaxies and Galactic center observations, particularly for the case of monochromatic γ-ray emission lines [30]. In order to address all these studies we can plan deep (in the order of 200 h) dedicated pointing, a typical exposure that could be accumulated during the first years, with the described ASTRI Mini-Array observing strategy.

3.3. Multi-Messenger and Time-Domain Astrophysics

Transients and multi-messenger studies such as γ-ray bursts (GRBs), gravitational waves (GWs), and neutrino emission (ν_s) from VHE sources are indeed the new frontiers of high-energy astrophysics.

γ-RAY BURSTS—GRBs have only been detected by IACTs starting from 2018 and, at the time of writing, we only count six[5] GRBs detected at energies in excess of 0.1 TeV: GRB 160821B ($z = 0.162$, MAGIC] [45]), GRB 180720B ($z = 0.653$, H.E.S.S. [46]), GRB 190114C ($z = 0.424$, MAGIC [47]), GRB 190829A ($z = 0.078$, H.E.S.S. [48]), GRB 201015A ($z = 0.42$, MAGIC [49]), GRB 201216C ($z = 1.1$, MAGIC [50]), GRB 221009A ($z = 0.151$, LHAASO] [51]). Two of them are particularly relevant for the ASTRI Mini-Array. GRB 190114C is the first GRB detected at VHE within one minute from the T_0, up to an energy of about 1 TeV. GRB 221009A, described as *"to be a once-in-10,000-year event"* [52,53] was detected by LHAASO up to 13 TeV [54], challenging the standard emission scenario for the canonical EBL absorption [55,56]. Taking into account the energetics observed in GRB 221009A, we can estimate that this event, placed at a different redshift ($z = 0.078, 0.25, 0.42$), can be detected up and above 10 TeV by the ASTRI Mini-Array within a few minutes from the event (L. Nava, Priv. Comm.).

NEUTRINOS—AGNs can be sources of extra-galactic ν_s. The IceCube data [57] seem to indicate that there could be an association between ν_s emission and a few AGNs: a Seyfert-2 galaxy (NGC 1068, $D = 14.4$ Mpc), and two BL Lac objects (TXS 0506 + 056, $z = 0.3365$; PKS 1424 + 240, $z = 0.16$). Although the two latter sources are known TeV emitters, NGC 1068 shows prominent emission only in the 0.1–300 GeV energy band [58,59]. NGC 1068 could show emission above 10 TeV under the assumption of a dominant contribution of relativistic particles accelerated by the AGN-driven wind, as discussed in Section 4.4.

3.4. Non γ-ray Astrophysics

STELLAR INTENSITY INTERFEROMETRY—Stellar intensity interferometry (SII) is based on the second-order coherence of light, which allows imaging sources at the level of 100 μas. This means that it is possible to reveal details on the surface and of the environment surrounding bright stars in the sky, which typically have angular diameters of 1–10 mas. The SII observing mode will take advantage of an additional, dedicated instrument that is being designed and will be installed on the ASTRI Mini-Array telescopes [60].

DIRECT MEASUREMENTS OF COSMIC RAYS—More than 99% of the signal acquired by the ASTRI Mini-Array is hadronic in nature. In particular, this hadronic component could be useful to investigate the cosmic ray composition in the TeV–PeV energy range and the measurement of the cosmic ray spectrum at energies characteristics of its "knee".

3.5. Synergies with Other Facilities

The ASTRI Mini-Array will be operating during a period when several facilities will cover the whole electromagnetic spectrum, from radio to PeV. The Sardinia radio telescope (SRT) will complement VHE observation with radio data at different frequencies, both for Galactic and extra-galactic objects. Galactic sources already observed with SRT are W 44, IC 433, and Tycho [61,62]. In the optical energy band, the Telescopio Nazionale Galileo [63] and the GASP/WEBT Consortium [64] can provide excellent coverage, as well as several facilities managed by Instituto de Astrofísica de Canarias (IAC) at the Canary Island. In the X-ray energy band, in addition to the long-standing ESA and NASA legacy Observatories, we can now exploit the eROSITA [65] surveys and, in particular, the IXPE [66] X-ray polarimetric data. At the extreme energy boundary, LHAASO, HAWC, and Tibet AS-γ will extend data at energies of a few PeV.

The ASTRI Mini-Array location at the Observatorio del Teide and its collaboration with the IAC will allow us to investigate sources synergically with both the MAGIC and CTAO−N arrays. In particular, they will be of paramount importance for their capability to investigate not only the local Universe but also to reach redshifts well beyond one and perform cosmological studies on extra-galactic sources. Moreover, both MAGIC and CTAO−N will allow us to extend the ASTRI Mini-Array spectral performance in the sub-TeV regime, with almost no breaks from a few tens of GeV up to hundreds of TeV.

4. The Observatory Phase

The ASTRI Mini-Array science program will gradually evolve from an experiment towards an Observatory Phase, built on the experience and results from the Core Science phase, and open to observational proposals from the scientific community at large. We foresee important synergies with the above-mentioned facilities to yield the best scientific return from the proposed observations.

An example of such synergies is illustrated in Figure 5, where we show, in Galactic coordinates and Aitoff projection, the First LHAASO Catalog of γ-ray Sources (1LHHASO, [31]) plotted as orange dots and, superimposed, the ASTRI Mini-Array Core Science target regions for both Pillar-1 (blue circles) and Pillar-2 (red circles) targets. Some 1LHAASO sources already overlap the ASTRI Mini-Array selected Pillar regions, in particular along the Galactic Plane. We also note that Mrk 501 and Mrk 421 are natural candidates for common variability studies.

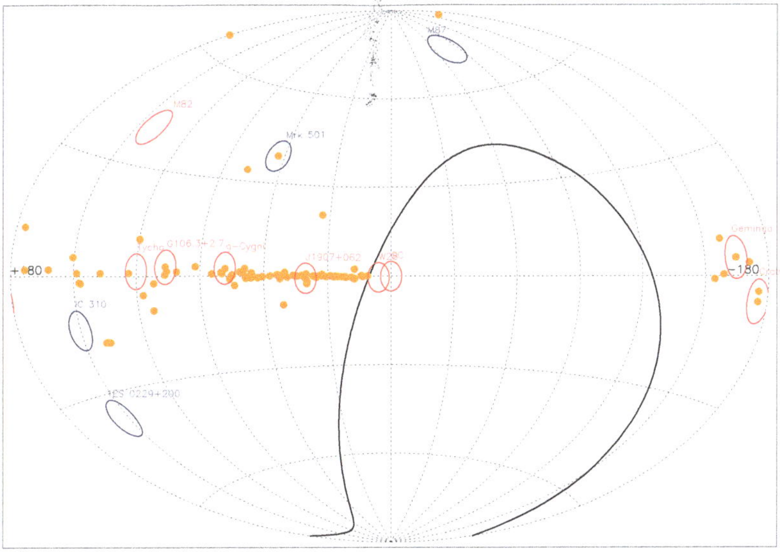

Figure 5. Pillar-1 and Pillar-2 target regions (red and blue circles, ≈10° in diameter) and 1LHAASO sources (orange dots) on the sky in Galactic coordinates (Aitoff projection). The black solid line shows the declination limit for the ASTRI Mini-Array pointings.

We can now discuss a few examples of foreseeable investigations that can be performed during the Observatory Phase. Since this phase will be open to the scientific community through competitive proposals, these examples represent ideas on how to best employ the ASTRI Mini-Array capabilities.

4.1. Cygnus Region Mini-Survey

The ASTRI Mini-Array wide FoV is well suited to perform mini-surveys of selected sky regions. One of the most important ones, as shown in Figure 5, is the Cygnus Region

($60° < l < 90°$), which contains several sources emitting above a few TeV, as reported in the First LHAASO Catalog. Figure 6 shows the results of a possible ASTRI Mini-Array mini-survey of this region with different exposure times, 50 h, 100 h, and 200 h from top to bottom, respectively.

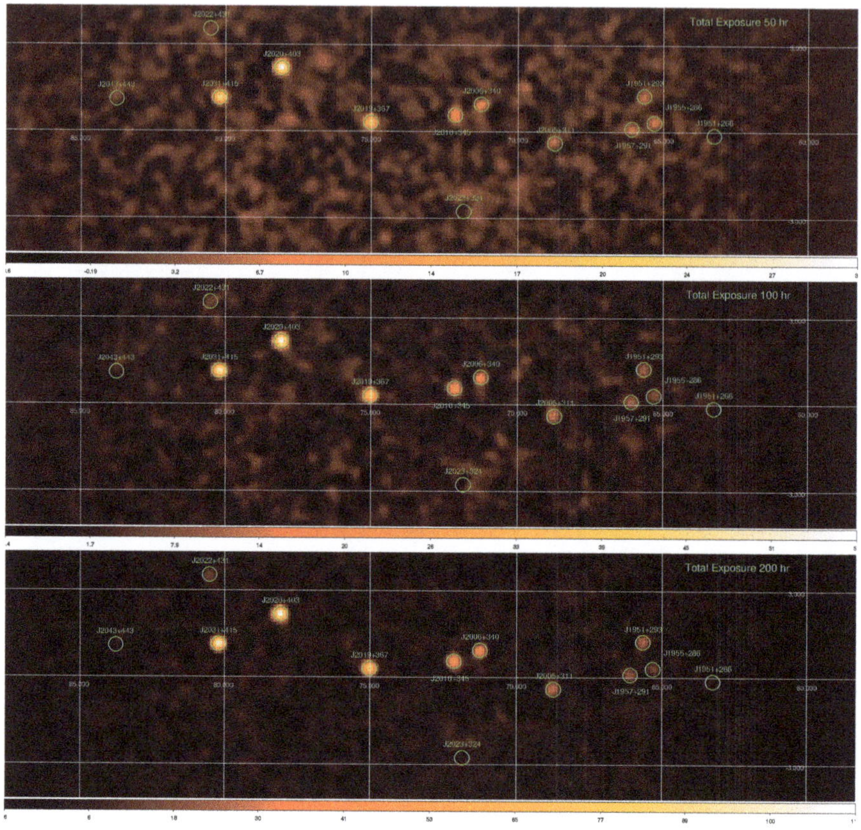

Figure 6. ASTRI Mini-Array simulations of the Cygnus region mini-survey. The count maps were produced assuming for each pointing an exposure of 1 h (**top panel**), 2 h (**middle panel**), and 4 h (**bottom panel**), respectively. Sky map units are counts/pixels. From [29].

The simulations, discussed in [29], combined fifty different pointings, at the same Galactic latitude and spaced by 0.4° in Galactic longitude, from (l, b) = (64, 0) to (l, b) = (84, 0), and lasting 1 h, 2 h, and 4 h hours each, respectively. The very high-energy simulated sources were drawn from the Third HAWC Catalog of very high-energy γ-ray sources (3HWC [67]). Thirteen of them fall inside the area considered and were simulated according to their published spectral parameters. Ten of these very high-energy sources are always significantly detected by the ASTRI Mini-Array, even at the shortest (50 h) exposure time. Recently, [68] reported the detection of an extend (about 6° in diameter) γ-ray emission centered on Cygnus-X (l,b)$\approx$(80°, 0°) with 66 photon-like events with energies greater than 400 TeV (see Figure 1 in [68]). The ASTRI Mini-Array wide FoV ($\sim$10° in diameter) and the stable off-axis performance (see Section 2) will allow us to investigate this region with a single pointing and prolonged exposure, performing a more accurate morphological measurement on the core region of the bubble discovered by LHAASO.

4.2. Gamma-Ray Binaries—LS 5039

A recent study [69] discusses the properties of γ-ray binaries. We currently know ten non-transient γ-ray binaries: seven have a compact source (six are located in our Galaxy and one, LMC P3, in the Large Magellanic Cloud), while three are colliding-wind binaries. We discuss the results of ASTRI Mini-Array simulations of LS 5039 to show the ASTRI Mini-Array capabilities of reproducing both the folded light-curve and the spectrum in different orbital phases for this source. We simulated 300 h of total exposure, 250 h in the low state, and 50 h in the high state (see [70] for a detailed description of the different flux levels). The orbit-averaged spectrum[6], described in [71], is a cut-off power-law with $\Gamma = 2.06 \pm 0.05$ and $E_{\rm cut} = 13.0 \pm 4.1$ TeV. We simulated a fixed exposure time of 10 h for each phase bin. Figure 7 shows the simulation results, the flux (left panel), and the 1 σ uncertainty ($\delta\Gamma$) on the spectral index (right panel) as a function of the orbital phase. The simulated source flux is fully consistent with the flux expected from the model. Moreover, for 90% of the orbital phase, the uncertainty on the photon index, $\delta\Gamma$, is between 0.1 and 0.25, while only in the case of the lowest-flux bin (phase range 0.1–0.2) its value rises up to about 0.4.

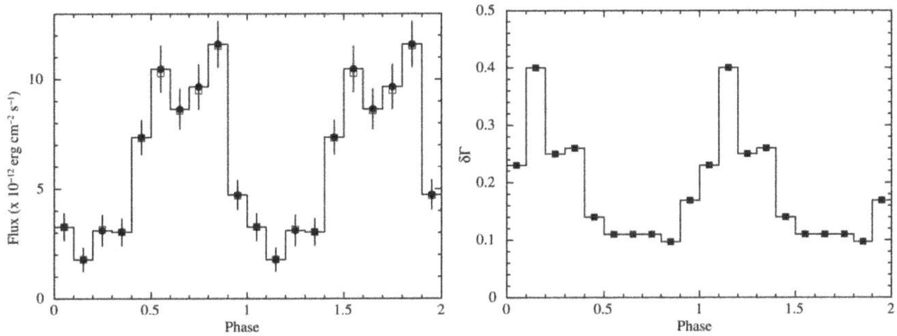

Figure 7. ASTRI Mini-Array simulations of LS 5039. **Left panel**: orbital modulation obtained with 10 h-long simulations per orbital phase bin. The open squares are the expected fluxes from the models, while the filled circles are the simulated fluxes in 0.8–200 TeV. Error bars are at 1 σ C.L. **Right panel**: 1 σ uncertainty ($\delta\Gamma$) on the spectral index obtained for 10 h-long simulations per orbital bin. From [29].

4.3. Spectral Features—Mrk 501

Mrk 501 ($z = 0.032983 \pm 0.00005$) is the second extra-galactic source detected at VHE [72]. It is classified as a high-synchrotron-peaked BL Lac object, which means that the synchrotron peak of the usual double-humped blazar spectral energy distribution reaches the ultra-violet energy band or even higher frequencies ($\nu_{\rm peak} \gtrsim 10^{15}$ Hz). This source is extremely variable, at almost all frequencies. Recently MAGIC detected a peculiar spectral feature during the highest X-ray ($E > 0.3$ KeV) flux ever recorded from this source [73].

The spectral feature, emerging during the highest X-ray flux state at about 3 TeV with a significance of $\approx 4\sigma$, can be modeled both as a curved narrow-band log-parabola or a Gaussian function superimposed to a broad-band simple log-parabola, respectively. The physical interpretation is still debated and three possible scenarios can be invoked: a two-zone emitting region model, a pile-up in the electron energy distribution, or a pair cascade from electrons accelerated in a black hole magnetospheric vacuum gap. Figure 8 shows the ASTRI Mini-Array simulations performed to investigate its capabilities in terms of energy resolution to detect such spectral features. Simulations (see [30] for a detailed discussion) were performed to investigate the percentage of number of detections of the spectral feature with respect to a broad-band log-parabola above 5σ confidence level for 200 realizations. In order to have at least a $\approx$50% probability of detection of the feature, 1.5 h of observation time would be required, increasing up to $\approx$80% probability for 2 h of exposure.

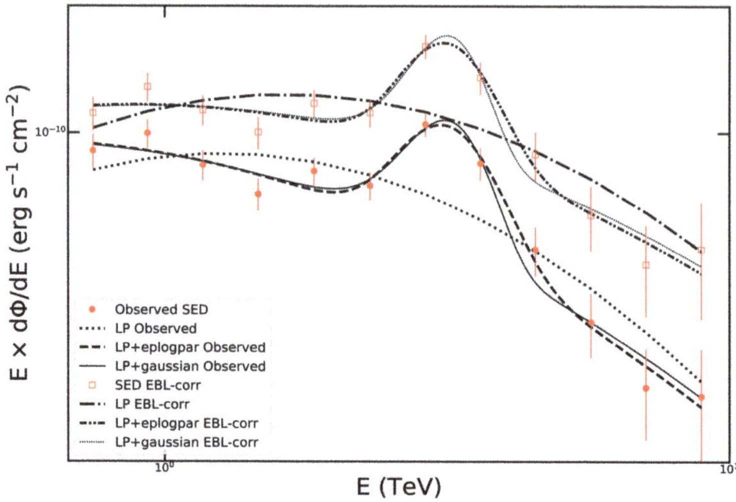

Figure 8. ASTRI Mini-Array simulations of the spectral feature emerging at ≈3 TeV in the spectrum on Mrk 501 during its highest ever-recorded X-ray flux state. `LP` = log-parabola; `eplogpar` = curved log-parabola; `EBL` = extra-galactic background light. From [30].

4.4. *Disentangling Spectral Models in Misaligned Jetted Sources—NGC 1068*

NGC 1068 ($z = 0.00379 \pm 0.00001$) is a powerful γ-ray Seyfert-2 galaxy detected by *Fermi*-LAT. It also hosts starburst activity in its central region and AGN-driven winds. The origin of the γ-ray emission is still debated because of the presence of different particle acceleration sites, such as the starburst ring, the circum-nuclear disk, and the jet [58,74]. Moreover, it has recently been associated with a possible source of neutrino emission [57]. While the canonical jet model does not extend above 10 TeV (see [74]), the AGN wind model predicts a hard spectrum that extends in the very high energy band. Figure 9 shows the results of a simulated deep observation (200 h) to test if we can detect VHE emission expected by the AGN wind model. The ASTRI Mini-Array is able to measure the source spectrum in the energy bins ∼2–5 TeV and ∼5–13 TeV at about 5 σ level.

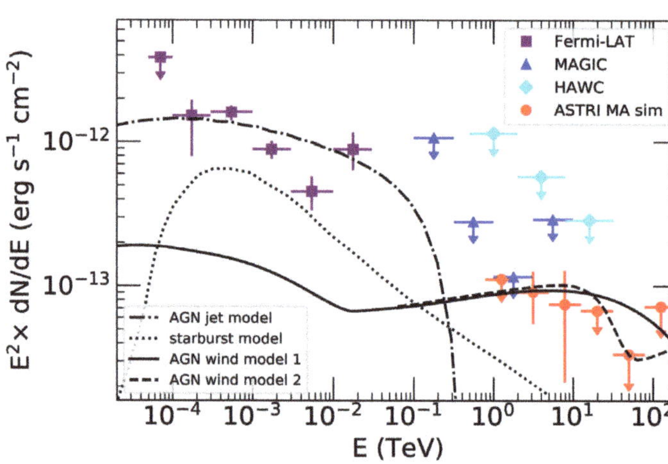

Figure 9. ASTRI Mini-Array simulations of the NGC 1068 VHE spectral energy distribution. Different emission models (see [74] for a detailed description) have been considered. From [30].

5. Conclusions

The ASTRI Mini-Array will commence scientific observations at the end of 2025 from the Observatorio del Teide, collecting data that will create a natural connection between current and future VHE facilities and other multi-wavelength observatories by providing light-curves, spectra, and high-resolution images of point-like and extended sources. Its 10° field of view will allow us to investigate both extended sources (e.g., supernova remnants) and crowded/rich fields (e.g., the Galactic Center) with a single pointing, while its $3'$ angular resolution at 10 TeV will allow us to perform detailed morphological studies of extended sources. Moreover, its sensitivity, extending above 100 TeV with a moderate degradation (about a factor of 2) up to the edge of the FoV, will make it the most sensitive IACT in the 5–200 TeV energy range in the Northern Hemisphere before the advent of CTAO−N. The ASTRI Mini-Array will join the energy domain typical of EASs with the precision domain (excellent angular and energy resolutions) typical of IACTs, allowing several synergies with LHAASO, HAWC, and Tibet AS-γ, investigating PeV-only sources, obtaining broad-band spectra, and detailed source morphology. For the first four years, the ASTRI Mini-Array will be run as an experiment with dedicated pointings in order to address specific Core Science Topics. Afterward, we expect a smooth transition toward an Observatory Phase open to observational proposals from the scientific community.

Funding: This work received no funding.

Data Availability Statement: This research has made use of the ASTRI Mini-Array Instrument Response Functions (IRFs) provided by the ASTRI Project [75] which are publicly available. All physical models are available through the cited literature. The ASTRI Mini-Array simulations were performed using the publicly available software `ctools` [v1.6.3] [76] and `gammapy` [v0.17] [77].

Acknowledgments: This work was conducted in the context of the ASTRI Project thanks to the support of the Italian Ministry of University and Research (MUR) as well as the Ministry for Economic Development (MISE), with funds explicitly assigned to the Italian National Institute of Astrophysics (INAF). We acknowledge the support of the Brazilian Funding Agency FAPESP (Grant 2013/10559-5) and the South African Department of Science and Technology through Funding Agreement 0227/2014 for the South African Gamma-Ray Astronomy Program. IAC is supported by the Spanish Ministry of Science and Innovation (MICIU). They are partially supported by H2020-ASTERICS, a project funded by the European Commission Framework Programme Horizon 2020 Research and Innovation action under grant agreement n. 653477. The ASTRI project is becoming a reality thanks to Giovanni "Nanni" Bignami, Nicolò "Nichi" D'Amico, two outstanding scientists who, in their capability as INAF Presidents, provided continuous support and invaluable guidance. Although Nanni was instrumental in starting the ASTRI telescope, Nichi transformed it into the Mini-Array in Tenerife. Now, the project is being built owing to the unfaltering support of Marco Tavani, the current INAF President. Paolo Vettolani and Filippo Zerbi, the past and current INAF Science Directors, and Massimo Cappi, the Coordinator of the High Energy branch of INAF, have been also very supportive of our work. We are very grateful to all of them. Unfortunately, Nanni and Nichi passed away, but their vision still guides us. This review went through the internal ASTRI review process. SV also wishes to thank the *Universe* Editor of the Special Issue "Recent Advances in Gamma Ray Astrophysics and Future Perspectives", P. Romano, for inviting him to write a review, and the referees for their comments that helped improve the manuscript.

Conflicts of Interest: The author declares no conflict of interest.

Abbreviations

The following abbreviations are used in this manuscript:

ALP	Axion-like particles
AS-γ	Air shower γ-ray array
ASTRI	Astrofisica con specchi a tecnologia replicante italiana
C.L.	Confidence Limit
CTAO	Čerenkov telescope array Observatory
DM	Dark matter

EAS	Extended air showers arrays
ESA	European space agency
EBL	Extra-galactic background light
E-HBL	Extreme high-peaked BL Lacs
eROSITA	Extended Roentgen survey with an imaging telescope array
FOV	Field of view
FR	Fanaroff–Riley galaxies
FSRQ	Flat-spectrum radio quasar
GASP	GLAST-AGILE support programme
GRB	Gamma-ray burst
GW	Gravitational wave
HAWC	High-altitude water Čerenkov observatory
HB	Hadron Beam
HBL	High-peaked BL Lacs
HE	High-energy
H.E.S.S.	High-energy stereoscopic system
IAC	Instituto de Astrofísica de Canarias
IACT	Imaging atmospheric Čerenkov telescope arrays
IBL	Intermediate-peaked BL Lacs
IGMF	Inter-galactic magnetic field
IR	Infra-red
IXPE	Imaging X-ray polarimetry explorer
LHAASO	Large high-altitude air shower observatory
LIV	Lorentz invariance violation
LST	Large-sized telescope
MAGIC	Major atmospheric gamma-ray imaging Čerenkov telescopes
MST	Medium-sized telescope
NASA	National aeronautics and space administration
SBG	Star-bursting galaxies
SC	Schwarzschild-Couder
SII	Stellar intensity interferometry
SRT	Sardinia radio telescope
SST	Small-sized telescope
TNG	Telescopio Nazionale Galileo
VERITAS	Very energetic radiation imaging telescope array system
VHE	Very high-energy
WEBT	Whole-Earth blazar telescope

Notes

1. http://tevcat.uchicago.edu/, accessed on 14 February 2024.
2. Large-sized telescopes.
3. Medium-sized telescopes.
4. As noted in [33], a 1.1 PeV photon requires a parent electron of energy $E{\sim}2.3$ PeV.
5. GRB 160821B and GRB 201015A have been detected at a significance of $\sim$3.1 σ and $\sim$3.5 σ, respectively, contrary to all the other GRBs whose detection significance exceeds $\sim$5 σ.
6. More detailed, phase-resolved spectra could not be investigated given the short exposure times in each phase bin.

References

1. Wakely, S.P.; Horan, D. TeVCat: An online catalog for Very High Energy Gamma-Ray Astronomy. *Int. Cosm. Ray Conf.* **2008**, *3*, 1341–1344.
2. Aharonian, F.; Akhperjanian, A.G.; Bazer-Bachi, A.R.; Beilicke, M.; Benbow, W.; Berge, D.; Bernlöhr, K.; Boisson, C.; Bolz, O.; Borrel, V.; et al. Observations of the Crab nebula with HESS. *A&A* **2006**, *457*, 899–915. [CrossRef]
3. Aleksić, J.; Alvarez, E.A.; Antonelli, L.A.; Antoranz, P.; Asensio, M.; Backes, M.; Barrio, J.A.; Bastieri, D.; Becerra González, J.; Bednarek, W.; et al. Performance of the MAGIC stereo system obtained with Crab Nebula data. *Astropart. Phys.* **2012**, *35*, 435–448. [CrossRef]
4. Weekes, T.C.; Badran, H.; Biller, S.D.; Bond, I.; Bradbury, S.; Buckley, J.; Carter-Lewis, D.; Catanese, M.; Criswell, S.; Cui, W.; et al. VERITAS: The Very Energetic Radiation Imaging Telescope Array System. *Astropart. Phys.* **2002**, *17*, 221–243. [CrossRef]

5. Abeysekara, A.U.; Albert, A.; Alfaro, R.; Alvarez, C.; Álvarez, J.D.; Arceo, R.; Arteaga-Velázquez, J.C.; Ayala Solares, H.A.; Barber, A.S.; Bautista-Elivar, N.; et al. Observation of the Crab Nebula with the HAWC Gamma-Ray Observatory. *Astrophys. J.* **2017**, *843*, 39. [CrossRef]
6. Cao, Z. A future project at tibet: The large high altitude air shower observatory (LHAASO). *Chin. Phys. C* **2010**, *34*, 249–252. [CrossRef]
7. Amenomori, M.; Bi, X.J.; Chen, D.; Chen, T.L.; Chen, W.Y.; Cui, S.W.; Danzengluobu; Ding, L.K.; Feng, C.F.; Feng, Z.; et al. Search for Gamma Rays above 100 TeV from the Crab Nebula with the Tibet Air Shower Array and the 100 m^2 muon Detector. *Astrophys. J.* **2015**, *813*, 98. [CrossRef]
8. Cherenkov Telescope Array Consortium; Acharya, B.S.; Agudo, I.; Al Samarai, I.; Alfaro, R.; Alfaro, J.; Alispach, C.; Alves Batista, R.; Amans, J.P.; Amato, E.; et al. *Science with the Cherenkov Telescope Array*; World Scientific: Singapore, 2019. [CrossRef]
9. Gueta, O. The Cherenkov Telescope Array: Layout, design and performance. In Proceedings of the 37th International Cosmic Ray Conference, Virtual, 12–23 July 2022; p. 885. [CrossRef]
10. Scuderi, S. The ASTRI Program. *Eur. Phys. J. Web Conf.* **2019**, *209*, 01001. [CrossRef]
11. Giro, E.; Canestrari, R.; Sironi, G.; Antolini, E.; Conconi, P.; Fermino, C.E.; Gargano, C.; Rodeghiero, G.; Russo, F.; Scuderi, S.; et al. First optical validation of a Schwarzschild Couder telescope: The ASTRI SST-2M Cherenkov telescope. *A&A* **2017**, *608*, A86. [CrossRef]
12. Lombardi, S.; Catalano, O.; Scuderi, S.; Antonelli, L.A.; Pareschi, G.; Antolini, E.; Arrabito, L.; Bellassai, G.; Bernlöhr, K.; Bigongiari, C.; et al. First detection of the Crab Nebula at TeV energies with a Cherenkov telescope in a dual-mirror Schwarzschild-Couder configuration: The ASTRI-Horn telescope. *A&A* **2020**, *634*, A22. [CrossRef]
13. Scuderi, S.; Giuliani, A.; Pareschi, G.; Tosti, G.; Catalano, O.; Amato, E.; Antonelli, L.A.; Becerra Gonzàles, J.; Bellassai, G.; Bigongiari, C.; et al. The ASTRI Mini-Array of Cherenkov telescopes at the Observatorio del Teide. *J. High Energy Astrophys.* **2022**, *35*, 52–68. [CrossRef]
14. Scuderi, S. The ASTRI Mini-Array: A new pathfinder for Imaging Cherenkov Telescope Arrays. *Universe* **2024**, *submitted*.
15. Vercellone, S.; Bigongiari, C.; Burtovoi, A.; Cardillo, M.; Catalano, O.; Franceschini, A.; Lombardi, S.; Nava, L.; Pintore, F.; Stamerra, A.; et al. ASTRI Mini-Array core science at the Observatorio del Teide. *J. High Energy Astrophys.* **2022**, *35*, 1–42. [CrossRef]
16. Aleksić, J.; Ansoldi, S.; Antonelli, L.A.; Antoranz, P.; Babic, A.; Bangale, P.; Barceló, M.; Barrio, J.A.; Becerra González, J.; Bednarek, W.; et al. The major upgrade of the MAGIC telescopes, Part II: A performance study using observations of the Crab Nebula. *Astropart. Phys.* **2016**, *72*, 76–94. [CrossRef]
17. Holder, J.; Atkins, R.W.; Badran, H.M.; Blaylock, G.; Bradbury, S.M.; Buckley, J.H.; Byrum, K.L.; Carter-Lewis, D.A.; Celik, O.; Chow, Y.C.K.; et al. The first VERITAS telescope. *Astropart. Phys.* **2006**, *25*, 391–401. [CrossRef]
18. De Naurois, M. H.E.S.S.-II—Gamma ray astronomy from 20 GeV to hundreds of TeV's. *Eur. Phys. J. Web Conf.* **2017**, *136*, 03001. [CrossRef]
19. Lombardi, S.; Antonelli, L.A.; Bigongiari, C.; Cardillo, M.; Gallozzi, S.; Green, J.G.; Lucarelli, F.; Saturni, F.G. Performance of the ASTRI Mini-Array at the Observatorio del Teide. In Proceedings of the 37th International Cosmic Ray Conference, Virtual, 12–23 July 2022; p. 884.
20. Zanin, R. (CTAO Observatory, Bologna, Italy) CTAO Project Scientist. Personal Communication, 2021.
21. Abeysekara, A.U.; Alfaro, R.; Alvarez, C.; Álvarez, J.D.; Arceo, R.; Arteaga-Velázquez, J.C.; Avila Rojas, D.; Ayala Solares, H.A.; Barber, A.S.; Bautista-Elivar, N.; et al. The HAWC Real-time Flare Monitor for Rapid Detection of Transient Events. *Astrophys. J.* **2017**, *843*, 116. [CrossRef]
22. Amenomori, M.; Bao, Y.W.; Bi, X.J.; Chen, D.; Chen, T.L.; Chen, W.Y.; Chen, X.; Chen, Y.; Cirennima; Cui, S.W.; et al. First Detection of Photons with Energy beyond 100 TeV from an Astrophysical Source. *Phys. Rev. Lett.* **2019**, *123*, 051101. [CrossRef]
23. di Sciascio, G.; Lhaaso Collaboration. The LHAASO experiment: From Gamma-Ray Astronomy to Cosmic Rays. *Nucl. Part. Phys. Proc.* **2016**, *279-281*, 166–173. [CrossRef]
24. Holler, M.; de Naurois, M.; Zaborov, D.; Balzer, A.; Chalmé-Calvet, R. Photon Reconstruction for H.E.S.S. Using a Semi-Analytical Model. *Int. Cosm. Ray Conf.* **2015**, *34*, 980. [CrossRef]
25. Abeysekara, A.U.; Albert, A.; Alfaro, R.; Alvarez, C.; Álvarez, J.D.; Arceo, R.; Arteaga-Velázquez, J.C.; Ayala Solares, H.A.; Barber, A.S.; Baughman, B.; et al. The 2HWC HAWC Observatory Gamma-Ray Catalog. *Astrophys. J.* **2017**, *843*, 40. [CrossRef]
26. Kawata, K.; Sako, T.K.; Ohnishi, M.; Takita, M.; Nakamura, Y.; Munakata, K. Energy determination of gamma-ray induced air showers observed by an extensive air shower array. *Exp. Astron.* **2017**, *44*, 1–9. [CrossRef]
27. Aharonian, F.; An, Q.; Axikegu; Bai, L.X.; Bai, Y.X.; Bao, Y.W.; Bastieri, D.; Bi, X.J.; Bi, Y.J.; Cai, H.; et al. Observation of the Crab Nebula with LHAASO-KM2A—A performance study. *Chin. Phys. C* **2021**, *45*, 025002. [CrossRef]
28. Cao, Z.; Aharonian, F.A.; An, Q.; Axikegu; Bai, Y.X.; Bao, Y.W.; Bastieri, D.; Bi, X.J.; Bi, Y.J.; Cai, H.; et al. Ultrahigh-energy photons up to 1.4 petaelectronvolts from 12 γ-ray Galactic sources. *Nature* **2021**, *594*, 33–36. [CrossRef]
29. D'Aì, A.; Amato, E.; Burtovoi, A.; Compagnino, A.A.; Fiori, M.; Giuliani, A.; La Palombara, N.; Paizis, A.; Piano, G.; Saturni, F.G.; et al. Galactic observatory science with the ASTRI Mini-Array at the Observatorio del Teide. *J. High Energy Astrophys.* **2022**, *35*, 139–175. [CrossRef]

30. Saturni, F.G.; Arcaro, C.H.E.; Balmaverde, B.; Becerra González, J.; Caccianiga, A.; Capalbi, M.; Lamastra, A.; Lombardi, S.; Lucarelli, F.; Alves Batista, R.; et al. Extragalactic observatory science with the ASTRI mini-array at the Observatorio del Teide. *J. High Energy Astrophys.* **2022**, *35*, 91–111. [CrossRef]
31. Cao, Z.; Aharonian, F.; An, Q.; Axikegu; Bai, Y.X.; Bao, Y.W.; Bastieri, D.; Bi, X.J.; Bi, Y.J.; Cai, J.T.; et al. The First LHAASO Catalog of Gamma-Ray Sources. *arXiv* **2023**, arXiv:2305.17030. [CrossRef].
32. Saturni, F.G.; Lombardi, S.; Antonelli, L.A.; Bigongiari, C.; Fedorova, E.; Giuliani, A.; Lucarelli, F.; Mastropietro, M.; Pareschi, G. Performance of the ASTRI Mini-Array with different levels of the Night Sky Background. In Proceedings of the 38th International Cosmic Ray Conference—PoS(ICRC2023), Nagoya, Japan, 26 July–3 August 2023; Volume 444, p. 717. [CrossRef]
33. Cao, Z.; Aharonian, F.A.; An, Q.; Axikegu; Bai, Y.X.; Bao, Y.W.; Bastieri, D.; Bi, X.J.; Bi, Y.J.; Cai, H.; et al. Peta,Äìelectron volt gamma-ray emission from the Crab Nebula. *Science* **2021**, *373*, 425–430. [CrossRef] [PubMed]
34. Aharonian, F.; An, Q.; Axikegu; Bai, Y.X.; Bao, Y.W.; Bastieri, D.; Bi, X.J.; Bi, Y.J.; Cai, H.; Tian, W.W.; et al. Performance of LHAASO-WCDA and observation of the Crab Nebula as a standard candle. *Chin. Phys. C* **2021**, *45*, 085002. [CrossRef]
35. Punch, M.; Akerlof, C.W.; Cawley, M.F.; Chantell, M.; Fegan, D.J.; Fennell, S.; Gaidos, J.A.; Hagan, J.; Hillas, A.M.; Jiang, Y.; et al. Detection of TeV photons from the active galaxy Markarian 421. *Nature* **1992**, *358*, 477–478. [CrossRef]
36. Ahnen, M.L.; Ansoldi, S.; Antonelli, L.A.; Arcaro, C.; Babić, A.; Banerjee, B.; Bangale, P.; Barres de Almeida, U.; Barrio, J.A.; Becerra González, J.; et al. First multi-wavelength campaign on the gamma-ray-loud active galaxy IC 310. *A&A* **2017**, *603*, A25. [CrossRef]
37. Aleksić, J.; Antonelli, L.A.; Antoranz, P.; Babic, A.; Barres de Almeida, U.; Barrio, J.A.; Becerra González, J.; Bednarek, W.; Berger, K.; Bernardini, E.; et al. Rapid and multiband variability of the TeV bright active nucleus of the galaxy IC 310. *A&A* **2014**, *563*, A91. [CrossRef]
38. MAGIC Collaboration; Acciari, V.A.; Ansoldi, S.; Antonelli, L.A.; Arbet Engels, A.; Arcaro, C.; Baack, D.; Babić, A.; Banerjee, B.; Bangale, P.; et al. Monitoring of the radio galaxy M 87 during a low-emission state from 2012 to 2015 with MAGIC. *MNRAS* **2020**, *492*, 5354–5365. [CrossRef]
39. Aharonian, F.; Akhperjanian, A.G.; Bazer-Bachi, A.R.; Beilicke, M.; Benbow, W.; Berge, D.; Bernlöhr, K.; Boisson, C.; Bolz, O.; Borrel, V.; et al. Fast Variability of Tera-Electron Volt γ Rays from the Radio Galaxy M87. *Science* **2006**, *314*, 1424–1427. [CrossRef] [PubMed]
40. Aliu, E.; Arlen, T.; Aune, T.; Beilicke, M.; Benbow, W.; Bouvier, A.; Bradbury, S.M.; Buckley, J.H.; Bugaev, V.; Byrum, K.; et al. VERITAS Observations of Day-scale Flaring of M 87 in 2010 April. *Astrophys. J.* **2012**, *746*, 141. [CrossRef]
41. Biteau, J.; Prandini, E.; Costamante, L.; Lemoine, M.; Padovani, P.; Pueschel, E.; Resconi, E.; Tavecchio, F.; Taylor, A.; Zech, A. Progress in unveiling extreme particle acceleration in persistent astrophysical jets. *Nat. Astron.* **2020**, *4*, 124–131. [CrossRef]
42. Galanti, G.; Roncadelli, M. Behavior of axionlike particles in smoothed out domainlike magnetic fields. *Phys. Rev. D* **2018**, *98*, 043018. [CrossRef]
43. Alves Batista, R.; Saveliev, A. The Gamma-Ray Window to Intergalactic Magnetism. *Universe* **2021**, *7*, 223. [CrossRef]
44. Tavecchio, F.; Romano, P.; Landoni, M.; Vercellone, S. Putting the hadron beam scenario for extreme blazars to the test with the Cherenkov Telescope Array. *MNRAS* **2019**, *483*, 1802–1807. [CrossRef]
45. Acciari, V.A.; Ansoldi, S.; Antonelli, L.A.; Arbet Engels, A.; Asano, K.; Baack, D.; Babić, A.; Baquero, A.; Barres de Almeida, U.; Barrio, J.A.; et al. MAGIC Observations of the Nearby Short Gamma-Ray Burst GRB 160821B. *Astrophys. J.* **2021**, *908*, 90. [CrossRef]
46. Abdalla, H.; Adam, R.; Aharonian, F.; Ait Benkhali, F.; Angüner, E.O.; Arakawa, M.; Arcaro, C.; Armand, C.; Ashkar, H.; Backes, M.; et al. A very-high-energy component deep in the γ-ray burst afterglow. *Nature* **2019**, *575*, 464–467. [CrossRef]
47. MAGIC Collaboration; Acciari, V.A.; Ansoldi, S.; Antonelli, L.A.; Arbet Engels, A.; Baack, D.; Babić, A.; Banerjee, B.; Barres de Almeida, U.; Barrio, J.A.; et al. Teraelectronvolt emission from the γ-ray burst GRB 190114C. *Nature* **2019**, *575*, 455–458. [CrossRef]
48. H. E. S. S. Collaboration; Abdalla, H.; Aharonian, F.; Ait Benkhali, F.; Angüner, E.O.; Arcaro, C.; Armand, C.; Armstrong, T.; Ashkar, H.; Backes, M.; et al. Revealing X-ray and gamma ray temporal and spectral similarities in the GRB 190829A afterglow. *Science* **2021**, *372*, 1081–1085. [CrossRef]
49. Suda, Y.; Artero, M.; Asano, K.; Berti, A.; Nava, L.; Noda, K.; Terauchi, K.; Acciari, V.A.; Ansoldi, S.; Antonelli, L.A.; et al. Observation of a relatively low luminosity long duration GRB 201015A by the MAGIC telescopes. In Proceedings of the 37th International Cosmic Ray Conference, Virtual, 12–23 July 2022; p. 797. [CrossRef]
50. Fukami, S.; Berti, A.; Loporchio, S.; Suda, Y.; Nava, L.; Noda, K.; Bošnjak, Ž.; Asano, K.; Longo, F.; The MAGIC Collaboration; et al. Very-high-energy gamma-ray emission from GRB 201216C detected by MAGIC. In Proceedings of the 37th International Cosmic Ray Conference, Virtual, 12–23 July 2022; p. 788. [CrossRef]
51. Huang, Y.; Hu, S.; Chen, S.; Zha, M.; Liu, C.; Yao, Z.; Cao, Z.; Experiment, T.L. LHAASO observed GRB 221009A with more than 5000 VHE photons up to around 18 TeV. *GRB Coord. Netw.* **2022**, *32677*, 1.
52. Burns, E.; Svinkin, D.; Fenimore, E.; Kann, D.A.; Agüí Fernández, J.F.; Frederiks, D.; Hamburg, R.; Lesage, S.; Temiraev, Y.; Tsvetkova, A.; et al. GRB 221009A: The Boat. *Astrophys. J. Lett.* **2023**, *946*, L31. [CrossRef]
53. LHAASO Collaboration; Cao, Z.; Aharonian, F.; An, Q.; Axikegu; Bai, L.X.; Bai, Y.X.; Bao, Y.W.; Bastieri, D.; Bi, X.J.; et al. A tera-electron volt afterglow from a narrow jet in an extremely bright gamma-ray burst. *Science* **2023**, *380*, 1390–1396. [CrossRef] [PubMed]

54. Cao, Z.; Aharonian, F.; An, Q.; Axikegu; Bai, Y.X.; Bao, Y.W.; Bastieri, D.; Bi, X.J.; Bi, Y.J.; Cai, J.T.; et al. Very high-energy gamma-ray emission beyond 10 TeV from GRB 221009A. *Sci. Adv.* **2023**, *9*, eadj2778. [CrossRef]
55. Galanti, G.; Roncadelli, M.; Tavecchio, F. Assessment of ALP scenarios for GRB 221009A. *arXiv* **2022**, arXiv:2211.06935. [CrossRef].
56. Galanti, G.; Nava, L.; Roncadelli, M.; Tavecchio, F.; Bonnoli, G. Observability of the Very-High-Energy Emission from GRB 221009A. *Phys. Rev. Lett.* **2023**, *131*, 251001. [CrossRef]
57. IceCube Collaboration; Abbasi, R.; Ackermann, M.; Adams, J.; Aguilar, J.A.; Ahlers, M.; Ahrens, M.; Alameddine, J.M.; Alispach, C.; Alves, A.A., Jr.; et al. Evidence for neutrino emission from the nearby active galaxy NGC 1068. *Science* **2022**, *378*, 538–543. [CrossRef]
58. Lamastra, A.; Fiore, F.; Guetta, D.; Antonelli, L.A.; Colafrancesco, S.; Menci, N.; Puccetti, S.; Stamerra, A.; Zappacosta, L. Galactic outflow driven by the active nucleus and the origin of the gamma-ray emission in NGC 1068. *A&A* **2016**, *596*, A68. [CrossRef]
59. Acciari, V.A.; Ansoldi, S.; Antonelli, L.A.; Arbet Engels, A.; Baack, D.; Babić, A.; Banerjee, B.; Barres de Almeida, U.; Barrio, J.A.; Becerra González, J.; et al. Constraints on Gamma-Ray and Neutrino Emission from NGC 1068 with the MAGIC Telescopes. *Astrophys. J.* **2019**, *883*, 135. [CrossRef]
60. Zampieri, L.; Bonanno, G.; Bruno, P.; Gargano, C.; Lessio, L.; Naletto, G.; Paoletti, L.; Rodeghiero, G.; Romeo, G.; Bulgarelli, A.; et al. A stellar intensity interferometry instrument for the ASTRI Mini-Array telescopes. In *Optical and Infrared Interferometry and Imaging VIII*; Mérand, A., Sallum, S., Sanchez-Bermudez, J., Eds.; Society of Photo-Optical Instrumentation Engineers (SPIE) Conference Series; SPIE: Bellingham, WA, USA, 2022; Volume 12183, p. 121830F. [CrossRef]
61. Egron, E.; Pellizzoni, A.; Iacolina, M.N.; Loru, S.; Marongiu, M.; Righini, S.; Cardillo, M.; Giuliani, A.; Mulas, S.; Murtas, G.; et al. Imaging of SNR IC443 and W44 with the Sardinia Radio Telescope at 1.5 and 7 GHz. *MNRAS* **2017**, *470*, 1329–1341. [CrossRef]
62. Loru, S.; Pellizzoni, A.; Egron, E.; Righini, S.; Iacolina, M.N.; Mulas, S.; Cardillo, M.; Marongiu, M.; Ricci, R.; Bachetti, M.; et al. Investigating the high-frequency spectral features of SNRs Tycho, W44, and IC443 with the Sardinia Radio Telescope. *MNRAS* **2019**, *482*, 3857–3867. [CrossRef]
63. Bortoletto, F.; Bonoli, C.; D'Alessandro, M.; Ragazzoni, R.; Conconi, P.; Mancini, D.; Pucillo, M. Commissioning of the Italian National Telescope Galileo. In *Advanced Technology Optical/IR Telescopes VI*; Stepp, L.M., Ed.; Society of Photo-Optical Instrumentation Engineers (SPIE) Conference Series; SPIE: Bellingham, WA, USA, 1998; Volume 3352, pp. 91–101. [CrossRef]
64. Villata, M.; Raiteri, C.M.; Larionov, V.M.; Kurtanidze, O.M.; Nilsson, K.; Aller, M.F.; Tornikoski, M.; Volvach, A.; Aller, H.D.; Arkharov, A.A.; et al. Multifrequency monitoring of the blazar 0716+714 during the GASP-WEBT-AGILE campaign of 2007. *A&A* **2008**, *481*, L79–L82. [CrossRef]
65. Predehl, P.; Andritschke, R.; Arefiev, V.; Babyshkin, V.; Batanov, O.; Becker, W.; Böhringer, H.; Bogomolov, A.; Boller, T.; Borm, K.; et al. The eROSITA X-ray telescope on SRG. *A&A* **2021**, *647*, A1. [CrossRef]
66. Weisskopf, M.C.; Ramsey, B.; O'Dell, S.; Tennant, A.; Elsner, R.; Soffitta, P.; Bellazzini, R.; Costa, E.; Kolodziejczak, J.; Kaspi, V.; et al. The Imaging X-ray Polarimetry Explorer (IXPE). In *Space Telescopes and Instrumentation 2016: Ultraviolet to Gamma Ray*; den Herder, J.W.A., Takahashi, T., Bautz, M., Eds.; SPIE: Bellingham, WA, USA, 2016; Society of Photo-Optical Instrumentation Engineers (SPIE) Conference Series; Volume 9905, p. 990517. [CrossRef]
67. Albert, A.; Alfaro, R.; Alvarez, C.; Camacho, J.R.A.; Arteaga-Velázquez, J.C.; Arunbabu, K.P.; Avila Rojas, D.; Ayala Solares, H.A.; Baghmanyan, V.; Belmont-Moreno, E.; et al. 3HWC: The Third HAWC Catalog of Very-high-energy Gamma-Ray Sources. *Astrophys. J.* **2020**, *905*, 76. [CrossRef]
68. LHAASO Collaboration. An ultrahigh-energy γ-ray bubble powered by a super PeVatron. *Sci. Bull.* **2024**, *69*, 449–457. [CrossRef] [PubMed]
69. Chernyakova, M.; Malyshev, D.; Paizis, A.; La Palombara, N.; Balbo, M.; Walter, R.; Hnatyk, B.; van Soelen, B.; Romano, P.; Munar-Adrover, P.; et al. Overview of non-transient γ-ray binaries and prospects for the Cherenkov Telescope Array. *A&A* **2019**, *631*, A177. [CrossRef]
70. Pintore, F.; Giuliani, A.; Belfiore, A.; Paizis, A.; Mereghetti, S.; La Palombara, N.; Crestan, S.; Sidoli, L.; Lombardi, S.; D'Aì, A.; et al. Scientific prospects for a mini-array of ASTRI telescopes: A γ-ray TeV data challenge. *J. High Energy Astrophys.* **2020**, *26*, 83–94. [CrossRef]
71. Aharonian, F.; Akhperjanian, A.G.; Bazer-Bachi, A.R.; Beilicke, M.; Benbow, W.; Berge, D.; Bernlöhr, K.; Boisson, C.; Bolz, O.; Borrel, V.; et al. 3.9 day orbital modulation in the TeV γ-ray flux and spectrum from the X-ray binary LS 5039. *A&A* **2006**, *460*, 743–749. [CrossRef]
72. Quinn, J.; Akerlof, C.W.; Biller, S.; Buckley, J.; Carter-Lewis, D.A.; Cawley, M.F.; Catanese, M.; Connaughton, V.; Fegan, D.J.; Finley, J.P.; et al. Detection of Gamma Rays with E > 300 GeV from Markarian 501. *Astrophys. J.* **1996**, *456*, L83. [CrossRef]
73. MAGIC Collaboration; Acciari, V.A.; Ansoldi, S.; Antonelli, L.A.; Babić, A.; Banerjee, B.; Barres de Almeida, U.; Barrio, J.A.; Becerra González, J.; Bednarek, W.; et al. Study of the variable broadband emission of Markarian 501 during the most extreme Swift X-ray activity. *A&A* **2020**, *637*, A86. [CrossRef]
74. Lamastra, A.; Tavecchio, F.; Romano, P.; Landoni, M.; Vercellone, S. Unveiling the origin of the gamma-ray emission in NGC 1068 with the Cherenkov Telescope Array. *Astropart. Phys.* **2019**, *112*, 16–23. [CrossRef]
75. ASTRI Project, ASTRI Mini-Array Instrument Response Functions (Prod2, v1.0). 2022. Available online: https://zenodo.org/records/6827882 (accessed on 21 January 2024). [CrossRef]

76. Knödlseder, J.; Mayer, M.; Deil, C.; Cayrou, J.B.; Owen, E.; Kelley-Hoskins, N.; Lu, C.C.; Buehler, R.; Forest, F.; Louge, T.; et al. GammaLib and ctools. A software framework for the analysis of astronomical gamma-ray data. *A&A* **2016**, *593*, A1. [CrossRef]
77. Donath, A.; Terrier, R.; Remy, Q.; Sinha, A.; Nigro, C.; Pintore, F.; Khélifi, B.; Olivera-Nieto, L.; Ruiz, J.E.; Brügge, K.; et al. Gammapy: A Python package for gamma-ray astronomy. *A&A* **2023**, *678*, A157. [CrossRef]

universe

MDPI

Review

Scientific Highlights of the AGILE Gamma-ray Mission

Stefano Vercellone [1,*,†], Carlotta Pittori [2,3,†] and Marco Tavani [4,5,†]

1 INAF Osservatorio Astronomico di Brera, Via E. Bianchi 46, 23807 Merate, Italy
2 INAF Osservatorio Astronomico di Roma, Via Frascati 33, 00078 Rome, Italy; carlotta.pittori@inaf.it
3 ASI Space Science Data Center, Via del Politecnico, 00133 Rome, Italy
4 INAF-IAPS Roma, Via del Fosso del Cavaliere 100, 00133 Rome, Italy; marco.tavani@inaf.it
5 Dipartimento di Fisica, Universitá di Roma Tor Vergata, Via della Ricerca Scientifica 1, 00133 Rome, Italy
* Correspondence: stefano.vercellone@inaf.it; Tel.: +39-02-72320-509
† These authors contributed equally to this work.

Abstract: The γ-ray sky above a few tens of megaelectronvolts (MeV) reveals some of the most powerful and energetic phenomena of our Universe. The *Astrorivelatore Gamma ad Immagini LEggero* (AGILE) Gamma-ray Mission was launched in 2007 with the aim of observing celestial sources by means of three instruments covering a wide range of energies, from hard X-rays up to 30 GeV. Thanks to its wide field of view, AGILE set to observe and detect emission from pulsars, pulsar wind nebulae, gamma-ray bursts, active galactic nuclei, fast radio bursts, terrestrial gamma-ray flashes, and the electromagnetic counterparts of neutrinos and gravitational waves. In particular, the fast on-ground processing and analysis chain allowed the AGILE team to promptly respond to transient events, and activate or participate in multiwavelength observing campaigns. Eventually, after 17 years of operations, the AGILE Italian scientific satellite re-entered the atmosphere on 14 February 2024, ending its intense activity as a hunter of some of the most energetic cosmic sources in the Universe that emit X and γ-rays. We will review the most relevant AGILE results to date and their impact on the advancements of theoretical models.

Keywords: AGILE; γ-ray astrophysics; astroparticle physics

check for updates

Citation: Vercellone, S.; Pittori, C.; Tavani, M. Scientific Highlights of the AGILE Gamma-ray Mission. *Universe* **2024**, *10*, 153. https://doi.org/10.3390/universe10040153

Academic Editor: Tina Kahniashvili

Received: 31 January 2024
Revised: 11 March 2024
Accepted: 21 March 2024
Published: 25 March 2024

1. Introduction

The γ-ray sky at high energies (HE, $E > 10\,\mathrm{MeV}$) has been investigated since the beginning of the 1960s. On 27 April 1961, the *EXPLORER XI* (1961) satellite [1] was launched with the aim of producing an all-sky survey at energies above a few tens of MeV.

Despite the very limited number of *fairly certain* γ-rays, as the authors note, and no certain sources detected, this may mark the beginning of studies of γ-ray astrophysics with space satellites[1]. Figure 1 shows the sky distribution of the 22 photons detected by the *EXPLORER XI* satellite in its early months.

The first confirmed γ-ray-emitting regions were detected by the *OSO-3* (1967–1969) satellite [2]. *OSO-3* reported γ-ray emission with energies above 50 MeV from the galactic disk with peak intensity towards the galactic center. The first two pointlike sources, Crab and Vela (both pulsar wind nebulae (PWNe)), were detected by the *SAS-2* (1972–1973) satellite [3].

A further improvement was achieved with the *COS-B* (1975–1982) satellite, which detected 25 sources, reported in [4]. We note that only 4 sources were identified: Crab, Vela, ρ Oph, and the first extragalactic one, the blazar 3C 273.

Among the *SAS-2* and *COS-B* unidentified γ-ray sources, one deserves a special mention, $\gamma\,195+5$, in the *Gemini* constellation. This may represent the first example of a multiwavelength campaign to identify a newly discovered γ-ray source, as described in [5]. Nowadays, this source is commonly known as *Geminga*.

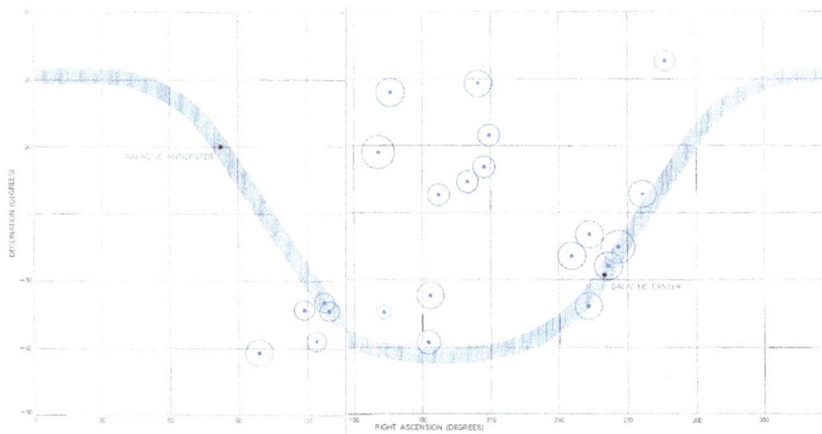

Figure 1. The γ-ray sky as seen by the *EXPLORER XI* satellite. The x-axis is the RA (deg); the y-axis is the Dec (deg). The shaded stripe represents the galactic plane in celestial coordinates. The blue dots mark the position of the 22 detected γ-rays. The circles are proportional to the scanned time length in that particular region of the sky. From [1].

A remarkable step ahead was obtained in the 1990s, with the launch of the Compton Gamma Ray Observatory, hosting on board the *EGRET* (1991–2000) telescope [6]. *EGRET* detected more than 270 sources, both galactic and extragalactic. Many of them were unidentified at the time, as reported in the *Third EGRET Catalog* (3EGC) [7]. 3EGC was instrumental in performing the first population studies of different source types and helped plan observations and target definitions for future HE space missions, e.g., [8] for the study of the γ-ray AGN duty cycle.

AGILE was a step forward with respect to previous γ-ray satellites under several aspects. Thanks to its silicon-based γ-ray tracker, it improved the angular resolution near 100 MeV by at least a factor of 2–3 compared with EGRET. The AGILE field of view (FoV) was about five times larger, improving the transient source detection and obtaining broad-band spectral information by including large FoV detectors in the MeV and X-ray energy ranges. These characteristics, combined with a rapid quick-look analysis of the γ-ray data and a fast dissemination of results, allowed the AGILE collaboration to provide alerts, stimulating efficient multifrequency programs.

In the next sections, we describe the AGILE Mission, its concepts, and the major scientific results obtained so far. We show how the AGILE Mission has been able to overcome some of the bottlenecks of the previous γ-ray missions by dramatically improving the ground segment efficiency, which allowed a rapid analysis of γ-ray data and dissemination of the results. Finally, we would like to note that this review cannot be a compilation of all the AGILE scientific results that the AGILE collaboration obtained during its 17 years of operation.

2. The AGILE Mission

The *Astrorivelatore Gamma ad Immagini LEggero* (AGILE[2]) satellite [9] (2007–2024) was a mission of the Italian Space Agency (ASI) devoted to high-energy astrophysics. It was launched on 23 April 2007 by the Indian PSLV-C8 rocket from the Sriharikota ISRO base (India). It ceased its operations on 18 January 2024 and subsequently re-entered the Earth's atmosphere on 14 February 2024 [10]. The AGILE scientific instrument combined four active detectors yielding broad-band coverage from hard X-ray to γ-ray energies: a silicon tracker [ST; 30 MeV–50 GeV] [11], a coaligned coded-mask hard X-ray imager, Super-AGILE [SA; 18–60 keV] [12], a nonimaging CsI mini-calorimeter [MCAL; 0.3–100 MeV] [13], and a segmented anticoincidence system [ACS] [14]. Any γ-ray detection was obtained by the combination of ST, MCAL, and ACS; these three detectors formed the AGILE gamma-ray imaging detector (GRID). The coaligned detectors, the silicon-strip-based tracker, a wide FoV γ-ray imager, and the fast-reaction ground segment were the AGILE innovative solutions with respect to the previous generation of γ-ray satellites.

Table 1 shows the main AGILE scientific performance. In addition to these figures, we note that AGILE had the possibility of jointly monitoring celestial sources both in the X-ray and in the γ-ray energy bands, as well as quickly reacting to fast transients.

The AGILE satellite was in a low-Earth equatorial orbit with an inclination of about 2.5 degrees, with an initial average altitude of about 500 km. AGILE raw telemetry level-0 (LV0) data were downlinked every ~100 min to the ASI Malindi ground station in Kenya and transmitted, through the fast ASINET network provided by ASI, first to the Telespazio Mission Control Center at Fucino in Italy and then to the AGILE Data Center[3] (ADC, Italy), part of the ASI multimission Space Science Data Center (SSDC), ~ 5 min after the end of each contact downlink. The ADC is in charge of all the scientific operations: data management, archiving, distribution of AGILE data and scientific software, and user support. Its main activities and architecture are described in [15].

Table 1. The AGILE scientific performance.

Gamma-ray Imaging Detector (GRID)	
Energy range	30 MeV–50 GeV
Field of view	~2.5 sr
Flux sensitivity ($E > 100$ MeV, 5σ in 3×10^6 s)	3×10^{-7} (ph cm^{-2} s^{-1})
Ang. resol. at 100 MeV (68% cont. radius)	3.5°
Ang. resol. at 400 MeV (68% cont. radius)	1.2°
Source localization acc. ($\lvert b \rvert > 10°$, 90% C.L.)	~15′
Energy resolution (at 400 MeV)	$\Delta E/E \sim 1$
Absolute time resolution	~2 μs
Deadtime	~100–200 μs
Hard X-ray Imaging Detector (SA)	
Energy range	18–60 keV
Single (1-dim.) detector FoV (FWZI)	107° × 68°
Combined (2-dim.) detector FoV (FWZI)	68° × 68°
Sensitivity (18–60 keV, 5σ in 1 day)	~15–30 mCrab
Angular resolution (pixel size)	6 arcmin
Source location accuracy (S/N~10)	~1–2 arcmin
Energy resolution (FWHM)	$\Delta E \sim 8$ keV
Absolute time resolution	~2 μs
Mini-Calorimeter (MCAL)	
Energy range	0.35–50 MeV
Energy resolution (at 1.3 MeV)	13% FWHM
Absolute time resolution	~3 μs
Deadtime (for each of the 30 CsI bars)	~20 μs

A ground segment alert system allowed the AGILE team to perform the full AGILE-GRID data reduction and the preliminary quick-look scientific analysis for a fast reaction to high-energy transients [15,16].

3. The Crab Nebula

The Crab Nebula (G184.6−5.8) is an expanding remnant of a supernova explosion (SN1054) recorded by Japanese and Chinese astronomers in 1054 A.D., located at an estimated distance of 2 kpc from Earth [17]. The Crab Nebula [18] is a complex pulsar wind nebula (PWN) system, powered by a powerful rotating neutron star, a pulsar of spin-down luminosity $L_{PSR} = 5 \times 10^{38}$ erg s^{-1}, and spin period $P = 33$ ms. Observations of the nebula have been carried out at every accessible wavelength, from radio up to very high energy (VHE) [19]. In the last decades, almost all high-energy (EGRET, *Fermi*-LAT, AGILE) and very-high-energy (H.E.S.S., MAGIC, VERITAS, HAWC, Tibet AS-γ, LHAASO) instruments provided invaluable information up to PeV energies. Moreover, in recent years, the first detection at TeV energies from the Crab Nebula by a dual-mirror Schwarzschild–Couder configuration Čherenkov telescope was reported by the 4 m ASTRI-Horn telescope, operated on Mt. Etna, Italy, and developed in the context of the Čherenkov Telescope Array Observatory (CTAO) preparatory phase [20]. Recently, the Crab was also observed with the Large-Sized Telescope Prototype of the Čherenkov Telescope Array LST-1 [21]. The results of the observations with high significance of the Crab Nebula in the energy ranges 10–100 TeV and >100 TeV were reported by the HAWC Collaboration [22] and by the LHAASO Collaboration using the first 5 months of the hybrid extensive air shower (EAS) half-array LHAASO-KM2A data [23].

Local variations in the inner nebula showing distinctive optical and X-ray features aligned with the pulsar jet ("wisps", "knots", and the "anvil") [24–27] have been attributed to enhancements of the synchrotron emission produced by instabilities or shocks in the pulsar wind outflow. However, when averaged over the whole inner region, the overall high-energy flux resulting from the unpulsed synchrotron radiation of the Crab Nebula has been considered essentially stable. Concerning the Crab Nebula variability, before 2010, only possible long-term nebular hard X-ray flux changes on a timescale of a few years have been reported. The X-ray variability has been observed to occur with an apparent relative amplitude of a few percent on timescales of $\sim$3 years [28–30]. The Crab was thus regarded as a nearly constant source at a level of a few percent from optical to γ-ray energies and used as a calibration source, a "standard candle" in astrophysics, up to very high energies [31].

In September 2010, thanks to its rapid alert system [15,16], AGILE detected a fast γ-ray flare above 100 MeV from the Crab Nebula over a daily timescale (see Flare F7 in Figure 2, left panel) and made the first public announcement on 22 September 2010 [32,33]. This finding was confirmed the following day by the Fermi satellite [34].

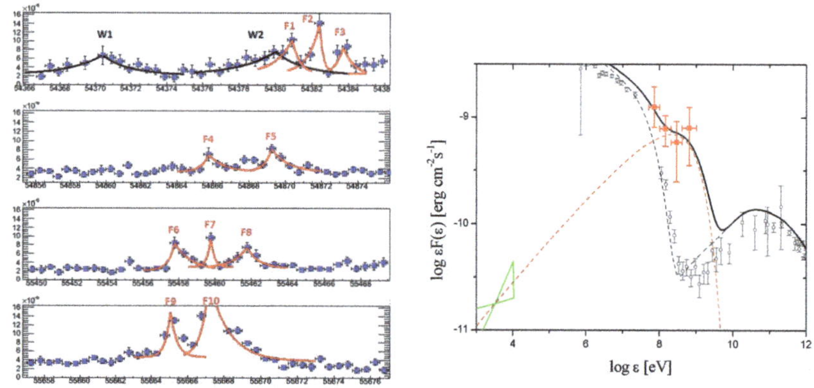

Figure 2. Left panel: decomposition of the multiyear Crab light curve in "waves" (W) and "flares" (F) according to [35], represented with exponential fits (see Table 1 in [35] for detailed fitting parameters). The x-axis is the date in MJD; the y-axis is the flux in units of photons cm^{-2} s^{-1}. **Right panel**: AGILE (filled squares) 48 h averaged data of the September 2010 γ-ray flare. The black line represents the 48 h averaged synchrotron emission model of the September 2010 flare summed with the standard nebular emission. The dashed red curve shows the flaring component averaged over 48 h. Open circles mark the standard average Crab Nebula spectrum, modeled by the dashed black curve. The green bow tie marks the X-ray data. Further details in [36].

The surprising discovery by AGILE of variable γ-ray emission from the Crab Nebula started a new era of investigation of this system. As a consequence, the 2012 Bruno Rossi International Prize was awarded to the principal investigator (PI), Marco Tavani, and the

AGILE team for this important and unexpected discovery. The detailed and exciting story that led to the AGILE discovery has been very well described in the *ScienceMagazine* article "The Crab that Roared" [37].

The interest in the discovery triggered several prompt multifrequency observations (from radio to TeV). However, no enhancement was detected at any of these energies, indicating that the flaring episode happened only in the γ-ray band. Moreover, no variations were detected in the pulsed emission, leaving out any interpretation in terms of magnetospheric origin near the Crab pulsar. AGILE had also previously detected a giant flare from the Crab in October 2007 during the initial science verification period of the satellite; furthermore, the first AGILE catalog paper [38] reported that anomalous episodic high-flux values observed from the Crab in 2007 were under investigation. Several major γ-ray flares from the Crab Nebula were detected by the AGILE-GRID and *Fermi*-LAT in the following years, with an approximate rate of one every few years (see Table 2), as well as more frequent enhanced γ-ray weeklong activity of lower intensity (the so-called "waves", W) [35], as shown in Figure 2, left panel, and as can be seen in [35], Figure 7.

Table 2. AGILE-GRID and *Fermi*-LAT Crab flares.

Flare Date	Duration (Days)	Peak γ-ray Flux $\times 10^{-6}$ Photons cm^{-2} s^{-1}	Instrument
2007 October	≈15	≈6	AGILE
2009 February	≈15	≈4	*Fermi*
2010 September	≈4	≈5	AGILE, *Fermi*
2011 April	≈2	≈0	*Fermi*, AGILE
2012 July	≈3	≈5	*Fermi*
2013 March	≈4	≈11	AGILE
2013 October	≈3	≈10	*Fermi*, AGILE
2014 August	≈4	≈7	*Fermi*
2016 October	≈13	≈7	AGILE, *Fermi*
2018 March	≈3	≈5	*Fermi*, AGILE
2018 October	≈10	≈11	*Fermi*, AGILE
2019 May	≈6	≈5	*Fermi*

During the flaring states, γ-ray spectra above 100 MeV have cutoff energies below a few GeVs, and the data are compatible with the observation of a new, almost monochromatic γ-ray spectral component evolving during the flare [36], as shown in Figure 2, right panel. The γ-ray flux can increase up to a factor of 8 over timescales of days [39], and drop below the average flux within a similar timescale [40]. γ-ray data provide evidence for particle acceleration mechanisms in nebular shock regions more efficient than previously expected from theoretical models, as discussed in [41] in a recent review.

4. Probes for Cosmic-Ray Acceleration

Understanding the escape of accelerated particles from expanding spherical shocks is a key ingredient to establish a connection between supernova remnants (SNRs) and the origin of galactic cosmic rays (CR). Moreover, SNRs are excellent laboratories to investigate the hadronic scenario as the main emission model in astrophysical sources [42–49]. In this section, we describe a few AGILE results on this topic.

4.1. W28

W28 is a middle-aged SNR with an age of at least 35,000 years. It is surrounded by molecular clouds [50], located at a projected distance of 10–20 pc from the SNR shell and detected by the High Energy Stereoscopic System (H.E.S.S.) up to several TeVs [51]. AGILE investigated W28 by means of γ-ray data accumulated during the "pointing observing mode" (2007–2009) [52], when it detected this source with a flux of $F_{E>100\,\mathrm{MeV}} = (40 \pm 11) \times 10^{-8}$ photons cm^{-2} s^{-1}. Two ingredients are fundamental in this investigation. The first one is an accurate modeling of the diffuse galactic γ-ray background, as presented in [53]. The second is the multiwavelength study of the region around this source. For this reason, the AGILE collaboration made use of ^{12}CO ($J = 1 \rightarrow 0$) molecular line observations taken by the NANTEN telescope, which detected a system of molecular clouds (cloud-N and cloud-S, respectively) associated with W28 [50], complemented with VHE observation by H.E.S.S. [51]. Figure 3 shows the AGILE-GRID data ($E > 400$ GeV, Gaussian-smoothed map) with the superimposed ^{12}CO contours (black lines) and the position of W28 (cyan circle).

A broad-band spectral energy distribution (SED) of the nonsimultaneous AGILE and H.E.S.S. data allowed us to model the observed data in terms of hadronic-induced interaction with the two molecular clouds adjacent to the SNR. This model explains the morphological and spectral features detected by both AGILE in the MeV–GeV energy range and H.E.S.S. in the TeV energy range, i.e., the different emission levels between cloud-N and cloud-S, the former being brighter than the latter at MeV–GeV energies, while the opposite occurs in the TeV energy band.

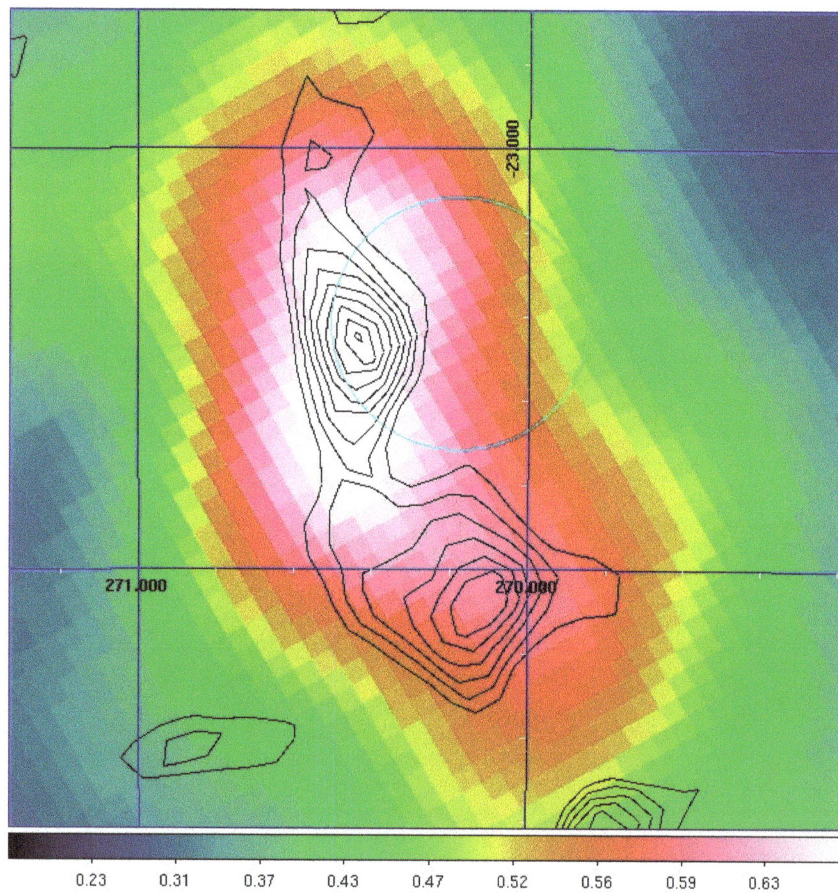

Figure 3. AGILE-GRID data ($E > 400$ GeV, Gaussian-smoothed map) with the superimposed ^{12}CO contours (black lines) and the position of W28 (cyan circle). The grid represents celestial (RA, Dec) coordinates. Cloud-N and cloud-S are located at (RA, Dec) = (270.4°, −23.4°) and (RA, Dec) = (270.2°, −24.1°), respectively. Further details are available in [52].

4.2. W44

W44 is a middle-aged (∼20,000 yr) SNR located in the galactic disk at a distance of ∼3 kpc from Earth at celestial coordinates (RA, Dec) = (284.04, 1.22)°. A direct proof that

SNRs are the origin of cosmic rays can be given by an unambiguous detection of the γ-ray emission expected from neutral pion decay in hadronic interactions, which is expected to be more evident at low energies (50–100 MeV), where it can be disentangled from the leptonic emission. In this respect, the AGILE-GRID is an excellent detector for investigating emission at such low energies. In [54] and subsequently in [55], the AGILE collaboration discussed the AGILE data accumulated in the period July 2007–April 2011. The γ-ray flux for energies above 400 MeV (where the AGILE-GRID angular resolution is optimized for the best performance) is $F_{400\,\mathrm{MeV}} = (16.0 \pm 1.2) \times 10^{-8}$ photons cm^{-2} s^{-1}. Figure 4 shows the first evidence of hadronic cosmic-ray acceleration in the 50–100 MeV energy range. Figure 4 shows the multiwavelength SED of W44 with observations from radio [and references therein] [56] up to TeV [57–59]. A model described by a synchrotron, bremsstrahlung, and inverse Compton contributions is shown and constitutes the first evidence of hadronic cosmic-ray acceleration in the 50–100 MeV energy band.

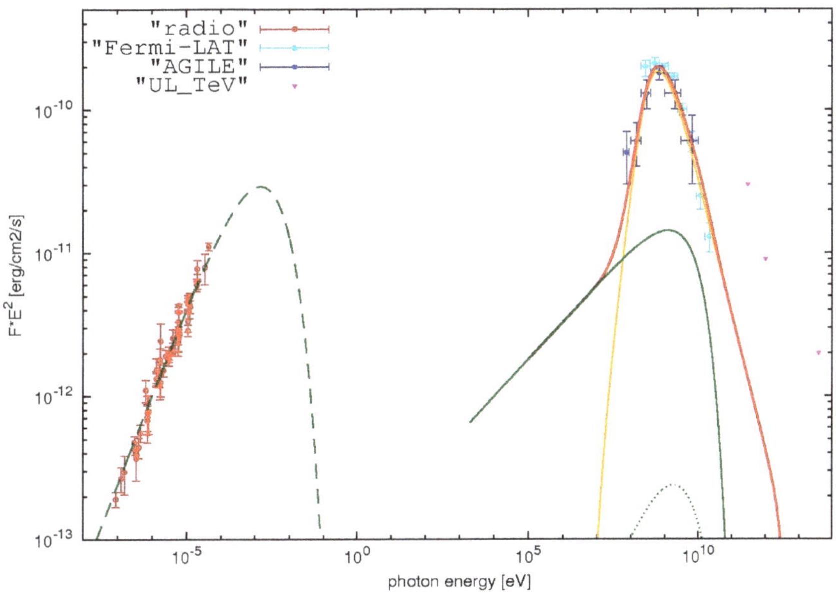

Figure 4. W44 SED. Red, blue, cyan, and pink points represent the radio, AGILE-GRID, Fermi-LAT, and TeV data, respectively. The yellow curve shows the neutral pion emission from the accelerated proton distribution. The green curves show the electron contribution by synchrotron (dashed curve), bremsstrahlung (solid curve), and inverse-Compton (dotted curve) emissions. The red curve marks the total γ-ray emission. See [54] for model details.

4.3. *IC 433*

The intermediate-age SNR IC 443, $\sim$(1–2) $\times$ 10^4 yr, is located near the galactic anticenter (l = 189°.1, b = 3°.0), and at a close distance from Earth, $\sim$1.5 kpc. During the first 2 years of operation, AGILE-GRID accumulated about 1 Ms of exposure towards this source [60].

Figure 5 shows the AGILE-GRID data and the different sources in the IC 433 region. The AGILE-GRID data show four possible excesses (A–D). The γ-ray flux above 100 MeV of excess-A is of $F_{E>100\,\mathrm{MeV}} = (47 \pm 10) \times 10^{-8}$ photons cm^{-2} s^{-1}.

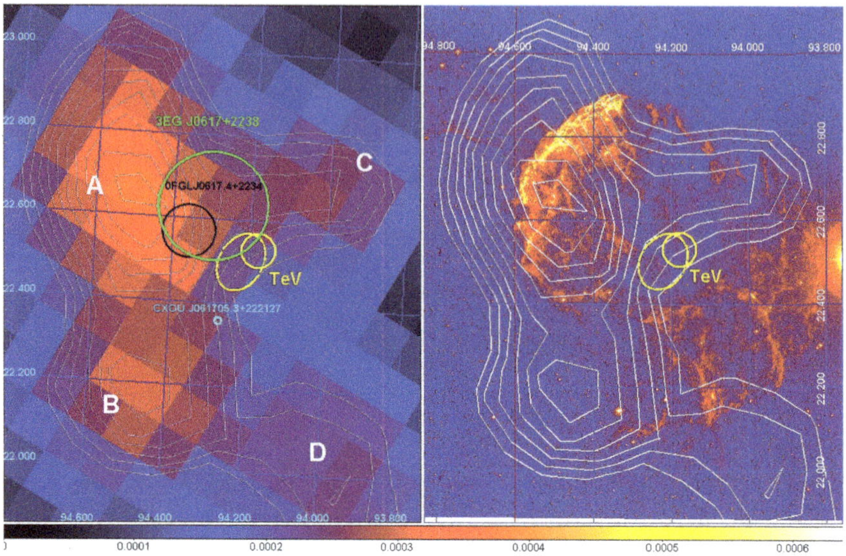

Figure 5. **Left panel**: AGILE γ-ray intensity map above 400 MeV centered on IC 443. White contour levels mark the γ-ray intensity. The green and black circles mark the 95% error boxes of the EGRET (3EG J0617+2238) and Fermi-LAT (0FGL J0617.4+2234) sources, respectively. The position of the TeV source associated with IC 443 is marked by a yellow circle and ellipse that give the 95% confidence level error boxes determined by MAGIC and VERITAS. The X-ray Chandra source CXOU J061705.3+222127 is marked by a cyan circle. **Right panel**: optical image of IC 443 (Palomar Digitized Sky Survey) superimposed with the AGILE γ-ray intensity contours above 400 MeV (same as the left panel). Further details are reported in [60].

From the AGILE-GRID observations, and from the lack of detectable diffuse TeV emission, it was demonstrated that electrons cannot be the main emitters of γ-rays in the range 0.1–10 GeV at the site of the strongest SNR shock. The intensity, spectral characteristics, and location of the most prominent γ-ray emission together with the absence of cospatial detectable TeV emission are consistent only with a hadronic model of cosmic-ray acceleration in the SNR.

5. The Cygnus Region

The Cygnus region ($60^\circ < l < 90^\circ$) is a site hosting bright diffuse emission, both transient and persistent pointlike and extended sources. The AGILE-GRID detected several sources in this region [38,61,62]. The most prominent persistent γ-ray pointlike sources are the three pulsars: PSR J2021+3651 (2AGL J2021+3654), PSR J2021+4026 (2AGL J2021+4029), and PSR J2032+4127 (2AGL J2032+4135). Moreover, three microquasars were detected in this region with variable γ-ray emission: Cygnus X-1 [63,64], Cygnus X-3 [65–68], and V404 Cygni [69]. Last but not least, the analysis of the AGILE-GRID data allowed the discovery of a Be-type star with a black hole companion in MWC 656 (AGL J2241+4454) [70]. This region is shining both above a few hundred of GeV [see the H.E.S.S. Galactic Plane Survey] [71] and in the TeV–PeV energy band, as reported in the first LHAASO catalog [72].

Figure 6 shows the AGILE-GRID count map ($E > 100$ MeV) centered on the Cygnus region. The positions of the three PSRs and of the three microquasars are marked with black and white crosses, respectively.

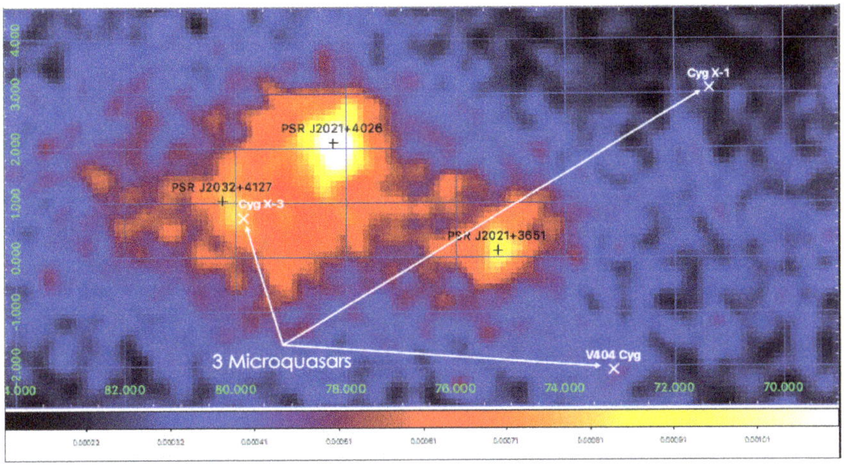

Figure 6. AGILE-GRID count map ($E > 100\,\text{MeV}$) centered on the Cygnus region. The positions of three PSRs and of the three microquasars are marked with black and white crosses, respectively. [G. Piano, Priv. Comm.]

Table 3 shows the principal properties of the three microquasars detected by the AGILE-GRID.

Table 3. Principal properties of the three microquasars detected by the AGILE-GRID. Data are drawn from [63–67,69] and references therein.

	Cygnus X-1	Cygnus X-3	V404 Cygni
Type	HMXB	HMXB	LMXB
Compact Object	BH (4.8-14.8) $M_\odot$	BH or NS	BH 9 $M_\odot$
Companion Star	O9.7 Iab (17–31) $M_\odot$	WR (>7 $M_\odot$)	K III 0.7 $M_\odot$
Distance	1.9 kpc	(7–10) kpc	2.39 kpc
Orbital Period	5.6 d	4.8 h	6.47 d

AGILE extensive monitoring of Cyg X-1 in the energy range 100 MeV–3 GeV during the period July 2007–October 2009 confirmed the existence of a spectral cutoff between 1 and 100 MeV during the typical hard spectral state of the source. However, on 15–16 October 2009, the AGILE-GRID detected Cyg X-1 at a flux of $F_\gamma = (232 \pm 66) \times 10^{-8}$ photons cm^{-2} s^{-1} in the energy range 100 MeV–3 GeV [63], which demonstrates that Cyg X-1 is capable of producing episodes of extreme particle acceleration on 1-day timescales.

On the contrary, the AGILE-GRID detected Cyg X-3 flaring events several times in the period February 2008–July 2009 [65,67]. The typical flare has a γ-ray flux of about one order of magnitude larger than its steady flux, $F^{\text{steady}}_{\text{E>100 MeV}} = (14 \pm 3) \times 10^{-8}$ photons cm^{-2} s^{-1}. The γ-ray flares seem to occur in anticorrelation with hard X-ray emission, during soft X-ray spectral state, and a few days before major radio flares. This picture seems to indicate that quenched states are a key condition for γ-ray flares [65].

The AGILE "spinning" observing mode[4] allowed the detection of a γ-ray flare from V404 Cygni, after more than a quarter of a century of quiescence, coincident with outbursts in radio, hard X-ray, and soft γ-ray (continuum and 511 keV annihilation line), as reported in [69] and references therein. The AGILE-GRID detected V404 Cygni on 24–26 June 2015 in the energy band (50–400) MeV with a flux of $F_\gamma = (4.6 \pm 1.5) \times 10^{-6}$ photons cm^{-2} s^{-1}. The AGILE observations are compatible with a microquasar scenario in which plasmoid ejections in a lepton-dominated transient jet are responsible for the high-energy γ-ray emission.

Finally, another transient source in the Cygnus region, namely, AGL J2241+4454, was detected by the AGILE-GRID on July 2010 at a flux of $F_{E>100\,\mathrm{MeV}} \approx 150 \times 10^{-8}$ photons cm^{-2} s^{-1} [73]. The combination of this detection with optical observations allowed the firm discovery of a black hole of (3.8–6.9) $M_{\odot}$, which orbits the Be star MWC 656, counterpart of AGL J2241+4454 [70].

6. The Crazy Diamond 3C 454.3

The renown flat-spectrum radio quasar (FSRQ) 3C 454.3 ($z = 0.859$) was originally detected in the γ-ray energy band by EGRET [74,75]. In 2005, 3C 454.3 underwent a major flaring activity at almost all energy bands and caused the start of a multiwavelength campaign, e.g., [76–78]. One interesting piece of evidence is a clear signature of the accretion disk in low states. 3C 454.3 was the first blazar detected in a flaring state by AGILE in 2007 [79]. Subsequently, it also became the brightest γ-ray source in the AGILE sky above 100 MeV, thus earning the nickname *Crazy Diamond* because of its prolonged and fast variability, resembling a diamond lit by light.

AGILE performed several multiwavelength campaigns on 3C 454.3, which enabled the study of the different SED and the discussion of innovative theoretical models to account for different flaring activity periods. Between July 2007 and October 2009, a long-term observing campaign [80] was performed during which fast γ-ray variability ($t^{\gamma}_{\mathrm{var}} \leq 1$ d) was recorded, with almost no time lag with respect to the optical one.

Figure 7 (left) shows the 18-month multiwavelength light curves that display how the radio band presents a markedly distinct behavior from the higher-frequency bands.

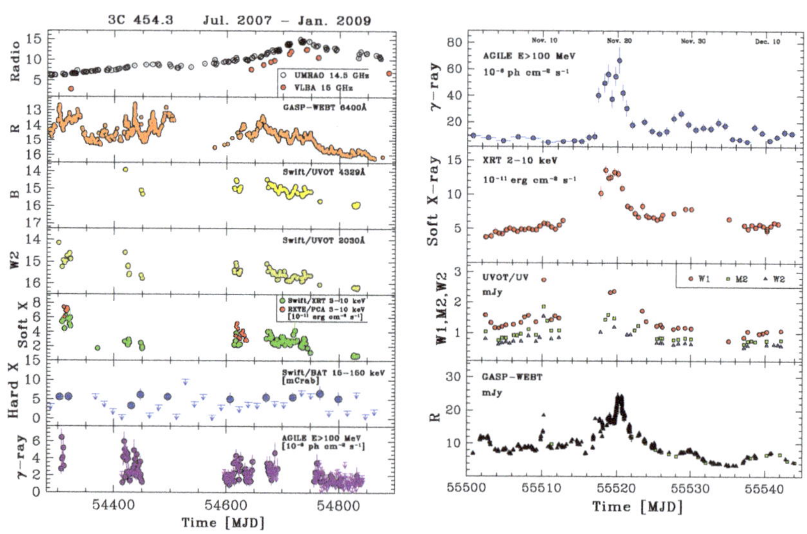

Figure 7. Left panel: 3C 454.3 light curves at different energies, covering about 18 months of monitoring. Data from [80]. **Right panel**: from top to bottom, AGILE ($E > 100$ MeV), *Swift*/XRT (2–10 keV), *Swift*/UVOT ($w1$, $m2$, $w2$), and GASP-WEBT (R) light curves obtained during the November 2010 flare. Data from [81].

Thanks to this long timescale multiwavelength coverage, the AGILE collaboration was able to uncover a slow, nearly constant increase of the 15 GHz flux uncorrelated with other wavebands. This different behavior of the light curves at different wavelengths could be interpreted in terms of a changing of the jet geometry between 2007 and 2008.

On the other hand, as shown in Figure 7 (right), the γ-ray flare that occurred on 20 November 2010 [81] showed some peculiar behaviors, among which was a γ-ray-orphan optical flare, which may challenge the model of a uniform external photon field responsible for the high-energy emission. Figure 8 shows three SEDs of 3C 454.3 obtained in November 2010, when the source was in a γ-ray-enhanced and flaring state [81]. In particular, preflare, flare, and postflare states can be distinguished, with remarkably different γ-ray flux levels and a Compton dominance reaching up to a factor of 20. The figure also shows, as a comparison, a SED for a very low γ-ray state for this source obtained 2 years earlier [80]. The AGILE team presented many observational and theoretical results on this source throughout the mission; see [79–86] and references therein.

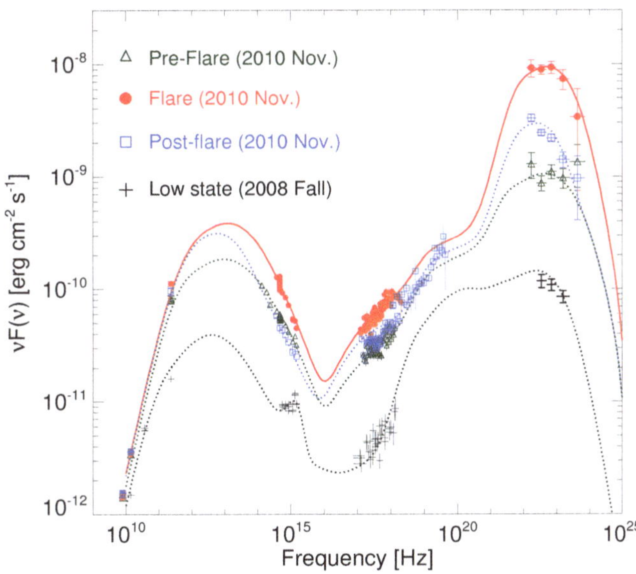

Figure 8. SED of the flat-spectrum radio quasar 3C 454.3 accumulated during the November 2010 flare in colors, data from [81] compared with a SED accumulated during a particularly low γ-ray state during fall 2008 in black, data from [80].

7. High-Redshift Sources

High-redshift ($z > 2$) blazars have a SED whose inverse Compton peak usually lies in the MeV–GeV energy range (see [87] for a seminal study of high-redshift γ-ray sources at $z > 2$). In a recent paper, [88] considered the extragalactic sources listed in the *Fermi*-LAT 4FGL-DR2 catalog [89], selected those outside the galactic plane ($|b| > 10°$), and reclassified them from the spectroscopic point of view. Out of 2980 sources, 1465 have spectroscopically determined redshift, and among them, 102 have $z > 2$. The majority of these high-redshift jetted sources are FSRQ (95, 93%), while BLLAC (5, 5%) and CLAGN[5] (2, 2%) are marginal. We focus on two high-redshift sources that have been investigated with AGILE.

7.1. 4C +71.07

The flat-spectrum radio quasar 4C+71.07 is a high-redshift ($z = 2.172$), γ-ray loud blazar whose optical emission is dominated by thermal radiation from the accretion disk. In [90], the AGILE collaboration reported on the high γ-ray activity of this source in the period October–November 2015. AGILE detected two separate flares (F1 and F2) with fluxes $F^1_{\mathrm{E}>100\,\mathrm{MeV}} = (1.2 \pm 0.3) \times 10^{-6}$ photons cm^{-2} s^{-1} between 26 October 2015 and 1 November 2015 and $F^2_{\mathrm{E}>100\,\mathrm{MeV}} = (3.1 \pm 0.6) \times 10^{-6}$ photons cm^{-2} s^{-1} between 7 November 2015 and 13 November 2015, respectively. The AGILE collaboration activated a multiwavelength campaign involving the *Neil Gehrels Swift* Observatory (*Swift*) [91] narrow field instruments, the X-ray Telescope [XRT] [92] and the UV/Optical Telescope [UVOT] [93], and the GLAST-

AGILE Support Program of the Whole Earth Blazar Telescope [GASP/WEBT] [94], with a coverage from the radio (5 GHz) up to GeV energy bands.

Figure 9 shows the multiwavelength campaign light curves, from the radio to the γ-ray energy band. In particular, the second and most prominent γ-ray flare (F2) has a much richer multiwavelength coverage, including the mm and the near-infrared wavelengths.

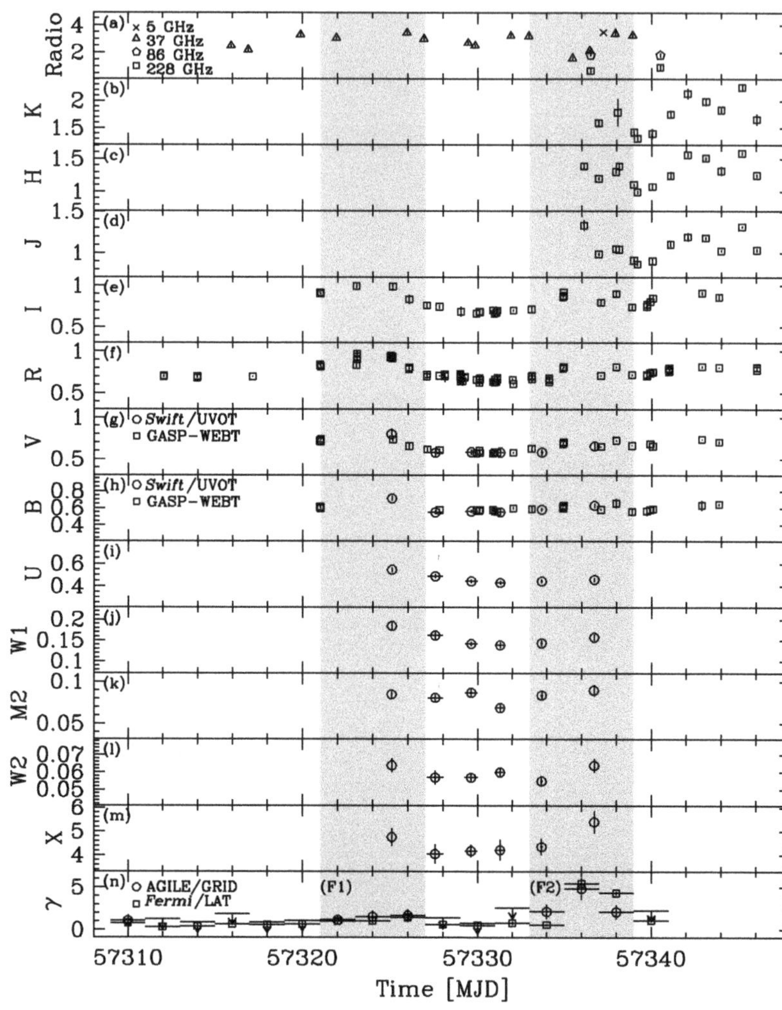

Figure 9. Multiwavelength light curves for the observing campaign on 4C+71.07. **Panel (a)**: GASP-WEBT 5 GHz (cross sign), 37 GHz (triangles), 86 GHz (diamonds), and 228 GHz (squares) data [Jy]. **Panels (b–h)**: K, H, J, I, R, V, B bands (open squares, [mJy]). **Panels (g–l)**: *Swift*/UVOT v, b, u, $w1$, $m2$, $w2$ bands (open circles, [mJy]). **Panel (m)**: *Swift*/XRT 0.3–10 keV observed flux [10^{-11} erg cm^{-2} s^{-1}]. **Panel (n)**: AGILE/GRID (circles) and *Fermi*-LAT (squares) data ($E > 100$ MeV, [10^{-6} photons cm^{-2} s^{-1}]). The gray-dashed areas mark the time intervals F1 (MJD 57,321.0–57,327.0) and F2 (MJD 57,333.0–57,339.0) used to accumulate the almost simultaneous SEDs. Data from [90].

7.2. PKS 1830−211

PKS 1830−211 is a γ-ray-emitting, high-redshift ($z = 2.507 \pm 0.002$), lensed flat-spectrum radio quasar (see [95]). Recently, Ref. [96] investigated the γ-ray flare detected by AGILE/GRID and *Fermi*/LAT of PKS 1830−211.

Figure 10 shows the multiwavelength light curves from radio (5 GHz) to γ-ray ($E > 100$ MeV). The source reached its maximum flux above 100 MeV ($F_{E>100\,\mathrm{MeV}} = (2.28 \pm 0.25) \times 10^{-5}$ photons cm^{-2} s^{-1}) around 24 April 2019 (MJD= 58597.25 ± 1.0), as shown in panel (d). This flux level is unprecedented for this source, and it is one of the largest ever detected in γ-rays from blazars at redshift $z > 2$.

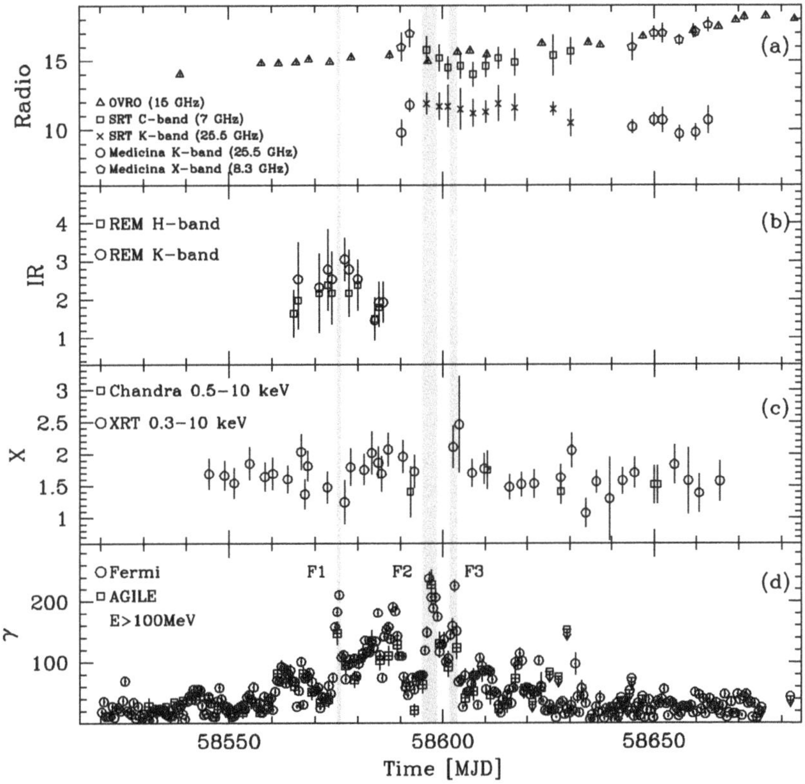

Figure 10. PKS 1830−211 multiwavelength light curves. **Panel** (**a**): radio (7, 8.3, 15, 25.5) GHz [Jy]. **Panel** (**b**): IR (H-band, K-band [10^{-13} erg cm^{-2} s^{-1}]). **Panel** (**c**): X-ray [10^{-11} erg cm^{-2} s^{-1}]. **Panel** (**d**): γ-ray (E > 100 MeV [10^{-7} photons cm^{-2} s^{-1}]) data. The shaded areas correspond to the major γ-ray flares F1, F2, and F3, when the spectral energy distributions were computed. Arrows mark 3σ upper limits. Data from [96].

7.3. SED Comparison

We can now compare the SEDs of these two sources. Figure 11 shows the SEDs for both 4C+71.07 and PKS 1830−211, the latter in different epochs and emission states.

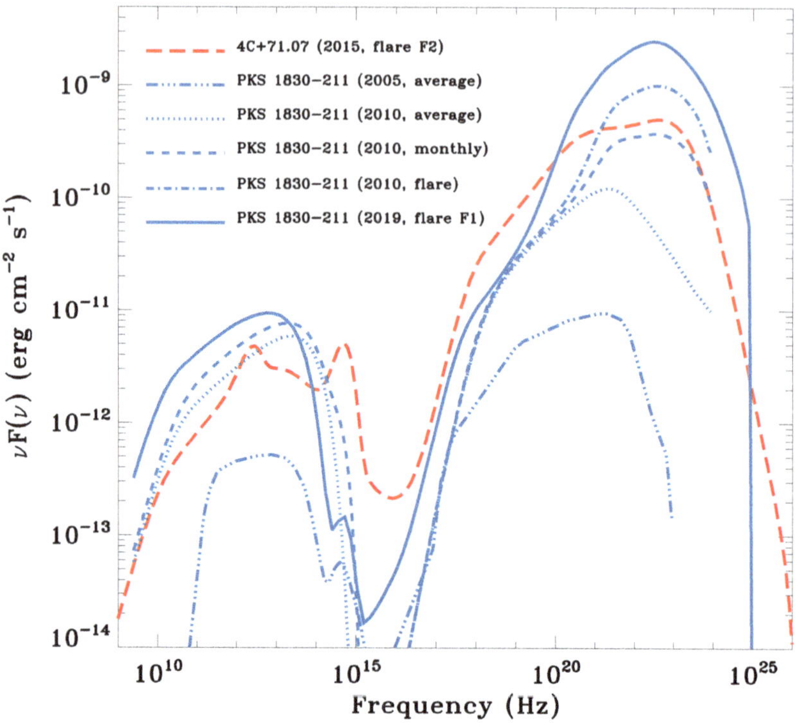

Figure 11. Comparison of the 4C+71.07 and PKS 1830−211 SEDs. Only the sum of different emission components is reported. Data drawn from 4C+71.07 (2015) [90], PKS 1830−211 (2005) [97], PKS 1830−211 (2010, three distinct states) [95], and PKS 1830−211 (2019) [96].

We note that, during the flaring states, both sources are within a factor of about 10 in flux. In particular, the isotropic γ-ray luminosity for 4C+71.07 at its maximum is $L^{\rm iso}_{\gamma,E>100\rm MeV} \approx 3 \times 10^{49}\,\rm erg\,s^{-1}$, while the Eddington luminosity is $L_{\rm Edd} \approx 6 \times 10^{47}\,\rm erg\,s^{-1}$, assuming a black hole mass of $M_{\rm BH} = 5 \times 10^9\,\rm M_\odot$, as computed in [98]. For PKS 1830−211, the AGILE collaboration similarly obtains $L^{\rm iso}_{\gamma,E>100\rm MeV} \approx 8.6 \times 10^{50}\,\rm erg\,s^{-1}$, while $L_{\rm Edd} \approx 6 \times 10^{46}\,\rm erg\,s^{-1}$, where for the latter, the value of the black hole mass reported in [99], $M_{\rm BH} = 5 \times 10^8\,\rm M_\odot$, was assumed. A notable difference between the flaring and the average state of PKS 1830−211 is the value of the Compton dominance (CD), i.e., the ratio between the inverse Compton and the synchrotron peaks. During the average 2005 state, the CD is of the order of $\approx$20, rising to $\approx$100 in 2010 and topping at >200 in 2019. Such high CD values may challenge the canonical one-component emission model, requiring alternative models to explain this remarkable SED, such as the "mirror model" [100,101] or the "jet-cloud interaction model" [86,102].

8. From MeV to TeV

The search for possible very-high-energy blazar candidates was also one of the *Beppo*SAX legacies, as discussed in [103] and recently updated, taking advantage of the *Fermi*-LAT all-sky γ-ray survey applied to the ROMA-BZCAT and sedentary survey samples [104]. Since the beginning of observations of the sky with both imaging atmospheric Čherenkov telescopes and extended air-shower arrays, jetted extragalactic sources started to emerge as a TeV-emitting class. In 1992, the Whipple telescope detected Mrk 421 [105], while a few years later, it also detected Mrk 501 [106]. An important recent result is the detection [107] of FSRQ OP 313 by means of the Čherenkov Telescope Array Observatory (CTAO) large-sized telescope (LST) prototype LST-1 [21]. This source, at a redshift of $z = 0.997$, is the most distant blazar ever detected by a Čherenkov telescope. LST-1 detected OP 313 during a target of opportunity repointing on 11–14 December 2023, with an integrated flux, above 100 GeV, of $\approx 15\%$ of the flux of the Crab Nebula in the same energy band.

Currently, the number of extragalactic jetted sources is about 90, as reported by the TeVCat [108] website[6]. Figure 12 shows the fraction of extragalactic jetted sources divided according to the TeVCat nomenclature.

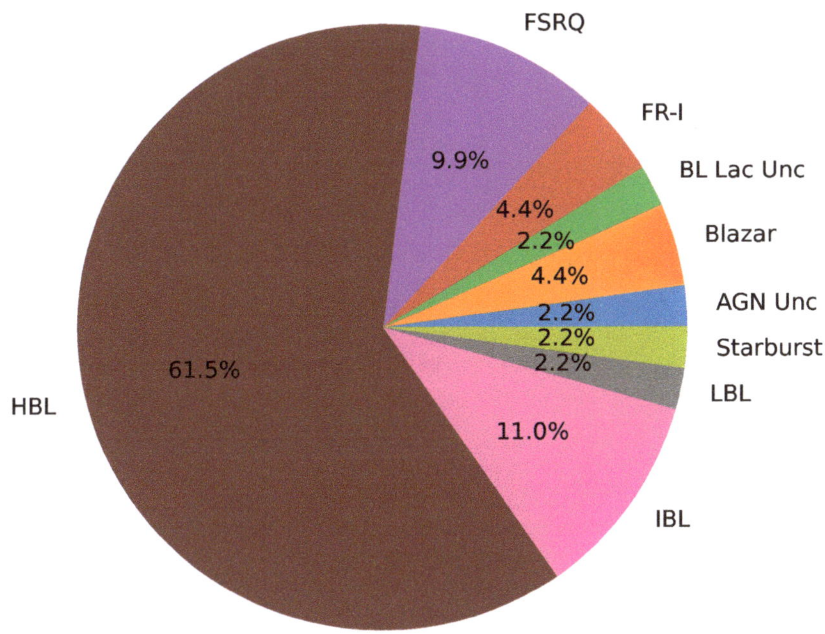

Figure 12. Fraction of jetted sources detected above $\approx$100 GeV by imaging atmospheric Čherenkov telescopes and extended air-shower arrays according to the TeVCat nomenclature. Data from the TeVCat website.

The majority of TeV-detected extragalactic jetted sources are high-peaked BL Lac objects (HBLs), accounting for more than 60% of the entire sample.

8.1. Search for MeV–GeV Counterparts of TeV Sources

A preliminary yet comprehensive analysis of the MeV-GeV search for AGILE counterparts of TeV extragalactic jetted sources is reported in [109]. The authors focused on the period 9 July 2007–18 October 2009, during which the AGILE satellite operated in the nominal pointing mode. During this period, the satellite was mainly pointed to observe two regions near the galactic plane: towards the Cygnus region ($l \approx 90°$) and the Vela region ($l \approx 270°$). These observations, therefore, are suboptimal to investigate extragalactic sources. The counterpart search was performed using the maximum likelihood estimator (MLE) described in [110] on the data accumulated in the period MJD (54,290.5–55,122.5). The AGILE collaboration used as input sources those reported in [38,61,66]. When the analysis was performed, TeVCat contained a lower number of sources than the current one, and the analysis was performed on 152 TeV sources.

Table 4 shows the extragalactic TeV sources listed in TeVCat that have been detected by AGILE in the period 9 July 2007–18 October 2009. Among them, there are five HBLs, two IBLs, two LBLs, two FSRQs, and two FR-I galaxies.

Table 4. TeVCat sources detected by AGILE in the period 9 July 2007–18 October 2009.

TeV Name	Counterpart	Class
TeV J0222+430	3C 66A	IBL
TeV J0232+202	1ES 0229+200	HBL
TeV J0319+415	NGC 1275	FR-I
TeV J0521+211	RGB J0521.8+211	IBL
TeV J0721+713	S5 0716+714	LBL
TeV J1104+382	Mrk 421	HBL
TeV J1256−057	3C 279	FSRQ
TeV J1325−430	Centaurus A	FR-I
TeV J1512−091	PKS 1510−089	FSRQ
TeV J2001+438	MAGIC J2001+435	HBL
TeV J2158−302	PKS 2155−304	HBL
TeV J2202+422	BL Lacertae	LBL
TeV J2359−306	H 2356−309	HBL

8.2. The Multiwavelength View of TeV Sources: W Comae and Mrk 421

In addition to the sources listed in Table 4, AGILE performed multiwavelength studies on some peculiar TeV sources and, among them, W Comae and Mrk 421.

W Comae ($z = 0.102$, IBL) was detected by AGILE on June 9–15 2008 as a follow-up observation [111] of the detection at VHE by VERITAS [112] on 7–8 June 2008 with a flux of $F_{E>200\,\mathrm{GeV}} = (5.7 \pm 0.6) \times 10^{-11}$ photons cm^{-2} s^{-1}. This value is about three times brighter than the flux detected by VERITAS in March 2008, when the sources were discovered as a VHE emitter [113]. AGILE detected W Comae with a flux of $F_{E>100\,\mathrm{MeV}} = (90 \pm 34) \times 10^{-8}$ photons cm^{-2} s^{-1}, which was, at that time, a factor of 1.5 higher than any γ-ray emission detected by both EGRET and *Fermi*-LAT. A multiwavelength campaign was immediately activated, as reported in [114]. Figure 13 (left panel) shows the almost-simultaneous multiwavelength light curves covering W Comae HE and VHE flare from the radio up to the GeV energy bands from MJD 54622 to MJD 54636. The SED was accumulated during the high-state period MJD 54624–54626, and fit with both a pure synchrotron self-Compton (SSC) model and an SSC plus external Compton (EC) one. The latter model fits the observed data better, in particular the near-infrared bump (see Figure 5 in [114]), suggesting a dusty molecular torus as a possible source of seed photons for the EC component.

Mrk 421 ($z = 0.03$, HBL) is a well-known TeV emitter. AGILE detected Mrk 421 both by means of its onboard hard X-ray detector, *Super-AGILE*, on 10 June 2008 [115] and in the γ-ray energy band [116], by integrating in the period 9–15 June 2008. In the period 9–15 June 2008, Mrk 421 reached a flux in the hard X-ray energy band of $F_{20-60\,\mathrm{keV}} =$ 55 mCrab ($\approx (4.9 \pm 0.5) \times 10^{-10}$ erg cm^{-2} s^{-1}) and a γ-ray flux of $F_{E>100\,\mathrm{MeV}} = 42^{+14}_{-12} \times 10^{-8}$ photons cm^{-2} s^{-1} [117]. Figure 13, right panel, illustrates the Mrk 421 light curves at different energies, showing the broad-band coverage for this flaring event. The X-ray and HE/VHE correlated variability points towards a possible SSC modeling of the SED. In particular, the decreasing trend in the R-band data, not followed by a similar trend in the X-ray energy band, may suggest a scenario in which the inner jet region would produce the X-rays while the outer region can only produce lower-frequency emission; see, e.g., [118,119].

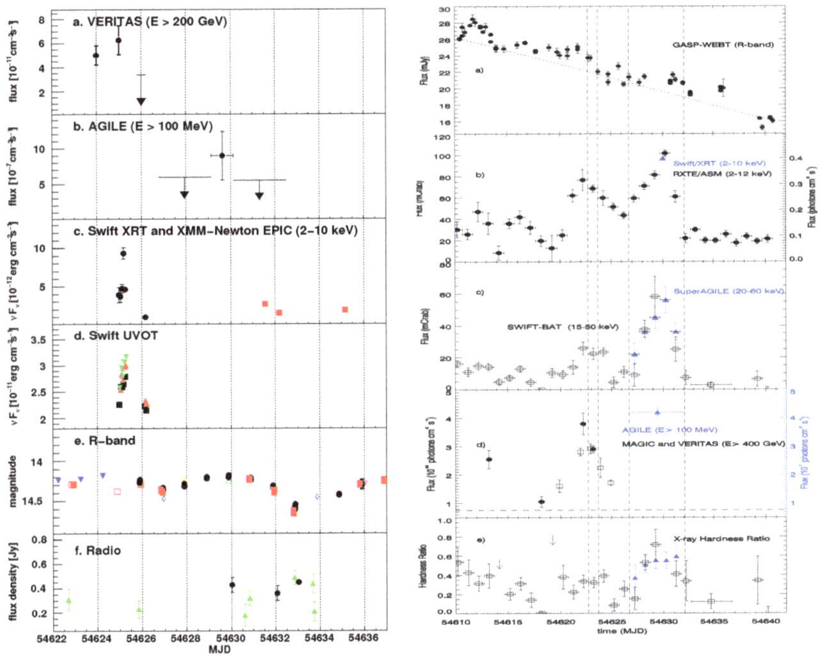

Figure 13. Left panel: W Comae multiwavelength light curve covering the HE and VHE flare from the radio up to the GeV energy bands. The x-axis is the date in MJD. Panel (**a**): VERITAS γ-ray ($E > 200\,\text{GeV}$) light curve. Panel (**b**): AGILE-GRID γ-ray ($E > 100\,\text{MeV}$) light curve. Panel (**c**): 2–10 keV *Swift*-XRT (circles) and *XMM-Newton* (squares) X-ray light curves. Panel (**d**): *Swift*-UVOT (UVW1: squares; UVM2: downward-pointing triangles; UVW2: upward-pointing triangles) light curves. Panel (**e**): GASP/WEBT *R*-band light curves. Panel (**f**): radio light curve (circles: UMRAO 14.5 GHz; triangles: Metsähovi 37 GHz). Downward-pointing arrows indicate 99% C.L. upper limits. **Right panel:** Mrk 421 multiwavelength light curve covering the HE and VHE flare from the optical up to the GeV energy bands. The x-axis is the date in MJD. Panel (**a**): R-band GASP/WEBT optical light curve. Panel (**b**): *RXTE*/ASM (2–12 keV) daily binned light curve and *Swift*/XRT (2–10 keV) light curve (blue triangle). Panel (**c**): Super-AGILE (20–60 keV, blue triangles; 1 Crab = 0.2 photons $\text{cm}^{-2}\,\text{s}^{-1}$) and the *Swift* Burst Alert Telescope [BAT,][15–50 keV] [120] (empty black squares; 1 Crab = 0.29 photons $\text{cm}^{-2}\,\text{s}^{-1}$). Panel (**d**): MAGIC, VERITAS ($E > 400\,\text{GeV}$, empty black squares, and black circles, respectively), and AGILE ($E > 100\,\text{MeV}$, blue triangles); the horizontal dashed line marks the 1 Crab flux level at $E > 400\,\text{GeV}$. Panel (**e**): hardness ratio computed using the Super-AGILE (or *Swift*/BAT) and *RXTE*/ASM data for each day. Vertical dashed lines represent two distinct time periods, P1 (6 June 2008) and P2 (9–15 June 2008). Further details are available in [W Comae] [114] and [Mrk 421] [117].

9. Gamma-ray Bursts and Multimessenger Astrophysics

An important contribution of the AGILE team has been to alert the community for short timescale (seconds/minutes/hours) transients, such as gamma-ray bursts (GRBs), through dedicated channels, such as the General Coordinates Network[7] (GCN) and the Astronomer's Telegram[8] (ATel), and to perform the follow-up of gravitational wave (GW) events, cosmic neutrinos, fast radio bursts (FRB), and other bursting events.

The efficient real-time analysis (RTA) system developed for the AGILE space mission to detect transient sources on short timescales has been presented in [121]. Two types of pipelines were implemented. The first one executed automated analyses as soon as new AGILE data were available, sharing the detection of sources with the community (more than 90 automated notices have been sent to the GCN since May 2019). The other pipeline reacted to external science alerts (GRBs from other missions, neutrinos, GW, etc.) to search for electromagnetic counterparts in the AGILE data. The AGILE alert system is also considered a heritage for the development of future RTA systems of the next generation of space and ground-based γ-ray observatories. Here, we present a short summary of selected results.

9.1. The AGILE-GRID View of GRBs and the Exceptional GRB 221009A

Following the EGRET seminal detections of GRBs above a few tens of MeVs in the early 1990s, in 2008, AGILE detected its first GRB with photons of energy above several tens of MeVs [GRB 080514B] [122]. The hard X-ray emission observed by Super-AGILE lasted about 7 s, while the emission observed by the AGILE-GRID above 30 MeV had almost twice the duration (at least 13 s). Prior to AGILE, such behavior regarding a possible longer-lasting high-energy component had only been hypothesized from a few other GRBs observed with EGRET. However, EGRET measurements were affected by instrumental deadtime effects, resulting in only lower limits to the GRB intensity. Thanks to the small deadtime of the AGILE-GRID and the unique simultaneous hard X-ray/γ-ray AGILE capability, for the first time, it could be assessed that the arrival times of the high-energy photons detected with the AGILE-GRID do not coincide with the brightest peaks seen in hard X-rays. Three high-energy photons are concentrated within 2 s at the beginning of the burst, while the next ones arrive only when the X-ray emission has returned to a level consistent with the background (7 s after the beginning of the burst). This implies a rapid time evolution of the γ-ray to X-ray flux ratio, although a quantitative assessment of this variability is hampered by the small statistics.

The AGILE and *Fermi* satellites have since then increased the sample of GRBs detected at γ-ray energies above 100 MeV to a couple of hundreds [123] with a rate of $\sim$10–17 GRBs per year. Most of them are are long bursts with typical prompt durations above 2 s, probably associated with stellar explosions of massive stars. However, they still represent a small fraction (<5%) of the GRB samples detected in the X-ray band.

AGILE and *Fermi* also contributed to the detection of short GRBs, characterized by durations below 2 s, which are usually spectrally hard compared to the average properties of GRBs and are believed to be associated with the coalescence of neutron star binaries. While *Fermi* detected the first short GRB in the γ-ray energy band, GRB 081024B [124,125], AGILE contributed to the science of short GRBs with the detection of GRB 090510 [126] (also observed by *Fermi* [127]), which provided the first case of a short GRB with delayed γ-ray emission. Indeed, Ref. [126] reported a delay of 0.2 s between the AGILE-GRID data with respect to the AGILE-MCAL ones (see their Figure 2). The short GRB 090510 is now considered a reference for potential electromagnetic γ-ray emission that could be associated with a gravitational wave event, and its light curve has been used as a possible high-energy template counterpart of GW events [128,129].

Now, this interesting and unexpected finding that for some GRBs the GeV emission starts with a delay after the MeV emission has become an often revealed trait. GRB 190114C was the first GRB with delayed emission ever detected above 300 GeV, a breakthrough discovery reported by MAGIC [130]. The sub-MeV/MeV data of the prompt and early afterglow emissions of GRB 190114C as detected by AGILE and Konus-Wind have been presented in [131]. In that AGILE paper, the first $\sim$200 s of the early afterglow of GRB 190114C have been carefully analyzed, and a previously unnoticed flux temporal break near T0+100 s was identified. Such a break is incompatible with the commonly assumed adiabatic evolution of a fireball in a constant-density medium, and it has been tentatively interpreted as a consequence of radiative evolution of the early afterglow from a fireball expanding in a wind-like circumburst medium.

Among the exceptional events observed, the long-duration GRB 221009A needs to be mentioned, as it was the brightest and most energetic GRB ever recorded, hence named *brightest of all time* or the "BOAT". During this unprecedented event, AGILE detected an extraordinary incoming flux of hard X-ray and high-energy γ-ray photons.

Figure 14 shows the AGILE-GRID sky count map accumulated during the first 48 h after the GRB 221009A onset. The BOAT clearly outshines all other γ-ray sources active at that time. The high-energy emission has been observed with an almost-continuous time coverage, from $\sim$200 s up to $\sim$20 ks after the GRB onset [132]. AGILE observations provide crucial flux and spectral γ-ray information regarding the early phases of GRB 221009A, during which emission in the TeV range was reported together with the first detection of photons above 10 TeV from GRBs [133]. The AGILE data suggest a dramatic transition between prompt and afterglow emission with a peculiar phase of coexistence of MeV and GeV emissions with very different spectral properties. A general review on the main discoveries about GRB science can be found in [134].

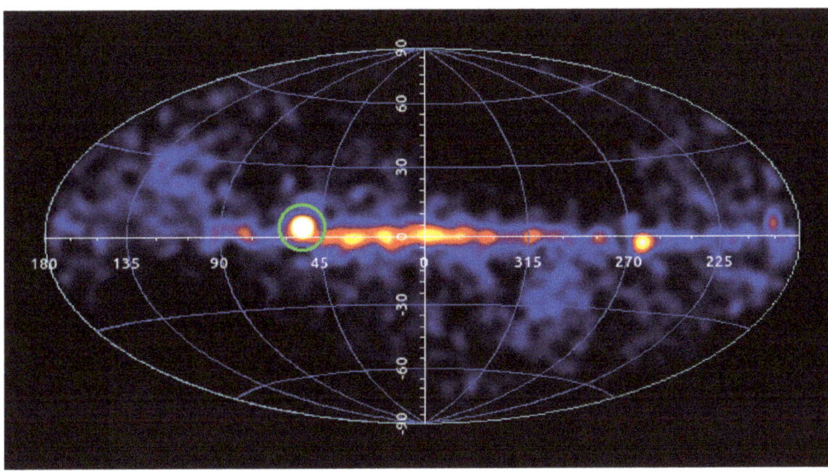

Figure 14. Sky count map above 100 MeV in galactic coordinates of the AGILE-GRID γ-ray detector during the time interval $[T_0, T_0 + 48\,\text{hr}]$. The γ-ray source associated with GRB 221009A is shown inside the green circle. The darkened sky regions are due to seasonal lack of exposure of the AGILE-GRID detector due to solar panel constraints. Further details in [132].

9.2. AGILE and Other Transients

The AGILE space mission, with its fast ground segment alert system and its unique observing capability to cover about 80% of the sky in $\sim$7 min in the so-called "spinning observing mode", provided crucial contribution in follow-up observations of multiwavelength and multimessenger transients, such as gravitational wave events, cosmic neutrinos, and fast radio bursts.

A detailed report on these significant results that involved all AGILE payload instruments (GRID, MCAL, and AC system) on timescales ranging from sub-milliseconds to tens to hundreds of seconds goes beyond the scope of this paper. In this section, we briefly present a list of AGILE main publications on these topics.

- AGILE and GWs: AGILE follow-up observations have provided in general the fastest response and the most significant upper limits above 100 MeV on all GW events detected by the Ligo–Virgo–Kagra Collaboration up to now [129,135–137].
- AGILE and neutrinos: AGILE published results from follow-up observations of IceCube neutrinos range from the first (still unconfirmed) tentative discovery of a γ-ray precursor in 2017 [138] to the systematic search for transient γ-ray sources temporally and spatially coincident with ten high-energy neutrino IceCube events published up to August 2018 [139], and the AGILE detection of the flaring blazar TXS 0506+056 in 2017, following the most interesting neutrino event detected to date [140].

- AGILE and FRBs: FRBs are millisecond radio pulses originating from powerful sources of unknown origin at extragalactic distances. AGILE observations in a multiwavelength context provide important constraints on the prompt (millisecond and hundreds of millisecond timescales) emission in the sub-MeV–MeV range. AGILE also studied the persistent long timescale γ-ray emission above 30 MeV from repeating FRBs [141–144]. A breakthrough in FRB science happened in 2020, with the AGILE detection of an X-ray burst from the galactic magnetar SGR 1935+2154 [145], an important finding that supports magnetar models and sheds light on the understanding of the physical mechanism of FRBs.

AGILE also produced important results on terrestrial gamma-ray flashes and solar flares, with the publication of dedicated catalogs, as described in Section 10.

10. The AGILE Legacy: The Catalogs

In this section, we present a summary of the main AGILE catalogs published at the time of writing (January 2024). Our goal is to provide a centralized source of information, which allows the reader to easily access online catalogs and their references, as reported in Table 5. The AGILE catalogs cover both celestial (including GRBs and solar flares) and terrestrial events (TGFs). This shows the AGILE versatility, which, thanks to the information collected by its detectors, could detect steady, flaring, and transient events from 20 keV up to 30 GeV.

Table 5. List of AGILE catalogs up to January 2024. References: (a) Pittori et al. [38]; (b) Feroci et al. [146]; (c) Galli et al. [147]; (d) Verrecchia et al. [61]; (e) Marisaldi et al. [148]; (f) Marisaldi et al. [149]; (g) Rappoldi et al. [109]; (h) Bulgarelli et al. [62]; (i) Marisaldi et al. [150]; (j) https://www.ssdc.asi.it/mcal3tgfcat/ (accessed on 18 March 2024); (k) Ursi et al. [151]; (l) Ursi et al. [152].

Catalog Title	Description	Reference	Link
The 1st AGILE-GRID Catalog of High Confidence Gamma-ray Sources	Jul. 2007–Jun. 2008 47 Sources	(a)	1AGL
Monitoring the hard X-ray sky with SuperAGILE	Jul. 2007–Apr. 2009 53 Sources	(b)	1SA
The AGILE MCAL Gamma-ray Burst Catalog	Apr. 2007–Oct. 2008 84 Sources	(c)	1GRB
An updated list of AGILE bright γ-ray sources and their variability in pointing mode	Jul. 2007–Oct. 2009 54 Sources	(d)	1AGLR
Properties of Terrestrial Gamma-ray Flashes detected by AGILE MCAL below 30 MeV	Mar. 2009–Jul. 2012 308 Events	(e)	1TGF
Enhanced detection of Terrestrial Gamma-ray Flashes by AGILE	Mar.–Jun. 2015 279 Events	(f)	2TGF
Search of MeV-GeV counterparts of TeV sources with AGILE in pointing mode	Jul. 2007–Oct. 2009 52 Sources	(g)	1ATEV
The 2nd AGILE Catalog of Gamma-ray sources AGILE in pointing mode	Jul. 2007–Oct. 2009 175 Sources	(h)	2AGL
On The High-Energy Spectral Component and Fine Time Structure of Terrestrial Gamma-ray Flashes	Mar–Jun. 2015 84 Events	(i)	1HETGF
The 3rd AGILE/MCAL TGF Catalog	Apr. 2007–Jun. 2022 5344 Events	(j)	3TGF
The 1st AGILE/MCAL GRB Catalog	Nov. 2007–Nov. 2020 503 Sources	(k)	2GRB
The 1st AGILE Solar Flare Catalog	May 2007–Aug. 2022 5003 Events	(l)	1SOL

11. Conclusions

The AGILE mission has been a *first of its kind*. It was selected by ASI as the *first* among the small mission programs in 1998. It was the *first* HE mission to have a hard X-ray monitor on board. For the *first* time in the γ-ray energy range, a fast ground segment allowed the dissemination of flaring events and the activation of the target of opportunity observations with other observatories. AGILE provided the *first* evidence of hadronic cosmic-ray acceleration in supernova remnants. For the *first* time, a γ-ray mission devoted to the observations of the sky became an asset in the study of terrestrial γ-ray flashes. Last, but not least, AGILE discovered, for the *first* time, flux variability in the Crab Nebula, previously considered to be a "standard candle" in the energy range (0.1–10) GeV. This discovery allowed the AGILE team to be awarded in 2012 with the Bruno Rossi Prize[9] of High Energy Astrophysics Division of the American Astronomical Society.

After 17 years of thriving operations, the AGILE Italian scientific satellite re-entered the atmosphere on 14 February 2024, thus ending its intense activity as a hunter of some of the most energetic cosmic sources in the Universe that emit X and γ-rays [10]. With AGILE's re-entry, the in-orbit operational phase comes to a close, but a new phase of scientific work on the satellite legacy data archive opens.

Currently, only *Fermi*-LAT is acquiring γ-ray data. In the future, ASTROGAM, a proposed observatory space mission dedicated to the study of the nonthermal Universe in the photon energy range from 0.3 MeV to 3 GeV [153,154], could provide important information not only above a few hundreds of MeV but also at lower energies. Furthermore, the Compton Spectrometer and Imager [COSI] [155] should be launched in 2027. It is designed as a soft γ-ray survey telescope sensitive in the 0.2–5 MeV energy band, performing studies of γ-ray polarization.

Author Contributions: Conceptualization, S.V., C.P. and M.T.; writing—original draft preparation, S.V. and C.P.; writing—review and editing, S.V., C.P. and M.T. All authors have read and agreed to the published version of the manuscript.

Funding: This research received no external funding.

Data Availability Statement: The data underlying this review are publicly available from their archives and processed with publicly available software.

Acknowledgments: This paper is written on behalf of the AGILE collaboration. Part of this work is based on archival data, software, or online services provided by the ASI–Space Science Data Center (SSDC). This research has made use of the TeVCat online source catalog (http://tevcat.uchicago.edu) (accessed on 18 March 2024). The scientific research carried out for the project has been partially supported under the grants ASI-I/R/045/04, ASI-I/089/06/0, and ASI-I/028/12/0 and subsequent grants and addenda. S.V. acknowledges partial financial contribution from the agreement ASI–INAF n.2017-14-H.0.

Conflicts of Interest: The authors declare no conflicts of interest.

Abbreviations

The following abbreviations are used in this manuscript:

3EG	third EGRET catalog
ACS	anticoincidence system
ADC	AGILE data center
AGILE	Astrorivelatore Gamma ad Immagini LEggero
AGN	active galactic nuclei
ASI	Agenzia spaziale Italiana
ATel	Astronomer's Telegram
BH	black hole
C.L.	confidence limit
CLAGN	changing-look AGN
COSI	Compton spectrometer and imager
CR	cosmic ray
CsI	cesium iodide
CTAO	Čherenkov Telescope Array Observatory

EAS	extensive air shower
EC	external Compton
EGRET	Energetic Gamma Ray Experiment Telescope
FR	Fanaroff–Riley galaxies
FoV	field of view
FRB	fast radio burst
FSRQ	flat-spectrum radio quasar
FWHM	full-width half maximum
FWZI	full-width zero intensity
GASP	GLAST-AGILE Support Program
GCN	general coordinates network
GRB	gamma-ray burst
GRID	gamma-ray imaging detector
GW	gravitational wave
HBL	high-peaked BL Lacs
HE	high energy
HMXB	high-mass X-ray binary
IBL	intermediate-peaked BL Lacs
LAT	large area telescope
LBL	low-peaked BL Lacs
LST	large-sized telescope
LMXB	low-mass X-ray binary
MCAL	mini-calorimeter
MJD	Modified Julian Day
NS	neutron star
PI	principal investigator
PSR	pulsar
RTA	real-time analysis
SA	Super-AGILE
SED	spectral energy distribution
S/N	signal-to-noise
SNR	supernova remnant
SSC	synchrotron self-Compton
SSDC	space science data center
ST	silicon tracker
TGF	terrestrial gamma-ray flashes
WEBT	whole-earth blazar telescope

Notes

1 We do not discuss here the large number of balloon-based instruments.

2 http://agile.rm.iasf.cnr.it/ (accessed on 18 March 2024).

3 https://agile.ssdc.asi.it/ (accessed on 18 March 2024).

4 On 2009 November 4 the AGILE scientific operations were reconfigured following a malfunction of the unique rotation wheel. See, https://agile.asdc.asi.it/news.html#115 and https://agile.asdc.asi.it/news.html#117 (accessed on 18 March 2024). Since then, the satellite started operating in a controlled "spinning observing mode", with the solar panels pointing at the Sun and the instrument axis sweeping the accessible sky with an angular speed of about $0.8\,\mathrm{deg\,s^{-1}}$. In spinning mode AGILE was able to survey a large fraction (about 80%) of the sky each day.

5 We note that changing-look AGN, CLAGN, are sources changing from one type to another, and vice versa.

6 http://tevcat.uchicago.edu/ (accessed on 18 March 2024).

7 https://gcn.nasa.gov/ (accessed on 18 March 2024).

8 https://www.astronomerstelegram.org/ (accessed on 18 March 2024).

9 https://head.aas.org/rossi/rossi.recip.html#AB (accessed on 18 March 2024).

References

1. Kraushaar, W.L.; Clark, G.W. GAMMA RAY ASTRONOMY. *Sci. Am.* **1962**, *206*, 52–61. [CrossRef]
2. Kraushaar, W.L.; Clark, G.W.; Garmire, G.P.; Borken, R.; Higbie, P.; Leong, V.; Thorsos, T. High-Energy Cosmic Gamma-ray Observations from the OSO-3 Satellite. *Astrophys. J.* **1972**, *177*, 341. [CrossRef]
3. Fichtel, C.E.; Hartman, R.C.; Kniffen, D.A.; Thompson, D.J.; Bignami, G.F.; Ögelman, H.; Özel, M.E.; Tümer, T. High-energy gamma-ray results from the second Small Astronomy Satellite. *Astrophys. J.* **1975**, *198*, 163–182. [CrossRef]
4. Swanenburg, B.N.; Bennett, K.; Bignami, G.F.; Buccheri, R.; Caraveo, P.; Hermsen, W.; Kanbach, G.; Lichti, G.G.; Masnou, J.L.; Mayer-Hasselwander, H.A.; et al. Second COS-B catalogue of high-energy gamma-ray sources. *Astrophys. J. Lett.* **1981**, *243*, L69–L73. [CrossRef]
5. Bignami, G.F.; Caraveo, P.A. Geminga: Its Phenomenology, Its Fraternity, and Its Physics. *Annu. Rev. Astron. Astrophys.* **1996**, *34*, 331–382. [CrossRef]
6. Kanbach, G.; Bertsch, D.L.; Fichtel, C.E.; Hartman, R.C.; Hunter, S.D.; Kniffen, D.A.; Hughlock, B.W.; Favale, A.; Hofstadter, R.; Hughes, E.B. The project EGRET (energetic gamma-ray experiment telescope) on NASA's Gamma-ray Observatory GRO. *Space Sci. Rev.* **1989**, *49*, 69–84. [CrossRef]
7. Hartman, R.C.; Bertsch, D.L.; Bloom, S.D.; Chen, A.W.; Deines-Jones, P.; Esposito, J.A.; Fichtel, C.E.; Friedlander, D.P.; Hunter, S.D.; McDonald, L.M.; et al. The Third EGRET Catalog of High-Energy Gamma-ray Sources. *Astrophys. J. Suppl. Ser.* **1999**, *123*, 79–202. [CrossRef]
8. Vercellone, S.; Soldi, S.; Chen, A.W.; Tavani, M. On the duty-cycle of γ-ray blazars. *Mon. Not. R. Astron. Soc.* **2004**, *353*, 890–902. [CrossRef]
9. Tavani, M.; Barbiellini, G.; Argan, A.; Boffelli, F.; Bulgarelli, A.; Caraveo, P.; Cattaneo, P.W.; Chen, A.W.; Cocco, V.; Costa, E.; et al. The AGILE Mission. *Astron. Astrophys.* **2009**, *502*, 995–1013. [CrossRef]
10. Tavani, M.; Addis, A.; Argan, A.; Antonelli, L.A.; Auricchio, N.; Barbiellini, G.; Baroncelli, L.; Basset, M.; Boffelli, F.; Bulgarelli, A.; et al. The AGILE satellite ceased operations and re-entered today into the atmosphere. *Astron. Telegr.* **2024**, *16450*, 1.
11. Prest, M.; Barbiellini, G.; Bordignon, G.; Fedel, G.; Liello, F.; Longo, F.; Pontoni, C.; Vallazza, E. The AGILE silicon tracker: An innovative /γ-ray instrument for space. *Nucl. Instruments Methods Phys. Res.* **2003**, *501*, 280–287. [CrossRef]
12. Feroci, M.; Costa, E.; Soffitta, P.; Del Monte, E.; di Persio, G.; Donnarumma, I.; Evangelista, Y.; Frutti, M.; Lapshov, I.; Lazzarotto, F.; et al. SuperAGILE: The hard X-ray imager for the AGILE space mission. *Nucl. Instruments Methods Phys. Res.* **2007**, *581*, 728–754. [CrossRef]
13. Labanti, C.; Marisaldi, M.; Fuschino, F.; Galli, M.; Argan, A.; Bulgarelli, A.; Di Cocco, G.; Gianotti, F.; Tavani, M.; Trifoglio, M. Design and construction of the Mini-Calorimeter of the AGILE satellite. *Nucl. Instruments Methods Phys. Res.* **2009**, *598*, 470–479. [CrossRef]
14. Perotti, F.; Fiorini, M.; Incorvaia, S.; Mattaini, E.; Sant'Ambrogio, E. The AGILE anticoincidence detector. *Nucl. Instruments Methods Phys. Res.* **2006**, *556*, 228–236. [CrossRef]
15. Pittori, C.; The Agile-Ssdc Team. The AGILE data center and its legacy. *Rend. Lincei. Sci. Fis. Nat.* **2019**, *30*, 217–223. [CrossRef]
16. Bulgarelli, A.; Trifoglio, M.; Gianotti, F.; Tavani, M.; Parmiggiani, N.; Fioretti, V.; Chen, A.W.; Vercellone, S.; Pittori, C.; Verrecchia, F.; et al. The AGILE Alert System for Gamma-ray Transients. *Astrophys. J.* **2014**, *781*, 19. [CrossRef]
17. Trimble, V. Motions and Structure of the Filamentary Envelope of the Crab Nebula. *Astron. J.* **1968**, *73*, 535. [CrossRef]
18. Hester, J.J. The Crab Nebula: An astrophysical chimera. *Annu. Rev. Astron. Astrophys.* **2008**, *46*, 127–155. [CrossRef]
19. Weekes, T.C.; Cawley, M.F.; Fegan, D.J.; Gibbs, K.G.; Hillas, A.M.; Kowk, P.W.; Lamb, R.C.; Lewis, D.A.; Macomb, D.; Porter, N.A.; et al. Observation of TeV Gamma Rays from the Crab Nebula Using the Atmospheric Cerenkov Imaging Technique. *Astrophys. J.* **1989**, *342*, 379. [CrossRef]
20. Lombardi, S.; Catalano, O.; Scuderi, S.; Antonelli, L.A.; Pareschi, G.; Antolini, E.; Arrabito, L.; Bellassai, G.; Bernlöhr, K.; Bigongiari, C.; et al. First detection of the Crab Nebula at TeV energies with a Cherenkov telescope in a dual-mirror Schwarzschild-Couder configuration: The ASTRI-Horn telescope. *Astron. Astrophys.* **2020**, *634*, A22. [CrossRef]
21. Abe, H.; Abe, K.; Abe, S.; Aguasca-Cabot, A.; Agudo, I.; Alvarez Crespo, N.; Antonelli, L.A.; Aramo, C.; Arbet-Engels, A.; Arcaro, C.; et al. Observations of the Crab Nebula and Pulsar with the Large-sized Telescope Prototype of the Cherenkov Telescope Array. *Astrophys. J.* **2023**, *956*, 980. [CrossRef]
22. Abeysekara, A.U.; Albert, A.; Alfaro, R.; Alvarez, C.; Álvarez, J.D.; Camacho, J.R.A.; Arceo, R.; Arteaga-Velázquez, J.C.; Arunbabu, K.P.; Avila Rojas, D.; et al. Measurement of the Crab Nebula Spectrum Past 100 TeV with HAWC. *Astrophys. J.* **2019**, *881*, 134. [CrossRef]
23. Aharonian, F.; An, Q.; Axikegu; Bai, L.X.; Bai, Y.X.; Bao, Y.W.; Bastieri, D.; Bi, X.J.; Bi, Y.J.; Cai, H.; et al. Observation of the Crab Nebula with LHAASO-KM2A - a performance study. *Chin. Phys. C* **2021**, *45*, 025002. [CrossRef]
24. Scargle, J.D. Activity in the Crab Nebula. *Astrophys. J.* **1969**, *156*, 401. [CrossRef]
25. Hester, J.J.; Scowen, P.A.; Sankrit, R.; Burrows, C.J.; Gallagher, J.S.; Holtzman, J.A.; Watson, A.; Trauger, J.T.; Ballester, G.E.; Casertano, S.; et al. WFPC2 Studies of the Crab Nebula. I. HST and ROSAT Imaging of the Synchrotron Nebula. *Astrophys. J.* **1995**, *448*, 240. [CrossRef]
26. Hester, J.J.; Mori, K.; Burrows, D.; Gallagher, J.S.; Graham, J.R.; Halverson, M.; Kader, A.; Michel, F.C.; Scowen, P. Hubble Space Telescope and Chandra Monitoring of the Crab Synchrotron Nebula. *Astrophys. J. Lett.* **2002**, *577*, L49–L52. [CrossRef]

27. Weisskopf, M.C.; Hester, J.J.; Tennant, A.F.; Elsner, R.F.; Schulz, N.S.; Marshall, H.L.; Karovska, M.; Nichols, J.S.; Swartz, D.A.; Kolodziejczak, J.J.; et al. Discovery of Spatial and Spectral Structure in the X-ray Emission from the Crab Nebula. *Astrophys. J. Lett.* **2000**, *536*, L81–L84. [CrossRef]
28. Greiveldinger, C.; Aschenbach, B. Temporal Variability of the X-ray Emission of the Crab Nebula Torus. *Astrophys. J.* **1999**, *510*, 305–311. [CrossRef]
29. Ling, J.C.; Wheaton, W.A. Gamma-ray Spectra and Variability of the Crab Nebula Emission Observed by BATSE. *Astrophys. J.* **2003**, *598*, 334–348. [CrossRef]
30. Wilson-Hodge, C.A.; Cherry, M.L.; Case, G.L.; Baumgartner, W.H.; Beklen, E.; Narayana Bhat, P.; Briggs, M.S.; Camero-Arranz, A.; Chaplin, V.; Connaughton, V.; et al. When a Standard Candle Flickers. *Astrophys. J. Lett.* **2011**, *727*, L40. [CrossRef]
31. Meyer, M.; Horns, D.; Zechlin, H.S. The Crab Nebula as a standard candle in very high-energy astrophysics. *Astron. Astrophys.* **2010**, *523*, A2. [CrossRef]
32. Tavani, M.; Striani, E.; Bulgarelli, A.; Gianotti, F.; Trifoglio, M.; Pittori, C.; Verrecchia, F.; Argan, A.; Trois, A.; de Paris, G.; et al. AGILE detection of enhanced gamma-ray emission from the Crab Nebula region. *Astron. Telegr.* **2010**, *2855*, 1.
33. Tavani, M.; Bulgarelli, A.; Vittorini, V.; Pellizzoni, A.; Striani, E.; Caraveo, P.; Weisskopf, M.C.; Tennant, A.; Pucella, G.; Trois, A.; et al. Discovery of Powerful Gamma-ray Flares from the Crab Nebula. *Science* **2011**, *331*, 736. [CrossRef]
34. Buehler, R.; D'Ammando, F.; Hays, E. Fermi LAT confirmation of enhanced gamma-ray emission from the Crab Nebula region. *Astron. Telegr.* **2010**, *2861*, 1.
35. Striani, E.; Tavani, M.; Vittorini, V.; Donnarumma, I.; Giuliani, A.; Pucella, G.; Argan, A.; Bulgarelli, A.; Colafrancesco, S.; Cardillo, M.; et al. Variable Gamma-ray Emission from the Crab Nebula: Short Flares and Long "Waves". *Astrophys. J.* **2013**, *765*, 52. [CrossRef]
36. Vittorini, V.; Tavani, M.; Pucella, G.; Striani, E.; Donnarumma, I.; Caraveo, P.; Giuliani, A.; Mereghetti, S.; Pellizzoni, A.; Trois, A.; et al. Spectral Evolution of the 2010 September Gamma-ray Flare from the Crab Nebula. *Astrophys. J. Lett.* **2011**, *732*, L22. [CrossRef]
37. Bhattacharje, Y. Crab That Roared. *Science* **2013**, *341*, 710–711. [CrossRef] [PubMed]
38. Pittori, C.; Verrecchia, F.; Chen, A.W.; Bulgarelli, A.; Pellizzoni, A.; Giuliani, A.; Vercellone, S.; Longo, F.; Tavani, M.; Giommi, P.; et al. First AGILE catalog of high-confidence gamma-ray sources. *Astron. Astrophys.* **2009**, *506*, 1563–1574. [CrossRef]
39. Buehler, R.; Scargle, J.D.; Blandford, R.D.; Baldini, L.; Baring, M.G.; Belfiore, A.; Charles, E.; Chiang, J.; D'Ammando, F.; Dermer, C.D.; et al. Gamma-ray Activity in the Crab Nebula: The Exceptional Flare of 2011 April. *Astrophys. J.* **2012**, *749*, 26. [CrossRef]
40. Yeung, P.K.H.; Horns, D. Fermi Large Area Telescope observations of the fast-dimming Crab Nebula in 60-600 MeV. *Astron. Astrophys.* **2020**, *638*, A147. [CrossRef]
41. Amato, E.; Olmi, B. The Crab Pulsar and Nebula as Seen in Gamma-rays. *Universe* **2021**, *7*, 448. [CrossRef]
42. Blandford, R.; Eichler, D. Particle acceleration at astrophysical shocks: A theory of cosmic ray origin. *Phys. Rep.* **1987**, *154*, 1–75. [CrossRef]
43. Malkov, M.A.; Drury, L.O. Nonlinear theory of diffusive acceleration of particles by shock waves. *Rep. Prog. Phys.* **2001**, *64*, 429–481. [CrossRef]
44. Reynolds, S.P. Supernova remnants at high energy. *Annu. Rev. Astron. Astrophys.* **2008**, *46*, 89–126. [CrossRef]
45. Aharonian, F.A.; Akhperjanian, A.G.; Aye, K.M.; Bazer-Bachi, A.R.; Beilicke, M.; Benbow, W.; Berge, D.; Berghaus, P.; Bernlöhr, K.; Bolz, O.; et al. High-energy particle acceleration in the shell of a supernova remnant. *Nature* **2004**, *432*, 75–77. [CrossRef]
46. Aharonian, F.; Akhperjanian, A.G.; Bazer-Bachi, A.R.; Beilicke, M.; Benbow, W.; Berge, D.; Bernlöhr, K.; Boisson, C.; Bolz, O.; Borrel, V.; et al. Primary particle acceleration above 100 TeV in the shell-type supernova remnant <ASTROBJ>RX J1713.7-3946</ASTROBJ> with deep HESS observations. *Astron. Astrophys.* **2007**, *464*, 235–243. [CrossRef]
47. Aharonian, F.; Akhperjanian, A.G.; Bazer-Bachi, A.R.; Beilicke, M.; Benbow, W.; Berge, D.; Bernlöhr, K.; Boisson, C.; Bolz, O.; Borrel, V.; et al. H.E.S.S. Observations of the Supernova Remnant RX J0852.0-4622: Shell-Type Morphology and Spectrum of a Widely Extended Very High Energy Gamma-ray Source. *Astrophys. J.* **2007**, *661*, 236–249. [CrossRef]
48. Dermer, C.D.; Powale, G. Gamma rays from cosmic rays in supernova remnants. *Astron. Astrophys.* **2013**, *553*, A34. [CrossRef]
49. Peron, G.; Aharonian, F. Probing the galactic cosmic-ray density with current and future γ-ray instruments. *Astron. Astrophys.* **2022**, *659*, A57. [CrossRef]
50. Mizuno, A.; Fukui, Y. Physical properties of molecular clouds as revealed by NANTEN CO survey: From the galactic center to the galactic warp. In Proceedings of the 5th Boston University Astrophysics Conference, Boston, MA, USA, 15–17 June 2003; p. 59.
51. Aharonian, F.; Akhperjanian, A.G.; Bazer-Bachi, A.R.; Behera, B.; Beilicke, M.; Benbow, W.; Berge, D.; Bernlöhr, K.; Boisson, C.; Bolz, O.; et al. Discovery of very high energy gamma-ray emission coincident with molecular clouds in the W 28 (G6.4-0.1) field. *Astron. Astrophys.* **2008**, *481*, 401–410. [CrossRef]
52. Giuliani, A.; Tavani, M.; Bulgarelli, A.; Striani, E.; Sabatini, S.; Cardillo, M.; Fukui, Y.; Kawamura, A.; Ohama, A.; Furukawa, N.; et al. AGILE detection of GeV γ-ray emission from the SNR W28. *Astron. Astrophys.* **2010**, *516*, L11. [CrossRef]
53. Giuliani, A.; Chen, A.; Mereghetti, S.; Pellizzoni, A.; Tavani, M.; Vercellone, S. Gamma-ray emission from the Galaxy: A new model for AGILE. *Mem. Della Soc. Astron. Ital. Suppl.* **2004**, *5*, 135.
54. Giuliani, A.; Cardillo, M.; Tavani, M.; Fukui, Y.; Yoshiike, S.; Torii, K.; Dubner, G.; Castelletti, G.; Barbiellini, G.; Bulgarelli, A.; et al. Neutral Pion Emission from Accelerated Protons in the Supernova Remnant W44. *Astrophys. J. Lett.* **2011**, *742*, L30. [CrossRef]

55. Cardillo, M.; Tavani, M.; Giuliani, A.; Yoshiike, S.; Sano, H.; Fukuda, T.; Fukui, Y.; Castelletti, G.; Dubner, G. The supernova remnant W44: Confirmations and challenges for cosmic-ray acceleration. *Astron. Astrophys.* **2014**, *565*, A74. [CrossRef]
56. Castelletti, G.; Dubner, G.; Brogan, C.; Kassim, N.E. The low-frequency radio emission and spectrum of the extended SNR <ASTROBJ>W44</ASTROBJ>: New VLA observations at 74 and 324 MHz. *Astron. Astrophys.* **2007**, *471*, 537–549. [CrossRef]
57. Buckley, J.H.; Akerlof, C.W.; Carter-Lewis, D.A.; Catanese, M.; Cawley, M.F.; Connaughton, V.; Fegan, D.J.; Finley, J.P.; Gaidos, J.A.; Hillas, A.M.; et al. Constraints on cosmic-ray origin from TeV gamma-ray observations of supernova remnants. *Astron. Astrophys.* **1998**, *329*, 639–658.
58. Aharonian, F.A.; Akhperjanian, A.G.; Beilicke, M.; Bernloehr, K.; Bojahr, H.; Bolz, O.; Boerst, H.; Coarasa, T.; Contreras, J.L.; Cortina, J.; et al. A search for TeV gamma-ray emission from SNRs, pulsars and unidentified GeV sources in the Galactic plane in the longitude range between -2 deg and 85 deg. *Astron. Astrophys.* **2002**, *395*, 803–811. [CrossRef]
59. Abdo, A.A.; Allen, B.T.; Aune, T.; Berley, D.; Chen, C.; Christopher, G.E.; DeYoung, T.; Dingus, B.L.; Ellsworth, R.W.; Gonzalez, M.M.; et al. Milagro Observations of Multi-TeV Emission from Galactic Sources in the Fermi Bright Source List. *Astrophys. J. Lett.* **2009**, *700*, L127–L131. [CrossRef]
60. Tavani, M.; Giuliani, A.; Chen, A.W.; Argan, A.; Barbiellini, G.; Bulgarelli, A.; Caraveo, P.; Cattaneo, P.W.; Cocco, V.; Contessi, T.; et al. Direct Evidence for Hadronic Cosmic-ray Acceleration in the Supernova Remnant IC 443. *Astrophys. J. Lett.* **2010**, *710*, L151–L155. [CrossRef]
61. Verrecchia, F.; Pittori, C.; Chen, A.W.; Bulgarelli, A.; Tavani, M.; Lucarelli, F.; Giommi, P.; Vercellone, S.; Pellizzoni, A.; Giuliani, A.; et al. An updated list of AGILE bright γ-ray sources and their variability in pointing mode. *Astron. Astrophys.* **2013**, *558*, A137. [CrossRef]
62. Bulgarelli, A.; Fioretti, V.; Parmiggiani, N.; Verrecchia, F.; Pittori, C.; Lucarelli, F.; Tavani, M.; Aboudan, A.; Cardillo, M.; Giuliani, A.; et al. Second AGILE catalogue of gamma-ray sources. *Astron. Astrophys.* **2019**, *627*, A13. [CrossRef]
63. Sabatini, S.; Tavani, M.; Striani, E.; Bulgarelli, A.; Vittorini, V.; Piano, G.; Del Monte, E.; Feroci, M.; de Pasquale, F.; Trifoglio, M.; et al. Episodic Transient Gamma-ray Emission from the Microquasar Cygnus X-1. *Astrophys. J. Lett.* **2010**, *712*, L10–L15. [CrossRef]
64. Sabatini, S.; Tavani, M.; Coppi, P.; Pooley, G.; Del Santo, M.; Campana, R.; Chen, A.; Evangelista, Y.; Piano, G.; Bulgarelli, A.; et al. Gamma-ray Observations of Cygnus X-1 above 100 MeV in the Hard and Soft States. *Astrophys. J.* **2013**, *766*, 83. [CrossRef]
65. Tavani, M.; Bulgarelli, A.; Piano, G.; Sabatini, S.; Striani, E.; Evangelista, Y.; Trois, A.; Pooley, G.; Trushkin, S.; Nizhelskij, N.A.; et al. Extreme particle acceleration in the microquasar CygnusX-3. *Nature* **2009**, *462*, 620–623. [CrossRef] [PubMed]
66. Bulgarelli, A.; Tavani, M.; Chen, A.W.; Evangelista, Y.; Trifoglio, M.; Gianotti, F.; Piano, G.; Sabatini, S.; Striani, E.; Pooley, G.; et al. AGILE detection of Cygnus X-3 γ-ray active states during the period mid-2009/mid-2010. *Astron. Astrophys.* **2012**, *538*, A63. [CrossRef]
67. Piano, G.; Tavani, M.; Vittorini, V.; Trois, A.; Giuliani, A.; Bulgarelli, A.; Evangelista, Y.; Coppi, P.; Del Monte, E.; Sabatini, S.; et al. The AGILE monitoring of Cygnus X-3: Transient gamma-ray emission and spectral constraints. *Astron. Astrophys.* **2012**, *545*, A110. [CrossRef]
68. Koljonen, K.I.I.; Maccarone, T.; McCollough, M.L.; Gurwell, M.; Trushkin, S.A.; Pooley, G.G.; Piano, G.; Tavani, M. The hypersoft state of Cygnus X-3. A key to jet quenching in X-ray binaries? *Astron. Astrophys.* **2018**, *612*, A27. [CrossRef]
69. Piano, G.; Munar-Adrover, P.; Verrecchia, F.; Tavani, M.; Trushkin, S.A. High-energy Gamma-ray Activity from V404 Cygni Detected by AGILE during the 2015 June Outburst. *Astrophys. J.* **2017**, *839*, 84. [CrossRef]
70. Casares, J.; Negueruela, I.; Ribó, M.; Ribas, I.; Paredes, J.M.; Herrero, A.; Simón-Díaz, S. A Be-type star with a black-hole companion. *Nature* **2014**, *505*, 378–381. [CrossRef]
71. H. E. S. S. Collaboration; Abdalla, H.; Abramowski, A.; Aharonian, F.; Ait Benkhali, F.; Angüner, E.O.; Arakawa, M.; Arrieta, M.; Aubert, P.; Backes, M.; et al. The H.E.S.S. Galactic plane survey. *Astron. Astrophys.* **2018**, *612*, A1. [CrossRef]
72. Cao, Z.; Aharonian, F.; An, Q.; Axikegu.; Bai, Y.X.; Bao, Y.W.; Bastieri, D.; Bi, X.J.; Bi, Y.J.; Cai, J.T.; et al. The First LHAASO Catalog of Gamma-ray Sources. *arXiv* **2023**, arXiv:2305.17030. [CrossRef]
73. Lucarelli, F.; Verrecchia, F.; Striani, E.; Pittori, C.; Tavani, M.; Vercellone, S.; Bulgarelli, A.; Gianotti, F.; Trifoglio, M.; Chen, A.; et al. AGILE detection of the new unidentified gamma-ray source AGL J2241+4454. *Astron. Telegr.* **2010**, *2761*, 1.
74. Hartman, R.C.; Bertsch, D.L.; Dingus, B.L.; Fichtel, C.E.; Hunter, S.D.; Kanbach, G.; Kniffen, D.A.; Lin, Y.C.; Mattox, J.R.; Mayer-Hasselwander, H.A.; et al. EGRET Detection of High-Energy Gamma Radiation from the OVV Quasar 3C 454.3. *Astrophys. J. Lett.* **1993**, *407*, L41. [CrossRef]
75. Aller, M.F.; Marscher, A.P.; Hartman, R.C.; Aller, H.D.; Aller, M.C.; Balonek, T.J.; Begelman, M.C.; Chiaberge, M.; Clements, S.D.; Collmar, W.; et al. Radio to γ-ray observations of 3C 454.3:1993-1995. In Proceedings of the Fourth Compton Symposium, Williamsburg, VA, USA, 27–30 April 1997; pp. 1423–1427. [CrossRef]
76. Giommi, P.; Blustin, A.J.; Capalbi, M.; Colafrancesco, S.; Cucchiara, A.; Fuhrmann, L.; Krimm, H.A.; Marchili, N.; Massaro, E.; Perri, M.; et al. Swift and infra-red observations of the blazar 3C 454.3 during the giant X-ray flare of May 2005. *Astron. Astrophys.* **2006**, *456*, 911–916. [CrossRef]
77. Villata, M.; Raiteri, C.M.; Balonek, T.J.; Aller, M.F.; Jorstad, S.G.; Kurtanidze, O.M.; Nicastro, F.; Nilsson, K.; Aller, H.D.; Arai, A.; et al. The unprecedented optical outburst of the quasar <ASTROBJ>3C 454.3</ASTROBJ>. The WEBT campaign of 2004-2005. *Astron. Astrophys.* **2006**, *453*, 817–822. [CrossRef]

78. Pian, E.; Foschini, L.; Beckmann, V.; Soldi, S.; Türler, M.; Gehrels, N.; Ghisellini, G.; Giommi, P.; Maraschi, L.; Pursimo, T.; et al. INTEGRAL observations of the blazar 3C 454.3 in outburst. *Astron. Astrophys.* **2006**, *449*, L21–L25. [CrossRef]
79. Vercellone, S.; Chen, A.W.; Giuliani, A.; Bulgarelli, A.; Donnarumma, I.; Lapshov, I.; Tavani, M.; Argan, A.; Barbiellini, G.; Caraveo, P.; et al. AGILE Detection of a Strong Gamma-ray Flare from the Blazar 3C 454.3. *Astrophys. J. Lett.* **2008**, *676*, L13. [CrossRef]
80. Vercellone, S.; D'Ammando, F.; Vittorini, V.; Donnarumma, I.; Pucella, G.; Tavani, M.; Ferrari, A.; Raiteri, C.M.; Villata, M.; Romano, P.; et al. Multiwavelength Observations of 3C 454.3. III. Eighteen Months of Agile Monitoring of the "Crazy Diamond". *Astrophys. J.* **2010**, *712*, 405–420. [CrossRef]
81. Vercellone, S.; Striani, E.; Vittorini, V.; Donnarumma, I.; Pacciani, L.; Pucella, G.; Tavani, M.; Raiteri, C.M.; Villata, M.; Romano, P.; et al. The Brightest Gamma-ray Flaring Blazar in the Sky: AGILE and Multi-wavelength Observations of 3C 454.3 During 2010 November. *Astrophys. J. Lett.* **2011**, *736*, L38. [CrossRef]
82. Vercellone, S.; Chen, A.W.; Vittorini, V.; Giuliani, A.; D'Ammando, F.; Tavani, M.; Donnarumma, I.; Pucella, G.; Raiteri, C.M.; Villata, M.; et al. Multiwavelength Observations of 3C 454.3. I. The AGILE 2007 November campaign on the "Crazy Diamond". *Astrophys. J.* **2009**, *690*, 1018–1030. [CrossRef]
83. Donnarumma, I.; Pucella, G.; Vittorini, V.; D'Ammando, F.; Vercellone, S.; Raiteri, C.M.; Villata, M.; Perri, M.; Chen, W.P.; Smart, R.L.; et al. Multiwavelength Observations of 3C 454.3. II. The AGILE 2007 December Campaign. *Astrophys. J.* **2009**, *707*, 1115–1123. [CrossRef]
84. Pacciani, L.; Vittorini, V.; Tavani, M.; Fiocchi, M.T.; Vercellone, S.; D'Ammando, F.; Sakamoto, T.; Pian, E.; Raiteri, C.M.; Villata, M.; et al. The 2009 December Gamma-ray Flare of 3C 454.3: The Multifrequency Campaign. *Astrophys. J. Lett.* **2010**, *716*, L170–L175. [CrossRef]
85. Striani, E.; Vercellone, S.; Tavani, M.; Vittorini, V.; D'Ammando, F.; Donnarumma, I.; Pacciani, L.; Pucella, G.; Bulgarelli, A.; Trifoglio, M.; et al. The Extraordinary Gamma-ray Flare of the Blazar 3C 454.3. *Astrophys. J.* **2010**, *718*, 455–459. [CrossRef]
86. Vittorini, V.; Tavani, M.; Cavaliere, A.; Striani, E.; Vercellone, S. The Blob Crashes into the Mirror: Modeling the Exceptional γ-ray Flaring Activity of 3C 454.3 in 2010 November. *Astrophys. J.* **2014**, *793*, 98. [CrossRef]
87. Ghisellini, G.; Tagliaferri, G.; Foschini, L.; Ghirlanda, G.; Tavecchio, F.; Della Ceca, R.; Haardt, F.; Volonteri, M.; Gehrels, N. High-redshift Fermi blazars. *Mon. Not. R. Astron. Soc.* **2011**, *411*, 901–914. [CrossRef]
88. Foschini, L.; Lister, M.L.; Andernach, H.; Ciroi, S.; Marziani, P.; Antón, S.; Berton, M.; Dalla Bontà, E.; Järvelä, E.; Marchã, M.J.M.; et al. A New Sample of Gamma-ray Emitting Jetted Active Galactic Nuclei. *Universe* **2022**, *8*, 587. [CrossRef]
89. Abdollahi, S.; Acero, F.; Ackermann, M.; Ajello, M.; Atwood, W.B.; Axelsson, M.; Baldini, L.; Ballet, J.; Barbiellini, G.; Bastieri, D.; et al. Fermi Large Area Telescope Fourth Source Catalog. *Astrophys. J. Suppl. Ser.* **2020**, *247*, 33. [CrossRef]
90. Vercellone, S.; Romano, P.; Piano, G.; Vittorini, V.; Donnarumma, I.; Munar-Adrover, P.; Raiteri, C.M.; Villata, M.; Verrecchia, F.; Lucarelli, F.; et al. AGILE, Fermi, Swift, and GASP/WEBT multi-wavelength observations of the high-redshift blazar 4C +71.07 in outburst. *Astron. Astrophys.* **2019**, *621*, A82. [CrossRef]
91. Gehrels, N.; Chincarini, G.; Giommi, P.; Mason, K.O.; Nousek, J.A.; Wells, A.A.; White, N.E.; Barthelmy, S.D.; Burrows, D.N.; Cominsky, L.R.; et al. The Swift Gamma-ray Burst Mission. *Astrophys. J.* **2004**, *611*, 1005–1020. [CrossRef]
92. Burrows, D.N.; Hill, J.E.; Nousek, J.A.; Kennea, J.A.; Wells, A.; Osborne, J.P.; Abbey, A.F.; Beardmore, A.; Mukerjee, K.; Short, A.D.T.; et al. The Swift X-ray Telescope. *Space Sci. Rev.* **2005**, *120*, 165–195. [CrossRef]
93. Roming, P.W.A.; Kennedy, T.E.; Mason, K.O.; Nousek, J.A.; Ahr, L.; Bingham, R.E.; Broos, P.S.; Carter, M.J.; Hancock, B.K.; Huckle, H.E.; et al. The Swift Ultra-Violet/Optical Telescope. *Space Sci. Rev.* **2005**, *120*, 95–142. [CrossRef]
94. Villata, M.; Raiteri, C.M.; Larionov, V.M.; Kurtanidze, O.M.; Nilsson, K.; Aller, M.F.; Tornikoski, M.; Volvach, A.; Aller, H.D.; Arkharov, A.A.; et al. Multifrequency monitoring of the blazar <ASTROBJ>0716+714</ASTROBJ> during the GASP-WEBT-AGILE campaign of 2007. *Astron. Astrophys.* **2008**, *481*, L79–L82. [CrossRef]
95. Donnarumma, I.; De Rosa, A.; Vittorini, V.; Miller, H.R.; Popović, L.Č.; Simić, S.; Tavani, M.; Eggen, J.; Maune, J.; Kuulkers, E.; et al. The Remarkable γ-ray Activity in the Gravitationally Lensed Blazar PKS 1830-211. *Astrophys. J. Lett.* **2011**, *736*, L30. [CrossRef]
96. Vercellone, S.; Donnarumma, I.; Pittori, C.; Capitanio, F.; De Rosa, A.; Di Gesu, L.; Kiehlmann, S.; Iacolina, M.N.; Pellizzoni, P.A.; Egron, E.; et al. Multiwavelength observations of the lensed quasar PKS 1830-211 during the 2019 γ-ray flare. *Mon. Not. R. Astron. Soc.* **2024**, *527*, 5717–5731. [CrossRef]
97. de Rosa, A.; Piro, L.; Tramacere, A.; Massaro, E.; Walter, R.; Bassani, L.; Malizia, A.; Bird, A.J.; Dean, A.J. The broad-band X-ray spectrum of the blazar PKS B1830-211 by Chandra and INTEGRAL. *Astron. Astrophys.* **2005**, *438*, 121–126. [CrossRef]
98. Tagliaferri, G.; Ghisellini, G.; Perri, M.; Hayashida, M.; Baloković, M.; Covino, S.; Giommi, P.; Madejski, G.M.; Puccetti, S.; Sbarrato, T.; et al. NuSTAR and Multifrequency Study of the Two High-redshift Blazars S5 0836+710 and PKS 2149-306. *Astrophys. J.* **2015**, *807*, 167. [CrossRef]
99. Nair, S.; Jin, C.; Garrett, M.A. Helical jet in the gravitationally lensed blazar PKS1830-211. *Mon. Not. R. Astron. Soc.* **2005**, *362*, 1157–1166. [CrossRef]
100. Tavani, M.; Vittorini, V.; Cavaliere, A. An Emerging Class of Gamma-ray Flares from Blazars: Beyond One-zone Models. *Astrophys. J.* **2015**, *814*, 51. [CrossRef]
101. Vittorini, V.; Tavani, M.; Cavaliere, A. Meeting the Challenge from Bright and Fast Gamma-ray Flares of 3C 279. *Astrophys. J. Lett.* **2017**, *843*, L23. [CrossRef]

102. Araudo, A.T.; Bosch-Ramon, V.; Romero, G.E. Gamma rays from cloud penetration at the base of AGN jets. *Astron. Astrophys.* **2010**, *522*, A97. [CrossRef]
103. Costamante, L.; Ghisellini, G. TeV candidate BL Lac objects. *Astron. Astrophys.* **2002**, *384*, 56–71. [CrossRef]
104. Costamante, L. TeV-peaked candidate BL Lac objects. *Mon. Not. R. Astron. Soc.* **2020**, *491*, 2771–2778. [CrossRef]
105. Punch, M.; Akerlof, C.W.; Cawley, M.F.; Chantell, M.; Fegan, D.J.; Fennell, S.; Gaidos, J.A.; Hagan, J.; Hillas, A.M.; Jiang, Y.; et al. Detection of TeV photons from the active galaxy Markarian 421. *Nature* **1992**, *358*, 477–478. [CrossRef]
106. Quinn, J.; Akerlof, C.W.; Biller, S.; Buckley, J.; Carter-Lewis, D.A.; Cawley, M.F.; Catanese, M.; Connaughton, V.; Fegan, D.J.; Finley, J.P.; et al. Detection of Gamma Rays with E > 300 GeV from Markarian 501. *Astrophys. J. Lett.* **1996**, *456*, L83. [CrossRef]
107. Cortina, J.; CTAO LST Collaboration. First detection of VHE gamma-ray emission from FSRQ OP 313 with LST-1. *Astron. Telegr.* **2023**, *16381*, 1.
108. Wakely, S.P.; Horan, D. TeVCat: An online catalog for Very High Energy Gamma-ray Astronomy. *Int. Cosm. Ray Conf.* **2008**, *3*, 1341–1344.
109. Rappoldi, A.; Lucarelli, F.; Pittori, C.; Longo, F.; Cattaneo, P.W.; Verrecchia, F.; Tavani, M.; Bulgarelli, A.; Chen, A.W.; Colafrancesco, S.; et al. Search of MeV-GeV counterparts of TeV sources with AGILE in pointing mode. *Astron. Astrophys.* **2016**, *587*, A93. [CrossRef]
110. Bulgarelli, A.; Chen, A.W.; Tavani, M.; Gianotti, F.; Trifoglio, M.; Contessi, T. Evaluating the maximum likelihood method for detecting short-term variability of AGILE γ-ray sources. *Astron. Astrophys.* **2012**, *540*, A79. [CrossRef]
111. Verrecchia, F.; Gasparrini, D.; Cutini, S.; Preger, B.; Pittori, C.; Colafrancesco, S.; Giommi, P.; Vercellone, S.; Chen, A.; Giuliani, A.; et al. AGILE detection of the blazar W Comae in the gamma-ray energy band. *Astron. Telegr.* **2008**, *1582*, 1.
112. Swordy, S. TeV Outburst from W Comae. *Astron. Telegr.* **2008**, *1565*, 1.
113. Swordy, S. VERITAS discovers TeV gamma rays from W Comae. *Astron. Telegr.* **2008**, *1422*, 1.
114. Acciari, V.A.; Aliu, E.; Aune, T.; Beilicke, M.; Benbow, W.; Böttcher, M.; Boltuch, D.; Buckley, J.H.; Bradbury, S.M.; Bugaev, V.; et al. Multiwavelength Observations of a TeV-Flare from W Comae. *Astrophys. J.* **2009**, *707*, 612–620. [CrossRef]
115. Costa, E.; Del Monte, E.; Donnarumma, I.; Evangelista, Y.; Feroci, M.; Lapshov, I.; Lazzarotto, F.; Pacciani, L.; Rapisarda, M.; Soffitta, P.; et al. SuperAGILE detects enhanced hard X-ray emission from Mkn 421. *Astron. Telegr.* **2008**, *1574*, 1.
116. Pittori, C.; Cutini, S.; Gasparrini, D.; Verrecchia, F.; Colafrancesco, S.; Giommi, P.; Vercellone, S.; Chen, A.; Giuliani, A.; Donnarumma, I.; et al. AGILE detection of the blazar Mrk 421. *Astron. Telegr.* **2008**, *1583*, 1.
117. Donnarumma, I.; Vittorini, V.; Vercellone, S.; del Monte, E.; Feroci, M.; D'Ammando, F.; Pacciani, L.; Chen, A.W.; Tavani, M.; Bulgarelli, A.; et al. The June 2008 Flare of Markarian 421 from Optical to TeV Energies. *Astrophys. J. Lett.* **2009**, *691*, L13–L19. [CrossRef]
118. Villata, M.; Raiteri, C.M. Helical jets in blazars. I. The case of MKN 501. *Astron. Astrophys.* **1999**, *347*, 30–36.
119. Villata, M.; Raiteri, C.M.; Kurtanidze, O.M.; Nikolashvili, M.G.; Ibrahimov, M.A.; Papadakis, I.E.; Tosti, G.; Hroch, F.; Takalo, L.O.; Sillanpää, A.; et al. The WEBT <ASTROBJ>BL Lacertae</ASTROBJ> Campaign 2001 and its extension. Optical light curves and colour analysis 1994-2002. *Astron. Astrophys.* **2004**, *421*, 103–114. [CrossRef]
120. Barthelmy, S.D.; Barbier, L.M.; Cummings, J.R.; Fenimore, E.E.; Gehrels, N.; Hullinger, D.; Krimm, H.A.; Markwardt, C.B.; Palmer, D.M.; Parsons, A.; et al. The Burst Alert Telescope (BAT) on the SWIFT Midex Mission. *Space Sci. Rev.* **2005**, *120*, 143–164. [CrossRef]
121. Parmiggiani, N.; Bulgarelli, A.; Ursi, A.; Addis, A.; Baroncelli, L.; Fioretti, V.; Di Piano, A.; Panebianco, G.; Tavani, M.; Pittori, C.; et al. The AGILE real-time analysis software system to detect short-transient events in the multi-messenger era. *Astron. Comput.* **2023**, *44*, 100726. [CrossRef]
122. Giuliani, A.; Mereghetti, S.; Fornari, F.; Del Monte, E.; Feroci, M.; Marisaldi, M.; Esposito, P.; Perotti, F.; Tavani, M.; Argan, A.; et al. AGILE detection of delayed gamma-ray emission from GRB 080514B. *Astron. Astrophys.* **2008**, *491*, L25–L28. [CrossRef]
123. Ajello, M.; Arimoto, M.; Axelsson, M.; Baldini, L.; Barbiellini, G.; Bastieri, D.; Bellazzini, R.; Bhat, P.N.; Bissaldi, E.; Blandford, R.D.; et al. A Decade of Gamma-ray Bursts Observed by Fermi-LAT: The Second GRB Catalog. *Astrophys. J.* **2019**, *878*, 52. [CrossRef]
124. Omodei, N. Fermi-LAT observation of GRB 081024B. *GRB Coord. Netw.* **2008**, *8407*, 1.
125. Connaughton, V.; Briggs, M. GRB 081024B: Fermi gamma-ray burst monitor detection. *GRB Coord. Netw.* **2008**, *8408*, 1.
126. Giuliani, A.; Fuschino, F.; Vianello, G.; Marisaldi, M.; Mereghetti, S.; Tavani, M.; Cutini, S.; Barbiellini, G.; Longo, F.; Moretti, E.; et al. AGILE Detection of Delayed Gamma-ray Emission From the Short Gamma-ray Burst GRB 090510. *Astrophys. J. Lett.* **2010**, *708*, L84–L88. [CrossRef]
127. Ackermann, M.; Asano, K.; Atwood, W.B.; Axelsson, M.; Baldini, L.; Ballet, J.; Barbiellini, G.; Baring, M.G.; Bastieri, D.; Bechtol, K.; et al. Fermi Observations of GRB 090510: A Short-Hard Gamma-ray Burst with an Additional, Hard Power-law Component from 10 keV TO GeV Energies. *Astrophys. J.* **2010**, *716*, 1178–1190. [CrossRef]
128. Ackermann, M.; Ajello, M.; Albert, A.; Anderson, B.; Arimoto, M.; Atwood, W.B.; Axelsson, M.; Baldini, L.; Ballet, J.; Barbiellini, G.; et al. Fermi-LAT Observations of the LIGO Event GW150914. *Astrophys. J. Lett.* **2016**, *823*, L2. [CrossRef]
129. Tavani, M.; Pittori, C.; Verrecchia, F.; Bulgarelli, A.; Giuliani, A.; Donnarumma, I.; Argan, A.; Trois, A.; Lucarelli, F.; Marisaldi, M.; et al. AGILE Observations of the Gravitational-wave Event GW150914. *Astrophys. J. Lett.* **2016**, *825*, L4. [CrossRef]
130. MAGIC Collaboration.; Acciari, V.A.; Ansoldi, S.; Antonelli, L.A.; Engels, A.A.; Baack, D.; Babić, A.; Banerjee, B.; Barres de Almeida, U.; Barrio, J.A.; et al. Observation of inverse Compton emission from a long γ-ray burst. *Nature* **2019**, *575*, 459–463. [CrossRef]

131. Ursi, A.; Tavani, M.; Frederiks, D.D.; Romani, M.; Verrecchia, F.; Marisaldi, M.; Aptekar, R.L.; Antonelli, L.A.; Argan, A.; Bulgarelli, A.; et al. AGILE and Konus-Wind Observations of GRB 190114C: The Remarkable Prompt and Early Afterglow Phases. *Astrophys. J.* **2020**, *904*, 133. [CrossRef]
132. Tavani, M.; Piano, G.; Bulgarelli, A.; Foffano, L.; Ursi, A.; Verrecchia, F.; Pittori, C.; Casentini, C.; Giuliani, A.; Longo, F.; et al. AGILE Gamma-ray Detection of the Exceptional GRB 221009A. *Astrophys. J. Lett.***2023**, *956*, L23. [CrossRef]
133. Cao, Z.; Aharonian, F.; An, Q.; Axikegu; Bai, Y.X.; Bao, Y.W.; Bastieri, D.; Bi, X.J.; Bi, Y.J.; Cai, J.T.; et al. Very high-energy gamma-ray emission beyond 10 TeV from GRB 221009A. *Sci. Adv.* **2023**, *9*, eadj2778. [CrossRef]
134. Vigliano, A.; Longo, F. Gamma-ray Bursts: 50 Years and Counting! *Universe* **2024**, *10*, 57. [CrossRef]
135. Verrecchia, F.; Tavani, M.; Ursi, A.; Argan, A.; Pittori, C.; Donnarumma, I.; Bulgarelli, A.; Fuschino, F.; Labanti, C.; Marisaldi, M.; et al. AGILE Observations of the Gravitational-wave Source GW170104. *Astrophys. J. Lett.* **2017**, *847*, L20. [CrossRef]
136. Verrecchia, F.; Tavani, M.; Donnarumma, I.; Bulgarelli, A.; Evangelista, Y.; Pacciani, L.; Ursi, A.; Piano, G.; Pilia, M.; Cardillo, M.; et al. AGILE Observations of the Gravitational-wave Source GW170817: Constraining Gamma-ray Emission from an NS-NS Coalescence. *Astrophys. J. Lett.* **2017**, *850*, L27. [CrossRef]
137. Ursi, A.; Verrecchia, F.; Piano, G.; Casentini, C.; Tavani, M.; Bulgarelli, A.; Cardillo, M.; Longo, F.; Lucarelli, F.; Morselli, A.; et al. AGILE Observations of the LIGO-Virgo Gravitational-wave Events of the GWTC-1 Catalog. *Astrophys. J.* **2022**, *924*, 80. [CrossRef]
138. Lucarelli, F.; Pittori, C.; Verrecchia, F.; Donnarumma, I.; Tavani, M.; Bulgarelli, A.; Giuliani, A.; Antonelli, L.A.; Caraveo, P.; Cattaneo, P.W.; et al. AGILE Detection of a Candidate Gamma-ray Precursor to the ICECUBE-160731 Neutrino Event. *Astrophys. J.* **2017**, *846*, 121. [CrossRef]
139. Lucarelli, F.; Tavani, M.; Piano, G.; Bulgarelli, A.; Donnarumma, I.; Verrecchia, F.; Pittori, C.; Antonelli, L.A.; Argan, A.; Barbiellini, G.; et al. AGILE Detection of Gamma-ray Sources Coincident with Cosmic Neutrino Events. *Astrophys. J.* **2019**, *870*, 136. [CrossRef]
140. IceCube Collaboration; Aartsen, M.G.; Ackermann, M.; Adams, J.; Aguilar, J.A.; Ahlers, M.; Ahrens, M.; Al Samarai, I.; Altmann, D.; Andeen, K.; et al. Multimessenger observations of a flaring blazar coincident with high-energy neutrino IceCube-170922A. *Science* **2018**, *361*, eaat1378. [CrossRef]
141. Casentini, C.; Verrecchia, F.; Tavani, M.; Ursi, A.; Antonelli, L.A.; Argan, A.; Barbiellini, G.; Bulgarelli, A.; Caraveo, P.; Cardillo, M.; et al. AGILE Observations of Two Repeating Fast Radio Bursts with Low Intrinsic Dispersion Measures. *Astrophys. J. Lett.* **2020**, *890*, L32. [CrossRef]
142. Tavani, M.; Verrecchia, F.; Casentini, C.; Perri, M.; Ursi, A.; Pacciani, L.; Pittori, C.; Bulgarelli, A.; Piano, G.; Pilia, M.; et al. Gamma-ray and X-ray Observations of the Periodic-repeater FRB 180916 during Active Phases. *Astrophys. J. Lett.* **2020**, *893*, L42. [CrossRef]
143. Pilia, M.; Burgay, M.; Possenti, A.; Ridolfi, A.; Gajjar, V.; Corongiu, A.; Perrodin, D.; Bernardi, G.; Naldi, G.; Pupillo, G.; et al. The Lowest-frequency Fast Radio Bursts: Sardinia Radio Telescope Detection of the Periodic FRB 180916 at 328 MHz. *Astrophys. J. Lett.* **2020**, *896*, L40. [CrossRef]
144. Verrecchia, F.; Casentini, C.; Tavani, M.; Ursi, A.; Mereghetti, S.; Pilia, M.; Cardillo, M.; Addis, A.; Barbiellini, G.; Baroncelli, L.; et al. AGILE Observations of Fast Radio Bursts. *Astrophys. J.* **2021**, *915*, 102. [CrossRef]
145. Tavani, M.; Casentini, C.; Ursi, A.; Verrecchia, F.; Addis, A.; Antonelli, L.A.; Argan, A.; Barbiellini, G.; Baroncelli, L.; Bernardi, G.; et al. An X-ray burst from a magnetar enlightening the mechanism of fast radio bursts. *Nat. Astron.* **2021**, *5*, 401–407. [CrossRef]
146. Feroci, M.; Costa, E.; Del Monte, E.; Donnarumma, I.; Evangelista, Y.; Lapshov, I.; Lazzarotto, F.; Pacciani, L.; Rapisarda, M.; Soffitta, P.; et al. Monitoring the hard X-ray sky with SuperAGILE. *Astron. Astrophys.* **2010**, *510*, A9. [CrossRef]
147. Galli, M.; Marisaldi, M.; Fuschino, F.; Labanti, C.; Argan, A.; Barbiellini, G.; Bulgarelli, A.; Cattaneo, P.W.; Colafrancesco, S.; Del Monte, E.; et al. AGILE mini-calorimeter gamma-ray burst catalog. *Astron. Astrophys.* **2013**, *553*, A33. [CrossRef]
148. Marisaldi, M.; Fuschino, F.; Tavani, M.; Dietrich, S.; Price, C.; Galli, M.; Pittori, C.; Verrecchia, F.; Mereghetti, S.; Cattaneo, P.W.; et al. Properties of terrestrial gamma ray flashes detected by AGILE MCAL below 30 MeV. *J. Geophys. Res.* **2014**, *119*, 1337–1355. [CrossRef]
149. Marisaldi, M.; Argan, A.; Ursi, A.; Gjesteland, T.; Fuschino, F.; Labanti, C.; Galli, M.; Tavani, M.; Pittori, C.; Verrecchia, F.; et al. Enhanced detection of terrestrial gamma-ray flashes by AGILE. *Geophys. Res. Lett.* **2015**, *42*, 9481–9487. [CrossRef] [PubMed]
150. Marisaldi, M.; Galli, M.; Labanti, C.; Østgaard, N.; Sarria, D.; Cummer, S.A.; Lyu, F.; Lindanger, A.; Campana, R.; Ursi, A.; et al. On the High-Energy Spectral Component and Fine Time Structure of Terrestrial Gamma Ray Flashes. *J. Geophys. Res.* **2019**, *124*, 7484–7497. [CrossRef]
151. Ursi, A.; Romani, M.; Verrecchia, F.; Pittori, C.; Tavani, M.; Marisaldi, M.; Galli, M.; Labanti, C.; Parmiggiani, N.; Bulgarelli, A.; et al. The Second AGILE MCAL Gamma-ray Burst Catalog: 13 yr of Observations. *Astrophys. J.* **2022**, *925*, 152. [CrossRef]
152. Ursi, A.; Parmiggiani, N.; Messerotti, M.; Pellizzoni, A.; Pittori, C.; Longo, F.; Verrecchia, F.; Argan, A.; Bulgarelli, A.; Tavani, M.; et al. The First AGILE Solar Flare Catalog. *Astrophys. J. Suppl. Ser.* **2023**, *267*, 9. [CrossRef]
153. De Angelis, A.; Tatischeff, V.; Tavani, M.; Oberlack, U.; Grenier, I.; Hanlon, L.; Walter, R.; Argan, A.; von Ballmoos, P.; Bulgarelli, A.; et al. The e-ASTROGAM mission. Exploring the extreme Universe with gamma rays in the MeV-GeV range. *Exp. Astron.* **2017**, *44*, 25–82. [CrossRef]

154. De Angelis, A.; Tatischeff, V.; Argan, A.; Brandt, S.; Bulgarelli, A.; Bykov, A.; Costantini, E.; Silva, R.C.d.; Grenier, I.A.; Hanlon, L.; et al. Gamma-ray astrophysics in the MeV range. *Exp. Astron.* **2021**, *51*, 1225–1254. [CrossRef]
155. Tomsick, J.A.; Boggs, S.E.; Zoglauer, A.; Hartmann, D.; Ajello, M.; Burns, E.; Fryer, C.; Karwin, C.; Kierans, C.; Lowell, A.; et al. The Compton Spectrometer and Imager. *arXiv* **2023**, arXiv:2308.12362. [CrossRef]

universe

MDPI

Review

Supernova Remnants in Gamma Rays

Andrea Giuliani [1,*] and Martina Cardillo [2]

1 INAF—Istituto di Astrofisica Spaziale e Fisica Cosmica, Via Alfonso Corti 12, 20133 Milano, Italy
2 INAF—Istituto di Astrofisica e Planetologia Spaziali, Via del Fosso del Cavaliere 100, 00133 Roma, Italy; martina.cardillo@inaf.it
* Correspondence: andrea.giuliani@inaf.it

Abstract: In the 1960s, the remnants of supernova explosions (SNRs) were indicated as a possible source of galactic cosmic rays through the Diffusive Shock Acceleration (DSA) mechanism. Since then, the observation of gamma-ray emission from relativistic ions in these objects has been one of the main goals of high-energy astrophysics. A few dozen SNRs have been detected at GeV and TeV photon energies in the last two decades. However, these observations have shown a complex phenomenology that is not easy to reduce to the standard paradigm based on DSA acceleration. Although the understanding of these objects has greatly increased, and their nature as efficient electron and proton accelerators has been observed, it remains to be clarified whether these objects are the main contributors to galactic cosmic rays. Here, we review the observations of γ-ray emission from SNRs and the perspectives for the future.

Keywords: supernova remnants; high energy; PeVatrons; galactic cosmic rays

check for updates

Citation: Giuliani, A.; Cardillo, M. Supernova Remnants in Gamma Rays. *Universe* **2024**, *10*, 203. https://doi.org/10.3390/universe10050203

Academic Editor: Fridolin Weber

Received: 29 February 2024
Revised: 21 April 2024
Accepted: 25 April 2024
Published: 1 May 2024

1. Supernova Remnants

1.1. Supernova Remnant Evolution

Supernova explosions can result from two main physical processes related to the life cycle of a star. The first one happens when a star with a mass greater than 8 $M_{\odot}$ reaches the end of its life, and the radiation produced within the nucleus is no longer able to counteract the weight of the star's envelope. The collapse of the star triggers the material to produce a supernova explosion. Alternatively, a white dwarf in a binary system can exceed the Chandrashekar equilibrium limit (approximately 1.4 $M_{\odot}$) due to accretion from its companions and explode in a type Ia supernova. In both cases, the supernova explosion expels a significant amount of matter, ranging from 1 to several solar masses, into the interstellar medium at a speed of 10^4 km s^{-1}. This corresponds to a kinetic energy of approximately 10^{51} erg. The main differences between the two scenarios concern the chemical composition of the ejected materials and the interstellar medium (ISM) into which the supernova remnant (SNR) expands.

Over the subsequent 100,000 years, the ejecta will expand, interact with the surrounding medium, and eventually dissolve into it. The evolutionary phases of an SNR can be defined by the type of interaction with the surrounding medium and the ratio between the ejecta mass (M_{ej}) and the mass of the swept-up material (M_{sw}).

During the free expansion phase, lasting a few hundred years as long as $M_{ej} > M_{sw}$, the star's ejecta expands freely into the surrounding medium with an expansion law linear in time, $R \sim t$. In this phase, the material's speed exceeds the speed of sound in the surrounding medium, thus creating a shock. The temperature decreases as the gas expands adiabatically: $T \sim -3(\gamma - 1)$, where γ is the specific heat ratio.

In the Sedov–Taylor phase, lasting between 20,000 and 40,000 years as long as $M_{ej} \leq M_{sw}$, the deceleration of the shell becomes significant, and a reverse shock begins to propagate toward the interior of the SNR, heating the gas, which thus becomes visible in the soft X-ray band. The evolution of the remnant is well described by the adiabatic blast-wave solution of Taylor and Sedov [1], and the radius increases with time as $R \propto t^{2/5}$.

The forward and reverse shocks formed during the free expansion and Sedov–Taylor phases can accelerate particles (see Section 1.2). The observation of non-thermal radio and γ-ray emissions from SNRs with ages up to a few tens of thousands of years confirms this aspect.

In the radiative phase ($M_{\rm ej} \ll M_{\rm sw}$), the shell expands at a sub-thermal speed. After this phase, the remnant of the supernova dissolves and becomes part of the interstellar medium.

The duration of each phase of an SNR is determined by the interaction between the ejecta and the ISM, which, in turn, depends on the type of circumstellar medium in which the SNR is situated. On the other hand, supernova explosions play a crucial role in shaping the evolution of the ISM and galaxies by injecting a significant amount of energy and momentum and then influencing the process of star formation [2]. Figure 1 shows the young SNR Cas A as seen in the X-ray.

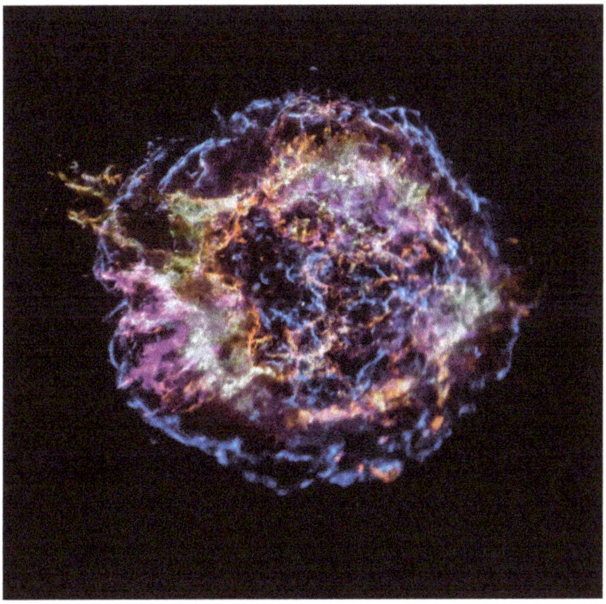

Figure 1. Cas A image in X-ray band as observed by the Chandra X-ray Observatory telescope [3].

1.2. Supernova Remnants as Cosmic Accelerators

In the initial two stages of an SNR's evolution, a shock forms between the ejecta (including swept material) and the local interstellar medium. Additionally, a second shock may develop and propagate toward the center.

Over the past decades, various works have shown how shocks can accelerate particles, including cosmic rays (CRs), and there is a consensus that the dominant process is Diffusive Shock Acceleration (DSA) [4,5]. This process is based on Fermi I-order acceleration [6]. The acceleration index provided by DSA, $\gamma = \frac{3R}{R-1}$, is strictly correlated with the compression ratio of the shock, $R = \frac{u_d}{u_u}$, where u_u and u_d are the upstream and downstream velocities, respectively, related to the shock Mach number. For strong shocks and in the test-particle limit, i.e., when the particle energy is negligible compared to that of the shock, DSA predicts a particle energy distribution with an energy spectral index close to 2. However, the backreaction of CRs on the shock and then also on the surrounding ambient can modify the acceleration spectrum, and if the feedback effects on the shock are not negligible, the spectral index of accelerated particles may become steeper than 2 (for a review of the DSA theory, see [7] and references therein).

If SNRs accelerate cosmic rays (CRs), they should be observed as sources of non-thermal emission. This emission results from the interaction of electrons and hadrons with the surrounding medium. The interaction between electrons and magnetic fields at the site produces synchrotron radiation. If the electron spectral distribution can be described as a power law with index α and maximum energy $E_{\rm max}$ (i.e., $f(E) \propto E^{-\alpha} exp(E/E_{\rm max})$), the emerging emission has a spectrum with slope $\gamma = -(\alpha+1)/2$, typically resulting in a hard spectrum. In the spectral energy distribution (SED, see Figure 2), the peak energy of synchrotron radiation can reach the X-ray band for an $E_{\rm max}$ of few TeV. The same electron population interacting with the radiation fields produces emission through the Inverse Compton process. The emission has a spectral shape similar to that of the synchrotron but shifted toward higher energies, typically in the γ-ray band. If a dense medium is present, electrons can also radiate efficiently through Bremsstrahlung, which concurs with the γ-ray emission of the objects.

If hadrons (protons or heavier nuclei) are also accelerated, they scatter inelastically against the nuclei of the medium in the so-called "pp interaction", producing neutral pions, which rapidly (8.5×10^{-17} s) decay into two γ-ray photons. In addition to protons, secondary electrons (produced by pp interactions) can produce γ-ray emission through IC and Bremsstrahlung (see [8] and references therein).

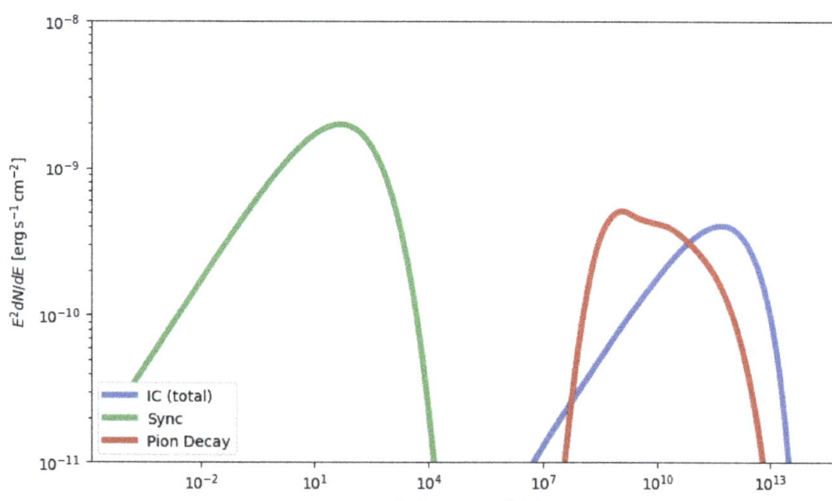

Figure 2. Example of SED given by synchrotron, IC and pion decay process.

SNRs have often been suggested as possible sources of CRs for two main reasons: their observed non-thermal emission (from radio to γ-ray) and a substantial match between the average power of supernova explosions in the Galaxy and the power needed to keep the galactic CR population stable. The energy density of CRs in the Galaxy disk is on the order of $w_{\rm CR} \simeq 1$ eV/cm^3. In the standard leaky-box approximation, we assume the disk of the Galaxy as a disk with a radius of 15 kpc and a height of 1 kpc, obtaining a total CR energy $E_{CR} = w_{\rm CR} \times V_{\rm disk} \simeq 3.34 \times 10^{55}$ erg. Isotope ratios in CRs can be used to estimate the typical permanence time of CRs in the plane of the Galaxy, which is approximately $t_{\rm disk} \approx 10^7$ years. Therefore, assuming that the galactic CR distribution has reached a steady-state condition, the power needed to maintain the observed CR density is $P_{CR} = \frac{E_{CR}}{t_{disk}} \simeq 10^{41}$ erg/s. Whichever process accelerates CRs in the Galaxy, it must therefore have a power of at least P_{CR}. A process with similar power is given by supernova explosions. Assuming that their average rate in our Galaxy is on the order of 1/30 years, we find that $P_{SN} \simeq 1 \simeq 10^{42}$ erg/s. The first to notice this similarity were Ginzburg and Syrovatskii [9], who concluded that "SuperNovae alone could maintain the CR population provided that about 10% of their kinetic energy is somehow converted into CRs". This was the first evidence pointing to supernovae (or, equivalently, SNRs) as prime candidates for CR sources.

The observation of non-thermal emission from SNRs strongly supports the hypothesis that they could be sources of galactic CRs. Indeed, the synchrotron spectra of these objects have been observed in the radio band since the 1960s, showing the presence of magnetic field amplification [10–12], and the modification of Balmer lines in the optical band due to CR presence was also observed [13].

The majority of cosmic rays are hadrons. Therefore, evidence supporting the idea that supernova remnants can sustain the galactic cosmic-ray population must come from observations of these objects in the gamma-ray band. This is the only band where a clear indication of their presence can be found. Moreover, the cosmic-ray spectrum suggests that some galactic sources can accelerate particles to at least 1 PeV. This should result in a γ-ray spectrum up to and beyond 100 TeV, with no cutoffs.

For this reason, SNRs have always been a primary focus of γ-ray observations. This includes both the GeV band (approximately 0.1–100 GeV), which is covered by γ-ray satellites, and the TeV band (approximately 0.1–100 TeV), which is covered by Cherenkov telescopes. On the one hand, we now know that even the first γ-ray instruments (SAS-2, COS-B in the 1970s, EGRET on board CGRO in the 1990s) were able to detect these objects. On the other hand, the limited angular resolution of these instruments (and the crowding of the galactic fields) did not allow the γ-ray emission to be associated with SNRs.

The first certain associations were made with Cherenkov instruments; HEGRA observed a source associated with SNR W28, while H.E.S.S. was able to resolve the shell morphology of SNR RX J1713.7-3946 [14].

In recent years, many other classes of sources have been shown to accelerate CRs ([15] and reference therein), and consequently, it is crucial to understand whether the contribution of SNRs is dominant or not.

In this new context, another fundamental channel is neutrino detection. This is an unquestionable hint of CR acceleration since neutrinos can be produced only by the decay of charged pions produced by p-p and p-γ interactions [16]. Looking for neutrino detection in correspondence with PeVatron candidates can confirm the nature of γ-ray-emitting sources [17].

2. SNR in Gamma-Ray Band

Although an exhaustive classification of SNRs in γ-rays is not possible, certain groups of objects have been recognized over time based on their spectrum morphology and multiwavelength behavior [18] (a list of firmly identified SNRs in gamma-ray can be found in Table 1). In young (few thousand years) shell-like SNRs, the γ-ray emission comes from the shell, and often, there is a good morphological correlation between X-ray and γ-ray morphology. Some objects of this class are RX J1713.7-3946, RX J0852.0-4622, RCW 86, and SN 1006. They have similar γ-ray spectra, composed of a hard component (index < 2) peaking around a few TeV, followed by a rapid spectrum decrease. They also show similar γ-ray luminosities [19]. It has been proposed that this class of SNRs is leptonic; in this scenario, the SED can be modeled quite easily with a single electron population emitting in X-rays through a synchrotron and in γ-rays through IC [19]. However, other authors proposed hadronic models for at least some of these objects. For example, the spectrum of RXJ 1713.7-3946 can be explained in terms of hadronic emission, taking into account the clumpiness of the surrounding medium [20].

Another class of sources includes the SNRs interacting with molecular clouds (MCs). A list of SNRs in this class can be found in [21]. In these objects, the γ-rays originate from a region of dense gas that is close to, or in contact with, the SNR. In these systems, typically, the target for the accelerating particles is the gas contained in large MCs (more than 10^3 $M_\odot$). Some famous examples are W 44, W 28, IC 443, and W 51C. The γ-ray spectra of these objects have a soft index (>2.5) and are more easily observed in the GeV than in the TeV band. The ages of these SNRs reach several tens of thousands of years, and their γ-ray luminosities are $\geq 10^{35}$ erg/s. These spectra can be interpreted as γ-ray emissions from particles that diffuse in a partially ionized interstellar medium [22]. The interaction between the SNR shock and the MC can be observed through enhanced CO(2-1)/CO(1-0) ratios, OH maser emission at 1720 MHz, and SiO emission [23]. Additionally, this interaction can be traced using the neutral iron line induced by MeV protons in dense gas [24].

Notably, the very young SNRs Tycho and Cas A exhibit properties that are not easily classified within these two categories, with a spectral index that is intermediate between the two.

Table 1. List of SNRs firmly identified with a gamma-ray source. The Columns GeV, TeV, and PeV indicate detection, respectively, in the bands 0.1–100 GeV, 0.1–100 TeV, and >0.1 PeV.

Name	Common Name	GeV	TeV	PeV
G004.5+06.8	Kepler	Yes [25]	Yes [26]	-
G006.4-00.1	W28	Yes [27]	Yes [28]	-
G008.7-00.1	W30	Yes [29]	-	-
G020.0-00.2	-	Yes [30]	-	-
G023.3-00.3	W41	Yes [31]	Yes [31]	-
G024.7+00.6	-	Yes [30]	Yes [32]	-
G034.7-00.4	W44	Yes [33]	-	-
G043.3-00.2	W49B	Yes [34]	Yes [35]	-
G045.7-00.4	-	Yes [30]	-	-
G049.2-00.7	W51C	Yes [36]	Yes [37]	Yes [38]
G074.0-08.5	CygnusLoop	Yes [39]	-	-
G078.2+02.1	γ-Cygni	Yes [40]	Yes [41]	Yes [38]
G089.0+04.7	HB21	Yes [42]	-	-
G106.3+02.7	-	Yes [43]	Yes [44]	Yes [45]
G109.1-01.0	-	Yes [46]	-	-
G111.7-02.1	CasA	Yes [47]	Yes [48]	-
G120.1+01.4	Tycho	Yes [49]	Yes [50]	-
G132.7+01.3	HB3	Yes [51]	-	-
G160.9+02.6	HB9	Yes [52]	-	-
G180.0-01.7	S147	Yes [53]	-	-
G189.1+03.0	IC443	Yes [54]	Yes [55]	-
G205.5+00.5	Monoceros Loop	Yes [56]	-	-
G260.4-03.4	PuppisA	Yes [57]	-	-
G266.2-01.2	Vela Jr	Yes [58]	Yes [59]	-
G291.0-00.1	-	Yes [30]	-	-
G292.0+01.8	-	Yes [60]	-	-
G296.5+10.0	-	Yes [61]	-	-
G298.6-00.0	-	Yes [60]	-	-
G315.4-02.3	RCW86	Yes [62]	Yes [63]	-
G321.9-00.3	-	Yes [64]	-	-
G323.7-01.0	HESSJ1534-571	-	Yes [65]	-
G326.3-01.8	-	Yes [66]	-	-
G327.6+14.6	SN1006	Yes [67]	Yes [68]	-
G347.3-00.5	RXJ1713.7-3946	Yes [69]	Yes [70]	-
G348.5+00.1	CTB37A	Yes [71]	Yes [72]	-
G348.7+00.3	CTB37B	Yes [73]	Yes [74]	-
G349.7+00.2	-	Yes [75]	Yes [76]	-
G353.6-00.7	HESSJ1731-347	Yes [77]	Yes [78]	-
G355.4+00.7	-	Yes [30]	-	-
G357.7-00.1	MSH 17-39	Yes [79]	-	-

2.1. Cas A

Cas A is one of the youngest known SNRs in the Galaxy, with an estimated age of about 350 years. It is the remnant of a Type IIb supernova [80] that likely passed through a red supergiant phase [81]. The remnant is located at a distance of 3.4 ± 0.4 kpc [82].

In the γ-ray band, this SNR was extensively observed by Fermi-LAT [47], MAGIC [83], and Veritas [84], covering an energy range from 100 MeV to about 10 TeV (see Figure 3). A single-zone hadronic component can accurately model the GeV-TeV spectrum (see Figure 3). The population of protons is described by a power law with an index of 2.17 and a cutoff energy of 17 TeV [85]. However, the contribution of a leptonic component cannot be ruled out.

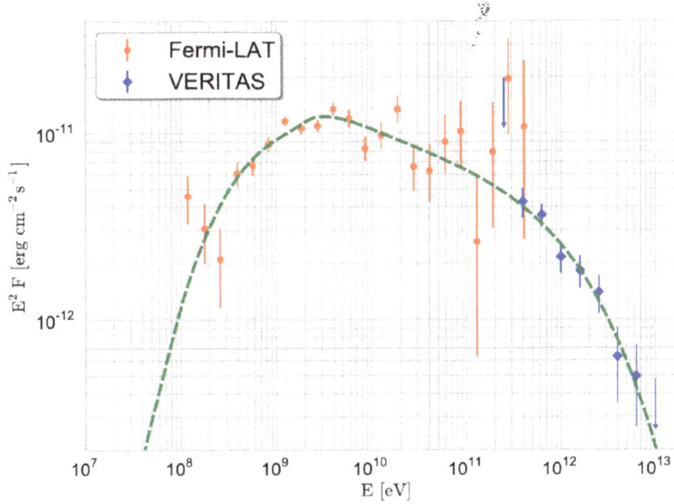

Figure 3. The Cas A spectrum, as seen by Fermi and Veritas (Figure from [85]). The dashed line represents a hadronic model for this source.

Given the observational evidence for a reverse shock in Cas A [86,87], a two-zone model with both forward and reverse shocks has been adopted for multiwavelength emission in many works. Moreover, multiwavelength observations indicate an asymmetric profile that could be modeled as a jet-like feature superposed onto an expanding spherical shell. The jet-like structure can produce a γ-ray emission detectable at 100 TeV by LHAASO, CTA, and ASTRI Mini-Array [88].

2.2. Tycho SNR

Tycho's SNR is the remnant of the Type Ia SN 1572 registered by Tycho Brahe, and it has a distance of about 3 kpc. It is one of the youngest known SNRs, with an age of 450 yrs, and it is still in its ejecta-dominated phase, with a shock velocity of about 4000 km s^{-1}.

X-ray observations by the Chandra Observatory have revealed synchrotron X-ray filaments at the shock location [89] with a very high variability [90], a clear indication of the presence of an amplified magnetic field, one of the main conditions to reach VHEs.

In the γ-ray observations, there is also no clear evidence of the presence of a cutoff because the last VERITAS results have large error bars that cannot confirm it [91,92]. Its power-law spectrum, with an index $\propto E^{-2.3}$, in agreement with the theoretical expectation [93], points toward a hadronic origin of this emission.

Very recent results obtained in X-ray polarization by IXPE [94] show a dominant radial magnetic field, in agreement with the radio polarization one and the one found in CasA [95] (see Figure 4). The polarization degree is about 12%, indicating the presence of such an order in the magnetic field that, however, cannot exclude the presence of turbulence.

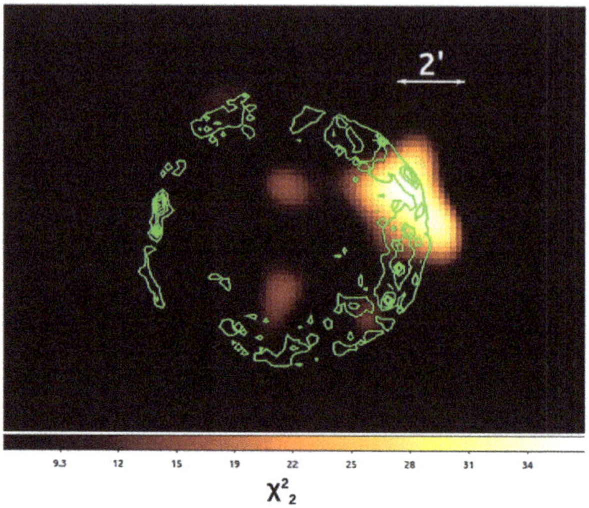

Figure 4. The polarization map of the Tycho SNR. The color scale gives the χ^2 values for the polarization signal in the 3–6 keV energy band smoothed with a Gaussian kernel. Superimposed in green are the Chandra 4–6 keV contours (figure from [94]).

These features make the Tycho SNR one of the main SNR candidate PeVatrons in the CR context and one of the best studied. However, LHAASO does not include this source in its list of candidate PeVatrons, and the most recent theoretical discussions [96,97] showed that the SNR could reach PeV energies only in the first 100 yrs of their evolution.

Only a future generation of IACTS with an effective area and a sensitivity better than those currently available will be able to better constrain the spectrum at VHEs and hence confirm or disprove the PeVatron nature of Tycho.

2.3. G 106.3+02.7

Among the first high-significance PeVatrons published by the LHAASO collaboration [15,45], there was only one source associated with an SNR: LHAASO J2226+6057, correlated with VER J2227+608/HAWC J2227+610 [98]. Its estimated age is about 10,000 years at a distance of 800 pc.

Its VHE/UHE emission could be explained by two kinds of sources: the SNR G106.3+2.7 with the associated MC in the "tail" of the TeV emission, and the Boomerang PWN, associated with the PSR J2229+6141, collocated in the "head". The low resolution of the UHE detection by HAWC [99], Tibet ASg [100], and finally, LHAASO [45] does not allow us to say whether the γ-ray emission is from the head or the tail region.

After the second LHAASO publication [38], in which more than one SNR is a PeVatron candidate, the chance that the VHE/UHE emission comes from the SNR G106.3+2.7 is taken into account more rigorously. This remnant was discovered by the Northern Galactic plane survey in the radio band by the DRAO [101] with a comet-shaped morphology. A recent Fermi-LAT GeV analysis showed that only the tail seems to emit at the highest energies detectable by the LAT instrument (10–500 GeV), explaining the whole emission with a hadronic model from the SNR/MC interaction [102]. A hadronic origin was also declared the most likely by the MAGIC collaboration, which resolved, for the first time, the VHE/UHE emission (see Figure 5), detecting $E > 10$ TeV only from the tail region [103,104]. Several studies and analyses are ongoing to disentangle the contribution from the two possible sources (for a review, see [15]). However, the real origin of the hadronic emission will only be understood with an in-depth analysis of the microphysics of the region.

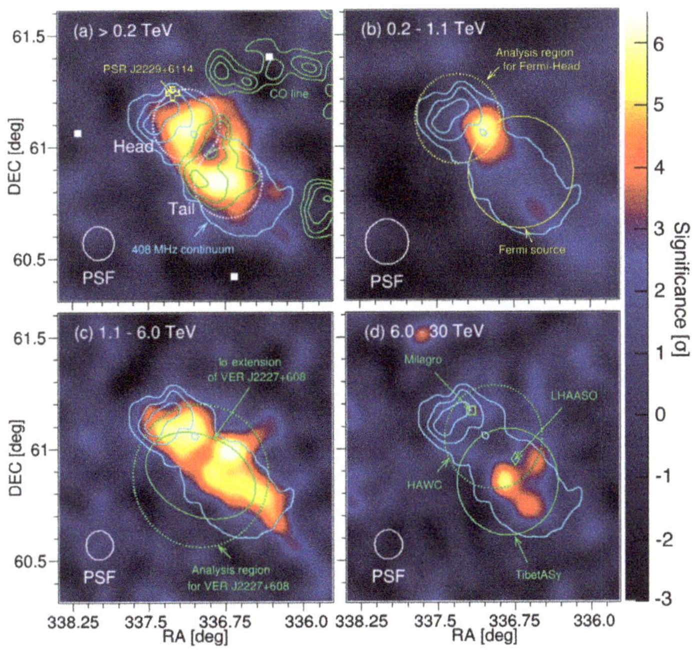

Figure 5. Energy-dependent pre-trial significance maps of the SNR G106.3+2.7 observed with the MAGIC telescopes at different energy ranges (from [104]).

In the future, the very good angular resolution of ASTRI Mini-Array [105,106] and CTA [107] could resolve the VHE emission location. Unfortunately, current estimations of the neutrino flux expected in the case of hadronic emission cannot be detected at the IceCube sensitivity [108].

2.4. RX J1713.7-3946

The "standard candle" in the debate about the hadronic or leptonic origin of the γ-ray emission from an SNR is the source RX J1713.7-3946 (G 347.3-0.5), because its GeV-TeV emission [14,69] can be reproduced with both types of models. Its age is about 1625 yrs with a distance of $d \simeq 1$ kpc [109,110] and an extension of $R_s \simeq 0.6$ deg [14].

The leptonic scenario is supported by the lack of thermal X-ray emission [111,112] and by the very good correlation between its X-ray shell [113,114] and TeV γ-ray emission. The very hard spectrum at GeV energy detected by Fermi-LAT [69] seemed a clear indication of an Inverse Compton-dominant emission (see Figure 6).

However, a more in-depth analysis of the environment in which the SNR is expanding stressed its non-homogeneous nature [115], particularly the presence of dense cloud cores in the northwestern part of the remnant, where there is enhanced X-ray emission. This could be a hint of the presence of an amplified magnetic field due to a shock–cloud interaction. The clumpy nature of the RX J1713.7-3946 environment was taken into account to explain the source γ-ray spectrum with a hadronic scenario [116–120].

For this ambiguity, RX J1713.7-3946 is the perfect candidate for the search for neutrino emission [121–123]. Indeed, after the recent results of LHAASO, it is now certain that the only way we have to finally distinguish hadronic from leptonic γ-ray emission is the detection of neutrinos correlated with VHE γ-ray.

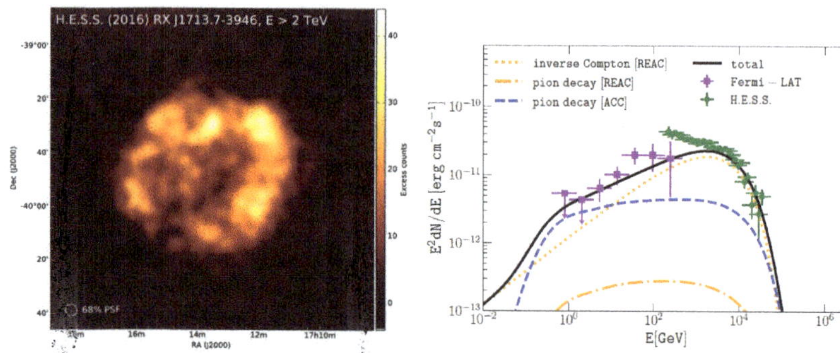

Figure 6. Left—The recent H.E.S.S. map of RXJ1713.7-3946, a glowing shell of γ-ray emission coincident with the outer shock of the SNR (from [124]). **Right**—The differential spectrum of the source obtained with H.E.S.S. and Fermi–LAT observations. The dotted (yellow) and dot-dashed (yellow) lines correspond to the gamma rays from re-accelerated electrons and protons. The dashed blue line corresponds to freshly accelerated protons. The solid black line is the sum of gamma rays from freshly accelerated protons and re-accelerated electrons (from [120]).

2.5. IC 443

IC 443 is a middle-aged SNR (the age is thought to be about 30,000 yr [125]) that belongs to the class of interacting SNRs; a system of MCs surrounds it [126].

In correspondence to the SNR-MC interaction region, γ-ray emission was detected in the TeV band by MAGIC [127] and VERITAS [55] and in the GeV band by AGILE [54] and Fermi [128] (see Figure 7). The association between γ-ray emission and MCs and the spectral features typical of pion decays indicate its hadronic origin [129].

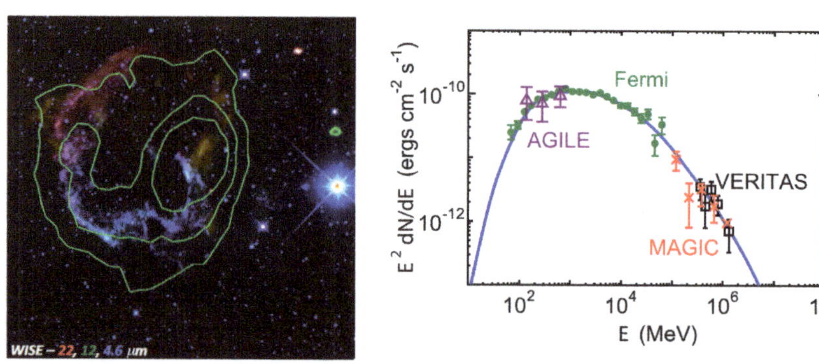

Figure 7. Morphology and spectrum of IC 443 (from [130]).

In the work in [131], an H+3 column density near IC 443 was measured, and a high ionization rate of 2×10^{15} s^{-1}, five times larger than the typical galactic values, was found. A bright blob-like enhancement of the Fe I Kα line located in the northwest and at the center of IC 443 was discovered [24], likely due to low-energy CR protons accelerated in the SNR leaking into the MCs and ionizing the Fe atoms therein.

A mass of about 1100 M$_\odot$ for the ambient gas in the [6.5, 1.5] km s^{-1} range was estimated. This estimation depends on the adopted ^{12}CO/^{13}CO isotopic ratio. Still, it is established that a total molecular mass of 0.9–3.1 × 10^3 M$_\odot$ is available to interact with CRs via pion decay in a more extended region.

2.6. W 44

W44 is an interacting SNR lying at a distance of $\sim$3 kpc with an estimated age of $\sim$20,000 years. Close to the remnant, there is a giant MC; the SNR-MC interaction is shown by the observation of OH maser emissions and a high ^{12}CO(J = 2-1)-to-^{12}CO(J = 1-0) ratio [132]. In the radio wavelength, W 44 appears as a bright source that extends tens of arcmins. The radio spectrum is a featureless power law in the range 0.02–10.7 GHz, with evidence of spectral variations across the shell [133,134] and with no evident polarization [135]. In X-ray, this SNR is filled with thermal emission, indicating the presence of ionized gas [136].

In the GeV band, W 44 is one of the brightest sources in the sky. The link between the γ-ray emission and the SNR was proposed, for the first time, in the work in [137] on the basis of EGRET data.

It was the first SNR in whose spectrum AGILE [33,138] and Fermi [129] detected a pion bump (see Figure 8). Beyond a few GeV, the spectrum becomes very soft. Moreover, Fermi-LAT has identified two hot spots outside the SNR, aligned with it. These hot spots are believed to be the result of the anisotropic diffusion of CRs from the SNR, taking the form of CR clouds [139,140].

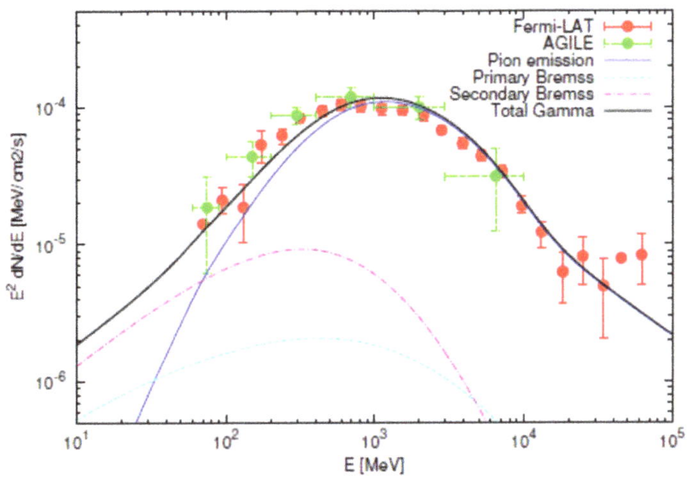

Figure 8. Gamma-ray spectrum of SNR W44 as seen by AGILE and Fermi. Hadronic and leptonic components are indicated.

2.7. Gamma Cygni

The SNR Gamma Cygni is an SNR in the constellation of Cygnus; it has an age of about 11,000 years and is characterized by a mixed morphology (a combination of a shell-like structure and a centrally filled interior) and a possible interaction with MCs. The γ-ray spectrum has been observed from a few tens of MeV to hundreds of Tev by AGILE [141], Fermi-LAT, Veritas [142], MAGIC [143] (see Figure 9), HAWC [144], and LHAASO [38].

The spectrum seems to classify this source as an interacting type of SNR. The target of the accelerated protons would be given by the gas contained in the associated MCs. These are observed in the radio band with a characteristic speed of 40–50 km s^{-1} and collect a mass of approximately 10^4 $M_\odot$. The recent LHAASO observation of emission up to energies of about 300 TeV makes this object interesting for understanding the ability of an SNR to accelerate particles up to VHEs.

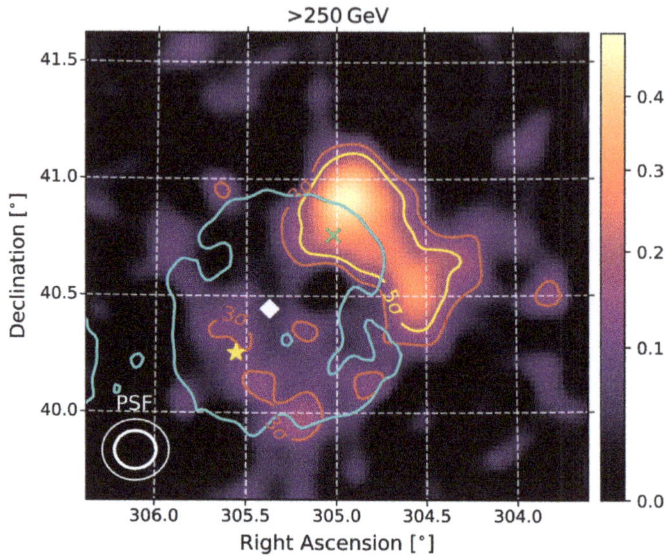

Figure 9. Gamma Cygni as seen by MAGIC [143]. The color map indicates the gamma-ray emission, while the radio shell is shown by the blue contour.

2.8. W 28

W 28 is an SNR that provides a unique opportunity to investigate the diffusion of CRs in an astrophysical environment. It is approximately 30,000 years old and lies at an estimated distance between 2 and 3.5 kpc. In continuous radio, it has a crescent shape, and the non-thermal spectrum indicates the presence of relativistic electrons in the shell [145]. W 28 provides a unique opportunity to investigate the diffusion of CRs in an astrophysical environment. Indeed, two giant MCs with masses of a few 10^4 $M_\odot$ appear near the SNR at compatible distances [146]. One cloud partially overlaps with the SNR shell, while the other is located approximately 0.5 degrees away from the shell. The system can be modeled as having one cloud in direct contact with the SNR shock and another cloud about 10–15 parsecs away, assuming that the three objects are at the same distance and interacting with each other. Evidence of an SNR-MC system interaction comes from observations of maser emission and high ionization levels [147].

W 28 was one of the first SNRs to be indicated as a γ-ray source in the Cos B data [148]. H.E.S.S. [28], AGILE [27], and Fermi-LAT [149] observations of this object measured a γ-ray spectrum that is typical of an interacting SNR, namely, a peak below 1 GeV and a very soft spectrum above it. It is interesting to note that the peak of the γ-ray spectrum in the cloud in direct contact with the shell is at different energies than the one estimated from the other clouds. This suggests that diffusion can have an important role in this system.

3. Conclusions

For the past fifty years, SNRs have been one of the main targets of γ-ray missions. Today, about thirty SNRs have been observed with certainty in γ-rays by satellites in the GeV band, by Cherenkov telescopes in the TeV band, and more recently, by hybrid detectors like LHAASO and HAWC in the PeV band. Their study has provided answers but also raised questions. For instance, their ability to accelerate CRs (electrons and hadrons) has been demonstrated. At the same time, their spectra and morphologies have proved to be non-trivial to interpret. Another open question concerns the maximum energy at which SNRs can accelerate particles. Some of these objects have an observed spectrum that extends beyond 100 TeV, confirming that these objects may be capable of accelerating particles up to about 1 PeV. However, they represent a small minority of both the LHAASO sources and the SNRs observed in the γ-ray regime. Overall, after almost half a century of study, it is still unclear whether or not their contribution to galactic CR acceleration is dominant compared to other sources. To answer these questions, a new generation of γ-ray instruments will come into operation in the next few years, in particular, instruments such as ASTRI Mini-Array and CTA, which will make it possible to better interpret the VHE/UHE observations that LHAASO, HAWC, and then the future SWGO will make of these objects and will finally allow us to know which astrophysical system supplies the Galaxy with its CR population.

Author Contributions: Conceptualization, A.G.; resources, M.C. and A.G.; writing—original draft preparation, A.G.; writing—review and editing, M.C. and A.G. All authors have read and agreed to the published version of the manuscript.

Funding: This research received no external funding.

Acknowledgments: We are grateful to the reviewers for their useful comments, which have enhanced the quality of our review. This research has made use of the TeVCat online source catalog (http://tevcat.uchicago.edu) and the SNRcat online catalog (http://snrcat.physics.umanitoba.ca), both accessed on 1 January 2024.

Conflicts of Interest: The authors declare no conflicts of interest.

Abbreviations

The following abbreviations are used in this manuscript:

ALPs	Axion-like particles
AS-γ	Air shower γ-ray array
ASTRI	Astrofisica con Specchi a Tecnologia Replicante Italiana
C.L.	Confidence Limit
CTAO	Čerenkov Telescope Array Observatory
DM	Dark matter
EAS	Extended air shower arrays
ESA	European Space Agency
EBL	Extra-galactic background light
E-HBL	Extreme high-peaked BL Lacs
eROSITA	Extended Roentgen survey with an imaging telescope array
FOV	Field of view
FR	Fanaroff–Riley galaxies
FSRQ	Flat-spectrum radio quasar
GASP	GLAST-AGILE Support Programme
GRB	Gamma-ray burst
GW	Gravitational wave
HAWC	High-Altitude Water Čerenkov Observatory
HB	Hadron Beam
HBL	High-peaked BL Lacs
HE	High energy
H.E.S.S.	High-Energy Stereoscopic System
IAC	Instituto de Astrofísica de Canarias
IACT	Imaging Atmospheric Čerenkov Telescope arrays
IBL	Intermediate-peaked BL Lacs
IGMF	Inter-galactic magnetic field
IR	Infra-red
IXPE	Imaging X-ray Polarimetry Explorer
LHAASO	Large High-Altitude Air Shower Observatory
LIV	Lorentz invariance violation

LST	Large-sized telescope
MAGIC	Major Atmospheric Gamma-Ray Imaging Čerenkov telescopes
MST	Medium-sized telescope
NASA	National Aeronautics and Space Administration
SBG	Star-bursting galaxies
SC	Schwarzschild–Couder
SII	Stellar intensity interferometry
SRT	Sardinia radio telescope
SST	Small-sized telescope
TNG	Telescopio Nazionale Galileo
VERITAS	Very Energetic Radiation Imaging Telescope Array System
VHE	Very high energy
WEBT	Whole-Earth Blazar Telescope

References

1. Sedov, L. *Similarity and Dimensional Methods in Mechanics*; Academic Press: Cambridge, MA, USA, 1959. [CrossRef]
2. Koo, B.C.; Kim, C.G.; Park, S.; Ostriker, E.C. Radiative Supernova Remnants and Supernova Feedback. *Astrophys. J.* **2020**, *905*, 35. [CrossRef]
3. Lee, J.J.; Park, S.; Hughes, J.P.; Slane, P.O. X-ray Observation of the Shocked Red Supergiant Wind of Cassiopeia A. *Astrophys. J.* **2014**, *789*, 7. [CrossRef]
4. Bell, A.R. The acceleration of cosmic rays in shock fronts—II. *Mon. Not. R. Astron. Soc.* **1978**, *182*, 443–455. [CrossRef]
5. Blandford, R.D.; Ostriker, J.P. Particle acceleration by astrophysical shocks. *Astrophys. J. Lett.* **1978**, *221*, L29–L32. [CrossRef]
6. Fermi, E. On the Origin of the Cosmic Radiation. *Phys. Rev.* **1949**, *75*, 1169–1174. [CrossRef]
7. Blasi, P. The origin of galactic cosmic rays. *Astron. Astrophys. Rev.* **2013**, *21*, 70. [CrossRef]
8. Huang, Y.; Li, Z.; Wang, W.; Zhao, X. Secondary-electron radiation accompanying hadronic GeV-TeV gamma-rays from supernova remnants. *Mon. Not. R. Astron. Soc.* **2020**, *492*, 4246–4253. [CrossRef]
9. Ginzburg, V.L.; Syrovatskii, S.I. *The Origin of Cosmic Rays*; Macmillan: New York, NY, USA, 1964.
10. Koyama, K.; Petre, R.; Gotthelf, E.V.; Hwang, U.; Matsuura, M.; Ozaki, M.; Holt, S.S. Evidence for shock acceleration of high-energy electrons in the supernova remnant SN1006. *Nature* **1995**, *378*, 255–258. [CrossRef]
11. Uchiyama, Y.; Aharonian, F.A.; Tanaka, T.; Takahashi, T.; Maeda, Y. Extremely fast acceleration of cosmic rays in a supernova remnant. *Nature* **2007**, *449*, 576–578. [CrossRef]
12. Vink, J. Supernova remnants: The X-ray perspective. *Astron. Astrophys. Rev.* **2012**, *20*, 49. [CrossRef]
13. Morlino, G.; Bandiera, R.; Blasi, P.; Amato, E. Collisionless Shocks in a Partially Ionized Medium. II. Balmer Emission. *Astrophys. J.* **2012**, *760*, 137. [CrossRef]
14. H. E. S. S. Collaboration; Abdalla, H.; Abramowski, A.; Aharonian, F.; Ait Benkhali, F.; Akhperjanian, A.G.; Andersson, T.; Angüner, E.O.; Arrieta, M.; Aubert, P.; et al. H.E.S.S. observations of RX J1713.7-3946 with improved angular and spectral resolution: Evidence for gamma-ray emission extending beyond the X-ray emitting shell. *Astron. Astrophys.* **2018**, *612*, A6. [CrossRef]
15. Cardillo, M.; Giuliani, A. The LHAASO PeVatron bright sky: What we learned. *Appl. Sci.* **2023**, *13*, 6433. [CrossRef]
16. Anchordoqui, L.A.; Barger, V.; Cholis, I.; Goldberg, H.; Hooper, D.; Kusenko, A.; Learned, J.G.; Marfatia, D.; Pakvasa, S.; Paul, T.C.; et al. Cosmic neutrino pevatrons: A brand new pathway to astronomy, astrophysics, and particle physics. *J. High Energy Astrophys.* **2014**, *1*, 1–30. [CrossRef]
17. Celli, S.; Aharonian, F.; Gabici, S. Spectral Signatures of PeVatrons. *Astrophys. J.* **2020**, *903*, 61. [CrossRef]
18. Funk, S. High-Energy Gamma Rays from Supernova Remnants. In *Handbook of Supernovae*; Alsabti, A.W., Murdin, P., Eds.; Springer: Cham, Switzerland, 2017; p. 1737. [CrossRef]
19. Acero, F.; Lemoine-Goumard, M.; Renaud, M.; Ballet, J.; Hewitt, J.W.; Rousseau, R.; Tanaka, T. Study of TeV shell supernova remnants at gamma-ray energies. *Astron. Astrophys.* **2015**, *580*, A74. [CrossRef]
20. Fukui, Y. Molecular and Atomic Gas in the Young TeV γ-Ray SNRs RX J1713.7-3946 and RX J0852.0-4622; Evidence for the Hadronic Production of γ-Rays. In *Cosmic Rays in Star-Forming Environments*; Torres, D.F., Reimer, O., Eds.; Astrophysics and Space Science Proceedings; Springer: Berlin/Heidelberg, Germany, 2013; Volume 34, p. 249. [CrossRef]
21. Jiang, B.; Chen, Y.; Wang, J.; Su, Y.; Zhou, X.; Safi-Harb, S.; DeLaney, T. Cavity of Molecular Gas Associated with Supernova Remnant 3C 397. *Astrophys. J.* **2010**, *712*, 1147–1156. [CrossRef]
22. Zirakashvili, V.N.; Ptuskin, V.S. Cosmic Rays and Nonthermal Radiation in Middle-Aged Supernova Remnants. *Astron. Lett.* **2018**, *44*, 769–776. [CrossRef]
23. Slane, P.; Bykov, A.; Ellison, D.C.; Dubner, G.; Castro, D. Supernova Remnants Interacting with Molecular Clouds: X-ray and Gamma-Ray Signatures. In *Multi-Scale Structure Formation and Dynamics in Cosmic Plasmas*; Balogh, A., Bykov, A., Eastwood, J., Kaastra, J., Eds.; Springer: Berlin/Heidelberg, Germany, 2016; Volume 51, p. 187. [CrossRef]

24. Nobukawa, K.K.; Nobukawa, M.; Koyama, K.; Yamauchi, S.; Uchiyama, H.; Okon, H.; Tanaka, T.; Uchida, H.; Tsuru, T.G. Evidence for a Neutral Iron Line Generated by MeV Protons from Supernova Remnants Interacting with Molecular Clouds. *Astrophys. J.* **2018**, *854*, 87. [CrossRef]
25. Acero, F.; Lemoine-Goumard, M.; Ballet, J. Characterization of the Gamma-ray Emission from the Kepler Supernova Remnant with Fermi-LAT. *Astron. Astrophys.* **2022**, *660*, A129. [CrossRef]
26. H. E. S. S. Collaboration; Aharonian, F.; Ait Benkhali, F.; Angüner, E.O.; Ashkar, H.; Backes, M.; Barbosa Martins, V.; Batzofin, R.; Becherini, Y.; Berge, D.; et al. Evidence for γ-ray emission from the remnant of Kepler's supernova based on deep H.E.S.S. observations (Corrigendum). *Astron. Astrophys.* **2024**, *683*, C1. [CrossRef]
27. Giuliani, A.; Tavani, M.; Bulgarelli, A.; Striani, E.; Sabatini, S.; Cardillo, M.; Fukui, Y.; Kawamura, A.; Ohama, A.; Furukawa, N.; et al. AGILE detection of GeV γ-ray emission from the SNR W28. *Astron. Astrophys.* **2010**, *516*, L11. [CrossRef]
28. Aharonian, F.; Akhperjanian, A.G.; Bazer-Bachi, A.R.; Behera, B.; Beilicke, M.; Benbow, W.; Berge, D.; Bernlöhr, K.; Boisson, C.; Bolz, O.; et al. Discovery of very high energy gamma-ray emission coincident with molecular clouds in the W 28 (G6.4-0.1) field. *Astron. Astrophys.* **2008**, *481*, 401–410. [CrossRef]
29. Ajello, M.; Allafort, A.; Baldini, L.; Ballet, J.; Barbiellini, G.; Bastieri, D.; Bechtol, K.; Bellazzini, R.; Berenji, B.; Blandford, R.D.; et al. Fermi Large Area Telescope Observations of the Supernova Remnant G8.7-0.1. *Astrophys. J.* **2012**, *744*, 80. [CrossRef]
30. Acero, F.; Ackermann, M.; Ajello, M.; Baldini, L.; Ballet, J.; Barbiellini, G.; Bastieri, D.; Bellazzini, R.; Bissaldi, E.; Blandford, R.D.; et al. The First Fermi LAT Supernova Remnant Catalog. *Astrophys. J. Suppl. Ser.* **2016**, *224*, 8. [CrossRef]
31. H. E. S. S. Collaboration; Abramowski, A.; Aharonian, F.; Ait Benkhali, F.; Akhperjanian, A.G.; Angüner, E.; Anton, G.; Backes, M.; Balenderan, S.; Balzer, A.; et al. Probing the gamma-ray emission from HESS J1834-087 using H.E.S.S. and Fermi LAT observations. *Astron. Astrophys.* **2015**, *574*, A27. [CrossRef]
32. MAGIC Collaboration; Acciari, V.A.; Ansoldi, S.; Antonelli, L.A.; Arbet Engels, A.; Arcaro, C.; Baack, D.; Babić, A.; Banerjee, B.; Bangale, P.; et al. Discovery of TeV γ-ray emission from the neighbourhood of the supernova remnant G24.7+0.6 by MAGIC. *Mon. Not. R. Astron. Soc.* **2019**, *483*, 4578–4585. [CrossRef]
33. Giuliani, A.; Cardillo, M.; Tavani, M.; Fukui, Y.; Yoshiike, S.; Torii, K.; Dubner, G.; Castelletti, G.; Barbiellini, G.; Bulgarelli, A.; et al. Neutral Pion Emission from Accelerated Protons in the Supernova Remnant W44. *Astrophys. J. Lett.* **2011**, *742*, L30. [CrossRef]
34. Abdo, A.A.; Ackermann, M.; Ajello, M.; Baldini, L.; Ballet, J.; Barbiellini, G.; Bastieri, D.; Bechtol, K.; Bellazzini, R.; Bloom, E.D.; et al. Fermi-LAT Study of Gamma-ray Emission in the Direction of Supernova Remnant W49B. *Astrophys. J.* **2010**, *722*, 1303–1311. [CrossRef]
35. Brun, F.; de Naurois, M.; Hofmann, W.; Carrigan, S.; Djannati-Ataï, A.; Ohm, S.; H. E. S. S. Collaboration. Discovery of VHE gamma-ray emission from the W49 region with H.E.S.S. In Proceedings of the 25th Texas Symposium on Relativistic Astrophysics, Heidelberg, Germany, 6–10 December 2010; Rieger, F.M., van Eldik, C., Hofmann, W., Eds.; p. 201. [CrossRef]
36. Abdo, A.A.; Ackermann, M.; Ajello, M.; Baldini, L.; Ballet, J.; Barbiellini, G.; Baring, M.G.; Bastieri, D.; Baughman, B.M.; Bechtol, K.; et al. Fermi LAT Discovery of Extended Gamma-Ray Emission in the Direction of Supernova Remnant W51C. *Astrophys. J. Lett.* **2009**, *706*, L1–L6. [CrossRef]
37. Aleksić, J.; Alvarez, E.A.; Antonelli, L.A.; Antoranz, P.; Asensio, M.; Backes, M.; Barres de Almeida, U.; Barrio, J.A.; Bastieri, D.; Becerra González, J.; et al. Morphological and spectral properties of the W51 region measured with the MAGIC telescopes. *Astron. Astrophys.* **2012**, *541*, A13. [CrossRef]
38. Cao, Z.; Aharonian, F.; An, Q.; Axikegu; Bai, Y.X.; Bao, Y.W.; Bastieri, D.; Bi, X.J.; Bi, Y.J.; Cai, J.T.; et al. The First LHAASO Catalog of Gamma-Ray Sources. *Astrophys. J. Suppl. Ser.* **2024**, *271*, 25. [CrossRef]
39. Katagiri, H.; Tibaldo, L.; Ballet, J.; Giordano, F.; Grenier, I.A.; Porter, T.A.; Roth, M.; Tibolla, O.; Uchiyama, Y.; Yamazaki, R. Fermi Large Area Telescope Observations of the Cygnus Loop Supernova Remnant. *Astrophys. J.* **2011**, *741*, 44. [CrossRef]
40. Lande, J.; Ackermann, M.; Allafort, A.; Ballet, J.; Bechtol, K.; Burnett, T.H.; Cohen-Tanugi, J.; Drlica-Wagner, A.; Funk, S.; Giordano, F.; et al. Search for Spatially Extended Fermi Large Area Telescope Sources Using Two Years of Data. *Astrophys. J.* **2012**, *756*, 5. [CrossRef]
41. Aliu, E.; Archambault, S.; Arlen, T.; Aune, T.; Beilicke, M.; Benbow, W.; Bird, R.; Bouvier, A.; Bradbury, S.M.; Buckley, J.H.; et al. Discovery of TeV Gamma-Ray Emission toward Supernova Remnant SNR G78.2+2.1. *Astrophys. J.* **2013**, *770*, 93. [CrossRef]
42. Pivato, G.; Hewitt, J.W.; Tibaldo, L.; Acero, F.; Ballet, J.; Brandt, T.J.; de Palma, F.; Giordano, F.; Janssen, G.H.; Jóhannesson, G.; et al. Fermi LAT and WMAP Observations of the Supernova Remnant HB 21. *Astrophys. J.* **2013**, *779*, 179. [CrossRef]
43. Xin, Y.; Zeng, H.; Liu, S.; Fan, Y.; Wei, D. VER J2227+608: A Hadronic PeVatron Pulsar Wind Nebula? *Astrophys. J.* **2019**, *885*, 162. [CrossRef]
44. Acciari, V.A.; Aliu, E.; Arlen, T.; Aune, T.; Bautista, M.; Beilicke, M.; Benbow, W.; Boltuch, D.; Bradbury, S.M.; Buckley, J.H.; et al. Detection of Extended VHE Gamma Ray Emission from G106.3+2.7 with Veritas. *Astrophys. J. Lett.* **2009**, *703*, L6–L9. [CrossRef]
45. Cao, Z.; Aharonian, F.A.; An, Q.; Axikegu; Bai, L.X.; Bai, Y.X.; Bao, Y.W.; Bastieri, D.; Bi, X.J.; Bi, Y.J.; et al. Ultrahigh-energy photons up to 1.4 petaelectronvolts from 12 γ-ray Galactic sources. *Nature* **2021**, *594*, 33–36. [CrossRef] [PubMed]
46. Castro, D.; Slane, P.; Ellison, D.C.; Patnaude, D.J. Fermi-LAT Observations and a Broadband Study of Supernova Remnant CTB 109. *Astrophys. J.* **2012**, *756*, 88. [CrossRef]
47. Abdo, A.A.; Ackermann, M.; Ajello, M.; Allafort, A.; Baldini, L.; Ballet, J.; Barbiellini, G.; Baring, M.G.; Bastieri, D.; Baughman, B.M.; et al. Fermi-Lat Discovery of GeV Gamma-Ray Emission from the Young Supernova Remnant Cassiopeia A. *Astrophys. J. Lett.* **2010**, *710*, L92–L97. [CrossRef]

48. Aharonian, F.; Akhperjanian, A.; Barrio, J.; Bernlöhr, K.; Börst, H.; Bojahr, H.; Bolz, O.; Contreras, J.; Cortina, J.; Denninghoff, S.; et al. Evidence for TeV gamma ray emission from Cassiopeia A. *Astron. Astrophys.* **2001**, *370*, 112–120. [CrossRef]
49. Giordano, F.; Naumann-Godo, M.; Ballet, J.; Bechtol, K.; Funk, S.; Lande, J.; Mazziotta, M.N.; Rainò, S.; Tanaka, T.; Tibolla, O.; et al. Fermi Large Area Telescope Detection of the Young Supernova Remnant Tycho. *Astrophys. J. Lett.* **2012**, *744*, L2. [CrossRef]
50. Acciari, V.A.; Aliu, E.; Arlen, T.; Aune, T.; Beilicke, M.; Benbow, W.; Bradbury, S.M.; Buckley, J.H.; Bugaev, V.; Byrum, K.; et al. Discovery of TeV Gamma-ray Emission from Tycho's Supernova Remnant. *Astrophys. J. Lett.* **2011**, *730*, L20. [CrossRef]
51. Katagiri, H.; Yoshida, K.; Ballet, J.; Grondin, M.H.; Hanabata, Y.; Hewitt, J.W.; Kubo, H.; Lemoine-Goumard, M. Fermi LAT Discovery of Extended Gamma-Ray Emissions in the Vicinity of the HB 3 Supernova Remnant. *Astrophys. J.* **2016**, *818*, 114. [CrossRef]
52. Araya, M. Fermi LAT observation of supernova remnant HB9. *Mon. Not. R. Astron. Soc.* **2014**, *444*, 860–865. [CrossRef]
53. Katsuta, J.; Uchiyama, Y.; Tanaka, T.; Tajima, H.; Bechtol, K.; Funk, S.; Lande, J.; Ballet, J.; Hanabata, Y.; Lemoine-Goumard, M.; et al. Fermi Large Area Telescope Observation of Supernova Remnant S147. *Astrophys. J.* **2012**, *752*, 135. [CrossRef]
54. Tavani, M.; Giuliani, A.; Chen, A.W.; Argan, A.; Barbiellini, G.; Bulgarelli, A.; Caraveo, P.; Cattaneo, P.W.; Cocco, V.; Contessi, T.; et al. Direct Evidence for Hadronic Cosmic-Ray Acceleration in the Supernova Remnant IC 443. *Astrophys. J. Lett.* **2010**, *710*, L151–L155. [CrossRef]
55. Acciari, V.A.; Aliu, E.; Arlen, T.; Aune, T.; Bautista, M.; Beilicke, M.; Benbow, W.; Bradbury, S.M.; Buckley, J.H.; Bugaev, V.; et al. Observation of Extended Very High Energy Emission from the Supernova Remnant IC 443 with VERITAS. *Astrophys. J. Lett.* **2009**, *698*, L133–L137. [CrossRef]
56. Liu, J.h.; Liu, B.; Yang, R.z. Diffuse gamma-ray emission around the Rosette Nebula. *Mon. Not. R. Astron. Soc.* **2023**, *526*, 175–180. [CrossRef]
57. Hewitt, J.W.; Grondin, M.H.; Lemoine-Goumard, M.; Reposeur, T.; Ballet, J.; Tanaka, T. Fermi-LAT and WMAP Observations of the Puppis A Supernova Remnant. *Astrophys. J.* **2012**, *759*, 89. [CrossRef]
58. Tanaka, T.; Allafort, A.; Ballet, J.; Funk, S.; Giordano, F.; Hewitt, J.; Lemoine-Goumard, M.; Tajima, H.; Tibolla, O.; Uchiyama, Y. Gamma-Ray Observations of the Supernova Remnant RX J0852.0-4622 with the Fermi Large Area Telescope. *Astrophys. J. Lett.* **2011**, *740*, L51. [CrossRef]
59. Aharonian, F.; Akhperjanian, A.G.; Bazer-Bachi, A.R.; Beilicke, M.; Benbow, W.; Berge, D.; Bernlöhr, K.; Boisson, C.; Bolz, O.; Borrel, V.; et al. H.E.S.S. Observations of the Supernova Remnant RX J0852.0-4622: Shell-Type Morphology and Spectrum of a Widely Extended Very High Energy Gamma-Ray Source. *Astrophys. J.* **2007**, *661*, 236–249. [CrossRef]
60. Saz Parkinson, P.M.; Xu, H.; Yu, P.L.H.; Salvetti, D.; Marelli, M.; Falcone, A.D. Classification and Ranking of Fermi LAT Gamma-ray Sources from the 3FGL Catalog using Machine Learning Techniques. *Astrophys. J.* **2016**, *820*, 8. [CrossRef]
61. Araya, M. Detection of gamma-ray emission in the region of the supernova remnants G296.5+10.0 and G166.0+4.3. *Mon. Not. R. Astron. Soc.* **2013**, *434*, 2202–2208. [CrossRef]
62. Yuan, Q.; Huang, X.; Liu, S.; Zhang, B. Fermi Large Area Telescope Detection of Supernova Remnant RCW 86. *Astrophys. J. Lett.* **2014**, *785*, L22. [CrossRef]
63. Aharonian, F.; Akhperjanian, A.G.; de Almeida, U.B.; Bazer-Bachi, A.R.; Behera, B.; Beilicke, M.; Benbow, W.; Bernlöhr, K.; Boisson, C.; Bochow, A.; et al. Discovery of Gamma-Ray Emission From the Shell-Type Supernova Remnant RCW 86 With Hess. *Astrophys. J.* **2009**, *692*, 1500–1505. [CrossRef]
64. Nolan, P.L.; Abdo, A.A.; Ackermann, M.; Ajello, M.; Allafort, A.; Antolini, E.; Atwood, W.B.; Axelsson, M.; Baldini, L.; Ballet, J.; et al. Fermi Large Area Telescope Second Source Catalog. *Astrophys. J. Suppl. Ser.* **2012**, *199*, 31. [CrossRef]
65. H. E. S. S. Collaboration; Abdalla, H.; Abramowski, A.; Aharonian, F.; Ait Benkhali, F.; Akhperjanian, A.G.; Andersson, T.; Angüner, E.O.; Arakawa, M.; Arrieta, M.; et al. A search for new supernova remnant shells in the Galactic plane with H.E.S.S. *Astron. Astrophys.* **2018**, *612*, A8. [CrossRef]
66. Devin, J.; Acero, F.; Ballet, J.; Schmid, J. Disentangling hadronic from leptonic emission in the composite SNR G326.3-1.8. *Astron. Astrophys.* **2018**, *617*, A5. [CrossRef]
67. Condon, B.; Lemoine-Goumard, M.; Acero, F.; Katagiri, H. Detection of Two TeV Shell-type Remnants at GeV Energies with FERMI LAT: HESS J1731-347 and SN 1006. *Astrophys. J.* **2017**, *851*, 100. [CrossRef]
68. Acero, F.; Aharonian, F.; Akhperjanian, A.G.; Anton, G.; Barres de Almeida, U.; Bazer-Bachi, A.R.; Becherini, Y.; Behera, B.; Beilicke, M.; Bernlöhr, K.; et al. First detection of VHE γ-rays from SN 1006 by HESS. *Astron. Astrophys.* **2010**, *516*, A62. [CrossRef]
69. Abdo, A.A.; Ackermann, M.; Ajello, M.; Allafort, A.; Baldini, L.; Ballet, J.; Barbiellini, G.; Baring, M.G.; Bastieri, D.; Bellazzini, R.; et al. Observations of the Young Supernova Remnant RX J1713.7-3946 with the Fermi Large Area Telescope. *Astrophys. J.* **2011**, *734*, 28. [CrossRef]
70. Aharonian, F.A.; Akhperjanian, A.G.; Aye, K.M.; Bazer-Bachi, A.R.; Beilicke, M.; Benbow, W.; Berge, D.; Berghaus, P.; Bernlöhr, K.; Bolz, O.; et al. High-energy particle acceleration in the shell of a supernova remnant. *Nature* **2004**, *432*, 75–77. [CrossRef] [PubMed]
71. Brandt, T.J.; Fermi-LAT Collaboration. A view of supernova remnant CTB 37A with the Fermi Gamma-ray Space Telescope. *Adv. Space Res.* **2013**, *51*, 247–252. [CrossRef]

72. Aharonian, F.; Akhperjanian, A.G.; Barres de Almeida, U.; Bazer-Bachi, A.R.; Behera, B.; Beilicke, M.; Benbow, W.; Bernlöhr, K.; Boisson, C.; Borrel, V.; et al. Discovery of a VHE gamma-ray source coincident with the supernova remnant CTB 37A. *Astron. Astrophys.* **2008**, *490*, 685–693. [CrossRef]
73. Xin, Y.L.; Liang, Y.F.; Li, X.; Yuan, Q.; Liu, S.M.; Wei, D.M. A GeV Source in the Direction of Supernova Remnant CTB 37B. *Astrophys. J.* **2016**, *817*, 64. [CrossRef]
74. Aharonian, F.; Akhperjanian, A.G.; Barres de Almeida, U.; Bazer-Bachi, A.R.; Behera, B.; Beilicke, M.; Benbow, W.; Bernlöhr, K.; Boisson, C.; Borrel, V.; et al. Chandra and HESS observations of the supernova remnant CTB 37B. *Astron. Astrophys.* **2008**, *486*, 829–836. [CrossRef]
75. Castro, D.; Slane, P. Fermi Large Area Telescope Observations of Supernova Remnants Interacting with Molecular Clouds. *Astrophys. J.* **2010**, *717*, 372–378. [CrossRef]
76. H. E. S. S. Collaboration; Abramowski, A.; Aharonian, F.; Ait Benkhali, F.; Akhperjanian, A.G.; Angüner, E.O.; Backes, M.; Balenderan, S.; Balzer, A.; Barnacka, A.; et al. H.E.S.S. detection of TeV emission from the interaction region between the supernova remnant G349.7+0.2 and a molecular cloud (Corrigendum). *Astron. Astrophys.* **2015**, *580*, C1. [CrossRef]
77. Yang, R.z.; Zhang, X.; Yuan, Q.; Liu, S. Fermi Large Area Telescope observations of the supernova remnant HESS J1731-347. *Astron. Astrophys.* **2014**, *567*, A23. [CrossRef]
78. H. E. S. S. Collaboration; Abramowski, A.; Acero, F.; Aharonian, F.; Akhperjanian, A.G.; Anton, G.; Balzer, A.; Barnacka, A.; Barres de Almeida, U.; Becherini, Y.; et al. A new SNR with TeV shell-type morphology: HESS J1731-347. *Astron. Astrophys.* **2011**, *531*, A81. [CrossRef]
79. Castro, D.; Slane, P.; Carlton, A.; Figueroa-Feliciano, E. Fermi-LAT Observations of Supernova Remnants Interacting with Molecular Clouds: W41, MSH 17-39, and G337.7-0.1. *Astrophys. J.* **2013**, *774*, 36. [CrossRef]
80. Krause, O.; Birkmann, S.M.; Usuda, T.; Hattori, T.; Goto, M.; Rieke, G.H.; Misselt, K.A. The Cassiopeia A Supernova Was of Type IIb. *Science* **2008**, *320*, 1195. [CrossRef]
81. Chevalier, R.A.; Oishi, J. Cassiopeia A and Its Clumpy Presupernova Wind. *Astrophys. J. Lett.* **2003**, *593*, L23–L26. [CrossRef]
82. Reed, J.E.; Hester, J.J.; Fabian, A.C.; Winkler, P.F. The Three-dimensional Structure of the Cassiopeia A Supernova Remnant. I. The Spherical Shell. *Astrophys. J.* **1995**, *440*, 706. [CrossRef]
83. Albert, J.; Aliu, E.; Anderhub, H.; Antoranz, P.; Armada, A.; Baixeras, C.; Barrio, J.A.; Bartko, H.; Bastieri, D.; Becker, J.K.; et al. Observation of VHE γ-rays from Cassiopeia A with the MAGIC telescope. *Astron. Astrophys.* **2007**, *474*, 937–940. [CrossRef]
84. Acciari, V.A.; Aliu, E.; Arlen, T.; Aune, T.; Bautista, M.; Beilicke, M.; Benbow, W.; Boltuch, D.; Bradbury, S.M.; Buckley, J.H.; et al. Observations of the Shell-type Supernova Remnant Cassiopeia A at TeV Energies with VERITAS. *Astrophys. J.* **2010**, *714*, 163–169. [CrossRef]
85. Abeysekara, A.U.; Archer, A.; Benbow, W.; Bird, R.; Brose, R.; Buchovecky, M.; Buckley, J.H.; Chromey, A.J.; Cui, W.; Daniel, M.K.; et al. Evidence for Proton Acceleration up to TeV Energies Based on VERITAS and Fermi-LAT Observations of the Cas A SNR. *Astrophys. J.* **2020**, *894*, 51. [CrossRef]
86. Gotthelf, E.V.; Koralesky, B.; Rudnick, L.; Jones, T.W.; Hwang, U.; Petre, R. Chandra Detection of the Forward and Reverse Shocks in Cassiopeia A. *Astrophys. J. Lett.* **2001**, *552*, L39–L43. [CrossRef]
87. Morse, J.A.; Fesen, R.A.; Chevalier, R.A.; Borkowski, K.J.; Gerardy, C.L.; Lawrence, S.S.; van den Bergh, S. Location of the Optical Reverse Shock in the Cassiopeia A Supernova Remnant. *Astrophys. J.* **2004**, *614*, 727–736. [CrossRef]
88. Zhan, S.; Wang, W.; Mou, G.; Li, Z. An asymmetrical model for high-energy radiation of Cassiopeia A. *Mon. Not. R. Astron. Soc.* **2022**, *513*, 2471–2477. [CrossRef]
89. Hwang, U.; Decourchelle, A.; Holt, S.S.; Petre, R. Thermal and Nonthermal X-ray Emission from the Forward Shock in Tycho's Supernova Remnant. *Astrophys. J.* **2002**, *581*, 1101–1115. [CrossRef]
90. Matsuda, M.; Uchida, H.; Tanaka, T.; Yamaguchi, H.; Tsuru, T.G. Discovery of Year-scale Time Variability from Thermal X-ray Emission in Tycho's Supernova Remnant. *Astrophys. J.* **2022**, *940*, 105. [CrossRef]
91. Park, N.; VERITAS Collaboration. Study of high-energy particle acceleration in Tycho with gamma-ray observations. In Proceedings of the 34th International Cosmic Ray Conference (ICRC2015), The Hague, The Netherlands, 30 July–6 August 2015; Volume 34, p. 769. [CrossRef]
92. Archambault, S.; Archer, A.; Benbow, W.; Bird, R.; Bourbeau, E.; Buchovecky, M.; Buckley, J.H.; Bugaev, V.; Cerruti, M.; Connolly, M.P.; et al. Gamma-Ray Observations of Tycho's Supernova Remnant with VERITAS and Fermi. *Astrophys. J.* **2017**, *836*, 23. [CrossRef]
93. Morlino, G.; Caprioli, D. Strong evidence for hadron acceleration in Tycho's supernova remnant. *Astron. Astrophys.* **2012**, *538*, A81. [CrossRef]
94. Ferrazzoli, R.; Slane, P.; Prokhorov, D.; Zhou, P.; Vink, J.; Bucciantini, N.; Costa, E.; Di Lalla, N.; Di Marco, A.; Soffitta, P.; et al. X-ray Polarimetry Reveals the Magnetic-field Topology on Sub-parsec Scales in Tycho's Supernova Remnant. *Astrophys. J.* **2023**, *945*, 52. [CrossRef]
95. Vink, J.; Prokhorov, D.; Ferrazzoli, R.; Slane, P.; Zhou, P.; Asakura, K.; Baldini, L.; Bucciantini, N.; Costa, E.; Di Marco, A.; et al. X-ray Polarization Detection of Cassiopeia A with IXPE. *Astrophys. J.* **2022**, *938*, 40. [CrossRef]
96. Bell, A.R.; Schure, K.M.; Reville, B.; Giacinti, G. Cosmic-ray acceleration and escape from supernova remnants. *Mon. Not. R. Astron. Soc.* **2013**, *431*, 415–429. [CrossRef]

97. Cardillo, M.; Amato, E.; Blasi, P. On the cosmic ray spectrum from type II supernovae expanding in their red giant presupernova wind. *Astropart. Phys.* **2015**, *69*, 1–10. [CrossRef]
98. Wakely, S.P.; Horan, D. TeVCat: An online catalog for Very High Energy Gamma-Ray Astronomy. *Int. Cosm. Ray Conf.* **2008**, *3*, 1341–1344.
99. Albert, A.; Alfaro, R.; Alvarez, C.; Camacho, J.R.A.; Arteaga-Velázquez, J.C.; Arunbabu, K.P.; Avila Rojas, D.; Ayala Solares, H.A.; Baghmanyan, V.; Belmont-Moreno, E.; et al. HAWC J2227+610 and Its Association with G106.3+2.7, a New Potential Galactic PeVatron. *Astrophys. J. Lett.* **2020**, *896*, L29. [CrossRef]
100. Amenomori, M.; Bao, Y.W.; Bi, X.J.; Chen, D.; Chen, T.L.; Chen, W.Y.; Chen, X.; Chen, Y.; Cirennima; Cui, S.W.; et al. Gamma-Ray Observation of the Cygnus Region in the 100-TeV Energy Region. *Phys. Rev. Lett.* **2021**, *127*, 031102. [CrossRef] [PubMed]
101. Joncas, G.; Higgs, L.A. The DRAO galactic-plane survey. II. Field at l = 105. *Astron. Astrophys. Suppl. Ser.* **1990**, *82*, 113–144.
102. Fang, K.; Kerr, M.; Blandford, R.; Fleischhack, H.; Charles, E. Evidence for PeV Proton Acceleration from Fermi-LAT Observations of SNR G 106.3 +2.7. *Phys. Rev. Lett.* **2022**, *129*, 071101. [CrossRef]
103. MAGIC Collaboration; Acciari, V.A.; Ansoldi, S.; Antonelli, L.A.; Arbet Engels, A.; Baack, D.; Babić, A.; Banerjee, B.; Barres de Almeida, U.; Barrio, J.A.; et al. Resolving the origin of very-high-energy gamma-ray emission from the PeVatron candidate SNR G106.3+2.7 using MAGIC telescopes. In Proceedings of the 37th International Cosmic Ray Conference (ICRC2021), Online, 12–23 July 2021; Volume 395, p. 796.
104. MAGIC Collaboration; Abe, H.; Abe, S.; Acciari, V.A.; Agudo, I.; Aniello, T.; Ansoldi, S.; Antonelli, L.A.; Arbet Engels, A.; Arcaro, C.; et al. MAGIC observations provide compelling evidence of hadronic multi-TeV emission from the putative PeVatron SNR G106.3+2.7. *Astron. Astrophys.* **2023**, *671*, A12. [CrossRef]
105. Cardillo, M. The ASTRI Mini-Array: In the search for hidden Pevatrons. *arXiv* **2023**, arXiv:2302.10059.
106. Cardillo, M.; Tutone, A. G106.3+2.7: A candidate PeVatron with ASTRI Mini-Array. In Proceedings of the 38th International Cosmic Ray Conference (ICRC2023), Nagoya, Japan, 26 July–3 August 2023; p. 965. [CrossRef]
107. Verna, G.; Cassol, F.; Costantini, H.; Consortium, C. HAWC J2227+610: A potential PeVatron candidate for the CTA in the northern hemisphere. In Proceedings of the 37th International Cosmic Ray Conference, Online, 12–23 July 2021; p. 904. [CrossRef]
108. Sarmah, P.; Chakraborty, S.; Joshi, J.C. Probing LHAASO galactic PeVatrons through gamma-ray and neutrino correspondence. *Mon. Not. R. Astron. Soc.* **2023**, *521*, 1144–1151. [CrossRef]
109. Fukui, Y.; Moriguchi, Y.; Tamura, K.; Yamamoto, H.; Tawara, Y.; Mizuno, N.; Onishi, T.; Mizuno, A.; Uchiyama, Y.; Hiraga, J.; et al. Discovery of Interacting Molecular Gas toward the TeV Gamma-Ray Peak of the SNR G 347.3–0.5. *Publ. Astron. Soc. Jpn.* **2003**, *55*, L61–L64. [CrossRef]
110. Moriguchi, Y.; Tamura, K.; Tawara, Y.; Sasago, H.; Yamaoka, K.; Onishi, T.; Fukui, Y. A Detailed Study of Molecular Clouds toward the TeV Gamma-Ray Supernova Remnant G347.3-0.5. *Astrophys. J.* **2005**, *631*, 947–963. [CrossRef]
111. Ellison, D.C.; Patnaude, D.J.; Slane, P.; Raymond, J. Efficient Cosmic Ray Acceleration, Hydrodynamics, and Self-Consistent Thermal X-ray Emission Applied to Supernova Remnant RX J1713.7-3946. *Astrophys. J.* **2010**, *712*, 287–293. [CrossRef]
112. Katz, B.; Waxman, E. In which shell-type SNRs should we look for gamma-rays and neutrinos from P-P collisions? *J. Cosmol. Astropart. Phys.* **2008**, *2008*, 018. [CrossRef]
113. Slane, P.; Gaensler, B.M.; Dame, T.M.; Hughes, J.P.; Plucinsky, P.P.; Green, A. Nonthermal X-ray Emission from the Shell-Type Supernova Remnant G347.3-0.5. *Astrophys. J.* **1999**, *525*, 357–367. [CrossRef]
114. Tanaka, T.; Uchiyama, Y.; Aharonian, F.A.; Takahashi, T.; Bamba, A.; Hiraga, J.S.; Kataoka, J.; Kishishita, T.; Kokubun, M.; Mori, K.; et al. Study of Nonthermal Emission from SNR RX J1713.7-3946 with Suzaku. *Astrophys. J.* **2008**, *685*, 988–1004. [CrossRef]
115. Sano, H.; Fukuda, T.; Yoshiike, S.; Sato, J.; Horachi, H.; Kuwahara, T.; Torii, K.; Hayakawa, T.; Tanaka, T.; Matsumoto, H.; et al. A Detailed Study of Non-thermal X-ray Properties and Interstellar Gas toward the γ-Ray Supernova Remnant RX J1713.7-3946. *Astrophys. J.* **2015**, *799*, 175. [CrossRef]
116. Zirakashvili, V.N.; Aharonian, F.A. Nonthermal Radiation of Young Supernova Remnants: The Case of RX J1713.7-3946. *Astrophys. J.* **2010**, *708*, 965–980. [CrossRef]
117. Gabici, S.; Aharonian, F.A. Hadronic gamma-rays from RX J1713.7-3946? *Mon. Not. R. Astron. Soc.* **2014**, *445*, L70–L73. [CrossRef]
118. Inoue, T.; Yamazaki, R.; Inutsuka, S.i.; Fukui, Y. Toward Understanding the Cosmic-Ray Acceleration at Young Supernova Remnants Interacting with Interstellar Clouds: Possible Applications to RX J1713.7-3946. *Astrophys. J.* **2012**, *744*, 71. [CrossRef]
119. Celli, S. *Gamma-Ray and Neutrino Signatures of Galactic Cosmic-Ray Accelerators*; Springer Nature: Berlin/Heidelberg, Germany, 2019. [CrossRef]
120. Cristofari, P.; Niro, V.; Gabici, S. Gamma-rays and neutrinos from RX J1713-3946 in a lepto-hadronic scenario. *Mon. Not. R. Astron. Soc.* **2021**, *508*, 2204–2209. [CrossRef]
121. Kappes, A.; Hinton, J.; Stegmann, C.; Aharonian, F.A. Potential neutrino signals from galactic γ-ray sources. In Proceedings of the Journal of Physics Conference Series, Madison, WI, USA, 28–31 August 2006; Volume 60, pp. 243–246. [CrossRef]
122. Morlino, G.; Amato, E.; Blasi, P. Gamma-ray emission from SNR RX J1713.7-3946 and the origin of galactic cosmic rays. *Mon. Not. R. Astron. Soc.* **2009**, *392*, 240–250. [CrossRef]
123. Villante, F.L.; Vissani, F. How precisely can neutrino emission from supernova remnants be constrained by gamma ray observations? *Phys. Rev.* **2008**, *78*, 103007. [CrossRef]
124. O'C. Drury, L. Galactic Cosmic Rays—Theory and Interpretation. *arXiv* **2017**, arXiv:1708.08858.

125. Petre, R.; Szymkowiak, A.E.; Seward, F.D.; Willingale, R. A Comprehensive Study of the X-ray Structure and Spectrum of IC 443. *Astrophys. J.* **1988**, *335*, 215. [CrossRef]
126. Cornett, R.H.; Chin, G.; Knapp, G.R. Observations of CO emission from a dense cloud associated with the supernova remnant IC 443. *Astron. Astrophys.* **1977**, *54*, 889–894.
127. Albert, J.; Aliu, E.; Anderhub, H.; Antoranz, P.; Armada, A.; Baixeras, C.; Barrio, J.A.; Bartko, H.; Bastieri, D.; Becker, J.K.; et al. Discovery of Very High Energy Gamma Radiation from IC 443 with the MAGIC Telescope. *Astrophys. J. Lett.* **2007**, *664*, L87–L90. [CrossRef]
128. Abdo, A.A.; Ackermann, M.; Ajello, M.; Baldini, L.; Ballet, J.; Barbiellini, G.; Bastieri, D.; Baughman, B.M.; Bechtol, K.; Bellazzini, R.; et al. Observation of Supernova Remnant IC 443 with the Fermi Large Area Telescope. *Astrophys. J.* **2010**, *712*, 459–468. [CrossRef]
129. Ackermann, M.; Ajello, M.; Allafort, A.; Baldini, L.; Ballet, J.; Barbiellini, G.; Baring, M.G.; Bastieri, D.; Bechtol, K.; Bellazzini, R.; et al. Detection of the Characteristic Pion-Decay Signature in Supernova Remnants. *Science* **2013**, *339*, 807–811. [CrossRef] [PubMed]
130. Humensky, T.B. The TeV morphology of the interacting supernova remnant IC 443. In Proceedings of the Supernova Remnants: An Odyssey in Space after Stellar Death, Chania, Greece, 6–11 June 2016; p. 21.
131. Indriolo, N.; Blake, G.A.; Goto, M.; Usuda, T.; Oka, T.; Geballe, T.R.; Fields, B.D.; McCall, B.J. Investigating the Cosmic-ray Ionization Rate Near the Supernova Remnant IC 443 through $H^+{}_3$ Observations. *Astrophys. J.* **2010**, *724*, 1357–1365. [CrossRef]
132. Yoshiike, S.; Fukuda, T.; Sano, H.; Ohama, A.; Moribe, N.; Torii, K.; Hayakawa, T.; Okuda, T.; Yamamoto, H.; Tajima, H.; et al. The Neutral Interstellar Gas toward SNR W44: Candidates for Target Protons in Hadronic γ-Ray Production in a Middle-aged Supernova Remnant. *Astrophys. J.* **2013**, *768*, 179. [CrossRef]
133. Egron, E.; Pellizzoni, A.; Iacolina, M.N.; Loru, S.; Marongiu, M.; Righini, S.; Cardillo, M.; Giuliani, A.; Mulas, S.; Murtas, G.; et al. Imaging of SNR IC443 and W44 with the Sardinia Radio Telescope at 1.5 and 7 GHz. *Mon. Not. R. Astron. Soc.* **2017**, *470*, 1329–1341. [CrossRef]
134. Loru, S.; Pellizzoni, A.; Egron, E.; Righini, S.; Iacolina, M.N.; Mulas, S.; Cardillo, M.; Marongiu, M.; Ricci, R.; Bachetti, M.; et al. Investigating the high-frequency spectral features of SNRs Tycho, W44, and IC443 with the Sardinia Radio Telescope. *Mon. Not. R. Astron. Soc.* **2019**, *482*, 3857–3867. [CrossRef]
135. Dokara, R.; Roy, N.; Menten, K.; Vig, S.; Dutta, P.; Beuther, H.; Pandian, J.D.; Rugel, M.; Rashid, M.; Brunthaler, A. Metrewave Galactic Plane with the uGMRT (MeGaPluG) Survey: Lessons from the pilot study. *Astron. Astrophys.* **2023**, *678*, A72. [CrossRef]
136. Okon, H.; Tanaka, T.; Uchida, H.; Yamaguchi, H.; Tsuru, T.G.; Seta, M.; Smith, R.K.; Yoshiike, S.; Orlando, S.; Bocchino, F.; et al. Deep XMM-Newton Observations Reveal the Origin of Recombining Plasma in the Supernova Remnant W44. *Astrophys. J.* **2020**, *890*, 62. [CrossRef]
137. Esposito, J.A.; Sreekumar, P.; Hunter, S.D.; Kanbach, G. EGRET observations of gamma-ray emission from supernova remnants. *Bull. Am. Astron. Soc.* **1996**, *28*, 861.
138. Cardillo, M.; Tavani, M.; Giuliani, A.; Yoshiike, S.; Sano, H.; Fukuda, T.; Fukui, Y.; Castelletti, G.; Dubner, G. The supernova remnant W44: Confirmations and challenges for cosmic-ray acceleration. *Astron. Astrophys.* **2014**, *565*, A74. [CrossRef]
139. Gabici, S.; Krause, J.; Morlino, G.; Nava, L. Acceleration of cosmic rays and gamma-ray emission from supernova remnant/molecular cloud associations. In Proceedings of the European Physical Journal Web of Conferences, SuGAR 2015 – Searching for the Sources of Galactic Cosmic Rays Geneva, Switzerland, 21–23 January 2015; Volume 105, p. 02001. [CrossRef]
140. Peron, G.; Aharonian, F.; Casanova, S.; Zanin, R.; Romoli, C. On the Gamma-Ray Emission of W44 and Its Surroundings. *Astrophys. J. Lett.* **2020**, *896*, L23. [CrossRef]
141. Piano, G.; Cardillo, M.; Pilia, M.; Trois, A.; Giuliani, A.; Bulgarelli, A.; Parmiggiani, N.; Tavani, M. AGILE Study of the Gamma-Ray Emission from the SNR G78.2+2.1 (Gamma Cygni). *Astrophys. J.* **2019**, *878*, 54. [CrossRef]
142. Abeysekara, A.U.; Archer, A.; Aune, T.; Benbow, W.; Bird, R.; Brose, R.; Buchovecky, M.; Bugaev, V.; Cui, W.; Daniel, M.K.; et al. A Very High Energy γ-Ray Survey toward the Cygnus Region of the Galaxy. *Astrophys. J.* **2018**, *861*, 134. [CrossRef]
143. MAGIC Collaboration; Acciari, V.A.; Ansoldi, S.; Antonelli, L.A.; Arbet Engels, A.; Baack, D.; Babić, A.; Banerjee, B.; Barres de Almeida, U.; Barrio, J.A.; et al. Study of the GeV to TeV morphology of the γ Cygni SNR (G 78.2+2.1) with MAGIC and Fermi-LAT. Evidence for cosmic ray escape. *Astron. Astrophys.* **2023**, *670*, A8. [CrossRef]
144. Albert, A.; Alfaro, R.; Alvarez, C.; Camacho, J.R.A.; Arteaga-Velázquez, J.C.; Arunbabu, K.P.; Avila Rojas, D.; Ayala Solares, H.A.; Baghmanyan, V.; Belmont-Moreno, E.; et al. 3HWC: The Third HAWC Catalog of Very-high-energy Gamma-Ray Sources. *Astrophys. J.* **2020**, *905*, 76. [CrossRef]
145. Dubner, G.M.; Velázquez, P.F.; Goss, W.M.; Holdaway, M.A. High-Resolution VLA Imaging of the Supernova Remnant W28 at 328 and 1415 MHZ. *Astron. J.* **2000**, *120*, 1933–1945. [CrossRef]
146. Mizuno, A.; Fukui, Y. Physical properties of molecular clouds as revealed by NANTEN CO survey: From the galactic center to the galactic warp. In Proceedings of the Milky Way Surveys: The Structure and Evolution of our Galaxy, Boston, MA, USA, 15–17 June 2003; Clemens, D., Shah, R., Brainerd, T., Eds.; Astronomical Society of the Pacific Conference Series; Astronomical Society of the Pacific: San Francisco, CA, USA, 2004; Volume 317, p. 59.
147. Frail, D.A.; Goss, W.M.; Slysh, V.I. Shock-excited maser emission from the supernova remnant W28. *Astrophys. J. Lett.* **1994**, *424*, L111–L113. [CrossRef]

148. Pollock, A.M.T. The probable identification of two COS-B gamma-ray sources with molecular clouds compressed by supernova remnants. *Astron. Astrophys.* **1985**, *150*, 339–342.
149. Abdo, A.A.; Ackermann, M.; Ajello, M.; Allafort, A.; Baldini, L.; Ballet, J.; Barbiellini, G.; Bastieri, D.; Bechtol, K.; Bellazzini, R.; et al. Fermi Large Area Telescope Observations of the Supernova Remnant W28 (G6.4-0.1). *Astrophys. J.* **2010**, *718*, 348–356. [CrossRef]

universe

MDPI

Review

Gamma-ray Bursts: 50 Years and Counting!

Alessandro Armando Vigliano [1,2,*,†] **and Francesco Longo** [2,3,†]

[1] Department of Mathematical, Computer and Physical Sciences, University of Udine, 33100 Udine, Italy
[2] Istituto Nazionale di Fisica Nucleare (INFN), Sezione di Trieste, 34149 Trieste, Italy; francesco.longo@ts.infn.it
[3] Department of Physics, University of Trieste, 34127 Trieste, Italy
* Correspondence: alessandro.armando.vigliano@ts.infn.it; Tel.: +39-3315875712
† These authors contributed equally to this work.

Abstract: Gamma-ray bursts were discovered by the Vela satellites in the late 1960s, but they were announced for the first time exactly 50 years ago, in 1973. The history of our understanding of gamma-ray bursts can be subdivided into several eras. We will highlight the main discoveries about GRBs, as well as the path toward the future that each GRB era could still indicate.

Keywords: gamma-ray astrophysics; gamma-ray burst; transients; gamma-ray instrumentation

check for updates

Citation: Vigliano, A.A.; Longo, F. Gamma-ray Bursts: 50 Years and Counting! *Universe* **2024**, *10*, 57. https://doi.org/10.3390/universe10020057

Academic Editor: Binbin Zhang

Received: 28 December 2023
Revised: 15 January 2024
Accepted: 18 January 2024
Published: 26 January 2024

1. Introduction

Gamma-ray bursts (GRBs) are extraordinary astrophysical events characterized by brief and intense flashes of gamma-ray radiation, representing some of the most energetic phenomena known in the universe. Their enigmatic nature has intrigued the scientific community for decades, prompting extensive research to decipher their origins and the underlying physical processes involved.

The serendipitous discovery of GRBs traces back to the late 1960s when the secret Vela satellites, designed for nuclear test detection, revealed anomalous gamma-ray emissions from deep space. Following meticulous study and analysis, the first public announcement of these mysterious gamma-ray bursts was made in 1973, marking the inception of a new era in astrophysical exploration.

This review paper provides a comprehensive historical account of the progression of GRB research, encompassing seven distinctive eras of study.

2. The "Dark" Era (1973(67)–1991)

The discovery of GRBs was made by the US military Vela satellites [1] in the late 1960s. These were a satellite constellation launched in order to monitor Soviet compliance with the Partial Test Ban Treaty of 1963, which prohibited all test detonations of nuclear weapons, except for those conducted underground.

In order to look for traces of nuclear tests in space, the Vela satellites were equipped with scintillation X-ray detectors, sensitive between 3 and 12 keV, and highly sensitive CsI gamma-ray detectors, sensitive between 150 and 750 keV. The satellites were launched in pairs, in a common circular orbit at an altitude of 118,000 km, in order to be well above the Van Allen radiation belts, drastically reducing the noise in the sensors and allowing some localization of the signal through time triangulation techniques.

At 14:19 UTC on 2 July 1967, the Vela 3 and Vela 4 satellites recorded an unprecedented gamma-radiation burst distinct from any previously identified astrophysical source. Despite the limited temporal resolution of the Vela satellites, GRB670702 (as it would be called with the modern-day naming convention) exhibited fundamental characteristics typical of GRBs (such as a duration of approximately 10 s, a two-pulse light curve and a peak emission around the MeV energy band), very different from the properties (duration, spectrum, variability) expected for a nuclear test in space (a short-duration, hard, nonstructured X-ray burst). Although the data were never classified (the emission was not of nuclear origin), they were only published in 1973 [1], after a total of 16 such signals had been observed. Around the same time, other instruments reported the observation of GRB signals, such as the American satellite IMP-6 [2] and the Soviet space satellite Konus/Venera [3].

The low number of signals made it impossible to estimate the spatial distribution of the events, and the lack of a method to determine the distance of their sources meant that the total energy and the location remained unknown. The poor localization capability of gamma-ray detectors made it very difficult to discover the electromagnetic counterparts of GRBs at lower frequencies. Still, a lot of information was gathered. It was clear that these signals, not thermal in origin, were not coming from Earth and not even from the Solar System plane. Spectral lines, not confirmed by subsequent missions, were reported with low significance in some GRBs detected with the Soviet Konus/Venera instruments [4] and with the Japanese Ginga satellite [5,6]. With so few constraints, theorists came up with all kinds of models based on a wide range of physics, with energy requirements varying over 20 orders of magnitude. By 1992, over 100 models existed [7], some proposed even before the official publication of the Vela results. The proposed models can be broadly divided into three categories: Earth and Solar System (for example, Terrestrial Gamma-ray Flashes, gamma rays produced by lightning at high altitudes in Earth's atmosphere), galactic, and cosmological models.

Galactic models, favored by the lower energies required in the processes, posited that the GRBs were produced either by the accretion/fall of material onto compact objects (supernova fallbacks, accretion from a binary companion onto white dwarfs or neutron stars, later observed as X-ray Novae) or through magnetic reconnection/quakes in the crusts of magnetars (neutron stars with magnetic fields reaching 10^{15} G). The first models came out even before the publication of the Vela results, and as early as 1968, Stirling Colgate proposed a model of GRBs as bremsstrahlung and inverse Compton emissions from Type II supernova shocks [8]. According to the current standard paradigm, long GRBs are indeed associated with supernovae, but only with one special type, namely, broad-line Type Ic, not Type II as envisaged by Colgate. A clear prediction of all these models was that, given the galactic nature of the phenomena, the spatial distribution of GRBs would be centered around the Galactic Equator. The hope was, therefore, that by placing more sensitive satellites with superior localization capabilities into orbit, a distribution along the plane of the Milky Way would be observed. We now note that a subclass of GRBs is definitively of galactic origin: Soft Gamma-ray Repeaters (SGRs), objects that emit repeated flashes of gamma and X-rays at irregular intervals. It is hypothesized that they could be a type of magnetar or neutron star surrounded by fossil disks [9].

Due to the large distances, extragalactic models in general had very high energy demands (with energies up to 10^{51}–10^{53} erg). This, in addition to implying a particularly low rate of events (10^{-6}–10^{-5} events per year in a galaxy with the luminosity of the Milky Way), has led to the proposal of exotic phenomena and/or objects (collapsing white dwarfs, stars accreting onto AGNs, white holes, accretion onto black holes in binary mergers or collapsing stars) among the possible origins of GRBs. The first extragalactic model of GRBs was probably the one proposed by Prilutskii and Usov [10], who proposed that GRBs are produced in the collapse of the cores of active galaxies. As early as 1975, Ruderman recognized that the electron–positron pair production condition is a limiting factor for the achievable GRB luminosity (the so-called compactness problem) [11]. He noted that this condition posed fewer challenges for galactic models compared to extragalactic ones and pointed out that relativistic motion had the potential to mitigate the compactness problem by enlarging the emission size. In 1978, using a more elaborate version of the two-photon pair production condition proposed by Ruderman to set general constraints on the luminosities of GRBs, Cavallo and Rees proposed, for the first time, a fireball model for modeling GRB emissions [12]. In 1989, Eichler and colleagues conducted an extensive examination of the neutron-star–neutron-star (NS-NS) merger model, presenting a detailed analysis of its potential relevance to gamma-ray bursts (GRBs). Their work introduced the notion that mergers of compact objects (either NS-NS, NS-black hole, or BH-BH) could serve as the progenitors for a specific subclass of observed GRBs [13]. In the paper, it was also suggested that these systems are important multi-messenger emitting sources, as, beyond gravitational waves, binary neutron star mergers are important sources of neutrino emissions, as well as fundamental heavy-element factories through the rapid neutron capture process (the r-process) of neutron-rich material ejected from the merger. Ref. [13] laid the foundation for the standard paradigm of the modern short-GRB models.

This era can be summarized as follows:

- GRBs were discovered in the 1960s by the Vela military satellites and later observed by other missions.
- The spatial distribution and total energy of GRBs remained unknown due to low instrumental capabilities and a low number of events.
- Galactic models suggested that GRBs were related to processes like supernovae, accretion onto compact objects and magnetic reconnection in magnetars.
- Galactic models predicted GRBs to be distributed on the galactic plane.
- Extragalactic models had high energy demands, and possible origins included collapsing white dwarfs, active galaxy cores, and neutron star mergers.
- The compactness problem and relativistic motion were already considered in these models.
- The relativistic fireball model was introduced to explain GRB emissions.

3. The BATSE Era (1991–1996)

On 5 April 1991, the US spacecraft Compton Gamma Ray Observatory (CGRO) was launched in low Earth orbit. On board, there were four instruments that covered a huge energy range, at the time unprecedented, of six orders of magnitude of the electromagnetic spectrum, ranging from 20 keV X-rays to 30 GeV gamma rays. Among these instruments, the Burst and Transient Source Experiment (BATSE) [14] has led to substantial improvements in our understanding of GRBs. It consisted of eight identical detector modules, each composed of a NaI (sodium iodide, i.e., crystal detector) scintillator Large Area Detector covering the energy range 20 keV–2 MeV to act as a trigger for the observations and a thick NaI Spectroscopy Detector covering the energy range 10 keV–100 MeV. Each module was placed on one of the corners of the satellite in order to obtain complete coverage of the sky, achieving a burst detection sensitivity of 3×10^{-8} erg cm^{-2} for a 1 s burst. BATSE was thus able to obtain approximate localizations of GRB signals based on the ratios of the count rates between the eight detectors.

GRB light curves have a broad range of characteristics and are quite irregular, ranging from single peak light curves to more complex multipeaked extended structures. However, in this period, enormous progress was made in their characterization. In 1996, Norris and colleagues found that the light curves of individual pulses are well fitted with a fast-rise exponential decay (FRED) profile with an average rise-to-decay ratio of 1:3 [15]:

$$I(t) = \begin{cases} A \exp(-(|t - t_{\max}| / \sigma_r)^{\nu}) & t \leq t_{\max} \\ A \exp(-(|t - t_{\max}| / \sigma_d)^{\nu}) & t > t_{\max} \end{cases}, \tag{1}$$

where $t_{\max}$ is the time of the pulse's maximum intensity; A is the peak value; σ_r and σ_d are the rise ($t \leq t_{\max}$) and decay ($t > t_{\max}$) time constants, respectively; and ν is the "peakedness", a measure of the pulse's sharpness. In the same paper, a delay between the emission at low energies and that at high energies was highlighted. Low/high energies are defined based on the low vs. high channels of BATSE: 20–50 keV, 50–100 keV, 100–300 keV, and >300 keV. The BATSE team also reported a width–symmetry–intensity correlation. In particular, high-intensity pulses were (statistically) more symmetric (i.e., with lower decay-to-rise ratios) and had shorter spectral delays. Furthermore, the low-energy light curves of individual pulses are generally wider than the high-energy ones, with a width of $\sim E^{-0.4}$ [16]. BATSE observed GRBs with very disparate durations, from $\sim$6 ms up to

~2000 s. To classify the light curves, a duration measurement standard was introduced, the T90 (T50), in which 90% (50%) of the total counts of the GRB arrive. After subtracting the background counts and normalizing the cumulative counts so that the total is 1, the T90 (T50) is defined as the time interval between the arrival of 5% (25%) and 95% (75%) of the counts. With this definition, Kouveliotou and colleagues showed that a GRB can be divided into two distinct groups: long bursts with T90 $>$ 2 s and short bursts with T90 $<$ 2 s [17]. The two groups also have different spectral characteristics, with short GRBs tending to be spectrally "harder" than long GRBs, where the hardness ratio is the ratio of photons observed in two BATSE channels: channel 3 (100–300 keV) counts divided by channel 2 (50–100 keV) counts.

Spectral studies of BATSE GRBs have shown that the emission is non-thermal, with spectra generally characterized by a double power law and a smooth break. Band and colleagues proposed a phenomenological function, now known as the Band function, to fit GRB spectra [18]:

$$F_{\text{Band}}(E) = \begin{cases} A\left(\frac{E}{100\,\text{keV}}\right)^{\alpha} \exp\left(-\frac{E(2+\alpha)}{E_{\text{peak}}}\right) & E < E_c \\ A\left[\frac{(\alpha-\beta)E_{\text{peak}}}{100\,\text{keV}(2+\alpha)}\right]^{\alpha-\beta} \exp(\beta-\alpha)\left(\frac{E}{100\,\text{keV}}\right)^{\beta} & E \geq E_c \end{cases}, \tag{2}$$

where

$$E_c = (\alpha-\beta)\frac{E_{\text{peak}}}{2+\alpha} \equiv (\alpha-\beta)E_0$$

α is the low-energy index, β is the high-energy index, E_c is the break energy, and E_{peak} is the peak energy of the νF_ν spectrum. No specific theoretical model predicts this spectral shape; however, the Band function provides an excellent fit to most observed spectra. Most GRBs show substantial spectral evolution with two typical behaviors: hard to soft throughout the whole GRB duration or hard to soft during each pulse. The value of E_{peak} generally follows the shape of the light curve. In the same year, CGRO/EGRET detected high-energy (HE) photons, from 30 MeV up to 18 GeV, for a few GRBs, with emission lasting up to thousands of seconds after the onset of the GRB. The average spectrum of the events detected by EGRET is well fitted by a single power law with an index of ~2, consistent with the extension of the low-energy spectra.

While the exact distances of GRB sources remained a topic of discussion, the data collected by BATSE offered pivotal insights strongly indicative of the cosmological nature of GRBs. GRB signals were found to be distributed isotropically in the sky, with no clustering near the Galactic Disk (as predicted by galactic models) [19]. However, this did not completely rule out the galactic models, which were recalibrated to shift the GRB sources to the galactic halo to smooth out the inconsistency with the observed distribution. The intensity distribution of the bursts was found to deviate from homogeneity at the faint end, indicating a deficit of weak bursts compared to the predictions [20]. The intensity distribution provides information on the radial distribution of the sources and is defined as the ratio $V/V_{\text{max}} = (C_{\text{max}}/C_{\text{min}})^{-3/2}$, where V is the volume contained by a sphere extending to the location of the burst, V_{max} is the volume of a sphere extending to the maximum distance at which the burst would still be detectable by the instrument, C_{max} is the peak count rate, C_{min} is the minimum rate required for triggering, and $-3/2$ is the power-law exponent predicted if the burst follows a homogeneous distribution in Euclidean space. For a homogeneous distribution of sources, the average ratio would be $\langle V/V_{\text{max}}\rangle = 1/2$. BATSE reported an average ratio of $\langle V/V_{\text{max}}\rangle = 0.348 \pm 0.024$, showing a deficit of low-fluence bursts and strongly hinting at a deviation from spatial homogeneity of the distribution of the sources. These findings imposed great constraints on current galactic models but could be easily explained if GRBs have an extragalactic origin. Despite great efforts to locate extragalactic GRBs with a certain precision, both through BATSE module counts and through temporal triangulation with other instruments

(BATSE Gamma-ray Burst Coordinates Distribution Network, BACODINE, now known as the GCN–GRB Coordinates Network), no host galaxy was identified, which is the so-called "No host problem".

During this era, the extragalactic models received a huge push in development. Peter Mészáros and Martin Rees proposed, in a series of papers, the now-standard fireball shock model, which already included the main ingredients of the current GRB theoretical framework, from the external shock of a relativistic fireball and synchrotron radiation as the main radiation mechanism to the external reverse shock and inverse Compton scattering. In 1997, two weeks before the discovery of the first X-ray and optical afterglows from BeppoSAX, they published a seminal paper in which they systematically predicted the multiwavelength afterglows of GRBs in a self-consistent manner [21]. In those years, new candidates for the sources of GRBs were also proposed. Woosley proposed that the collapse of a rapidly rotating Wolf–Rayet star (massive star whose outer hydrogen envelope is stripped away by a stellar wind), leading to a supernova, now known as the collapsar model for long GRBs, could be an appropriate progenitor of long-duration GRBs with a complex time profile [22]. The first mention of a millisecond magnetar (a newborn, rapidly spinning, highly magnetized neutron star) as a possible central engine [23] was made for the first time as an alternative to the standard hyper-accreting black hole scenario adopted by most modelers of the time.

The tension between extragalactic models, favored by the observation of the spatial distribution of GRBs but with surprisingly high energy demands (at a distance of 1 Gpc, the energy required would be $\sim 10^{51}$ erg, with a flow of $\sim 10^{-7}$ erg cm^{-2} s^{-1}), and galactic models, with decidedly more "reasonable" energy budgets ($E \sim 10^{43}$ erg at a distance of 100 kpc) but disfavored by observations, culminated in the so-called "Great Debate" of 1995, where Lamb and Paczynski showed evidence and arguments as to why a scale distance should be preferred to the other. As we will see, the solution to this debate later came thanks to precise localization measurements from the BeppoSAX satellite.

BATSE reported 2704 GRBs in its last catalog [24]. The failure of one of the three gyroscopes responsible for the maneuvering of CGRO determined the conclusion of the mission, and the satellite was de-orbited on 4 June 2000.

This era can be summarized as follows:

- BATSE observed a large number of GRBs, leading to better modeling of GRB properties.
- GRBs come in all shapes and sizes, but two subgroups exist: short hard bursts and long soft bursts.
- GRBs are distributed isotropically in the sky. This marked the beginning of the end of galactic models. Still, they were not ruled out completely, and this led to the Great Debate in 1995.

4. The BeppoSAX Era (1996–2000/2004)

As we have seen, much progress was made during the BATSE era in the investigations of GRBs. Still, regardless of the large amount of data acquired, these were not sufficient to single out a clear model for the production of these signals, and two vastly different classes, galactic and extragalactic models, were discussed in the community. The lack of multiwavelength observations limited the possibilities to understand the underlying mechanism of the production of these signals. At the same time, the poor localization capabilities, together with the impossibility of estimating the distance of the sources, meant that it was not yet possible to understand the energy scales of these phenomena and completely rule out any of the two model paradigms that were proposed. This was aggravated by the no-host problem, i.e., the impossibility of identifying the host galaxies of such signals. In order to move forward and solve these problems, it was therefore necessary to carry out a new type of observation. The key to making multiwavelength observations and estimating distances was to obtain the very precise localizations of the GRBs.

BeppoSAX was an Italian–Dutch satellite for X-ray astronomy [25]. Launched from the US space base at Cape Canaveral on 30 April 1996 and built largely by Italian companies, the satellite called SAX ("Satellite per Astronomia a raggi X", Italian for "Satellite for X-ray Astronomy"), once in orbit, was renamed BeppoSAX, from the nickname of Professor Giuseppe "Beppo" Occhialini, a pioneer of Italian high-energy astrophysics.

BeppoSAX hosted five scientific instruments:

- A Low-Energy Concentrator Spectrometer (LECS);
- A Medium-Energy Concentrator Spectrometer (MECS);
- A High-Pressure Gas-Scintillation-Proportional Counter (HPGSPC);
- A Phoswich Detector System (PDS);
- Two Wide-Field Cameras (WFCs).

Of these, the two X-ray focusing telescopes (LECS and MECS), the HPGSPC and the PDS (collectively called Narrow-Field Instruments, NFIs) pointed in the same direction, thus allowing observations in a large energy range, from 0.1 to 300 keV, while the two WFCs pointed orthogonally to the other instruments' axis, covering a large area of the sky. The LECS, MECS, and HPGSPC detectors were all based on gas-scintillation-proportional counters operating in a total energy window from 0.1 keV (LECS) to 120 keV (HPGSPC) with high sensitivity. The PDS detector was a crystal (sodium iodide/cesium iodide) scintillator detector capable of collecting photons from 15 to 300 keV. The PDS detector was covered on its sides by four CsI scintillators that acted as anticoincidence shields for background rejection. With sensitivities from 40 to 700 keV, these panels had large fields of view that were aligned with the two WFCs. It was thus possible to use them to trigger GRB observations, obtaining the first localization with the WFCs in order to then slew the satellite and observe with the NFIs. For this reason, they were referred to as the Gamma-Ray Burst Monitor (GRBM). The WFCs were two units each composed of a proportional counter and a coded mask to obtain excellent angular resolutions covering extended fields of view of $20^\circ \times 20^\circ$ in the energy range 2–28 keV. The simultaneous detection of GRBs by the GRBM and WFCs therefore made it possible to obtain the localizations of these signals with excellent precision, reaching accuracies of a few arcminutes. The estimated coordinates were quickly sent to the International Astronomical Union (IAU) and the Gamma-ray burst Coordinate Network Circular. Subsequently, immediate follow-up observations with the onboard NFIs and ground-based optical observatories allowed the accurate localization and detailed observations of GRBs.

The BeppoSAX mission achieved a breakthrough in February 1997, within a year of its launch. The gamma-ray burst GRB 970228 was initially detected by BeppoSAX's WFCs, and subsequently, when the X-ray NFIs aboard BeppoSAX were directed toward the source of the burst, they captured a diminishing X-ray emission [26]. Ground-based telescopes corroborated the findings by identifying a fading optical counterpart [27]. These later, fading emissions were called the afterglow of the GRB, to distinguish them from the initial prompt emission, which was the only signal observed up to that moment. Once the event's localization was pinpointed, subsequent deep imaging revealed a faint, very distant host galaxy in the GRB's location. This discovery had far-reaching implications. It swiftly resolved the longstanding debate about the distance scale of GRBs, confirming them as extragalactic events originating within faint galaxies situated at staggering distances. The establishment of this distance scale not only concluded a significant debate but also provided crucial insights into the environments in which GRBs unfold. A few months later, the observation of GRB970508 led to two further breakthroughs. In less than 4 h after its discovery, it was possible to obtain the accurate localization of this event, allowing follow-up observations to begin much earlier than with any previous burst, detecting the optical transient while still on the rise. In addition to the optical component, a radio component of the afterglow emission was observed for the first time [28]. Thanks to the analysis of the absorption lines in the optical spectrum, it was also possible to estimate, for the first time, the distance of the GRB, with a redshift of $z = 0.835$ [29], proving the extragalactic origin of these phenomena.

The possibility of measuring the distances of the origins of these signals allowed the discovery of the empirical correlation today known as the Amati relation [30] between the GRB isotropic bolometric emission energy ($E_{\gamma,\mathrm{iso}}$) and the rest-frame peak energy ($E_{p,z} = (1+z)E_p$):

$$\frac{E_{p,z}}{100\,\mathrm{keV}} = C\left(\frac{E_{\gamma,\mathrm{iso}}}{10^{52}\,\mathrm{erg}}\right)^m \quad (3)$$

with $C \sim 0.8$–1 and $m \sim 0.4$–0.6 [31]. This correlation holds for long GRBs with known redshifts, and its discovery hinted at the possibility of using GRBs as cosmological probes. In 2004, a correlation (now known as the Yonetoku relation) similar to the Amati relation was highlighted between the isotropic, bolometric peak luminosity $L_{\gamma,p,\mathrm{iso}}$ and the rest-frame peak energy $E_{p,z} = (1+z)E_p$ [32]:

$$\frac{E_{p,z}}{100\,\mathrm{keV}} \approx 1.8\left(\frac{L_{\gamma,p,\mathrm{iso}}}{10^{52}\,\mathrm{erg\,s^{-1}}}\right)^{0.52} \quad (4)$$

Although it has been suggested by several groups that the Amati and Yonetoku relations may be due to observational selection effects (e.g., [33]), this conclusion is not widely accepted, with other groups stating that the presence of such selection effects, despite lowering the correlation, are not enough to destroy it completely (e.g., [34]). Another piece of evidence supporting the existence of such correlations could be the existence of a similar correlation between L and E_p within some bursts themselves [35]. It is interesting to note that short and long GRBs do not share the same Amati relation, with short GRBs having a parallel and higher relation compared to that for the long GRBs, suggesting that they are systematically less energetic than the long GRBs given the same $E_{p,z}$. This could be attributed to the shorter durations, which would tighten the relationship between luminosity and $E_{p,z}$ for such phenomena. On the other hand, short and long GRBs are no longer well distinguishable in the $E_{p,z} - L_{\gamma,p,iso}$ plane, suggesting that the radiative processes responsible for the emissions are the same [36,37].

The abundant multiwavelength afterglow data allowed an in-depth understanding of the physics of GRBs. The spectral data were found to be generally in agreement with the power-law decay behavior predicted by the fireball forward shock model [38,39], while early optical flashes were interpreted as reverse shock emission [21,40]. The possibility of analyzing the evolution of the afterglow spectrum over time led to the proposal that GRB emission occurs in highly collimated jets: the discovery of an achromatic break (optical and radio bands) in the afterglow of GRB990510 (later observed in other GRBs) of around $t \sim 1$ day led, for the first time, to the hypothesis that the emission could occur in the jet rather than isotropically [41]. Analyzing the relatively small sample of afterglows observed until then, it was discovered that, surprisingly, the total jet-corrected energy of the GRB sample is essentially constant, narrowly clustered around 5×10^{50} erg [42]. This finding was at first interpreted as a hypothesis that different GRBs can collimate a standard energy reservoir into different jet angles. A different explanation was proposed in parallel in [43,44]: all GRBs have an (almost) universal, structured jet. The differences between the various GRBs may correspond to the different viewing angles of this universal jet such that the measured "jet angle" is not the opening angle of a uniform jet, but rather the observer's viewing angle from the jet axis.

Thanks to multiwavelength data and the ability to measure the jet angle, it was possible to discover another tight correlation between $E_{p,z}$ and the beaming-corrected bolometric emission energy E_γ (Ghirlanda's relation [45]):

$$\frac{E_{p,z}}{100\,\mathrm{keV}} \approx 4.8\left(\frac{E_{\gamma}}{10^{51}\,\mathrm{erg\,s^{-1}}}\right)^{0.7} \quad (5)$$

The discovery of these robust correlations led to the proposal to use GRBs as standard rulers for cosmological measurements (e.g., [46]).

Thanks to the possibility of carrying out multiband observations of these signals, particularly in the optical range, with excellent localization, it became possible to look for further clues about the nature of the progenitors. The detection of the long-lasting, fading, multiwavelength afterglow emissions allowed the identification of the host galaxies, solving the "no host problem" and firmly establishing the origin of GRBs as extragalactic. The ability to obtain accurate localizations showed that the GRBs are located toward the edges of the host galaxies [47,48], i.e., in the star formation zones [49], and not toward the center. This fitted well with one of the most investigated scenarios, i.e., the supernova–GRB association. The first hint of such an association was the discovery of SN 1998bw, a type Ic supernova in a galaxy near $z = 0.0085$, in the error box of the BeppoSAX burst GRB980425 [50,51]. Furthermore, a supernova red bump was discovered in the optical light curves of several other GRBs [52]. These findings led to the development of various models for the progenitors. Refinements of the collapsar model, first proposed by Woosley in 1993 [22], have been proposed by various groups. Woosley's group performed the first detailed numerical simulations of a jet launched from a collapsing Wolf–Rayet star, giving robustness to the collapsar model [53]. Many observational features were elegantly explained with this model, such as the presence of a less energetic "cocoon" surrounding the jet, as shown in [54]. In 1998, in the same article where he located GRBs within the star-forming regions of their host galaxies, Paczynski proposed a variant of Woosley's collapsar model, the so-called hypernova model: a rapidly rotating star collapses into a black hole surrounded by a thick accretion disk (or "torus"), which magnetically launches a relativistic jet, which, in turn, powers the observed GRB [49]. Another variation of this scenario was proposed in the Black-Hole Accretion Disk (BHAD) Models, where the binary merger of two compact objects or the collapse of a rotating star produces a rapidly accreting disk (>0.1 solar masses per second) around a "hyperaccreating" black hole [55]. Other models went in different directions. In particular, the supranova model posited that GRBs are emitted when a supramassive neutron star (a neutron star with a mass greater than the Tolman–Oppenheimer–Volkoff limit, supported by fast rotation) loses so much angular momentum that centrifugal support against self-gravity becomes impossible, and the star implodes into a black hole [56,57]. The interaction of the material ejected from the collapse with an almost baryon-clean environment would produce the GRB signals. A more radical proposal was put forward by Dar and De Rujula with the Cannonball model, an alternative to the paradigm of the forward-and-backward-shocks model, where each GRB pulse is given by the expulsion of "cannonballs" of relativistic plasmas from the supernova shell rather than by the interactions of the fronts in a collimated jet [58].

This era can be summarized as follows:

- In 1996, the BeppoSAX mission was launched with large-FoV gamma-ray instruments and X-ray instruments with great sensitivity, which allowed GRB signals to be localized with unprecedented precision.
- The discovery of afterglows led to the first multiwavelength observations, the measurement of redshifts (and therefore distances and energies), and the identification of host galaxies.
- The Amati, Yonetoku, and Ghirlanda correlations for long GRBs were discovered.
- In this era were the first indications of the GRB–supernova association.
- Since the X-ray telescopes and WFCs were not aligned, BeppoSAX had to be slewed after locating each GRB. The procedure had to be carried out from the ground, leading to slow replacements. For this reason, in this era were predominantly (almost exclusively) observed long GRBs.

5. The HETE-2 Era (2000–2004)

After the first High Energy Transient Explorer 1 (HETE-1) satellite was lost during its launch on 4 November 1996, a second mission was designed, and HETE-2 was then successfully launched on 9 October 2000. HETE-2 was a NASA science mission with international collaboration (mainly with Japan and France), with the main objective of making multiwavelength observations of GRB counterparts for the first time [59]. The satellite carried on board three main instruments: the French Gamma Telescope (FREGATE), the Wide-Field X-Ray Monitor (WXM), and the Soft X-ray Camera (SXC). FREGATE was composed of wide-field ($\sim\pi$ FoV) gamma-ray spectrometers sensitive between 6 and 400 keV, designed to conduct spectroscopy of GRBs thanks to its high spectral and timing resolutions. The WXM consisted of a pair of orthogonal, one-dimensional X-ray detectors sensitive in an energy range of 2 to 25 keV, allowing excellent localizations of $\sim$10 arcminutes. The SXC consisted of two orthogonal sets of one-dimensional coded-aperture X-ray imagers sensitive in the 0.5–14 keV energy range. These instruments had an identical field of view of 1.5 steradians and were able to communicate with each other in order to coordinate observations.

On 29 March 2003, HETE observed three different GRBs. The first of them, GRB030329, was fundamental to our understanding of GRBs by definitively confirming the association between long GRBs and supernovae [60,61]. GRB030329's distance from Earth allowed its afterglow to be studied extensively. On 6 April 2003, a spectroscopic analysis of the burst's optical afterglow revealed discernible peaks at approximately 570 nm and 470 nm. This spectral profile was effectively reproduced through a composite model, integrating a power-law distribution with the spectral characteristics derived from SN1998bw. Noteworthy was the sustained evolution of supernova-like features during the subsequent weeks following the initial burst. Optical observations conducted at the Kitt Peak National Observatory indicated a luminosity in the burst's optical afterglow that surpassed predictions based on a power-law decay. This observed departure from the expected decay pattern could be rationalized by postulating the influence of additional luminosity originating from a concurrent supernova event (later named SN2003dh) [62].

A further fundamental contribution of HETE was the discovery of X-ray Flashes (XRFs). XRFs are intense and transient cosmic events characterized by a rapid release of X-ray radiation. Although very similar to GRBs, the distinction between them lies in the energy distribution of the emitted radiation. Originally defined operationally as those X-ray signals detected by BeppoSAX's WFCs (energies between 2 and 25 keV), which were not triggered and not detected by the GRBM (in the energy range 40–700 keV) [63], they are now considered a subclass of gamma-ray bursts (GRBs), sharing similarities with traditional GRBs but with a spectral peak in the X-ray range. Even though the nature of XRFs is largely unknown, the most credited model posits that XRFs could be given by off-axis GRB emission: it is thus postulated that gamma rays are indeed emitted, but their trajectories are directed away from our instruments [64]. Consequently, the initially observable phenomenon is confined to lower-energy X-rays emitted in a divergent beam, exhibiting greater dispersion compared to the more narrowly focused gamma-ray beam.

This era can be summarized as follows:

- The common spectral properties of SN1998bw and SN2003dh provided a "smoking gun" for their common origin, proving once and for all the association between long-duration GRBs and supernovae.
- Discovery of X-ray Flashes (XRFs), a subpopulation of GRBs.
- Together with BeppoSAX, HETE-2 enabled accurate localization (of the order of arcminutes) that kick-started the era of multiwavelength observations of GRBs before the advent of *Swift*.

6. The Swift Era (2004–Ongoing)

In the early 2000s, the GRB phenomenon seemed almost completely decoded. BATSE observations had led to excellent modeling of gamma emissions, identifying two subpopulations of GRBs depending on their duration. Subsequent observations by BeppoSAX and HETE-2 led to the identification of multiwavelength afterglows and, through accurate localization, allowed long GRBs to be correlated with supernova emissions. These discoveries were accompanied by impressive and rapid theoretical advances that led to the development of, among others, the internal/external shock fireball model and the possible central engines of GRBs. Still, there were some "details" of these phenomena that were not yet well understood. First of all, even though BeppoSAX had been able to observe a large number of GRB afterglows, these were just from long-duration bursts due to the requirement to slew the telescopes to observe the signals with the main cameras once the wide-field cameras were triggered. This operation, carried out from the ground, was very slow and therefore limited the observation possibilities only to long-duration GRBs and meant that there was a gap in the data acquired between the prompt-emission signals that triggered the wide-field cameras and afterglow observations after slewing. To fill these observational gaps, the Neil Gehrels Swift Observatory mission [65] was launched on 20 November 2004.

Swift is an international space observatory with five main objectives: to quickly (automatically, in $\sim$1–2 min) repoint its NFIs in order to study the early afterglow emissions (and better understand both the afterglow onset and the connection with the prompt-emission phase), to observe GRBs across a large spectral band, to collect a large sample of GRBs and afterglows, and to study short GRBs and to trigger ground-based follow-up observations in order to perform accurate spectroscopy measurements.

To achieve these objectives, *Swift* hosts three instruments on board. The Burst Alert Telescope (BAT) [66] is a coded-aperture-mask telescope sensitive between 15 and 150 keV capable of localizing the position of the signals with an accuracy of 1 to 4 arcminutes within 15 s over a 1.4-steradian field of view. This instrument is used to quickly localize the positions of the GRBs to slew the telescope in order to point the NFIs and simultaneously send the localization to the ground through the GCN system to trigger follow-up observations. The X-ray Telescope (XRT) [67] is an X-ray focusing telescope sensitive between 0.3 and 10 keV capable of taking images and light curves and performing a spectral analysis of the GRB afterglows. This instrument also allows the refinement of the BAT localizations of GRBs, with a typical error radius of approximately 2 arcseconds. The Ultraviolet/Optical Telescope (UVOT) [68] is a 30 cm Ritchie–Chretien UV/optical telescope used to detect GRBs' optical afterglows, providing localization with $<$arcsecond precision and performing optical and ultraviolet photometry through the use of lenticular filters and low-resolution spectra (170–650 nm) thanks to optical and UV prisms. UVOT can also provide long-term follow-up of the afterglow light curves of GRBs. The combination of these instruments allowed a progressive refinement of the localizations of the signals. First, BAT is triggered by the GRB signal and calculates the position to $<$4 arcmin. Then, while the first localization is sent to the ground through the GCN network, the spacecraft autonomously slews to the GRB position in 20–70 s. XRT and UVOT then start their observations, determining the position to $<$5 arcseconds, and then transmit their data to the ground. In this way, *Swift* manages to identify on average 100 new GRBs every year, of which the vast majority are long GRBs. *Swift* has obtained (and continues to obtain) exceptional results in this way, which have revolutionized our understanding of GRBs in several ways thanks to some fundamental discoveries.

Thanks to the fast automatic slewing, *Swift* is able to detect the faint X-ray afterglow of short-duration GRBs. This was impossible before *Swift*. One year after the launch, in 2005, it was already possible to identify the host galaxies of some of these short GRBs (e.g., GRB050509B and GRB050709 detected with HETE-2 and GRB050724) and obtain their precise positions within them. The observations turned out to be rather different than expected: while long GRBs are generally found in blue, regular, and highly star-forming host galaxies and are located precisely in the star-forming regions of these galaxies, short GRBs, on the other hand, are hosted mainly by elliptical or irregular galaxies, far from star-forming regions [69–71]. This led to the suggestion that short GRBs may have different progenitors than long GRBs, not being associated with the death of massive stars but with the coalescence of compact objects (such as neutron stars or black holes).

The ability to quickly repoint the telescope and observe the initial stages of the afterglows has led to a true revolution in our understanding of these emissions thanks to the abundant available data [72,73]. Before *Swift*, GRB afterglows were simply modeled with two power laws, one for the "initial" phase and one for the "post jet-break" phase. It was therefore expected that by observing the initial phases of the afterglow, the same simple power-law trend would be found to connect up to the peak given by the prompt emission. In addition to the two phases already known, it was found that most GRBs have a complex light curve, with two additional phases connecting the prompt-emission phase with that of the afterglow. The early afterglows show, in addition to the two already-known decay phases, an "early steep decay" phase directly connected to the prompt emission [74,75] and a "plateau" phase (or "shallow decay") before the onset of the normal decay phase [76,77]. In nearly half of the GRBs, additional X-ray flares, following the prompt gamma-ray emission, were also discovered [78–80]. The abundance of data has allowed the development of a phenomenological model for all afterglow light curves, the "canonical X-ray afterglow lightcurve" [81,82]. In addition to these more widespread characteristics, some afterglows present even more particular behaviors, such as the presence in some GRBs of a "chromatic afterglow" behavior, where the optical light curve has no break coinciding with the X-ray break time (or vice versa), and in particular, there is no association between the temporal variations in the light curve in the optical and in the X-rays [83]. This evidence, which challenged the standard external shock fireball model, led to the proposal that the afterglow emission could be generated by the superposition of different physical processes. The initial steep-decay phase is the continuation of the prompt emission after it has ended due to emission from high latitudes relative to the observer's line of sight. The plateau phase could be given by an external shock emission with continuous energy injection, which requires either a long-duration central engine or a jet with stratified Lorentz factors. In particular, to justify this plateau phase, the models seem to favor the notion that the central engine could be a (supramassive) millisecond magnetar that, thanks to the rotation, manages to survive the collapse long enough to emit the signal in the early afterglow and then collapses into a black hole (with an extremely rapid decay of the light curve) [84]. And finally, X-ray flares may be internal emissions due to the tail end of central engine activity, similar to how prompt gamma-ray emission is produced. Various models have been proposed to explain such emissions as intermittent activity (or re-ignition) of the GRB's central engine. Among the proposed models, we mention the fragmentation of the star during the collapse [85] and/or the fragmentation of the accretion disk due to gravitational instability [86] in the black hole–torus/accretion disk model. Alternatively, these spikes could be produced by the magnetic activity of a rapidly spinning neutron star (magnetars) [87,88].

Thanks to its high performance, *Swift* was able to greatly expand the redshift range in which GRBs were observed: at low redshifts, several "low-luminosity" GRBs were observed, leading to the hypothesis that they may form a distinct population compared to "high-luminosity" GRBs [89], while at high redshifts, *Swift* broke the record several times over the years with different observations. The first record-breaking GRB observed was GRB050904 at a redshift $z = 6.29$ [90,91], discovered about a year after the mission's launch. Other notable GRBs were GRB080913 at $z = 6.7$ [92] and GRB090423 at $z = 8.2$ [93,94] (still the most distant GRB with strictly spectroscopic redshift estimation) and GRB090429B, which, at $z = 9.4$, remains the most distant GRB ever observed (with photometric redshift estimate) [95]. The increasing number of GRBs with known redshifts allowed the parameters of the Amati relation to be finely calibrated [31] and led to the discovery of two new empirical correlations between the observational parameters.

Some "anomalous" signals observed by *Swift* led, for the first time, to the question of whether the distinction between long and short GRBs is intrinsically linked to different progenitor populations. In 2006, *Swift* had already detected two long-duration GRBs (GRB060614 and GRB060505, [96]) that had temporal lags and peak luminosities that did not fit with those expected for long GRBs but were perfectly reasonable for short GRBs. Moreover, deep optical observations never found any supernovae associated with these emissions, putting very strong constraints on the possibility of an association. These events challenged both the collapsar and the merging neutron star theoretical models for long and short GRBs, suggesting that these long-duration events could be part of a subgroup of the category of short GRBs [96,97]. This first suggestion of the blurring of the two classes of GRBs was reinforced in the following years by the detection of several short-duration (or "rest-frame" short) bursts found to be more consistent with the long-duration population (e.g., GRB080913, GRB090423, GRB090426) [92,98]. All these findings led to the now-accepted conclusion that the duration criterion alone is insufficient to determine the physical category of a particular GRB, and other criteria must be taken into consideration [36].

Other exceptional events have, in time, appeared, challenging our models and understanding with unexpected characteristics. Some notable examples follow. GRB060218 (and SN2006aj) was a GRB with unusual characteristics never seen before: in fact, it had a duration of almost 2000 s, much longer than the typical gamma-ray bursts seen previously, and its host galaxy was identified at a distance of only 440 million light years, much closer than all the bursts previously observed. Furthermore, despite its proximity, the burst was much dimmer than the usual high-luminosity GRBs. Moreover, a smooth light curve, a thermal X-ray component in the time-resolved spectra, and a puzzling UV emission posed strong challenges to the theoretical models [99]. The aforementioned GRBs 060505, 060614, 080913, 090423, and 090426 have led to the understanding that the simple long–short classification scheme is not enough to completely describe the physical origin of GRBs. GRB080319B, also known as the "Naked eye Burst", was the first GRB to have a prompt optical emission visible to the naked eye: with a peak visibility at an apparent magnitude of 5.3, it remained visible to the human eye for about 30 s, breaking the record for being the object observable with the naked eye coming from the largest distance [100]. Some bursts, such as GRB101225 (the "Christmas Burst") or GRB111209A, were extremely long (with GRB111209A still holding the record as the longest GRB observed, with a duration of more than 7 h) with observational properties difficult to interpret with the models of the time; these represented a prototype of a separate class of ultralong GRBs with a possible progenitor different from the others [101]. GRB110328, now called "Sw J1644+57", was a totally anomalous event, later recognized as the gamma-ray emission from a tidal disruption event (TDE), the destruction of a star by a supermassive black hole [102–105]. This remarkable GRB marked the first time such type of events was ever observed.

This era can be summarized as follows:

- *Swift* was able to detect the X-ray afterglow of short-duration GRBs, and thanks to multiwavelength follow-up campaigns, it was possible to identify their host galaxies: short GRBs are hosted in different galaxies (and outside of star formation regions) with respect to long GRBs. This was the first strong hint that the two categories might come from different precursors.
- The observation of the early GRB afterglows allowed the discovery of a complex multi-phase structure given by a superposition of processes, at odds with the theoretical models of the time.
- The detection of some "anomalous" signals challenged the use of the temporal criterion (short vs. long) alone for the classification of GRBs.
- *Swift* is able to detect the highest-redshift GRBs, which led to their use to study the Early Universe.

7. The High-Energy (AGILE and Fermi) Era (2007–Ongoing)

Until the early 2000s, most GRB observations had been carried out exclusively in the hard-X-ray, dim-gamma-ray regimes in order to exploit the technological advances then available to optimize the capabilities (localization, spectral and temporal) of the detectors. In fact, until that time, only one mission had an instrument on board that was sensitive beyond a few MeV: CGRO. As seen in Section 3, CGRO had four instruments on board that covered an unprecedented energy range of 6 orders of magnitude, from 20 keV to 30 GeV. The Energetic Gamma Ray Experiment Telescope (EGRET) [106] was an instrument covering the higher-energy bands, sensitive between 30 MeV and 30 GeV. To achieve such high-energy detections, EGRET used the pair-conversion technique. The detector consisted of three components: a tracker, a calorimeter, and an anticoincidence detector. The tracker, a multilayer thin-plate spark chamber, was used to convert gamma-ray photons into electron–positron pairs, whose trajectories, reconstructed through the multiple interactions and the crossing of the different layers of the detector, were used to estimate the direction of arrival of the photon, reaching high angular resolutions. The NaI(Tl) calorimeter was used to collect the electron–positron shower and measure its energy. Finally, in order to reject unwanted signals from charged particles (cosmic rays), the telescope was covered by a plastic anticoincidence scintillation dome with a high-enough efficiency to reject charged particles but not to veto gamma rays. During its active life, EGRET allowed the observation of high-energy emission from some GRBs [107–109], not all of which could be explained by a simple extension of the models of the time [110], thus showing the need for optimal timing and fast broadband detectors to extend the studies of such phenomena to higher energies. In 2007 and 2008, two new missions for the observation of the gamma-ray sky were launched that substantially changed our way of understanding GRBs, starting a new high-energy era of observation characterized by the possibility of observing energies that, up to that moment, had been inaccessible ($>GeV$).

7.1. AGILE (2007–Ongoing)

AGILE ("Astro rivelatore Gamma a Immagini LEggero", Italian for "Lightweight imaging gamma astro detector") [111] is an Italian high-energy astrophysics mission launched on 23 April 2007 dedicated to the observation of the high-energy gamma-ray sky. With a weight of just ∼100 kg and the dimensions of a cube with sides of about 60 cm, AGILE is the lightest and most compact instrument for high-energy astrophysics flying today. AGILE combines, for the first time, a silicon–tungsten tracker gamma-ray imager ("GRID", sensitive in the 30 MeV–30 GeV range) [112] with a hard X-ray imager ("SuperAGILE", sensitive in the 18–60 keV energy range) [113] with a large FoV (∼1–2.5 steradians) and optimal angular resolution. AGILE is also equipped with a non-imaging gamma-ray scintillation mini-calorimeter (MCAL) sensitive between 350 keV and 100 MeV and a plastic scintillator anticoincidence system. AGILE operated in "pointing" mode until April 2009 when, following the failure of the inertial pointing system, the satellite was placed in "Sun Pointing Spinning" (scanning) mode to continue operations despite the failure. Since then, AGILE has operated in this scanning mode, rotating at a speed of $\sim 1^{\circ}\ s^{-1}$ and thus sweeping approximately 70% of the sky every day. This operating mode was found to be optimal for studying the variability of celestial sources. Central to AGILE's success is its world's fastest gamma-ray alarm monitoring system. AGILE is, in fact, equipped with two independent pipelines for monitoring gamma-ray alarms that process data with different data-quality results in order to optimize the response time [114].

Using data from the AGILE mini-calorimeter (MCAL), it was possible to publish two GRB catalogs, the first in 2013 [115] and the second, more recently, in 2022 [116]. The amount of data obtained made it possible to calculate the counts and flux upper limits pertaining to GRB high-energy emissions between 30 MeV and 3 GeV (the AGILE-GRID energy range), thereby imposing strong constraints on high-energy radiation originating from both the afterglow emission and synchrotron self-Compton emission within internal shocks [117].

Two exemplary GRBs seen by AGILE were GRB080514B and GRB090510. The first, GRB080514B [118], was the first GeV-bright GRB observed after EGRET. This event also had an afterglow phase, the observation of which allowed a photometric redshift measurement (z∼1.8). Remarkably, the high-energy tail of the spectrum of this GRB fell right on the continuation of the Band spectrum used at lower energies, extending it from 20 keV up to ∼50 MeV. GRB090510 was first localized by *Swift* and then detected both by AGILE [119] and by *Fermi*/LAT [120]. This nearby burst (estimated redshift z∼0.903) was unprecedented, as it was the first time that a short GRB was observed simultaneously by three different instruments, managing to extend its spectrum up to more than 300 GeV. The light curve was also anomalous compared to those observed up to that point for short GRBs, as it presented a delayed emission at high energies compared to the prompt phase and a substantial spectral evolution. This temporal behavior and the evolution of the power-law spectrum challenged the models based on synchrotron emission/synchrotron self-Compton in external shocks and on the hadronic models [121].

More recently, AGILE has had huge success in the application of modern machine-learning techniques for the identification of GRB signals in data from both the gamma-ray imaging detector (GRID) [122] and the anticoincidence system (ACS) [123], leading to the identification of 72 GRB signals with significance $\geq 3\sigma$, 15 of which are not present in the second MCAL GRB catalog, as they were not identified before when using traditional methods for the data analysis.

7.2. *Fermi* (2008–Ongoing)

In those same years, another NASA gamma-ray mission, the Gamma-ray Large Area Space Telescope (GLAST), later renamed the *Fermi* Gamma-Ray Space Telescope (FGST), was launched on 11 June 2008. *Fermi* carries two scientific instruments on board, the Large Area Telescope (LAT) [124] and the Gamma-ray Burst Monitor (GBM) [125]. The LAT is a gamma-ray imaging detector that detects photons with energies from about 20 MeV to about 300 GeV, with a field of view of about 20% of the sky, capable of exposing all parts of the sky for about 30 min every 3 h in sky survey mode. The GBM consists of 14 scintillation detectors, namely, 12 NaI crystals for the range from 8 keV to 1 MeV, primarily used for onboard triggering, onboard and ground localization, and spectroscopy, and two bismuth germanate (BGO) crystals with sensitivity from 150 keV to 30 MeV. The GBM is used for spectroscopy and is capable of detecting GRBs in this energy range throughout the entire sky not occluded by the Earth. Thanks to these instruments, *Fermi* offers numerous advantages compared to all previous instruments and, in particular, an enormous range of energies, spanning more than 7 orders of magnitude, including the largely unexplored 10 GeV–100 GeV band and a huge field of view. These allow *Fermi* to observe $\sim$200 GRB/year with observations from 8 keV to 40 MeV and $\sim$15 GRB/year with observations from 8 keV to 300 GeV with excellent spectral and timing capabilities, setting *Fermi* as one of the leading instruments in the current study of the gamma-ray sky. These two instruments have, in many ways, revolutionized our understanding of GRBs.

In its years of operation, *Fermi* observed a lot of GRBs, both with the GBM and with the LAT, and produced various catalogs. The latest ones are dated 2018 (though regularly updated) for both instruments, boasting as many as 2356 GRBs detected by the GBM [126] and 186 (approximately 7–8% of those seen by the GBM) above 100 MeV seen by the LAT [127]. For those detected by both instruments, the LAT band emission usually lasts much longer than the GBM band emission, and it decays as a power law [128,129]. This has been interpreted as an external-shock origin of the observed $>$ 100 MeV emission of the GRB, after the prompt-emission phase has ended [128–131]. Another, yet unexplained, feature that was found confronting the observations of the GBM and LAT is that, in some GRBs, the LAT GeV emission has a delayed onset with respect to the MeV emission. This was not predicted by theoretical models, and there is still no agreement on the origin of such emission, even though many mechanisms have been proposed. Still, the presence (or lack) of delays between the emissions at different energies has been used to put strong constraints on Lorentz Invariance Violations (LIVs), a feature allowed by many models of Quantum Gravity (e.g., GRB090510 [132]). The LAT was able to detect the presence of photons with rest-frame energies exceeding 100 GeV in several bright GRBs. Some examples are GRB080916C [133], i.e., the very first bright LAT burst; GRB090510 [132]; and GRB130427A [134], which, at the time, was a record-setting bright GRB detected both by *Fermi* and by *Swift*. The existence of these photons was not consistent with the predictions of the standard fireball synchrotron internal-shock model of the time and placed strong constraints on GRB physics, including the minimum bulk Lorentz factor (from opacity arguments) [135], particle acceleration mechanisms in relativistic shocks, and relativistic particle radiation mechanisms. These photons can also be used to study the extragalactic background light (EBL) via the expected attenuation of high-energy photons due to two-photon pair production (e.g., [136]).

The ability to carry out highly accurate spectral analysis over such a wide range of energies (7 orders of magnitude, including the hitherto unexplored range > 10 GeV) has provided fundamental information for understanding the composition of the GRB jet and the emission mechanisms of the prompt emission, triggering a real theoretical renaissance in trying to understand the large number of unexplained observational effects. The observation of the LAT's first bright GRB, GRB080916C, already showed almost-featureless, time-resolved spectra and spectral features divergent from the predictions put forward by the conventional fireball internal-shock model, thus necessitating a reevaluation of the fundamentals of the theoretical framework [137]. Subsequent observations of other GRBs (e.g., GRB090510, 090902B, and 090926A) have introduced a layer of complexity into the observed GRB spectra: these analyses have shown that the spectra of the observed GRBs are composed of the superposition of at least three distinct spectral components [138,139]. In addition to the conventional non-thermal component of the Band function, in some GRBs, a further quasi-thermal component is present, assuming a dominant [140] or sub-dominant role [141]. The observation of other signals (e.g., GRB090902B, 090510, and 090926A), moreover, suggested the existence of a third component of the power-law spectrum, which extends to both high and low (GBM) energies [142]. The origins of this additional spectral component remain elusive. Recently, more in-depth analyses of low- energy spectra have been carried out, going down to the optical, highlighting the existence of a spectral break in the low-energy part of prompt spectra. This is consistent with emission from synchrotron radiation in the moderately rapid cooling regime, identifying the spectral break with the cooling frequency [143,144].The Band spectrum thus modified is in better agreement with the data compared to the addition of ad hoc thermal and non-thermal components to the total spectrum.

This era can be summarized as follows:

- By fully exploiting the production of electron–positron pairs in the detectors, it was possible to extend the energy range to hundreds of GeV.
- Comparing GRBs observed by both the GBM and LAT, it was found that the LAT band emission usually lasts much longer than the GBM band emission.
- Comparing the GBM and LAT data also shows that the GeV emission has a delayed start compared to the MeV emission.
- The possibility of performing unprecedentedly detailed spectral analysis in a wide spectral window, including HE, has provided very important information for understanding the composition of GRB jets and the prompt-emission mechanisms.
- The use of modern machine-learning techniques applied to AGILE detectors led to the identification of previously unidentified GRB signals.

8. The Birth of the Multi-Messenger Era (2017–Ongoing)

GRBs have been believed to be multi-messenger emitters since 1989.

In 1995, three independent groups proposed that GRBs could be a dominant source of Ultra-High-Energy Cosmic Rays (UHECRs) through different mechanisms. Two groups proposed that the origin of UHECRs is in the internal shocks [145] or external shocks [146] of a GRB fireball. The latter group suggested a GRB-UHECR association after noticing that each of the error boxes of the two highest-energy cosmic-ray-shower events known at the time overlapped very well with that of a strong GRB [147]. GRB-accelerated cosmic rays can interact with background photons or other baryons through hadronic processes ($p\gamma$) to produce high-energy neutrinos, and as early as 1997, it was suggested that these neutrinos could reach PeV energies [148]. Predictions of the neutrino fluxes from GRBs are still a matter of debate, as they depend on many unknown parameters and are highly model-dependent [149,150]. Even though the theoretical evidence is particularly convincing, to date, no detection of neutrinos in conjunction with GRB signals has been observed, and progressively more stringent constraints on the neutrino flux from GRBs have been reported from both the IceCube neutrino observatory in the South Pole [151] and the ANTARES neutrino observatory in the Mediterranean Sea [152,153].

Ever since it was suggested that the merger of compact objects, such as two neutron stars, could be the progenitors of some GRBs [13], it was clear that this would make them ideal candidates for emitting gravitational waves together with gamma rays. In fact, these systems were already known sources of gravitational waves (GWs) [154]. Over the years, the possible association between short GRBs and GWs was revisited several times and consolidated. On 17 August 2017, GW170817 was detected by the gravitational wave observatories Advanced LIGO in the USA and Advanced Virgo in Italy, identifying it as an NS-NS merger event and obtaining an accurate localization [155]. Then, 1.7 s later, the low-luminosity short GRB GRB170817A was observed by *Fermi*/GBM and INTEGRAL, with a localization compatible with that obtained by LIGO-Virgo [156,157]. The joint detection thus confirmed the association between short GRBs and mergers of compact objects (NS-NS) [158]. At the same time, this detection triggered a massive observational follow-up campaign, which led to the observation of the signal in the optical, radio, and X-ray bands and allowed the localization of the signal in a nearby galaxy, NGC4993, at $\sim$40 Mpc [159–165]. Particularly remarkable is the observation of the simultaneous emission of a "kilonova", a short-lived (days) IR-UV signal powered by the radioactive decay of heavy elements synthesized in the ejected outflow, predicted by many models [88,166–168]. Other than these, various electromagnetic counterparts originating from gravitational wave (GW) sources have been linked to compact binary mergers, such as a faint radio afterglow ("radio flare"), arising from the interaction between the ejected material and the ambient medium [168], and an X-ray counterpart due to magnetic dissipation if the neutron-star–neutron-star (NS–NS) merger product takes the form of a millisecond magnetar or a black hole. It is speculated that collapsars might exhibit strong GW burst emissions, presenting the intriguing prospect of long GRBs and core-collapse hypernovae as potential multi-messenger targets [169].

The joint detection of GRB-GW170817 had enormous importance for fundamental physics, as it allowed a constraint to be imposed on the difference between the speed of gravity and the speed of light:

$$-3 \times 10^{-15} \leq \frac{\Delta\nu}{\nu_{EM}} \leq 7 \times 10^{-16} \tag{6}$$

This allowed new bounds to placed on Lorentz Invariance Violations and allowed a new test of the equivalence principle by constraining the Shapiro delay between gravitational and electromagnetic radiation [157]. These unprecedented results had huge repercussions for some dark matter (DM) and dark energy (DE) models, completely ruling out some classes of modified gravity theories that had been perfectly viable up to that detection [170].

This detection has renewed momentum in the development of strategies and synergies between different observatories to try to obtain repeat observations of this impact. To date, no other joint detections have been carried out, underscoring that the field of multi-messenger astronomy is still in its infancy.

This era can be summarized as follows:

- GRBs have been believed to be multi-messenger emitters since 1989.
- To date, no neutrino detections have been observed in conjunction with GRB signals, placing progressively more stringent constraints on the flux of neutrinos from GRBs.
- GRB-GW170717A was the first-ever multi-messenger observation of a GRB, allowing for some unprecedented results in GRB and fundamental physics.

9. The Very High Energy Era (2019–Ongoing)

The use of the photon-pair-conversion technique for the detection of gamma photons in space is severely limited by the volumes of the detectors. In order to observe photons with $\geq$TeV energies (VHE), the dimensions of the detectors are such that they make it impossible to observe this radiation from space. In order to observe photons of such energies, it is therefore necessary to use different techniques, exploiting the Earth's atmosphere as the "medium" of the telescope and observing the effect of the interaction of such photons with it. Gamma rays (as well as primary cosmic rays) crossing the atmosphere collide with atmospheric nuclei, producing secondary particles that, in turn, generate a cascade of particles as they traverse the atmosphere. The intricate interplay of these particles cascading down through the atmosphere creates an extensive spread of ionization and fluorescence, offering a unique signature for detection called an "Air Shower". There exist two types of ground observatory for VHE: Imaging Air Cherenkov Telescopes (IACTs) and Extensive Air Shower (EAS) experiments. Cherenkov experiments consist of almost-optical telescopes devoted to detecting the Cherenkov light emitted by particles produced in air showers and moving superluminally in the atmosphere. EAS experiments are huge arrays or carpets of particle detectors that directly collect the particles in air showers. Cherenkov experiments have lower energy thresholds but also a lower duty cycle, as well as a smaller field of view, when compared to EAS experiments.

Since as early as 1994, various theoretical models have predicted the possibility of the existence of a very high TeV energy component of the GRB emission [171–175]. This stimulated a search for this component of GRB emission with various instruments, however, with poor results. Some first tentative hints of TeV emissions from GRBs came with the claims of the detection of GRB920925C by the AIRshower Observation By angle Integrating Cherenkov Counters (AIROBICC) [176] and GRB970417A by Milagrito [177,178]. The HEGRA AIROBICC array was an array of 7×7 40 cm diameter PMT-based air Cherenkov integrating telescopes located in La Palma (Canary Islands) and was sensitive up to a few tens of TeV. After searching for GRBs above 20 TeV within AIROBIC's field of view, using data gathered between March 1992 and March 1993, evidence of a $\sim 2.7\,\sigma$ detection from GRB920925c was reported. However, since the "signal" preceded the activation of WATCH (an all-sky X-ray monitor on board the GRANAT satellite [179,180]) by <1 min and was located approximately 9° away from the position identified by this instrument, no evidence was claimed. Milagrito, a prototype of the Milagro EAS experiment, consisted of a flat array of 228 photomultiplier tubes (PMTs) submerged in a light-proof water pool with a size of $\sim 42 \times 42$ m^2. In 2000, Milagrito reported evidence for an emission above 650 GeV from GRB970417A, with a (post-trial) probability of 1.5×10^{-3} of being a background fluctuation. GRB970417A was a weak, soft GRB first observed by BATSE. Still, this was not enough to claim evidence of VHE emission from GRBs with full certainty.

Despite enormous and growing efforts, no evidence was found, and only upper limits from various collaborations were set [181–185] until 2019, with the observation of GRB190114C by the Major Atmospheric Gamma-ray Imaging Florian Goebel Cherenkov Telescopes (MAGIC) [186,187]. On 15 January 2019, MAGIC, a system of two IACTs located on the island of La Palma, reported the first observation of 0.2–1 TeV photons from GRB190114C. Triggered by the *Swift*-BAT alert, the MAGIC telescopes, thanks to their fast repointing capabilities, were able to start the observations of the GRB after just 57 s, "revealing a distinct emission component of the afterglow with power comparable to that of the synchrotron component" [187]. The observed emission, associated with the afterglow, has been explained as an emission from the inverse Compton scattering of synchrotron photons from high-energy electrons. The announcement of this unprecedented event triggered a renewed push in the theoretical research to understand VHE GRB emission. Moreover, the same year, 5 months later, the observation of the TeV afterglow of GRB180720B by the High Energy Stereoscopic System (H.E.S.S.) telescopes was announced [188]. Located in Namibia, H.E.S.S. is a system of five IACTs that are sensitive in the energy range of 0.03 to 100 TeV. It is currently the only VHE photon observatory in the Southern Hemisphere. The emission was also interpreted in this case as synchrotron photons accelerated by inverse Compton scattering with electrons (synchrotron self-Compton radiation, SSC). That same year, H.E.S.S. observed another TeV GRB, GRB190829A [189], bringing the total of such signals ever observed to three. The following year, the observations of GRB201015A [190], with a signal at the $>3\sigma$ level, and GRB201216C [191,192], with a $>5\sigma$ level, were announced by the MAGIC collaboration. After an in-depth analysis of the data taken for GRB160821B, an extremely close short GRB with a redshift of $z = 0.162$, MAGIC unveiled $\sim 3\sigma$ evidence pointing toward a gamma-ray signal exceeding approximately 0.5 TeV. The observations lasted approximately 4 h after the initial burst [193]. The presence of this signal poses a challenge to the most straightforward interpretation provided by one-zone models of the synchrotron self-Compton emission originating from the external direct shock. These models face difficulties in adequately explaining the reconstructed TeV flux associated with the observed gamma-ray signal.

On 9 October 2022, the last (to date) GRB with a TeV emission was observed. The Large High Altitude Air Shower Observatory (LHAASO) [194,195], the largest High Altitude Water Cherenkov Experiment in the world, located in China, reported the detection of the early onset of the afterglow of GRB221009A, with 64,000 photons (above $\sim$0.2 TeV) detected within the first 3000 s [196]. GRB221009A broke every record and remains to this day the Brightest (GRB) Of All Time (B.O.A.T.).

This era can be summarized as follows:

- Ground-based observations are needed to observe photons at TeV. There are two types of instruments, IACTs and EASs, with different performance and merits.
- Despite some claims (AIROBICC and Milagrito), no detection of GRB at TeV energies occurred until GRB190114C.
- To date, only five GRBs have been observed to have TeV emission with significance $> 5\sigma$, and two more have been observed with significance $> 3\sigma$. Despite enormous efforts, the detection of GRB emissions at TeV remains extremely difficult.

10. The Record Breaker: GRB221009A, the B.O.A.T.

On 19 October 2022, an unprecedented GRB event was observed by several instruments: GRB221009A. This event was so extreme that it broke the record as the brightest and most energetic GRB ever observed [197–199]. First detected by both the *Fermi* [200] and *Swift* [201] satellites, the GRB lasted around seven minutes, but the multiwavelength afterglow remained detectable for more than a month. This led to one of the biggest and most successful follow-up campaigns ever conducted, which led to the observation of the emission on an unprecedented 15 orders of magnitude on the electromagnetic spectrum, from radio emissions to VHE gamma rays. The burst completely saturated the detectors aboard *Fermi*, which captured gamma-ray photons with energies above 100 GeV [200]. GRB221009A produced the largest number of very high energy (VHE) photons ever observed by scientific instrumentation: before GRB221009A, the number of VHE photons detected in the entire history of GRB astronomy amounted to only a few hundred. For GRB221009A, however, the LHAASO alone saw more than 5000 VHE photons, some of these having a record energy of 18 TeV [196]. This claim is still debated, though. Still, photons of these energies are difficult to explain within the modern standard paradigms and prompted a new impulse in the development of theoretical models.

At the moment, there is still no consensus about the mechanism for generating these highly energetic gamma photons, which are difficult to explain with the standard synchrotron- and synchrotron-self-Compton-based leptonic models for the afterglow, and many different mechanisms and jet geometries are continuously being proposed to explain them (e.g., [202–205]). Some proposals have been put forward in which GRB221009A would be a huge accelerator of UHECRs, whose propagation would induce an electromagnetic cascade in the extragalactic medium. The line-of-sight component of this flow could explain the detected > 10 TeV emission [206,207].

The presence of photons of such energies, coming from an estimated distance of about $z = 0.151$ [208], is at odds with theoretical models because of the expected attenuation due to the extragalactic background light (EBL). This has led to much speculation on the presence or absence of exotic effects to explain the detection, with different groups proposing different mechanisms, such as the presence of LIV effects [209,210] or of axion-like particles (ALPs) [211,212]. Other groups have, however, highlighted how these observations, although exceptional, can still fit into standard paradigms.

Despite the extensive follow-up campaign, no multi-messenger detection occurred, managing to place stringent constraints on the emission mechanisms thanks to the non-observation of neutrino emission [213–215].

An important claim was made following an analysis of the spectral evolution of GRB221009A, as seen by *Fermi*/GBM, reporting the discovery of a narrow emission feature at around 10 MeV with very high significance ($>6\sigma$), interpreted in the paper as a blue-shifted annihilation line of relatively cold electron–positron pairs ($k_B T \ll m_e c^2$) [216]. This feature has never been observed in any other GRB, and its detection would be made possible by the exceptional brightness of the event, which dwarfs even the next-brightest bursts (such as GRB130427A). The presence of an excess around 10 MeV was also reported in the spectral analysis by the *Fermi* team [200].

In [197], it is argued that, while GRB2201009A might not be the intrinsically brightest GRB ever, its proximity to Earth makes it so that it is the brightest ever observed. In the same paper, they report an estimated timescale of approximately 10,000 years for any such event to happen again. These estimates underline the exceptional nature of this event, so much so that the authors themselves write in the acknowledgments at the end of the article:

> We acknowledge the universe for timing this burst to arrive at Earth after the invention of GRB monitors but during our active research careers. Our token optical astronomer would like to complain about the alignment with the Galactic plane and requests that the next one avoid this issue.

To this, we would also add the following:

> Our VHE colleagues would also request the next one not to arrive on a full moon night.

This exceptional event continues to surprise us by challenging our understanding of such phenomena, and there is no doubt that it will be a matter of study and debate for many years.

This era can be summarized as follows:

- GRB221009A was by far the brightest and most energetic burst ever observed by humans.
- The observation of photons at energies > 10 TeV challenges current theoretical models. Many mechanisms to explain such an emission have been proposed, but there is no consensus, yet.
- The observation of photons at energies > 10 TeV from a distance of $z = 0.151$ is at odds with predictions about attenuation by the EBL. Among the possible explanations, exotic/fundamental physics effects (LIV vs. ALPs) have been proposed.
- There is evidence of a highly significant narrow emission feature at around 10 MeV. However, this feature has never been observed in other GRBs and was observed in GRB221009A only due to its extreme brightness.

11. Future Prospects

As we have seen, the evolution of our understanding of the GRB phenomenon is intimately linked to the technological–scientific advances that have allowed us to carry out new measurements, thus revealing new pieces of this complex puzzle from time to time. Since a complete review of all the very interesting proposals for new instruments that could lead to a better understanding of these phenomena is an enormous undertaking, well beyond the scope of the present article, we mention here just four examples of instruments designed specifically for the observation of GRBs (among other things) that are guaranteed to be able to make a very important contribution to this field in the near future.

The High Energy Modular Ensemble of Satellites (HERMES) is a project for a constellation of nanosatellites (CubeSats) sensitive in the 50–300 keV range, with full-sky coverage, that will exploit the "temporal triangulation" technique to obtain the accurate localizations of GRB signals (and transient signals in general). In particular, the complete HERMES constellation should be able to obtain a localization accuracy better than 15′ (approx. an error area < 0.2 deg^2) for long GRBs and a localization better than 1° (error area ~ 3 deg^2) for short GRBs [217]. This is thanks to the very high precision of the timing (better than a few hundred nanoseconds), which allows the measurement of the delays between the arrival times of the signal at the different satellites at less than a few dozen μs, thus managing to accurately estimate the direction of arrival of the signal relative to the constellation. As we have seen, the ability to obtain the accurate and timely localizations of GRBs is essential to organizing a proper observational follow-up strategy. HERMES will therefore be extremely important in the future for the study of GRBs (and, in general, of gamma-ray transient events) thanks to its unprecedented localization capabilities. It should also be noted that a major advantage of CubeSats over typical larger gamma-ray space missions is that it is relatively cheap, greatly limiting construction and launch costs. A great milestone for this project was achieved on 1 December 2023 with the launch of the Space Industry Responsive Intelligent Thermal nanosatellite (SpIRIT). SpIRIT, an Australian–Italian collaborative project, will be the first satellite of the HERMES Scientific Pathfinder Constellation.

The Space Variable Objects Monitor (SVOM) [218,219] is a Chinese–French-approved mission, with a launch scheduled for spring 2024. SVOM will host four instruments on board and will be complemented by a ground segment composed of three other instruments. SVOM's payload is composed of the following four main instruments: a wide-field coded-mask camera sensitive between 4 and 120 keV to quickly localize GRBs with an accuracy of a few arcmin called ECLAIRs; a gamma-ray non-imaging spectrophotometer (GRM) sensitive in the 50 keV to 5 MeV energy range for monitoring the FoV of ECLAIRs; and two narrow-FoV telescopes for the study of GRB afterglows, one operating in the soft X-rays (MXT) and one in the optical band (VT). The ground segment includes three dedicated instruments: a set of cameras for follow-up observation of the ECLAIR FoV in the visible (GWAC) and two robotic telescopes (GFT), sensitive in the visible and NIR, for studying the GRB afterglow [218]. The ground segment will be fundamental for coordinated and automated follow-up observations of satellite alerts. This design was developed in order to better study two categories of GRBs specifically: very distant GRBs with redshifts $z > 5$ (in order to be able to use them as cosmological probes) and faint/soft nearby GRBs, effectively expanding on the scientific goals borrowed from *Swift* [219].

The Transient High-Energy Sky and Early Universe Surveyor (THESEUS) is a mission proposal to the European Space Agency for a satellite telescope with the main objective of using GRBs to study the Early Universe and will be invaluable for multi-messenger and time-domain astrophysics in general [220,221]. To achieve these goals, THESEUS will host three instruments on board whose combination will allow GRB and X-ray transient detection over an extensive field of view with accurate localization ($\sim$0.5–1 arcmin) over a wide energy range. These instruments will be a Soft X-ray Imager (SXI) sensitive between 0.3 and 6 keV with a wide FoV ($\sim$1 sr) and a great angular resolution ($\sim$1–2 arcmin), an InfraRed Telescope (IRT) operating in 0.7–1.8 μm with both imaging and (some) spectroscopic capabilities, and an X/Gamma-ray Imaging Spectrometer (XGIS), a set of coded-mask cameras sensitive in an unprecedented 2 keV–20 MeV energy range. With this instrumentation, THESEUS will be perfect for studying the most distant high-redshift GRBs, offering a unique opportunity to delve into pivotal unresolved questions within modern cosmology. These inquiries encompass understanding the population of low-mass and low-luminosity primordial galaxies and exploring the origins and the evolution of cosmic re-ionization, and the results will contribute to our comprehension of the evolution of the Star Formation Rate (SFR) and metallicity up to the "cosmic dawn" and through Pop-III stars [222]. Furthermore, THESEUS, by enabling the precise localization, identification, and in-depth study of the electromagnetic counterparts of GW and neutrino sources, will bring fundamental advances in multi-messenger and time-domain astrophysics.

We also mention the GAmma-ray burst Localizing Instrument (GALI) proposal [223]. GALI is a new concept for identifying the positions of GRBs by using numerous small scintillators in a 3D array using their mutual shielding. GALI can be thought of as an omnidirectional coded-mask detector, where the mask itself, rather than being applied externally to the detector, is made up of the grid of the sensor elements themselves. Furthermore, the detector (and therefore the mask) does not have a preferred direction and therefore provides coverage of the entire sky, unlike traditional coded-mask instruments. The GALI concept of using mutual masking of modular detector elements can be scaled to any size, thus adapting to a multitude of different missions. In particular, larger versions of the detector, with more elements, in addition to having greater sensitivity, will obtain a great improvement in angular resolution compared to smaller versions. However, simulations show how even a small detector, with a total volume of just $\sim$1 L, could identify the direction of a burst down to approximately $\pm$2 for a 1 s GRB in the 10 keV–1 MeV range and a flux of $\sim$10 ph cm^{-2}s^{-1}. While GALI is capable of functioning as a standalone instrument aboard a single satellite, its versatility extends to its integration within a distributed satellite architecture. This incorporation enhances both the sky coverage and the localization capabilities of the entire satellite constellation.

On 9 January 2024, the Einstein Probe (EP), a joint mission of the Chinese Academy of Sciences, the European Space Agency, and the Max Planck Institute for Extraterrestrial Physics [224], was successfully launched. The EP is a space satellite for the observation and monitoring of transient X-ray phenomena, which will exploit the combination of two instruments: a Wide-field X-ray Telescope (WXT) [225], which uses lobster-eye optics to obtain an FoV of approximately 3600 square degrees in the range 0.5 $\sim$ 4.0 keV, and the Follow-up X-ray Telescope (FXT) [226], a Wolter-I type of X-ray telescope highly sensitive between 0.3 and 10 keV, for observations of transients identified by WXT.

Finally, we mention some NASA mission proposals at various stages of development and financing.

BurstCube is a mission under development by NASA with a launch scheduled for March 2024, with the main objective of studying short GRBs and their gravitational wave counterparts [227]. BurstCube will be a cubesat equipped with four cesium iodide (CsI) detectors, each coupled to an array of silicon photomultipliers (SiPMs) sensitive in the 50 keV to 1 MeV energy range, thus expanding the coverage of the sky in this energy window.

Another smallsat project under development by NASA is the StarBurst Multimessenger Pioneer [228], also developed for the observation of the prompt-emission phase of short GRBs and the electromagnetic counterparts of gravitational waves, which will exploit the combination of scintillator arrays (12 NaI(TI) detectors) with SiPMs to achieve high sensitivities in the energy band between 30 keV and 1 MeV.

The Moon Burst Energetics All-sky Monitor (MoonBEAM) [229,230] proposal expands the concept of using the difference in light travel time between different spacecraft to precisely estimate the position of GRBs (time triangulation) by proposing to place two cubesats in Earth and cislunar orbit, thus maximizing the flight time (>1 s) of light between the two satellites and achieving a great improvement in localization capabilities.

Last but not least is the proposal of the LEAP GRB polarimeter [231,232]: a single wide-FoV Compton polarimeter ($\sim 1.5\pi$ steradians) capable of precisely measuring the polarization of GRBs between 50 and 500 keV and, at the same time, carrying out spectroscopy measurements between 20 keV and 5 MeV, to be mounted on board the International Space Station (ISS). The possibility of reliably measuring the polarization of GRB signals will allow an enormous advance in our understanding of these events, thanks to the possibility of clearly distinguishing the different emission mechanisms (synchrotron, inverse Compton, etc.) as well as directly probing the geometries of the magnetic fields in GRB jets, opening a completely new window for the study of these phenomena.

The proposals for new missions for the observation of GRBs are many and very varied, pushing in various directions (depth, rapid repointing, localization capability, polarimetry). It is interesting to note how, for the first time, a substantial methodological difference is emerging between large-single-satellite missions, with very high capabilities and costs, and distributed architectures, where constellations of smaller and cheaper instruments are used to obtain exceptional results. What the future of the sector will be remains to be seen:

"Ai posteri l'ardua sentenza!" (Il Cinque Maggio, Manzoni).

12. Conclusions and Open Questions

GRBs are extreme phenomena, pushing the limits of our physical theories. Even 50 years after the first publication about their discovery, they continue to surprise us and elude our complete understanding. Given their extreme nature, GRBs are (almost) unique in astrophysics in their multi-disciplinary nature. As discussed in [233], the study of GRBs has connections with many other major branches of astrophysics, with important consequences for stellar astronomy, the study of the interstellar medium, galactic astronomy, and cosmology. The connections do not end here, and indeed, the study of the GRB phenomenon has connections with various branches of physics (and not only). These connections go far beyond just physics; e.g., some groups have studied the possible links between GRBs and mass-extinction phenomena. Focusing more precisely on the study of the basic mechanisms of these phenomena, it is evident that every step forward in our understanding has been intimately linked to a technical–scientific advancement, which, through the conception and implementation of increasingly targeted measures, has allowed, from time to time, the discovery of more and more details of these fascinating and varied phenomena.

Without any pretense of completeness, we can summarize the open questions and the lines of research to follow as follows:

- The Standard Model:
  - The standard fireball (plus internal/external shocks) model can explain many features observed in prompt GRBs and afterglows ... before the *Swift*, HE, and VHE eras. More realistic assumptions are needed (ejecta, environment).
  - What is the nature of GRB jets?
- Long GRBs:
  - Collapsar vs. magnetar: which is dominant?
  - Where are the GRB remnants?
  - The nature of subluminous vs. ultraluminous GRBs: what determines the difference?
- Short GRBs:
  - Are all short-hard GRBs compact binary mergers? NS-NS or NS-BH mergers?
  - How can hard long GRB "spikes" be distinguished from short GRBs?
- GRBs as probes of the Early Universe:
  - What is reionizing the IGM?
  - Did the reionization begin at $z \sim 12$?
  - Do Pop III stars make luminous GRBs?
- GRB221009A and other extreme events:
  - How are $\geq$10 TeV photons produced?
  - Are UHECRs accelerated in high numbers in GRBs? Why are there still no associated detections of GRBs and neutrinos (only upper limits)?
  - How is it possible that $\geq$10 TeV photons arrived from $z \sim 0.151$ despite the theoretically predicted EBL absorption? Is the signal of exotic or fundamental physics effects (LIV, ALPs, other)?
  - Does the narrow $\sim$10 MeV emission component seen in GRB221009A really exist? Is this a characteristic of this burst only or of all GRBs?
- Experimental development: how can we develop new instruments to make new and better observations?
  - Large sky coverage with high localization capabilities/better angular resolutions to observe more signals (better statistics). Large-single-satellite missions vs. distributed geometries?

- Better spectroscopic capabilities over wider energy ranges.
- Better timing capabilities to characterize the light curves.
- Ability to measure gamma-ray polarization to distinguish emission mechanisms.
- Better follow-up strategies (and collaborations) for reliable multi-messenger observations.

After 50 years of research on GRBs, enormous progress has been made in the study of these phenomena. However, we are still far from a complete understanding of them. There is no doubt that this research area is one of the liveliest at the moment and that, thanks to the enormous effort of experimental improvement and theoretical understanding of the scientific community, it will still yield great surprises and lessons in the future.

Author Contributions: Writing, A.A.V.; review and editing, F.L. and A.A.V.; supervision, F.L. All authors have read and agreed to the published version of the manuscript.

Funding: This research received no external funding.

Data Availability Statement: No new data were created or analyzed in this study. Data sharing is not applicable to this article.

Acknowledgments: We acknowledge Lorenzo Amati for the original idea of dividing the history of GRBs into eras based both on the development of innovative instruments and the main discoveries of physical properties of GRBs they made.

Conflicts of Interest: The authors declare no conflicts of interest.

Abbreviations

The following abbreviations are used in this manuscript:

AGNs	Active Galactic Nuclei
ALP	Axion-Like Particle
BH	Black Hole
EAS	Extensive Air Shower experiment
EBL	Extragalactic Background Light
FOV	Field Of View
GCN	GRB Coordinate Network
GRB	Gamma-Ray Burst
GW	Gravitational Wave
IACT	Imaging Atmospheric Cherenkov Telescope
LIV	Lorentz Invariance Violation
NFI	Near-Field Instrument
NS	Neutron Star
SGR	Soft Gamma Repeater
SN	Supernovae
UHECR	Ultra-High-Energy Cosmic Ray

References

1. Klebesadel, R.W.; Strong, I.B.; Olson, R.A. Observations of Gamma-Ray Bursts of Cosmic Origin. *Astrophys. J.* **1973**, *182*, L85. [CrossRef]
2. Cline, T.L.; Desai, U.D.; Klebesadel, R.W.; Strong, I.B. Energy Spectra of Cosmic Gamma-Ray Bursts. *Astrophys. J.* **1973**, *185*, L1. [CrossRef]
3. Mazets, E.P.; Golenetskii, S.V.; Il'Inskii, V.N. Flare of cosmic gamma radiation as observed with "Cosmos-461" satellite. *Sov. J. Exp. Theor. Phys. Lett.* **1974**, *19*, 77.
4. Mazets, E.P.; Golenetskii, S.V.; Ilinskii, V.N.; Panov, V.N.; Aptekar, R.L.; Gurian, I.A.; Proskura, M.P.; Sokolov, I.A.; Sokolova, Z.I.; Kharitonova, T.V. Catalog of cosmic gamma-ray bursts from the KONUS experiment data. *Astrophys. Space Sci.* **1981**, *80*, 3–83. [CrossRef]
5. Murakami, T.; Fujii, M.; Hayashida, K.; Itoh, M.; Nishimura, J. Evidence for cyclotron absorption from spectral features in gamma-ray bursts seen with Ginga. *Nature* **1988**, *335*, 234–235. [CrossRef]
6. Fenimore, E.E.; Conner, J.P.; Epstein, R.I.; Klebesadel, R.W.; Laros, J.G.; Yoshida, A.; Fujii, M.; Hayashida, K.; Itoh, M.; Murakami, T.; et al. Interpretation of Multiple Absorption Features in a Gamma-Ray Burst Spectrum. *Astrophys. J.* **1988**, *335*, L71. [CrossRef]

7. Nemiroff, R.J. A Century of Gamma Ray Burst Models. *Comments Astrophys.* **1994**, *17*, 189. [CrossRef]
8. Colgate, S.A. Prompt gamma rays and X-rays from supernovae. *Can. J. Phys. Suppl.* **1968**, *46*, 476. [CrossRef]
9. Zhang, B.; Xu, R.X.; Qiao, G.J. Nature and Nurture: A Model for Soft Gamma-Ray Repeaters. *Astrophys. J.* **2000**, *545*, L127–L130. [CrossRef]
10. Prilutskii, O.F.; Usov, V.V. On the Nature of γ-ray Bursts. *Astrophys. Space Sci.* **1975**, *34*, 395–401. [CrossRef]
11. Ruderman, M. Theories of gamma γ-ray bursts. *Seventh Tex. Symp. Relativ. Astrophys.* **1975**, *262*, 164–180. [CrossRef]
12. Cavallo, G.; Rees, M.J. A qualitative study of cosmic fireballs and γ-ray bursts. *Mon. Not. R. Astron. Soc.* **1978**, *183*, 359–365. [CrossRef]
13. Eichler, D.; Livio, M.; Piran, T.; Schramm, D.N. Nucleosynthesis, neutrino bursts and γ-rays from coalescing neutron stars. *Nature* **1989**, *340*, 126–128. [CrossRef]
14. Fishman, G.J.; Meegan, C.A.; Wilson, R.B.; Paciesas, W.S.; Pendleton, G.N. The BATSE experiment on the Compton Gamma Ray Observatory: Status and some early results. *Nasa Conf. Publ.* **1992**, *3137*, 26–34.
15. Norris, J.P.; Nemiroff, R.J.; Bonnell, J.T.; Scargle, J.D.; Kouveliotou, C.; Paciesas, W.S.; Meegan, C.A.; Fishman, G.J. Attributes of Pulses in Long Bright Gamma-Ray Bursts. *Astrophys. J.* **1996**, *459*, 393. [CrossRef]
16. Fenimore, E.E.; in 't Zand, J.J.M.; Norris, J.P.; Bonnell, J.T.; Nemiroff, R.J. Gamma-Ray Burst Peak Duration as a Function of Energy. *Astrophys. J.* **1995**, *448*, L101. [CrossRef]
17. Kouveliotou, C.; Meegan, C.A.; Fishman, G.J.; Bhat, N.P.; Briggs, M.S.; Koshut, T.M.; Paciesas, W.S.; Pendleton, G.N. Identification of Two Classes of Gamma-Ray Bursts. *Astrophys. J.* **1993**, *413*, L101. [CrossRef]
18. Band, D.; Matteson, J.; Ford, L.; Schaefer, B.; Palmer, D.; Teegarden, B.; Cline, T.; Briggs, M.; Paciesas, W.; Pendleton, G.; et al. BATSE Observations of Gamma-Ray Burst Spectra. I. Spectral Diversity. *Astrophys. J.* **1993**, *413*, 281. [CrossRef]
19. Briggs, M.S.; Paciesas, W.S.; Pendleton, G.N.; Meegan, C.A.; Fishman, G.J.; Horack, J.M.; Brock, M.N.; Kouveliotou, C.; Hartmann, D.H.; Hakkila, J. BATSE Observations of the Large-Scale Isotropy of Gamma-Ray Bursts. *Astrophys. J.* **1996**, *459*, 40. [CrossRef]
20. Meegan, C.A.; Fishman, G.J.; Wilson, R.B.; Paciesas, W.S.; Pendleton, G.N.; Horack, J.M.; Brock, M.N.; Kouveliotou, C. Spatial distribution of γ-ray bursts observed by BATSE. *Nature* **1992**, *355*, 143–145. [CrossRef]
21. Mészáros, P.; Rees, M.J. Optical and Long-Wavelength Afterglow from Gamma-Ray Bursts. *Astrophys. J.* **1997**, *476*, 232–237. [CrossRef]
22. Woosley, S.E. Gamma-Ray Bursts from Stellar Mass Accretion Disks around Black Holes. *Astrophys. J.* **1993**, *405*, 273. [CrossRef]
23. Usov, V.V. Millisecond pulsars with extremely strong magnetic fields as a cosmological source of γ-ray bursts. *Nature* **1992**, *357*, 472–474. [CrossRef]
24. Paciesas, W.S.; Meegan, C.A.; Pendleton, G.N.; Briggs, M.S.; Kouveliotou, C.; Koshut, T.M.; Lestrade, J.P.; McCollough, M.L.; Brainerd, J.J.; Hakkila, J.; et al. The Fourth BATSE Gamma-Ray Burst Catalog (Revised). *ApJS* **1999**, *122*, 465–495. [CrossRef]
25. Piro, L.; Scarsi, L.; Butler, R.C. SAX: The wideband mission for x-ray astronomy. In Proceedings of the X-ray and EUV/FUV Spectroscopy and Polarimetry, San Diego, CA, USA, 11–12 July 1995; Society of Photo-Optical Instrumentation Engineers (SPIE) Conference Series; Fineschi, S., Ed.; Volume 2517, pp. 169–181. [CrossRef]
26. Costa, E.; Frontera, F.; Heise, J.; Feroci, M.; in't Zand, J.; Fiore, F.; Cinti, M.N.; Dal Fiume, D.; Nicastro, L.; Orlandini, M.; et al. Discovery of an X-ray afterglow associated with the γ-ray burst of 28 February 1997. *Nature* **1997**, *387*, 783–785. [CrossRef]
27. van Paradijs, J.; Groot, P.J.; Galama, T.; Kouveliotou, C.; Strom, R.G.; Telting, J.; Rutten, R.G.M.; Fishman, G.J.; Meegan, C.A.; Pettini, M.; et al. Transient optical emission from the error box of the γ-ray burst of 28 February 1997. *Nature* **1997**, *386*, 686–689. [CrossRef]
28. Frail, D.A.; Kulkarni, S.R.; Nicastro, L.; Feroci, M.; Taylor, G.B. The radio afterglow from the γ-ray burst of 8 May 1997. *Nature* **1997**, *389*, 261–263. [CrossRef]
29. Metzger, M.R.; Djorgovski, S.G.; Kulkarni, S.R.; Steidel, C.C.; Adelberger, K.L.; Frail, D.A.; Costa, E.; Frontera, F. Spectral constraints on the redshift of the optical counterpart to the γ-ray burst of 8 May 1997. *Nature* **1997**, *387*, 878–880. [CrossRef]
30. Amati, L.; Frontera, F.; Tavani, M.; in't Zand, J.J.M.; Antonelli, A.; Costa, E.; Feroci, M.; Guidorzi, C.; Heise, J.; Masetti, N.; et al. Intrinsic spectra and energetics of BeppoSAX Gamma-Ray Bursts with known redshifts. *Astron. Astrophys.* **2002**, *390*, 81–89. [CrossRef]
31. Amati, L. The $E_{p,i}$-E_{iso} correlation in gamma-ray bursts: Updated observational status, re-analysis and main implications. *Mon. Not. R. Astron. Soc.* **2006**, *372*, 233–245. [CrossRef]
32. Yonetoku, D.; Murakami, T.; Nakamura, T.; Yamazaki, R.; Inoue, A.K.; Ioka, K. Gamma-Ray Burst Formation Rate Inferred from the Spectral Peak Energy-Peak Luminosity Relation. *Astrophys. J.* **2004**, *609*, 935–951. [CrossRef]
33. Nakar, E.; Piran, T. Outliers to the peak energy-isotropic energy relation in gamma-ray bursts. *Mon. Not. R. Astron. Soc.* **2005**, *360*, L73–L76. [CrossRef]
34. Ghirlanda, G.; Nava, L.; Ghisellini, G.; Firmani, C.; Cabrera, J.I. The E_{peak}-E_{iso} plane of long gamma-ray bursts and selection effects. *Mon. Not. R. Astron. Soc.* **2008**, *387*, 319–330. [CrossRef]
35. Liang, E.W.; Dai, Z.G.; Wu, X.F. The Luminosity-E_p Relation within Gamma-Ray Bursts and the Implications for Fireball Models. *Astrophys. J.* **2004**, *606*, L29–L32. [CrossRef]
36. Zhang, B.; Zhang, B.B.; Virgili, F.J.; Liang, E.W.; Kann, D.A.; Wu, X.F.; Proga, D.; Lv, H.J.; Toma, K.; Mészáros, P.; et al. Discerning the Physical Origins of Cosmological Gamma-ray Bursts Based on Multiple Observational Criteria: The Cases of z = 6.7 GRB 080913, z = 8.2 GRB 090423, and Some Short/Hard GRBs. *Astrophys. J.* **2009**, *703*, 1696–1724. [CrossRef]

37. Ghirlanda, G.; Nava, L.; Ghisellini, G.; Celotti, A.; Firmani, C. Short versus long gamma-ray bursts: Spectra, energetics, and luminosities. *Astron. Astrophys.* **2009**, *496*, 585–595. [CrossRef]
38. Rees, M.J.; Meszaros, P. Relativistic fireballs—Energy conversion and time-scales. *Mon. Not. R. Astron. Soc.* **1992**, *258*, 41. [CrossRef]
39. Sari, R.; Piran, T.; Narayan, R. Spectra and Light Curves of Gamma-Ray Burst Afterglows. *Astrophys. J.* **1998**, *497*, L17–L20. [CrossRef]
40. Sari, R.; Piran, T. Predictions for the Very Early Afterglow and the Optical Flash. *Astrophys. J.* **1999**, *520*, 641–649. [CrossRef]
41. Harrison, F.A.; Bloom, J.S.; Frail, D.A.; Sari, R.; Kulkarni, S.R.; Djorgovski, S.G.; Axelrod, T.; Mould, J.; Schmidt, B.P.; Wieringa, M.H.; et al. Optical and Radio Observations of the Afterglow from GRB 990510: Evidence for a Jet. *Astrophys. J.* **1999**, *523*, L121–L124. [CrossRef]
42. Frail, D.A.; Kulkarni, S.R.; Sari, R.; Djorgovski, S.G.; Bloom, J.S.; Galama, T.J.; Reichart, D.E.; Berger, E.; Harrison, F.A.; Price, P.A.; et al. Beaming in Gamma-Ray Bursts: Evidence for a Standard Energy Reservoir. *Astrophys. J.* **2001**, *562*, L55–L58. [CrossRef]
43. Zhang, B.; Mészáros, P. Gamma-Ray Burst Beaming: A Universal Configuration with a Standard Energy Reservoir? *Astrophys. J.* **2002**, *571*, 876–879. [CrossRef]
44. Rossi, E.; Lazzati, D.; Rees, M.J. Afterglow light curves, viewing angle and the jet structure of γ-ray bursts. *Mon. Not. R. Astron. Soc.* **2002**, *332*, 945–950. [CrossRef]
45. Ghirlanda, G.; Ghisellini, G.; Lazzati, D. The Collimation-corrected Gamma-Ray Burst Energies Correlate with the Peak Energy of Their νF_ν Spectrum. *Astrophys. J.* **2004**, *616*, 331–338. [CrossRef]
46. Ghirlanda, G.; Ghisellini, G.; Lazzati, D.; Firmani, C. Gamma-Ray Bursts: New Rulers to Measure the Universe. *Astrophys. J.* **2004**, *613*, L13–L16. [CrossRef]
47. Fruchter, A.S.; Thorsett, S.E.; Metzger, M.R.; Sahu, K.C.; Petro, L.; Livio, M.; Ferguson, H.; Pian, E.; Hogg, D.W.; Galama, T.; et al. Hubble Space Telescope and Palomar Imaging of GRB 990123: Implications for the Nature of Gamma-Ray Bursts and Their Hosts. *Astrophys. J.* **1999**, *519*, L13–L16. [CrossRef]
48. Fryer, C.L.; Woosley, S.E.; Hartmann, D.H. Formation Rates of Black Hole Accretion Disk Gamma-Ray Bursts. *Astrophys. J.* **1999**, *526*, 152–177. [CrossRef]
49. Paczyński, B. Are Gamma-Ray Bursts in Star-Forming Regions? *Astrophys. J.* **1998**, *494*, L45–L48. [CrossRef]
50. Galama, T.J.; Vreeswijk, P.M.; van Paradijs, J.; Kouveliotou, C.; Augusteijn, T.; Böhnhardt, H.; Brewer, J.P.; Doublier, V.; Gonzalez, J.F.; Leibundgut, B.; et al. An unusual supernova in the error box of the γ-ray burst of 25 April 1998. *Nature* **1998**, *395*, 670–672. [CrossRef]
51. Kulkarni, S.R.; Frail, D.A.; Wieringa, M.H.; Ekers, R.D.; Sadler, E.M.; Wark, R.M.; Higdon, J.L.; Phinney, E.S.; Bloom, J.S. Radio emission from the unusual supernova 1998bw and its association with the γ-ray burst of 25 April 1998. *Nature* **1998**, *395*, 663–669. [CrossRef]
52. Bloom, J.S.; Kulkarni, S.R.; Djorgovski, S.G.; Eichelberger, A.C.; Côté, P.; Blakeslee, J.P.; Odewahn, S.C.; Harrison, F.A.; Frail, D.A.; Filippenko, A.V.; et al. The unusual afterglow of the γ-ray burst of 26 March 1998 as evidence for a supernova connection. *Nature* **1999**, *401*, 453–456. [CrossRef]
53. MacFadyen, A.I.; Woosley, S.E.; Heger, A. Supernovae, Jets, and Collapsars. *Astrophys. J.* **2001**, *550*, 410–425. [CrossRef]
54. Ramirez-Ruiz, E.; Celotti, A.; Rees, M.J. Events in the life of a cocoon surrounding a light, collapsar jet. *Mon. Not. R. Astron. Soc.* **2002**, *337*, 1349–1356. [CrossRef]
55. Fryer, C.L.; Woosley, S.E.; Herant, M.; Davies, M.B. Merging White Dwarf/Black Hole Binaries and Gamma-Ray Bursts. *Astrophys. J.* **1999**, *520*, 650–660. [CrossRef]
56. Salgado, M.; Bonazzola, S.; Gourgoulhon, E.; Haensel, P. High precision rotating neutron star models. I. Analysis of neutron star properties. *Astron. Astrophys.* **1994**, *291*, 155–170.
57. Vietri, M.; Stella, L. A Gamma-Ray Burst Model with Small Baryon Contamination. *Astrophys. J.* **1998**, *507*, L45–L48. [CrossRef]
58. Dar, A.; De Rújula, A. The Cannonball Model of Gamma Ray Bursts: Spectral and Temporal Properties of the Gamma Rays. *arXiv* **2000**, arXiv:astro-ph/0012227. [CrossRef]
59. Ricker, G.R.; Atteia, J.L.; Crew, G.B.; Doty, J.P.; Fenimore, E.E.; Galassi, M.; Graziani, C.; Hurley, K.; Jernigan, J.G.; Kawai, N.; et al. The High Energy Transient Explorer (HETE): Mission and Science Overview. In Proceedings of the Gamma-Ray Burst and Afterglow Astronomy 2001: A Workshop Celebrating the First Year of the HETE Mission, Woods Hole, MA, USA, 5–9 November 2001; American Institute of Physics Conference Series; Ricker, G.R., Vanderspek, R.K., Eds.; Volume 662, pp. 3–16. [CrossRef]
60. Hjorth, J.; Sollerman, J.; Møller, P.; Fynbo, J.P.U.; Woosley, S.E.; Kouveliotou, C.; Tanvir, N.R.; Greiner, J.; Andersen, M.I.; Castro-Tirado, A.J.; et al. A very energetic supernova associated with the γ-ray burst of 29 March 2003. *Nature* **2003**, *423*, 847–850. [CrossRef] [PubMed]
61. Matheson, T.; Garnavich, P.M.; Stanek, K.Z.; Bersier, D.; Holland, S.T.; Krisciunas, K.; Caldwell, N.; Berlind, P.; Bloom, J.S.; Bolte, M.; et al. Photometry and Spectroscopy of GRB 030329 and Its Associated Supernova 2003dh: The First Two Months. *Astrophys. J.* **2003**, *599*, 394–407. [CrossRef]
62. Stanek, K.Z.; Matheson, T.; Garnavich, P.M.; Martini, P.; Berlind, P.; Caldwell, N.; Challis, P.; Brown, W.R.; Schild, R.; Krisciunas, K.; et al. Spectroscopic Discovery of the Supernova 2003dh Associated with GRB 030329. *Astrophys. J.* **2003**, *591*, L17–L20. [CrossRef]

63. Heise, J.; Zand, J.I.; Kippen, R.M.; Woods, P.M. X-ray Flashes and X-ray Rich Gamma Ray Bursts. In Proceedings of the Gamma-Ray Bursts in the Afterglow Era, Rome, Italy, 17–20 October 2000; Costa, E.; Frontera, F.; Hjorth, J., Eds.; p. 16. [CrossRef]
64. Yamazaki, R.; Ioka, K.; Nakamura, T. X-ray Flashes from Off-Axis Gamma-ray Bursts. *Astrophys. J.* **2002**, *571*, L31–L35. [CrossRef]
65. Gehrels, N.; Chincarini, G.; Giommi, P.; Mason, K.O.; Nousek, J.A.; Wells, A.A.; White, N.E.; Barthelmy, S.D.; Burrows, D.N.; Cominsky, L.R.; et al. The Swift Gamma-Ray Burst Mission. *Astrophys. J.* **2004**, *611*, 1005–1020. [CrossRef]
66. Barthelmy, S.D.; Barbier, L.M.; Cummings, J.R.; Fenimore, E.E.; Gehrels, N.; Hullinger, D.; Krimm, H.A.; Markwardt, C.B.; Palmer, D.M.; Parsons, A.; et al. The Burst Alert Telescope (BAT) on the SWIFT Midex Mission. *Space Sci. Rev.* **2005**, *120*, 143–164. [CrossRef]
67. Burrows, D.N.; Hill, J.E.; Nousek, J.A.; Kennea, J.A.; Wells, A.; Osborne, J.P.; Abbey, A.F.; Beardmore, A.; Mukerjee, K.; Short, A.D.T.; et al. The Swift X-Ray Telescope. *Space Sci. Rev.* **2005**, *120*, 165–195. [CrossRef]
68. Roming, P.W.A.; Kennedy, T.E.; Mason, K.O.; Nousek, J.A.; Ahr, L.; Bingham, R.E.; Broos, P.S.; Carter, M.J.; Hancock, B.K.; Huckle, H.E.; et al. The Swift Ultra-Violet/Optical Telescope. *Space Sci. Rev.* **2005**, *120*, 95–142. [CrossRef]
69. Gehrels, N.; Sarazin, C.L.; O'Brien, P.T.; Zhang, B.; Barbier, L.; Barthelmy, S.D.; Blustin, A.; Burrows, D.N.; Cannizzo, J.; Cummings, J.R.; et al. A short γ-ray burst apparently associated with an elliptical galaxy at redshift z = 0.225. *Nature* **2005**, *437*, 851–854. [CrossRef] [PubMed]
70. Bloom, J.S.; Prochaska, J.X.; Pooley, D.; Blake, C.H.; Foley, R.J.; Jha, S.; Ramirez-Ruiz, E.; Granot, J.; Filippenko, A.V.; Sigurdsson, S.; et al. Closing in on a Short-Hard Burst Progenitor: Constraints from Early-Time Optical Imaging and Spectroscopy of a Possible Host Galaxy of GRB 050509b. *Astrophys. J.* **2006**, *638*, 354–368. [CrossRef]
71. Barthelmy, S.D.; Chincarini, G.; Burrows, D.N.; Gehrels, N.; Covino, S.; Moretti, A.; Romano, P.; O'Brien, P.T.; Sarazin, C.L.; Kouveliotou, C.; et al. An origin for short γ-ray bursts unassociated with current star formation. *Nature* **2005**, *438*, 994–996. [CrossRef]
72. Evans, P.A.; Beardmore, A.P.; Page, K.L.; Osborne, J.P.; O'Brien, P.T.; Willingale, R.; Starling, R.L.C.; Burrows, D.N.; Godet, O.; Vetere, L.; et al. Methods and results of an automatic analysis of a complete sample of Swift-XRT observations of GRBs. *Mon. Not. R. Astron. Soc.* **2009**, *397*, 1177–1201. [CrossRef]
73. O'Brien, P.T.; Willingale, R.; Osborne, J.; Goad, M.R.; Page, K.L.; Vaughan, S.; Rol, E.; Beardmore, A.; Godet, O.; Hurkett, C.P.; et al. The Early X-Ray Emission from GRBs. *Astrophys. J.* **2006**, *647*, 1213–1237. [CrossRef]
74. Tagliaferri, G.; Goad, M.; Chincarini, G.; Moretti, A.; Campana, S.; Burrows, D.N.; Perri, M.; Barthelmy, S.D.; Gehrels, N.; Krimm, H.; et al. An unexpectedly rapid decline in the X-ray afterglow emission of long γ-ray bursts. *Nature* **2005**, *436*, 985–988. [CrossRef]
75. Barthelmy, S.D.; Cannizzo, J.K.; Gehrels, N.; Cusumano, G.; Mangano, V.; O'Brien, P.T.; Vaughan, S.; Zhang, B.; Burrows, D.N.; Campana, S.; et al. Discovery of an Afterglow Extension of the Prompt Phase of Two Gamma-Ray Bursts Observed by Swift. *Astrophys. J.* **2005**, *635*, L133–L136. [CrossRef]
76. Campana, S.; Antonelli, L.A.; Chincarini, G.; Covino, S.; Cusumano, G.; Malesani, D.; Mangano, V.; Moretti, A.; Pagani, C.; Romano, P.; et al. Swift Observations of GRB 050128: The Early X-ray Afterglow. *Astrophys. J.* **2005**, *625*, L23–L26. [CrossRef]
77. Vaughan, S.; Goad, M.R.; Beardmore, A.P.; O'Brien, P.T.; Osborne, J.P.; Page, K.L.; Barthelmy, S.D.; Burrows, D.N.; Campana, S.; Cannizzo, J.K.; et al. Swift Observations of the X-ray-Bright GRB 050315. *Astrophys. J.* **2006**, *638*, 920–929. [CrossRef]
78. Burrows, D.N.; Romano, P.; Falcone, A.; Kobayashi, S.; Zhang, B.; Moretti, A.; O'Brien, P.T.; Goad, M.R.; Campana, S.; Page, K.L.; et al. Bright X-ray Flares in Gamma-Ray Burst Afterglows. *Science* **2005**, *309*, 1833–1835. [CrossRef] [PubMed]
79. Romano, P.; Campana, S.; Chincarini, G.; Cummings, J.; Cusumano, G.; Holland, S.T.; Mangano, V.; Mineo, T.; Page, K.L.; Pal'Shin, V.; et al. Panchromatic study of GRB 060124: From precursor to afterglow. *Astron. Astrophys.* **2006**, *456*, 917–927. [CrossRef]
80. Falcone, A.D.; Burrows, D.N.; Lazzati, D.; Campana, S.; Kobayashi, S.; Zhang, B.; Mészáros, P.; Page, K.L.; Kennea, J.A.; Romano, P.; et al. The Giant X-ray Flare of GRB 050502B: Evidence for Late-Time Internal Engine Activity. *Astrophys. J.* **2006**, *641*, 1010–1017. [CrossRef]
81. Zhang, B.; Fan, Y.Z.; Dyks, J.; Kobayashi, S.; Mészáros, P.; Burrows, D.N.; Nousek, J.A.; Gehrels, N. Physical Processes Shaping Gamma-Ray Burst X-ray Afterglow Light Curves: Theoretical Implications from the Swift X-ray Telescope Observations. *Astrophys. J.* **2006**, *642*, 354–370. [CrossRef]
82. Nousek, J.A.; Kouveliotou, C.; Grupe, D.; Page, K.L.; Granot, J.; Ramirez-Ruiz, E.; Patel, S.K.; Burrows, D.N.; Mangano, V.; Barthelmy, S.; et al. Evidence for a Canonical Gamma-Ray Burst Afterglow Light Curve in the Swift XRT Data. *Astrophys. J.* **2006**, *642*, 389–400. [CrossRef]
83. Panaitescu, A.; Mészáros, P.; Burrows, D.; Nousek, J.; Gehrels, N.; O'Brien, P.; Willingale, R. Evidence for chromatic X-ray light-curve breaks in Swift gamma-ray burst afterglows and their theoretical implications. *Mon. Not. R. Astron. Soc.* **2006**, *369*, 2059–2064. [CrossRef]
84. Zhang, B.; Mészáros, P. Gamma-Ray Burst Afterglow with Continuous Energy Injection: Signature of a Highly Magnetized Millisecond Pulsar. *Astrophys. J.* **2001**, *552*, L35–L38. [CrossRef]
85. King, A.; O'Brien, P.T.; Goad, M.R.; Osborne, J.; Olsson, E.; Page, K. Gamma-Ray Bursts: Restarting the Engine. *Astrophys. J.* **2005**, *630*, L113–L115. [CrossRef]
86. Perna, R.; Armitage, P.J.; Zhang, B. Flares in Long and Short Gamma-Ray Bursts: A Common Origin in a Hyperaccreting Accretion Disk. *Astrophys. J.* **2006**, *636*, L29–L32. [CrossRef]

87. Dai, Z.G.; Wang, X.Y.; Wu, X.F.; Zhang, B. X-ray Flares from Postmerger Millisecond Pulsars. *Science* **2006**, *311*, 1127–1129. [CrossRef] [PubMed]
88. Metzger, B.D.; Quataert, E.; Thompson, T.A. Short-duration gamma-ray bursts with extended emission from protomagnetar spin-down. *Mon. Not. R. Astron. Soc.* **2008**, *385*, 1455–1460. [CrossRef]
89. Liang, E.; Zhang, B.; Virgili, F.; Dai, Z.G. Low-Luminosity Gamma-Ray Bursts as a Unique Population: Luminosity Function, Local Rate, and Beaming Factor. *Astrophys. J.* **2007**, *662*, 1111–1118. [CrossRef]
90. Totani, T.; Kawai, N.; Kosugi, G.; Aoki, K.; Yamada, T.; Iye, M.; Ohta, K.; Hattori, T. Implications for Cosmic Reionization from the Optical Afterglow Spectrum of the Gamma-Ray Burst 050904 at z = 6.3*. *Publ. Astron. Soc. Jpn.* **2006**, *58*, 485–498. [CrossRef]
91. Cusumano, G.; Mangano, V.; Chincarini, G.; Panaitescu, A.; Burrows, D.N.; La Parola, V.; Sakamoto, T.; Campana, S.; Mineo, T.; Tagliaferri, G.; et al. GRB 050904: The oldest cosmic explosion ever observed in the Universe. In Proceedings of the Gamma-Ray Bursts in the Swift Era: Sixteenth Maryland Astrophysics Conference, Washington, DC, USA, 29 November–2 December 2005; American Institute of Physics Conference Series; Holt, S.S., Gehrels, N., Nousek, J.A., Eds.; Volume 836, pp. 564–569. [CrossRef]
92. Greiner, J.; Krühler, T.; Fynbo, J.P.U.; Rossi, A.; Schwarz, R.; Klose, S.; Savaglio, S.; Tanvir, N.R.; McBreen, S.; Totani, T.; et al. GRB 080913 at Redshift 6.7. *Astrophys. J.* **2009**, *693*, 1610–1620. [CrossRef]
93. Tanvir, N.R.; Fox, D.B.; Levan, A.J.; Berger, E.; Wiersema, K.; Fynbo, J.P.U.; Cucchiara, A.; Krühler, T.; Gehrels, N.; Bloom, J.S.; et al. A γ-ray burst at a redshift of z$\sim$8.2. *Nature* **2009**, *461*, 1254–1257. [CrossRef]
94. Salvaterra, R.; Della Valle, M.; Campana, S.; Chincarini, G.; Covino, S.; D'Avanzo, P.; Fernández-Soto, A.; Guidorzi, C.; Mannucci, F.; Margutti, R.; et al. GRB090423 at a redshift of z$\sim$8.1. *Nature* **2009**, *461*, 1258–1260. [CrossRef]
95. Cucchiara, A.; Levan, A.J.; Fox, D.B.; Tanvir, N.R.; Ukwatta, T.N.; Berger, E.; Krühler, T.; Küpcü Yoldaş, A.; Wu, X.F.; Toma, K.; et al. A Photometric Redshift of z$\sim$9.4 for GRB 090429B. *Astrophys. J.* **2011**, *736*, 7. [CrossRef]
96. Gehrels, N.; Norris, J.P.; Barthelmy, S.D.; Granot, J.; Kaneko, Y.; Kouveliotou, C.; Markwardt, C.B.; Mészáros, P.; Nakar, E.; Nousek, J.A.; et al. A new γ-ray burst classification scheme from GRB060614. *Nature* **2006**, *444*, 1044–1046. [CrossRef]
97. Zhang, B.; Zhang, B.B.; Liang, E.W.; Gehrels, N.; Burrows, D.N.; Mészáros, P. Making a Short Gamma-Ray Burst from a Long One: Implications for the Nature of GRB 060614. *Astrophys. J.* **2007**, *655*, L25–L28. [CrossRef]
98. Levesque, E.M.; Kewley, L.J.; Berger, E.; Zahid, H.J. The Host Galaxies of Gamma-ray Bursts. II. A Mass-metallicity Relation for Long-duration Gamma-ray Burst Host Galaxies. *Astron. J.* **2010**, *140*, 1557–1566. [CrossRef]
99. Campana, S.; Mangano, V.; Blustin, A.J.; Brown, P.; Burrows, D.N.; Chincarini, G.; Cummings, J.R.; Cusumano, G.; Della Valle, M.; Malesani, D.; et al. The association of GRB 060218 with a supernova and the evolution of the shock wave. *Nature* **2006**, *442*, 1008–1010. [CrossRef] [PubMed]
100. Racusin, J.L.; Karpov, S.V.; Sokolowski, M.; Granot, J.; Wu, X.F.; Pal'Shin, V.; Covino, S.; van der Horst, A.J.; Oates, S.R.; Schady, P.; et al. Broadband observations of the naked-eye γ-ray burst GRB080319B. *Nature* **2008**, *455*, 183–188. [CrossRef] [PubMed]
101. Levan, A.J.; Tanvir, N.R.; Starling, R.L.C.; Wiersema, K.; Page, K.L.; Perley, D.A.; Schulze, S.; Wynn, G.A.; Chornock, R.; Hjorth, J.; et al. A New Population of Ultra-long Duration Gamma-Ray Bursts. *Astrophys. J.* **2014**, *781*, 13. [CrossRef]
102. Bloom, J.S.; Giannios, D.; Metzger, B.D.; Cenko, S.B.; Perley, D.A.; Butler, N.R.; Tanvir, N.R.; Levan, A.J.; O'Brien, P.T.; Strubbe, L.E.; et al. A Possible Relativistic Jetted Outburst from a Massive Black Hole Fed by a Tidally Disrupted Star. *Science* **2011**, *333*, 203. [CrossRef] [PubMed]
103. Burrows, D.N.; Kennea, J.A.; Ghisellini, G.; Mangano, V.; Zhang, B.; Page, K.L.; Eracleous, M.; Romano, P.; Sakamoto, T.; Falcone, A.D.; et al. Relativistic jet activity from the tidal disruption of a star by a massive black hole. *Nature* **2011**, *476*, 421–424. [CrossRef] [PubMed]
104. Levan, A.J.; Butler, N.; Bloom, J.; Tanvir, N.R.; Fruchter, A.S. GRB 110328/Swift J164449.3+573451: Chandra observations. *GRB Coord. Netw.* **2011**, *11886*, 1.
105. Zauderer, B.A.; Berger, E.; Soderberg, A.M.; Loeb, A.; Narayan, R.; Frail, D.A.; Petitpas, G.R.; Brunthaler, A.; Chornock, R.; Carpenter, J.M.; et al. Birth of a relativistic outflow in the unusual γ-ray transient Swift J164449.3+573451. *Nature* **2011**, *476*, 425–428. [CrossRef]
106. Kanbach, G.; Bertsch, D.L.; Fichtel, C.E.; Hartman, R.C.; Hunter, S.D.; Kniffen, D.A.; Hughlock, B.W.; Favale, A.; Hofstadter, R.; Hughes, E.B. The project EGRET (energetic gamma-ray experiment telescope) on NASA's Gamma-Ray Observatory GRO. *Space Sci. Rev.* **1989**, *49*, 69–84. [CrossRef]
107. Hurley, K.; Dingus, B.L.; Mukherjee, R.; Sreekumar, P.; Kouveliotou, C.; Meegan, C.; Fishman, G.J.; Band, D.; Ford, L.; Bertsch, D.; et al. Detection of a γ-ray burst of very long duration and very high energy. *Nature* **1994**, *372*, 652–654. [CrossRef]
108. Kouveliotou, C.; Preece, R.; Bhat, N.; Fishman, G.J.; Meegan, C.A.; Horack, J.M.; Briggs, M.S.; Paciesas, W.S.; Pendleton, G.N.; Band, D.; et al. BATSE Observations of the Very Intense Gamma-Ray Burst GRB 930131. *Astrophys. J.* **1994**, *422*, L59. [CrossRef]
109. Sommer, M.; Bertsch, D.L.; Dingus, B.L.; Fichtel, C.E.; Fishman, G.J.; Harding, A.K.; Hartman, R.C.; Hunter, S.D.; Hurley, K.; Kanbach, G.; et al. High-Energy Gamma Rays from the Intense 1993 January 31 Gamma-Ray Burst. *Astrophys. J.* **1994**, *422*, L63. [CrossRef]
110. González, M.M.; Dingus, B.L.; Kaneko, Y.; Preece, R.D.; Dermer, C.D.; Briggs, M.S. A γ-ray burst with a high-energy spectral component inconsistent with the synchrotron shock model. *Nature* **2003**, *424*, 749–751. [CrossRef]
111. Tavani, M.; Barbiellini, G.; Argan, A.; Boffelli, F.; Bulgarelli, A.; Caraveo, P.; Cattaneo, P.W.; Chen, A.W.; Cocco, V.; Costa, E.; et al. The AGILE Mission. *Astron. Astrophys.* **2009**, *502*, 995–1013. [CrossRef]

112. Barbiellini, G.; Bordignon, G.; Fedel, G.; Liello, F.; Longo, F.; Pontoni, C.; Prest, M.; Vallazza, E. The next generation of high-energy gamma-ray detectors for satellites: The AGILE silicon tracker. In Proceedings of the Gamma 2001: Gamma-Ray Astrophysics, Baltimore, MD, USA, 4–6 April 2001; American Institute of Physics Conference Series; Ritz, S., Gehrels, N., Shrader, C.R., Eds.; Volume 587, pp. 754–758. [CrossRef]
113. Costa, E.; Barbanera, L.; Feroci, M.; Frutti, M.; Lapshov, I.; Martino, B.; Mastropietro, M.; Morelli, E.; Rapisarda, M.; Rubini, A.; et al. Super-agile-The X-ray detector for the gamma-ray mission agile. In Proceedings of the X-ray Astronomy: Stellar Endpoints, AGN, and the Diffuse X-ray Background, Bologna, Italy, 6–10 September 1999; American Institute of Physics Conference Series; White, N.E., Malaguti, G., Palumbo, G.G.C., Eds.; Volume 599, pp. 582–585. [CrossRef]
114. Parmiggiani, N.; Bulgarelli, A.; Ursi, A.; Addis, A.; Baroncelli, L.; Fioretti, V.; Di Piano, A.; Panebianco, G.; Tavani, M.; Pittori, C.; et al. The AGILE real-time analysis software system to detect short-transient events in the multi-messenger era. *Astron. Comput.* **2023**, *44*, 100726. [CrossRef]
115. Galli, M.; Marisaldi, M.; Fuschino, F.; Labanti, C.; Argan, A.; Barbiellini, G.; Bulgarelli, A.; Cattaneo, P.W.; Colafrancesco, S.; Del Monte, E.; et al. AGILE mini-calorimeter gamma-ray burst catalog. *Astron. Astrophys.* **2013**, *553*, A33. [CrossRef]
116. Ursi, A.; Romani, M.; Verrecchia, F.; Pittori, C.; Tavani, M.; Marisaldi, M.; Galli, M.; Labanti, C.; Parmiggiani, N.; Bulgarelli, A.; et al. The Second AGILE MCAL Gamma-Ray Burst Catalog: 13 yr of Observations. *Astrophys. J.* **2022**, *925*, 152. [CrossRef]
117. Longo, F.; Moretti, E.; Nava, L.; Desiante, R.; Olivo, M.; Del Monte, E.; Rappoldi, A.; Fuschino, F.; Marisaldi, M.; Giuliani, A.; et al. Upper limits on the high-energy emission from gamma-ray bursts observed by AGILE-GRID. *Astron. Astrophys.* **2012**, *547*, A95. [CrossRef]
118. Giuliani, A.; Mereghetti, S.; Fornari, F.; Del Monte, E.; Feroci, M.; Marisaldi, M.; Esposito, P.; Perotti, F.; Tavani, M.; Argan, A.; et al. AGILE detection of delayed gamma-ray emission from GRB 080514B. *Astron. Astrophys.* **2008**, *491*, L25–L28. [CrossRef]
119. Giuliani, A.; Fuschino, F.; Vianello, G.; Marisaldi, M.; Mereghetti, S.; Tavani, M.; Cutini, S.; Barbiellini, G.; Longo, F.; Moretti, E.; et al. AGILE Detection of Delayed Gamma-ray Emission From the Short Gamma-Ray Burst GRB 090510. *Astrophys. J.* **2010**, *708*, L84–L88. [CrossRef]
120. Ackermann, M.; Asano, K.; Atwood, W.B.; Axelsson, M.; Baldini, L.; Ballet, J.; Barbiellini, G.; Baring, M.G.; Bastieri, D.; Bechtol, K.; et al. Fermi Observations of GRB 090510: A Short-Hard Gamma-ray Burst with an Additional, Hard Power-law Component from 10 keV TO GeV Energies. *Astrophys. J.* **2010**, *716*, 1178–1190. [CrossRef]
121. De Pasquale, M.; Schady, P.; Kuin, N.P.M.; Page, M.J.; Curran, P.A.; Zane, S.; Oates, S.R.; Holland, S.T.; Breeveld, A.A.; Hoversten, E.A.; et al. Swift and Fermi Observations of the Early Afterglow of the Short Gamma-Ray Burst 090510. *Astrophys. J.* **2010**, *709*, L146–L151. [CrossRef]
122. Parmiggiani, N.; Bulgarelli, A.; Fioretti, V.; Di Piano, A.; Giuliani, A.; Longo, F.; Verrecchia, F.; Tavani, M.; Beneventano, D.; Macaluso, A. A Deep Learning Method for AGILE-GRID Gamma-Ray Burst Detection. *Astrophys. J.* **2021**, *914*, 67. [CrossRef]
123. Parmiggiani, N.; Bulgarelli, A.; Ursi, A.; Macaluso, A.; Di Piano, A.; Fioretti, V.; Aboudan, A.; Baroncelli, L.; Addis, A.; Tavani, M.; et al. A Deep-learning Anomaly-detection Method to Identify Gamma-Ray Bursts in the Ratemeters of the AGILE Anticoincidence System. *Astrophys. J.* **2023**, *945*, 106. [CrossRef]
124. Atwood, W.B.; Abdo, A.A.; Ackermann, M.; Althouse, W.; Anderson, B.; Axelsson, M.; Baldini, L.; Ballet, J.; Band, D.L.; Barbiellini, G.; et al. The Large Area Telescope on the Fermi Gamma-Ray Space Telescope Mission. *Astrophys. J.* **2009**, *697*, 1071–1102. [CrossRef]
125. Meegan, C.; Lichti, G.; Bhat, P.N.; Bissaldi, E.; Briggs, M.S.; Connaughton, V.; Diehl, R.; Fishman, G.; Greiner, J.; Hoover, A.S.; et al. The Fermi Gamma-ray Burst Monitor. *Astrophys. J.* **2009**, *702*, 791–804. [CrossRef]
126. von Kienlin, A.; Meegan, C.A.; Paciesas, W.S.; Bhat, P.N.; Bissaldi, E.; Briggs, M.S.; Burns, E.; Cleveland, W.H.; Gibby, M.H.; Giles, M.M.; et al. The Fourth Fermi-GBM Gamma-Ray Burst Catalog: A Decade of Data. *Astrophys. J.* **2020**, *893*, 46. [CrossRef]
127. Ajello, M.; Arimoto, M.; Axelsson, M.; Baldini, L.; Barbiellini, G.; Bastieri, D.; Bellazzini, R.; Bhat, P.N.; Bissaldi, E.; Blandford, R.D.; et al. A Decade of Gamma-Ray Bursts Observed by Fermi-LAT: The Second GRB Catalog. *Astrophys. J.* **2019**, *878*, 52. [CrossRef]
128. Ghisellini, G.; Ghirlanda, G.; Nava, L.; Celotti, A. GeV emission from gamma-ray bursts: A radiative fireball? *Mon. Not. R. Astron. Soc.* **2010**, *403*, 926–937. [CrossRef]
129. Nava, L.; Vianello, G.; Omodei, N.; Ghisellini, G.; Ghirlanda, G.; Celotti, A.; Longo, F.; Desiante, R.; Barniol Duran, R. Clustering of LAT light curves: A clue to the origin of high-energy emission in gamma-ray bursts. *Mon. Not. R. Astron. Soc.* **2014**, *443*, 3578–3585. [CrossRef]
130. Kumar, P.; Barniol Duran, R. On the generation of high-energy photons detected by the Fermi Satellite from gamma-ray bursts. *Mon. Not. R. Astron. Soc.* **2009**, *400*, L75–L79. [CrossRef]
131. Kumar, P.; Barniol Duran, R. External forward shock origin of high-energy emission for three gamma-ray bursts detected by Fermi. *Mon. Not. R. Astron. Soc.* **2010**, *409*, 226–236. [CrossRef]
132. Abdo, A.A.; Ackermann, M.; Ajello, M.; Asano, K.; Atwood, W.B.; Axelsson, M.; Baldini, L.; Ballet, J.; Barbiellini, G.; Baring, M.G.; et al. A limit on the variation of the speed of light arising from quantum gravity effects. *Nature* **2009**, *462*, 331–334. [CrossRef]
133. Abdo, A.A.; Ackermann, M.; Arimoto, M.; Asano, K.; Atwood, W.B.; Axelsson, M.; Baldini, L.; Ballet, J.; Band, D.L.; Barbiellini, G.; et al. Fermi Observations of High-Energy Gamma-Ray Emission from GRB 080916C. *Science* **2009**, *323*, 1688. [CrossRef] [PubMed]

134. Ackermann, M.; Ajello, M.; Asano, K.; Atwood, W.B.; Axelsson, M.; Baldini, L.; Ballet, J.; Barbiellini, G.; Baring, M.G.; Bastieri, D.; et al. Fermi-LAT Observations of the Gamma-Ray Burst GRB 130427A. *Science* **2014**, *343*, 42–47. [CrossRef]
135. Nava, L.; Desiante, R.; Longo, F.; Celotti, A.; Omodei, N.; Vianello, G.; Bissaldi, E.; Piran, T. Constraints on the bulk Lorentz factor of gamma-ray burst jets from Fermi/LAT upper limits. *Mon. Not. R. Astron. Soc.* **2017**, *465*, 811–819. [CrossRef]
136. Razzaque, S.; Dermer, C.D.; Finke, J.D. The Stellar Contribution to the Extragalactic Background Light and Absorption of High-Energy Gamma Rays. *Astrophys. J.* **2009**, *697*, 483–492. [CrossRef]
137. Zhang, B.; Pe'er, A. Evidence of an Initially Magnetically Dominated Outflow in GRB 080916C. *Astrophys. J.* **2009**, *700*, L65–L68. [CrossRef]
138. Zhang, B.; Yan, H. The Internal-collision-induced Magnetic Reconnection and Turbulence (ICMART) Model of Gamma-ray Bursts. *Astrophys. J.* **2011**, *726*, 90. [CrossRef]
139. Guiriec, S.; Kouveliotou, C.; Daigne, F.; Zhang, B.; Hascoët, R.; Nemmen, R.S.; Thompson, D.J.; Bhat, P.N.; Gehrels, N.; Gonzalez, M.M.; et al. Toward a Better Understanding of the GRB Phenomenon: A New Model for GRB Prompt Emission and its Effects on the New L_i^{NT}- $E_{peak,i}^{rest,NT}$ Relation. *Astrophys. J.* **2015**, *807*, 148. [CrossRef]
140. Ryde, F.; Axelsson, M.; Zhang, B.B.; McGlynn, S.; Pe'er, A.; Lundman, C.; Larsson, S.; Battelino, M.; Zhang, B.; Bissaldi, E.; et al. Identification and Properties of the Photospheric Emission in GRB090902B. *Astrophys. J.* **2010**, *709*, L172–L177. [CrossRef]
141. Guiriec, S.; Connaughton, V.; Briggs, M.S.; Burgess, M.; Ryde, F.; Daigne, F.; Mészáros, P.; Goldstein, A.; McEnery, J.; Omodei, N.; et al. Detection of a Thermal Spectral Component in the Prompt Emission of GRB 100724B. *Astrophys. J.* **2011**, *727*, L33. [CrossRef]
142. Ackermann, M.; Ajello, M.; Atwood, W.B.; Baldini, L.; Ballet, J.; Barbiellini, G.; Bastieri, D.; Baughman, B.M.; Bechtol, K.; Bellardi, F.; et al. Fermi LAT observations of cosmic-ray electrons from 7 GeV to 1 TeV. *Phys. Rev. D* **2010**, *82*, 092004. [CrossRef]
143. Oganesyan, G.; Nava, L.; Ghirlanda, G.; Melandri, A.; Celotti, A. Prompt optical emission as a signature of synchrotron radiation in gamma-ray bursts. *Astron. Astrophys.* **2019**, *628*, A59. [CrossRef]
144. Oganesyan, G.; Nava, L.; Ghirlanda, G.; Celotti, A. Characterization of gamma-ray burst prompt emission spectra down to soft X-rays. *Astron. Astrophys.* **2018**, *616*, A138. [CrossRef]
145. Waxman, E. Cosmological Gamma-Ray Bursts and the Highest Energy Cosmic Rays. *Phys. Rev. Lett.* **1995**, *75*, 386–389. [CrossRef]
146. Vietri, M. The Acceleration of Ultra–High-Energy Cosmic Rays in Gamma-Ray Bursts. *Astrophys. J.* **1995**, *453*, 883. [CrossRef]
147. Milgrom, M.; Usov, V. Possible Association of Ultra–High-Energy Cosmic-Ray Events with Strong Gamma-Ray Bursts. *Astrophys. J.* **1995**, *449*, L37. [CrossRef]
148. Waxman, E.; Bahcall, J. High Energy Neutrinos from Cosmological Gamma-Ray Burst Fireballs. *Phys. Rev. Lett.* **1997**, *78*, 2292–2295. [CrossRef]
149. Murase, K.; Ioka, K.; Nagataki, S.; Nakamura, T. High-energy cosmic-ray nuclei from high- and low-luminosity gamma-ray bursts and implications for multimessenger astronomy. *Phys. Rev. D* **2008**, *78*, 023005. [CrossRef]
150. Zhang, B.; Kumar, P. Model-Dependent High-Energy Neutrino Flux from Gamma-Ray Bursts. *Phys. Rev. Lett.* **2013**, *110*, 121101. [CrossRef] [PubMed]
151. Aartsen, M.G.; Ackermann, M.; Adams, J.; Aguilar, J.A.; Ahlers, M.; Ahrens, M.; Samarai, I.A.; Altmann, D.; Andeen, K.; Anderson, T.; et al. Extending the Search for Muon Neutrinos Coincident with Gamma-Ray Bursts in IceCube Data. *Astrophys. J.* **2017**, *843*, 112. [CrossRef]
152. Albert, A.; André, M.; Anghinolfi, M.; Anton, G.; Ardid, M.; Aubert, J.J.; Avgitas, T.; Baret, B.; Barrios-Martí, J.; Basa, S.; et al. Search for high-energy neutrinos from bright GRBs with ANTARES. *Mon. Not. R. Astron. Soc.* **2017**, *469*, 906–915. [CrossRef]
153. Albert, A.; André, M.; Anghinolfi, M.; Anton, G.; Ardid, M.; Aubert, J.J.; Aublin, J.; Baret, B.; Basa, S.; Belhorma, B.; et al. Constraining the contribution of Gamma-Ray Bursts to the high-energy diffuse neutrino flux with 10 yr of ANTARES data. *Mon. Not. R. Astron. Soc.* **2021**, *500*, 5614–5628. [CrossRef]
154. Taylor, J.H.; Weisberg, J.M. Further Experimental Tests of Relativistic Gravity Using the Binary Pulsar PSR 1913+16. *Astrophys. J.* **1989**, *345*, 434. [CrossRef]
155. Abbott, B.P.; Abbott, R.; Abbott, T.D.; Acernese, F.; Ackley, K.; Adams, C.; Adams, T.; Addesso, P.; Adhikari, R.X.; Adya, V.B.; et al. GW170817: Observation of Gravitational Waves from a Binary Neutron Star Inspiral. *Phys. Rev. Lett.* **2017**, *119*, 161101. [CrossRef]
156. Abbott, B.P.; Abbott, R.; Abbott, T.D.; Acernese, F.; Ackley, K.; Adams, C.; Adams, T.; Addesso, P.; Adhikari, R.X.; Adya, V.B.; et al. Multi-messenger Observations of a Binary Neutron Star Merger. *Astrophys. J.* **2017**, *848*, L12. [CrossRef]
157. Abbott, B.P.; Abbott, R.; Abbott, T.D.; Acernese, F.; Ackley, K.; Adams, C.; Adams, T.; Addesso, P.; Adhikari, R.X.; Adya, V.B.; et al. Gravitational Waves and Gamma-Rays from a Binary Neutron Star Merger: GW170817 and GRB 170817A. *Astrophys. J.* **2017**, *848*, L13. [CrossRef]
158. Goldstein, A.; Veres, P.; Burns, E.; Briggs, M.S.; Hamburg, R.; Kocevski, D.; Wilson-Hodge, C.A.; Preece, R.D.; Poolakkil, S.; Roberts, O.J.; et al. An Ordinary Short Gamma-Ray Burst with Extraordinary Implications: Fermi-GBM Detection of GRB 170817A. *Astrophys. J.* **2017**, *848*, L14. [CrossRef]
159. Coulter, D.A.; Foley, R.J.; Kilpatrick, C.D.; Drout, M.R.; Piro, A.L.; Shappee, B.J.; Siebert, M.R.; Simon, J.D.; Ulloa, N.; Kasen, D.; et al. Swope Supernova Survey 2017a (SSS17a), the optical counterpart to a gravitational wave source. *Science* **2017**, *358*, 1556–1558. [CrossRef]

160. Pian, E.; D'Avanzo, P.; Benetti, S.; Branchesi, M.; Brocato, E.; Campana, S.; Cappellaro, E.; Covino, S.; D'Elia, V.; Fynbo, J.P.U.; et al. Spectroscopic identification of r-process nucleosynthesis in a double neutron-star merger. *Nature* **2017**, *551*, 67–70. [CrossRef] [PubMed]
161. Evans, P.A.; Cenko, S.B.; Kennea, J.A.; Emery, S.W.K.; Kuin, N.P.M.; Korobkin, O.; Wollaeger, R.T.; Fryer, C.L.; Madsen, K.K.; Harrison, F.A.; et al. Swift and NuSTAR observations of GW170817: Detection of a blue kilonova. *Science* **2017**, *358*, 1565–1570. [CrossRef] [PubMed]
162. Shappee, B.J.; Simon, J.D.; Drout, M.R.; Piro, A.L.; Morrell, N.; Prieto, J.L.; Kasen, D.; Holoien, T.W.S.; Kollmeier, J.A.; Kelson, D.D.; et al. Early spectra of the gravitational wave source GW170817: Evolution of a neutron star merger. *Science* **2017**, *358*, 1574–1578. [CrossRef]
163. Smartt, S.J.; Chen, T.W.; Jerkstrand, A.; Coughlin, M.; Kankare, E.; Sim, S.A.; Fraser, M.; Inserra, C.; Maguire, K.; Chambers, K.C.; et al. A kilonova as the electromagnetic counterpart to a gravitational-wave source. *Nature* **2017**, *551*, 75–79. [CrossRef]
164. Nicholl, M.; Berger, E.; Kasen, D.; Metzger, B.D.; Elias, J.; Briceño, C.; Alexander, K.D.; Blanchard, P.K.; Chornock, R.; Cowperthwaite, P.S.; et al. The Electromagnetic Counterpart of the Binary Neutron Star Merger LIGO/Virgo GW170817. III. Optical and UV Spectra of a Blue Kilonova from Fast Polar Ejecta. *Astrophys. J.* **2017**, *848*, L18. [CrossRef]
165. Chornock, R.; Berger, E.; Kasen, D.; Cowperthwaite, P.S.; Nicholl, M.; Villar, V.A.; Alexander, K.D.; Blanchard, P.K.; Eftekhari, T.; Fong, W.; et al. The Electromagnetic Counterpart of the Binary Neutron Star Merger LIGO/Virgo GW170817. IV. Detection of Near-infrared Signatures of r-process Nucleosynthesis with Gemini-South. *Astrophys. J.* **2017**, *848*, L19. [CrossRef]
166. Kulkarni, S.R. Modeling Supernova-like Explosions Associated with Gamma-ray Bursts with Short Durations. *arXiv* **2005**, arXiv:astro-ph/0510256. [CrossRef]
167. Li, L.X.; Paczyński, B. Transient Events from Neutron Star Mergers. *Astrophys. J.* **1998**, *507*, L59–L62. [CrossRef]
168. Piran, T.; Nakar, E.; Rosswog, S. The electromagnetic signals of compact binary mergers. *Mon. Not. R. Astron. Soc.* **2013**, *430*, 2121–2136. [CrossRef]
169. Kobayashi, S.; Mészáros, P. Gravitational Radiation from Gamma-Ray Burst Progenitors. *Astrophys. J.* **2003**, *589*, 861–870. [CrossRef]
170. Ezquiaga, J.M.; Zumalacárregui, M. Dark Energy After GW170817: Dead Ends and the Road Ahead. *Phys. Rev. Lett.* **2017**, *119*, 251304. [CrossRef] [PubMed]
171. Meszaros, P.; Rees, M.J.; Papathanassiou, H. Spectral Properties of Blast-Wave Models of Gamma-Ray Burst Sources. *Astrophys. J.* **1994**, *432*, 181. [CrossRef]
172. Mészáros, P.; Razzaque, S.; Zhang, B. GeV-TeV emission from γ-ray bursts. *New A Rev.* **2004**, *48*, 445–451. [CrossRef]
173. Fan, Y.Z.; Piran, T. High-energy γ-ray emission from gamma-ray bursts—Before GLAST. *Front. Phys. China* **2008**, *3*, 306–330. [CrossRef]
174. Inoue, S.; Granot, J.; O'Brien, P.T.; Asano, K.; Bouvier, A.; Carosi, A.; Connaughton, V.; Garczarczyk, M.; Gilmore, R.; Hinton, J.; et al. Gamma-ray burst science in the era of the Cherenkov Telescope Array. *Astropart. Phys.* **2013**, *43*, 252–275. [CrossRef]
175. Nava, L. High-energy emission from gamma-ray bursts. *Int. J. Mod. Phys. D* **2018**, *27*, 1842003. [CrossRef]
176. Padilla, L.; Funk, B.; Krawczynski, H.; Contreras, J.L.; Moralejo, A.; Aharonian, F.; Akhperjanian, A.G.; Barrio, J.A.; Beteta, J.G.; Cortina, J.; et al. Search for gamma-ray bursts above 20 TeV with the HEGRA AIROBICC Cherenkov array. *Astron. Astrophys.* **1998**, *337*, 43–50.
177. Atkins, R.; Benbow, W.; Berley, D.; Chen, M.L.; Coyne, D.G.; Dingus, B.L.; Dorfan, D.E.; Ellsworth, R.W.; Evans, D.; Falcone, A.; et al. Evidence for TEV Emission from GRB 970417A. *Astrophys. J.* **2000**, *533*, L119–L122. [CrossRef]
178. Atkins, R.; Benbow, W.; Berley, D.; Blaufuss, E.; Bussons, J.; Coyne, D.G.; Delay, R.S.; De Young, T.; Dingus, B.L.; Dorfan, D.E.; et al. Observation of TeV Gamma Rays from the Crab Nebula with Milagro Using a New Background Rejection Technique. *Astrophys. J.* **2003**, *595*, 803–811. [CrossRef]
179. Lund, N. The WATCH gamma-burst detector for EURECA-1. In Proceedings of the X-ray Instrumentation in Astronomy, Cannes, France, 2–4 December 1985; p. 55.
180. Castro-Tirado, A.J.; Brandt, S.; Lund, N.; Guziy, S.S. Optical Follow-Up of Gamma-ray Bursts Observed by WATCH. In Proceedings of the Gamma-Ray Bursts, Huntsville, AL, USA, 20–22 October 1993; American Institute of Physics Conference Series; Fishman, G.J., Ed.; Volume 307, p. 404. [CrossRef]
181. Albert, J.; Aliu, E.; Anderhub, H.; Antoranz, P.; Armada, A.; Asensio, M.; Baixeras, C.; Barrio, J.A.; Bartelt, M.; Bartko, H.; et al. Flux Upper Limit on Gamma-Ray Emission by GRB 050713a from MAGIC Telescope Observations. *Astrophys. J.* **2006**, *641*, L9–L12. [CrossRef]
182. Aleksić, J.; Ansoldi, S.; Antonelli, L.A.; Antoranz, P.; Babic, A.; Barres de Almeida, U.; Barrio, J.A.; Becerra González, J.; Bednarek, W.; Berger, K.; et al. MAGIC upper limits on the GRB 090102 afterglow. *Mon. Not. R. Astron. Soc.* **2014**, *437*, 3103–3111. [CrossRef]
183. Abramowski, A. et al. [H. E. S. S. Collaboration]. Search for TeV Gamma-ray Emission from GRB 100621A, an extremely bright GRB in X-rays, with H.E.S.S. *Astron. Astrophys.* **2014**, *565*, A16. [CrossRef]
184. Aliu, E.; Aune, T.; Barnacka, A.; Beilicke, M.; Benbow, W.; Berger, K.; Biteau, J.; Buckley, J.H.; Bugaev, V.; Byrum, K.; et al. Constraints on Very High Energy Emission from GRB 130427A. *Astrophys. J.* **2014**, *795*, L3. [CrossRef]
185. Abeysekara, A.U.; Alfaro, R.; Alvarez, C.; Álvarez, J.D.; Arceo, R.; Arteaga-Velázquez, J.C.; Ayala Solares, H.A.; Barber, A.S.; Baughman, B.M.; Bautista-Elivar, N.; et al. Search for Gamma-Rays from the Unusually Bright GRB 130427A with the HAWC Gamma-Ray Observatory. *Astrophys. J.* **2015**, *800*, 78. [CrossRef]

186. Mirzoyan, R. First time detection of a GRB at sub-TeV energies; MAGIC detects the GRB 190114C. *Astron. Telegr.* **2019**, *12390*, 1.
187. Acciari, V.A. et al. [MAGIC Collaboration]. Teraelectronvolt emission from the γ-ray burst GRB 190114C. *Nature* **2019**, *575*, 455–458. [CrossRef]
188. Abdalla, H.; Adam, R.; Aharonian, F.; Ait Benkhali, F.; Angüner, E.O.; Arakawa, M.; Arcaro, C.; Armand, C.; Ashkar, H.; Backes, M.; et al. A very-high-energy component deep in the γ-ray burst afterglow. *Nature* **2019**, *575*, 464–467. [CrossRef]
189. Abdalla, H. et al. [H. E. S. S. Collaboration]. Revealing X-ray and gamma ray temporal and spectral similarities in the GRB 190829A afterglow. *Science* **2021**, *372*, 1081–1085. [CrossRef]
190. Blanch, O.; Gaug, M.; Noda, K.; Berti, A.; Moretti, E.; Miceli, D.; Gliwny, P.; Ubach, S.; Schleicher, B.; Cerruti, M.; et al. MAGIC observations of GRB 201015A: Hint of very high energy gamma-ray signal. *GRB Coord. Netw.* **2020**, *28659*, 1.
191. Blanch, O. GRB 201216C: MAGIC detection in very high energy gamma rays. *Astron. Telegr.* **2020**, *14275*, 1.
192. Blanch, O.; Longo, F.; Berti, A.; Fukami, S.; Suda, Y.; Loporchio, S.; Micanovic, S.; Green, J.G.; Pinter, V.; Takahashi, M.; et al. GRB 201216C: MAGIC detection in very high energy gamma rays. *GRB Coord. Netw.* **2020**, *29075*, 1.
193. Acciari, V.A.; Ansoldi, S.; Antonelli, L.A.; Arbet Engels, A.; Asano, K.; Baack, D.; Babić, A.; Baquero, A.; Barres de Almeida, U.; Barrio, J.A.; et al. MAGIC Observations of the Nearby Short Gamma-Ray Burst GRB 160821B. *Astrophys. J.* **2021**, *908*, 90. [CrossRef]
194. Cao, Z. A future project at tibet: The large high altitude air shower observatory (LHAASO). *Chin. Phys. C* **2010**, *34*, 249–252. [CrossRef]
195. Cao, Z.; della Volpe, D.; Liu, S.; Bi, X.; Chen, Y.; D'Ettorre Piazzoli, B.; Feng, L.; Jia, H.; Li, Z.; Ma, X.; et al. The Large High Altitude Air Shower Observatory (LHAASO) Science Book (2021 Edition). *arXiv* **2019**, arXiv:1905.02773. [CrossRef]
196. Cao, Z. et al. [LHAASO Collaboration]. A tera-electron volt afterglow from a narrow jet in an extremely bright gamma-ray burst. *Science* **2023**, *380*, 1390–1396. [CrossRef]
197. Burns, E.; Svinkin, D.; Fenimore, E.; Kann, D.A.; Agüí Fernández, J.F.; Frederiks, D.; Hamburg, R.; Lesage, S.; Temiraev, Y.; Tsvetkova, A.; et al. GRB 221009A: The Boat. *Astrophys. J.* **2023**, *946*, L31. [CrossRef]
198. An, Z.H.; Antier, S.; Bi, X.Z.; Bu, Q.C.; Cai, C.; Cao, X.L.; Camisasca, A.E.; Chang, Z.; Chen, G.; Chen, L.; et al. Insight-HXMT and GECAM-C observations of the brightest-of-all-time GRB 221009A. *arXiv* **2023**, arXiv:2303.01203. [CrossRef]
199. Frederiks, D.; Svinkin, D.; Lysenko, A.L.; Molkov, S.; Tsvetkova, A.; Ulanov, M.; Ridnaia, A.; Lutovinov, A.A.; Lapshov, I.; Tkachenko, A.; et al. Properties of the Extremely Energetic GRB 221009A from Konus-WIND and SRG/ART-XC Observations. *Astrophys. J.* **2023**, *949*, L7. [CrossRef]
200. Lesage, S.; Veres, P.; Briggs, M.S.; Goldstein, A.; Kocevski, D.; Burns, E.; Wilson-Hodge, C.A.; Bhat, P.N.; Huppenkothen, D.; Fryer, C.L.; et al. Fermi-GBM Discovery of GRB 221009A: An Extraordinarily Bright GRB from Onset to Afterglow. *Astrophys. J.* **2023**, *952*, L42. [CrossRef]
201. Williams, M.A.; Kennea, J.A.; Dichiara, S.; Kobayashi, K.; Iwakiri, W.B.; Beardmore, A.P.; Evans, P.A.; Heinz, S.; Lien, A.; Oates, S.R.; et al. GRB 221009A: Discovery of an Exceptionally Rare Nearby and Energetic Gamma-Ray Burst. *Astrophys. J.* **2023**, *946*, L24. [CrossRef]
202. Ren, J.; Wang, Y.; Zhang, L.L.; Dai, Z.G. The Possibility of Modeling the Very High Energy Afterglow of GRB 221009A in a Wind Environment. *Astrophys. J.* **2023**, *947*, 53. [CrossRef]
203. O'Connor, B.; Troja, E.; Ryan, G.; Beniamini, P.; van Eerten, H.; Granot, J.; Dichiara, S.; Ricci, R.; Lipunov, V.; Gillanders, J.H.; et al. A structured jet explains the extreme GRB 221009A. *Sci. Adv.* **2023**, *9*, eadi1405. [CrossRef] [PubMed]
204. Zhang, B.T.; Murase, K.; Ioka, K.; Song, D.; Yuan, C.; Mészáros, P. External Inverse-compton and Proton Synchrotron Emission from the Reverse Shock as the Origin of VHE Gamma Rays from the Hyper-bright GRB 221009A. *Astrophys. J.* **2023**, *947*, L14. [CrossRef]
205. Sato, Y.; Murase, K.; Ohira, Y.; Yamazaki, R. Two-component jet model for multiwavelength afterglow emission of the extremely energetic burst GRB 221009A. *Mon. Not. R. Astron. Soc.* **2023**, *522*, L56–L60. [CrossRef]
206. Das, S.; Razzaque, S. Ultrahigh-energy cosmic-ray signature in GRB 221009A. *Astron. Astrophys.* **2023**, *670*, L12. [CrossRef]
207. Alves Batista, R. GRB 221009A: A potential source of ultra-high-energy cosmic rays. *arXiv* **2022**, arXiv:2210.12855. [CrossRef]
208. de Ugarte Postigo, A.; Izzo, L.; Pugliese, G.; Xu, D.; Schneider, B.; Fynbo, J.P.U.; Tanvir, N.R.; Malesani, D.B.; Saccardi, A.; Kann, D.A.; et al. GRB 221009A: Redshift from X-shooter/VLT. *GRB Coord. Netw.* **2022**, *32648*, 1.
209. Li, H.; Ma, B.Q. Lorentz invariance violation induced threshold anomaly versus very-high energy cosmic photon emission from GRB 221009A. *Astropart. Phys.* **2023**, *148*, 102831. [CrossRef]
210. Finke, J.D.; Razzaque, S. Possible Evidence for Lorentz Invariance Violation in Gamma-Ray Burst 221009A. *Astrophys. J.* **2023**, *942*, L21. [CrossRef]
211. Galanti, G.; Nava, L.; Roncadelli, M.; Tavecchio, F.; Bonnoli, G. Observability of the very-high-energy emission from GRB 221009A. *arXiv* **2022**, arXiv:2210.05659. [CrossRef]
212. Nakagawa, S.; Takahashi, F.; Yamada, M.; Yin, W. Axion dark matter from first-order phase transition, and very high energy photons from GRB 221009A. *Phys. Lett. B* **2023**, *839*, 137824. [CrossRef]
213. Murase, K.; Mukhopadhyay, M.; Kheirandish, A.; Kimura, S.S.; Fang, K. Neutrinos from the Brightest Gamma-Ray Burst? *Astrophys. J.* **2022**, *941*, L10. [CrossRef]
214. Ai, S.; Gao, H. Model Constraints Based on the IceCube Neutrino Nondetection of GRB 221009A. *Astrophys. J.* **2023**, *944*, 115. [CrossRef]

215. Abbasi, R.; Ackermann, M.; Adams, J.; Agarwalla, S.K.; Aggarwal, N.; Aguilar, J.A.; Ahlers, M.; Alameddine, J.M.; Amin, N.M.; Andeen, K.; et al. Limits on Neutrino Emission from GRB 221009A from MeV to PeV Using the IceCube Neutrino Observatory. *Astrophys. J.* **2023**, *946*, L26. [CrossRef]
216. Edvige Ravasio, M.; Sharan Salafia, O.; Oganesyan, G.; Mei, A.; Ghirlanda, G.; Ascenzi, S.; Banerjee, B.; Macera, S.; Branchesi, M.; Jonker, P.G.; et al. A bright megaelectronvolt emission line in γ-ray burst GRB 221009A. *arXiv* **2023**, arXiv:2303.16223. [CrossRef]
217. Fiore, F.; Burderi, L.; Lavagna, M.; Bertacin, R.; Evangelista, Y.; Campana, R.; Fuschino, F.; Lunghi, P.; Monge, A.; Negri, B.; et al. The HERMES-technologic and scientific pathfinder. *arXiv* **2020**, arXiv:2101.03078. [CrossRef]
218. Paul, J.; Wei, J.; Basa, S.; Zhang, S.N. The Chinese-French SVOM mission for gamma-ray burst studies. *Comptes Rendus Phys.* **2011**, *12*, 298–308. [CrossRef]
219. Wei, J.; Cordier, B.; Antier, S.; Antilogus, P.; Atteia, J.L.; Bajat, A.; Basa, S.; Beckmann, V.; Bernardini, M.G.; Boissier, S.; et al. The Deep and Transient Universe in the SVOM Era: New Challenges and Opportunities—Scientific prospects of the SVOM mission. *arXiv* **2016**, arXiv:1610.06892. [CrossRef]
220. Amati, L.; O'Brien, P.; Götz, D.; Bozzo, E.; Tenzer, C.; Frontera, F.; Ghirlanda, G.; Labanti, C.; Osborne, J.P.; Stratta, G.; et al. The THESEUS space mission concept: Science case, design and expected performances. *Adv. Space Res.* **2018**, *62*, 191–244. [CrossRef]
221. Stratta, G.; Ciolfi, R.; Amati, L.; Bozzo, E.; Ghirlanda, G.; Maiorano, E.; Nicastro, L.; Rossi, A.; Vinciguerra, S.; Frontera, F.; et al. THESEUS: A key space mission concept for Multi-Messenger Astrophysics. *Adv. Space Res.* **2018**, *62*, 662–682. [CrossRef]
222. Amati, L.; O'Brien, P.T.; Götz, D.; Bozzo, E.; Santangelo, A.; Tanvir, N.; Frontera, F.; Mereghetti, S.; Osborne, J.P.; Blain, A.; et al. The THESEUS space mission: Science goals, requirements and mission concept. *Exp. Astron.* **2021**, *52*, 183–218. [CrossRef]
223. Rahin, R.; Salh, J.; Tarem, S.; Behar, E. GALI (Gamma-ray Burst Localizing Instrument) A New GRB Detector Concept. *AAS/High Energy Astrophys. Div.* **2022**, *54*, 111.12.
224. Yuan, W.M. Special Topic of Einstein Probe: Exploring the Dynamic X-ray Universe. *Sci. Sin. Phys. Mech. Astron.* **2018**, *48*, 039501. [CrossRef]
225. Chen, Y.; Sun, X.; Li, Z.; Wang, C.; Zhang, C.; Sun, S. Detection system of the lobster eye telescope with large field of view. *Appl. Opt.* **2022**, *61*, 8813. [CrossRef] [PubMed]
226. Chen, Y.; Cui, W.; Han, D.; Wang, J.; Yang, Y.; Wang, Y.; Li, W.; Ma, J.; Xu, Y.; Lu, F.; et al. Status of the follow-up x-ray telescope onboard the Einstein Probe satellite. *Soc. Photo-Opt. Instrum. Eng. (SPIE) Conf. Ser.* **2020**, *11444*, 114445B. [CrossRef]
227. Racusin, J.; Perkins, J.S.; Briggs, M.S.; de Nolfo, G.; Krizmanic, J.; Caputo, R.; McEnery, J.E.; Shawhan, P.; Morris, D.; Connaughton, V.; et al. BurstCube: A CubeSat for Gravitational Wave Counterparts. *arXiv* **2017**, arXiv:1708.09292. [CrossRef]
228. Kocevski, D. The StarBurst Multimessenger Pioneer. *AAS/High Energy Astrophys. Div.* **2023**, *55*, 115.23.
229. Hui, C.M.; Briggs, M.S.; Goldstein, A.; Jenke, P.; Kocevski, D.; Wilson-Hodge, C.A. MoonBEAM: A Beyond-LEO Gamma-Ray Burst Detector for Gravitational-Wave Astronomy. In Proceedings of the Deep Space Gateway Concept Science Workshop, Denver, CO, USA, 27 February–1 March 2018; LPI Contributions; LPI Editorial Board, Ed.; Volume 2063, p. 3060.
230. Fletcher, C.; Hui, C.M.; Goldstein, A.; The MoonBEAM Team. The Scientific Performance of the MoonBurst Energetics All-sky Monitor(MoonBEAM). *arXiv* **2023**, arXiv:2308.16293. [CrossRef]
231. McConnell, M.L.; Baring, M.G.; Bloser, P.F.; Briggs, M.S.; Connaughton, V.; Dwyer, J.; Gaskin, J.; Grove, J.E.; Gunji, S.; Hartmann, D.; et al. LEAP—A Large Area GRB Polarimeter for the ISS. *AAS/High Energy Astrophys. Div.* **2017**, *16*, 103.20.
232. Gaskin, J.; Baring, M.; Bloser, P.; Briggs, M.; Dwyer, J.; Ertley, C.; Fletcher, G.; Galloway, P.; Gelmis, K.; Goldstein, A.; et al. The LEAP Gamma-Ray Burst Polarimeter: An Overview. *Am. Astron. Soc. Meet. Abstr.* **2022**, *54*, 127.01.
233. Zhang, B. *The Physics of Gamma-Ray Bursts*; Cambridge University Press: Cambridge, UK, 2018. [CrossRef]

universe

MDPI

Review

Future Perspectives for Gamma-ray Burst Detection from Space

Enrico Bozzo [1,*,†], Lorenzo Amati [2], Wayne Baumgartner [3], Tzu-Ching Chang [4], Bertrand Cordier [5], Nicolas De Angelis [6,7], Akihiro Doi [8], Marco Feroci [7], Cynthia Froning [9], Jessica Gaskin [3], Adam Goldstein [10], Diego Götz [5], Jon E. Grove [11], Sylvain Guiriec [12,13], Margarita Hernanz [14,15], C. Michelle Hui [3], Peter Jenke [16], Daniel Kocevski [3], Merlin Kole [6], Chryssa Kouveliotou [12], Thomas Maccarone [17], Mark L. McConnell [18,19], Hideo Matsuhara [8], Paul O'Brien [20], Nicolas Produit [1], Paul S. Ray [11], Peter Roming [9], Andrea Santangelo [21], Michael Seiffert [4], Hui Sun [22], Alexander van der Horst [12], Peter Veres [16], Jianyan Wei [22], Nicholas White [12], Colleen Wilson-Hodge [3], Daisuke Yonetoku [8,23], Weimin Yuan [22,24], Shuang-Nan Zhang [24,25]

1. Department of Astronomy, University of Geneva, Chemin d'Ecogia 16, 1290 Versoix, Switzerland; nicolas.produit@unige.ch (N.P.)
2. INAF-OAS Bologna, via P. Gobetti 93/3, 40129 Bologna, Italy
3. NASA Marshall Space Flight Center, Huntsville, AL 35812, USA; daniel.kocevski@nasa.gov (D.K.)
4. Jet Propulsion Lab, 4800 Oak Grove Dr, Pasadena, CA 91109, USA
5. IRFU/Département d'Astrophysique, CEA, Université Paris-Saclay, F-91191 Gif-sur-Yvette, France
6. DPNC, University of Geneva, quai Ernest-Ansermet 24, 1205 Geneva, Switzerland
7. INAF-IAPS Roma, via del Fosso del Cavaliere 100, I-00133 Rome, Italy
8. Institute of Space and Astronautical Science (ISAS), 3-1-1 Yoshinodai, Chuo-ku, Sagamihara, Kanagawa 229-8510, Japan
9. Southwest Research Institute Space Science and Engineering Division, 6220 Culebra Road, San Antonio, TX 78238, USA
10. Science and Technology Institute, Universities Space Research Association, Huntsville, AL 35805, USA
11. Space Science Division, U.S. Naval Research Laboratory, Washington, DC 20375, USA
12. Department of Physics, George Washington University, Corcoran Hall, 725 21st Street NW, Washington, DC 20052, USA
13. NASA Goddard Space Flight Center, Greenbelt, MD 20771, USA
14. Institute of Space Sciences (ICE-CSIC), Carrer de can Magrans, s/n, Campus UAB, 08193 Bellaterra (Barcelona), Spain; hernanz@ice.csic.es
15. Institut d'Estudis Espacials de Catalunya (IEEC), Barcelona, Spain
16. University of Alabama in Huntsville, Huntsville, AL 35805, USA
17. Department of Physics & Astronomy, Texas Tech University Box 41051, Lubbock, TX 79409-1051, USA
18. Space Science Center, University of New Hampshire, Durham, NH 03824, USA; mark.mcconnell@unh.edu
19. Department of Earth, Oceans, and Space, Southwest Research Institute, Durham, NH 03824, USA
20. School of Physics and Astronomy, University of Leicester, Leicester LE1 7RH, UK
21. Institut für Astronomie und Astrophysik, Kepler Center for Astro and Particle Physics, Eberhard Karls, Universität, Sand 1, D-72076 Tübingen, Germany
22. National Astronomical Observatories, Chinese Academy of Sciences, Beijing 100101, China
23. College of Science and Engineering, School of Mathematics and Physics, Kanazawa University, Kakuma, Kanazawa, Ishikawa 920-1192, Japan
24. University of Chinese Academy of Sciences, Chinese Academy of Sciences, Beijing 100049, China
25. Key Laboratory of Particle Astrophysics, Institute of High Energy Physics, Chinese Academy of Sciences, Beijing 100049, China

* Correspondence: enrico.bozzo@unige.ch

† Most authors provided inputs on behalf of a larger collaboration.

check for updates

Citation: Bozzo, E.; Amati, L.; Baumgartner, W.; Chang, T.-C.; Cordier, B.; De Angelis, N.; Doi, A.; Feroci, M.; Froning, C.; Gaskin, J.; et al Future Perspectives for Gamma-ray Burst Detection from Space. *Universe* **2024**, *10*, 187. https://doi.org/10.3390/universe10040187

Academic Editor: Tina Kahniashvili

Received: 13 February 2024
Revised: 11 April 2024
Accepted: 17 April 2024
Published: 19 April 2024

Abstract: Since their first discovery in the late 1960s, gamma-ray bursts have attracted an exponentially growing interest from the international community due to their central role in the most highly debated open questions of the modern research of astronomy, astrophysics, cosmology, and fundamental physics. These range from the intimate nuclear composition of high-density material within the core of ultra-dense neuron stars, to stellar evolution via the collapse of massive stars, the production and propagation of gravitational waves, as well as the exploration of the early universe by unveiling the first stars and galaxies (assessing also their evolution and cosmic re-ionization). GRBs in the past ~50 years have stimulated the development of cutting-edge technological instruments for observations of high-energy celestial sources from space, leading to the launch and successful operations of many different scientific missions (several of them still in data-taking mode currently). In this review, we provide a brief description of the GRB-dedicated missions from space being designed and developed for the future. The list of these projects, not meant to be exhaustive, shall serve as a reference to interested readers to understand what is likely to come next to lead the further development of GRB research and the associated phenomenology.

Keywords: γ-ray astrophysics; Gamma-ray bursts; X-ray polarimetry; X-ray surveys; X-ray instrumentation

---

1. Introduction

About 57 years have already gone by since the first discovery of a gamma-ray burst (GRB) in 1967. This occurred genuinely by chance, while the U.S. *Vela* military satellites were looking for evidence of nuclear bomb testing in space following the 1963 partial test-ban treaty (which prohibited testing in the atmosphere, outer space, and underwater). The existence of GRBs was later revealed to the scientific community, stimulating a fast-growing excitement for this fortunate discovery [1]. A number of experiments were quickly developed and launched into space, mainly by the Soviet Union and the United States, to boost the detection of similar events and understand their true nature. The inevitable requirement of space-based instrumentation to catch the unpredictable short and bright flash of high-energy radiation from these events largely contributed to boosting the international race in the development of cutting-edge technological detectors with exponentially increasing sensitivity, as well as the provision of more and more advanced platforms to ensure their most proficient operations (see, e.g., [2] for recent historical reviews on GRBs).

A major leap forward in GRB research was first provided by the Konus-WIND satellite [3,4] and the BATSE experiment on-board CGRO [5,6], which allowed the determination of the dichotomy into short and long GRBs, established the isotropic distribution of the events in the sky, and led to the first characterization of their non-thermal emission (see, e.g., [7] and the references therein). Subsequently, the Italian–Dutch Beppo-SAX satellite, launched in 1996, unveiled the cosmological nature of these events by identifying for the first time the GRB X-ray afterglows and providing arcsec-accurate localizations of their host galaxies [8] (enabling also multi-wavelength observations with ground-based observatories). This was possible thanks to an on-board powerful suite of instruments covering a large energy-band (2–700 keV) and providing simultaneously a high sensitivity to faint sources via dedicated pointed observations and monitoring of a large fraction of the sky at once via large field-of-view (FoV) instruments [9] (achieving up to 20×20 deg with full imaging and good localization capabilities of typically a few arcminutes).

The size of the GRB community and its efforts to probe and understand the physics of these uniquely powerful cosmological explosions have been growing ever since. GRBs are also widely recognized to have a key role in many of the most debated aspects of research in astronomy and astrophysics. These range from the problems of fundamental physics, such as the equation of state of supra-density matter, to stellar evolution, via the collapse of massive stars and the production of both kilonovae and supernovae, to jet formation and dissipation, to the generation and propagation of gravitational waves, up to the quest for key cosmological parameters (see, e.g., [10,11] and the references therein). In the most recent years, GRBs have also gained renewed interest as key sources in the field of multi-messenger astrophysics due to their association with kilonovae and gravitational wave (GW) sources (see, e.g., [12–14], the well-known cases of GRB 170817A, GRB 211211A, and GRB 230307A, and the references therein).

Following the decommissioning of *Beppo-SAX* in 2002, a number of successful missions have provided dramatic advancements in the fields related to GRBs. Among these, AGILE [15], Astro-SAT [16], Fermi [17], INTEGRAL [18], and the Neil Gehrels Swift Observatory [19] are still operational and work in synergy also with ground-based multi-wavelength observatories to boost the detection and characterization of bright impulsive cosmological events (in addition to transients of different kinds). While an appraisal of the most relevant achievements of these missions can be found in recent reviews, the aim of the current paper is to provide interested readers an overview of the world-wide efforts from different teams to develop the GRB instruments of the future. These are expected to lead the scientific community toward the next leaps forward in the understanding of GRBs and associated phenomena. Different missions are presented in alphabetical order in the following sections. The list is not meant to be exhaustive, and given the heavily dynamic and fast-evolving nature of GRB research, it is likely (and desirable) that additional relevant missions set to provide important contributions to GRB research are being developed by other teams.

2. Einstein Probe

2.1. Mission Overview

The Einstein Probe (https://ep.bao.ac.cn) (EP) is a mission designed to monitor the sky in the soft X-ray band. It will perform systematic surveys and the characterization of high-energy transients and the monitoring of variable objects at unprecedented sensitivity and monitoring cadences. It has a large instantaneous FoV (3600 sq. deg.), which is achieved by using the established technology of novel lobster-eye micro-pore optics (MPOs). The EP also carries a conventional X-ray-focusing telescope with a larger effective area to perform follow-up observations and precise localization of newly discovered transients. Public transient alerts will be issued rapidly to trigger multi-wavelength follow-up observations from the world-wide community.

The primary science objectives of the EP mission are as follows: (1) Discover and characterize cosmic X-ray transients, particularly faint, distant, and rare X-ray transients, in large numbers. (2) Discover and characterize X-ray outbursts from otherwise normally dormant black holes. (3) Search for X-ray sources associated with gravitational wave events and precisely locate them. The EP is an international mission led by the Chinese Academy of Sciences (CAS) in collaboration with the European Space Agency (ESA), the Max Planck Institute for Extraterrestrial Physics (MPE) in Germany, and the Centre National d'Etudes Spatiales (CNES) in France. The mission was successfully launched on 9 January 2024 with a nominal lifetime of 3 years (5 years as a goal).

2.2. Instrument Design

The EP carries a wide-field X-ray telescope (WXT) with a large instantaneous FoV, which adopts a novel lobster-eye MPO technology. Complementary to this wide-field instrument is a follow-up X-ray telescope (FXT) with a large effective area and a narrow FoV. Figure 1 shows the configuration of the EP payload:

- Wide-field X-ray telescope (WXT): There are 12 almost identical WXT modules, each with an FoV of $\sim$300 sq. deg. Each WXT module includes a lobster-eye MPO mirror assembly with a focal length of 375 mm, a focal plane detector array, and an electronics unit. The mirror assembly of each module comprises 36 MPOs mounted on a supporting alloy frame. An MPO is made of a thin plate with millions of square micro-pores perpendicular to the surface, slumped into a spherical shape. Incoming X-rays at a grazing-incidence angle are reflected off the walls of the pores and are brought onto a focal sphere with a radius of half the curvature of the optic. It produces true imaging with a characteristic cruciform point-spread function. The detector array of each module is composed of four black-illuminated CMOS sensors, each 6 cm by 6 cm in size and 4 k by 4 k pixels. An aluminum layer of a 200 nm thickness is coated on the surface to block incident optical and UV light. The nominal detection bandpass of the WXT is 0.5–4 keV.
- Follow-up X-ray telescope (FXT): The FXT is a set of two telescopes of the Wolter-I optics, which have almost the same design and are co-aligned. The design of the

mirror assembly is similar to that of the eROSITA telescopes, which has a focal length of 1.6 m. The focal plane detectors are built from pn-CCDs, and a set of thin and thick filters of aluminum layers are mounted on a filter wheel.

Figure 1. Preliminary configuration of the EP payloads' alignment, with twelve WXT modules surrounding two FXTs.

2.3. *Expected Performance*

The angular resolution of the WXT is about 5 arcmin (FWHM) for the central focal spot, and the effective area is in the range of 2–3 cm^2. The grasp parameter of the WXT is shown in Figure 2. In a 1000 s exposure, a sensitivity of approximately (2–3) $\times 10^{-11}$ $ergss^{-1}cm^{-2}$ in the 0.5–4 keV band can be achieved at the 5-σ level. Such sensitivity and spatial resolution improve by one order of magnitude or more upon the previous and current wide-field X-ray monitors.

Operating in the 0.3–10 keV energy range, the FXT has a narrow field of view (60 arcmin in diameter) and an effective area of about 300 cm^2 at 1 keV (for one unit). The spatial resolution (PSF) is about 23 arcsec in half-power diameter (HPD), which gives a source localization precision of 5–15 arcsec (90% c.l.) depending on the source intensity. The FXT is responsible for the quick follow-up observations (within 5 min) of the triggered sources from the WXT, and will also observe other targets of interest as target of opportunity (ToO) observations.

Once a transient is detected with the WXT, the spacecraft will slew to point the FXT to the target for quick follow-up observations. Meanwhile, the alert information of the transient will be down-linked quickly to the ground segment and made public to trigger follow-up observations. Quick command up-link for time-critical ToO observations is also possible.

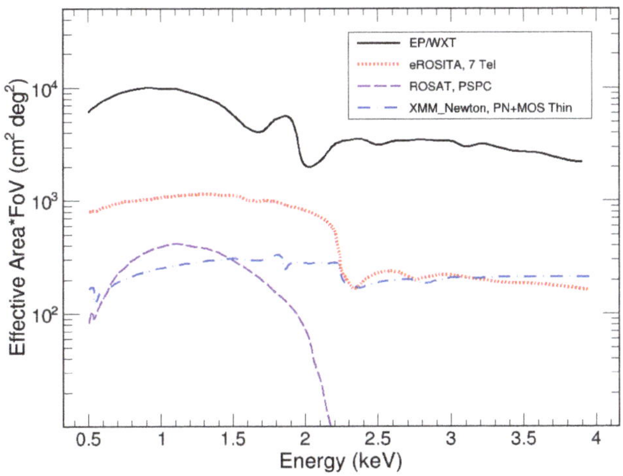

Figure 2. Representative grasp (effective area times FoV) of the WXT as a function of photon energy (black). As a comparison, the grasp parameters of several X-ray-focusing telescopes are overplotted. Adapted from Figure 10 in [20].

With the unprecedented capability of the WXT and FXT, the EP is expected to characterize the cosmic high-energy transients over wide time scales and at high cadences, revealing new insights into a diverse set of systems including dormant black holes, neutron stars, supernova shock breakouts, active galactic nuclei, X-ray binaries, gamma-ray bursts, stellar coronal activity, and electromagnetic-wave sources and gravitational-wave events. In particular, EP will provide valuable data for the prompt emission of GRBs in the soft X-ray band, which has not been common in previous GRB detections. Meanwhile, EP will also monitor the variability of several types of X-ray sources in large samples all over the sky.

3. eXTP

3.1. Mission Overview

The enhanced X-ray Timing and Polarimetry mission (https://www.isdc.unige.ch/extp/) (eXTP) is designed to study mainly the state of matter under extreme conditions of density, gravity, and magnetic field [21]. The core science objectives of the mission are focused on the determination of the equation of state of matter at supra-nuclear density and the study of the dynamics of accretion/ejection flows under the influence of strong gravitational fields. Thanks to an extensive suite of innovative instruments, the eXTP is designed also to be a general observatory for astrophysics, providing broad capabilities in the X-ray domain (0.5–50 keV) to conduct timing, spectral, and polarimetric observations of a wide variety of galactic and extra-galactic sources. Although the mission is not focusing on GRB science, the availability of a large FoV and sensitive instrument, the WFM (see below), makes the eXTP an important contributor to the possible detection and study of these events in the future.

The eXTP is planned to be the next flagship-class mission led by China, and it is currently in an advanced design phase, where virtually all payload elements have reached a relatively mature technology and could be ready for implementation (see Figure 3). The mission is being studied by a large collaboration including a wide scientific community in China and many European member states. At present, a launch of opportunity has still to be identified, and the programmatics are being cleared in order to possibly bring the eXTP into space in 2028–2029.

Figure 3. An artist impression of the eXTP satellite. In the image, we show the 13 nested shell optics of the SFA (9 telescopes) and PFA (4 telescopes), as well as the 6 cameras with coded masks composing the WFM. The LAD is placed on deployable panels equipped with solar shades. There are a total of 40 modules composing the LAD, and each module hosts 16 SDDs.

3.2. *Instrument Design*

The eXTP scientific payload consists of four main instruments: the spectroscopic-focusing array (SFA), the large-area detector (LAD), the polarimetry-focusing array (PFA), and the wide-field monitor (WFM). The SFA comprises nine identical Wolter-I grazing-incidence X-ray telescopes and is mainly used for spectral and timing observations of celestial sources in the energy range of 0.5–10 keV. In its current design, the SFA total effective area is 7400 cm^2 (at 2 keV) and the instrument FoV is circular with a diameter of 12 arcmin. The instrument achieves an energy resolution of 180 eV at 6 keV and a timing resolution of 10 μs in the whole operational energy band. The LAD is designed to perform photon-by-photon observations of X-ray sources in the 2–30 keV energy range. The LAD is exploiting the technology of innovative large-area Silicon Drift Detectors [22] (SDDs), to provide an unprecedented large effective area, reaching 3.4 m^2 at 8 keV. The instrument is not designed for imaging purposes, and the FoV is limited to 1 deg using micro-channel collimator plates [23] to simultaneously reduce source confusion and background. The LAD achieves an energy resolution of 260 eV at 6 keV and a timing resolution of 10 μs in the whole operating energy range. A similar instrument is also planned on-board the STROBE-X mission (see Section 10). The WFM exploits the same SDD technology coupled with coded masks to provide a coverage of about 5.5 sr of the sky at any time in the energy range of 2–50 keV. The instrument is capable of localizing X-ray sources within an accuracy of $<$1 arcmin and performing timing (spectral) investigations on these objects with a time (energy) resolution of 10 μs ($<$300 eV at 6 keV). The WFM is equipped with an automatic alert system to promptly broadcast to the ground (within a few tens of seconds at the most) the onset time and position of bright impulsive events detected on-board, including GRBs. A similar instrument is also planned on-board the STROBE-X mission (see Section 10). The PFA provides polarimetric capabilities in the energy range of 2–8 keV, exploiting X-ray optics optimized for polarimetric observations and a similar detector technology compared to that currently flying on the IXPE mission (the so-called gas pixel detectors (GPDs); see, e.g., [24] and the references therein).

3.3. *Expected Performance*

Given the capabilities of the eXTP/WFM (see Figure 4), a detection rate of about 100 GRB/year has been estimated. The WFM will be able to measure with good accuracy the spectral shape of the detected events and follow the hard to soft evolution of the prompt emission in the 2–50 keV band, where different GRB models are known to make different predictions [25,26]. This could provide information on the composition and magnetization of the emitting plasma, the geometry of the emission, and the structure of the jet and surrounding material. The combination of sensitivity and soft energy coverage of the WFM provides the opportunity to detect a few high redshift ($z > 6$) GRBs per year and many of the so-far elusive absorption features in tens of medium bright GRBs, probing (among others) the surroundings of GRB progenitors (see [27,28]). The WFM is also expected to detect up to 40 X-ray flashes each year (see [29]).

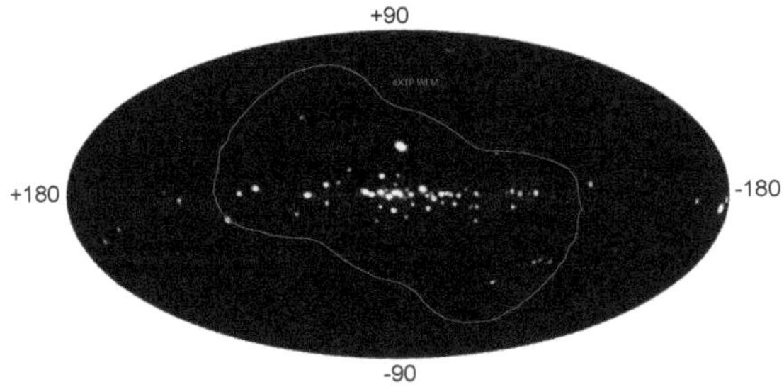

Figure 4. The large FoV of the eXTP/WFM (red line) overlapped on a background map of the high-energy sky, which has been provided as a courtesy by the MAXI team (T. Mihara, RIKEN, JAXA). The red line corresponds to the total FoV covered by the WFM in the configuration used for the eXTP, as illustrated by in't Zand et al. [30].

4. Gamow

4.1. Mission Overview

The Gamow Explorer mission [31] is optimized to: (1) probe the high-redshift universe ($z > 6$) when the first stars were born, galaxies formed, and Hydrogen was re-ionized and (2) enable multi-messenger astrophysics (MMA) by rapidly identifying electro-magnetic (EM) counterparts to gravitational wave (GW) events. GRBs have been detected out to $z{\sim}9$ (see, e.g., [32,33] and the references therein), and their afterglows are bright beacons lasting a few days, which can be used to observe the spectral fingerprints of the host galaxy and intergalactic medium. Gamow is designed to detect and rapidly identify high-redshift events. Rapid follow-up spectroscopy with the James Webb Space Telescope (JWST) and >6 m ground-based telescopes provide NIR $R{\sim}$ 2500 spectra to determine the IGM neutral fraction versus redshift using the damping wing of the Lyman-α absorption line (see Figure 5 from [34]).

GRB afterglows are particularly advantageous for this measurement versus, e.g., using QSOs. GRBs have a featureless power-law spectrum, ideal for fitting the Ly-α absorption profile, are hosted by low-mass galaxies directly tracing the typical ionization state of the IGM, and can be seen out to redshifts of 20, whereas the abundance of QSOs drops steeply with redshift. The same spectra will be used to determine metallicities from absorption lines to trace the early chemical enrichment and measure the escape fraction of ionizing photons that escape the galaxies to ionize atoms in the IGM. The co-moving GRB rate generally follows the star-formation rate (SFR). At higher redshifts, the GRB rate exceeds the SFR derived by other means. The greatly improved measurements of the high-redshift GRB rate by Gamow will provide crucial information to probe potential changes of the initial mass function of massive stars and the star-formation rate at high redshift and constrain GRB progenitors and their properties (e.g., luminosity distributions, progenitor binary fractions, etc.) in the high-redshift universe [33].

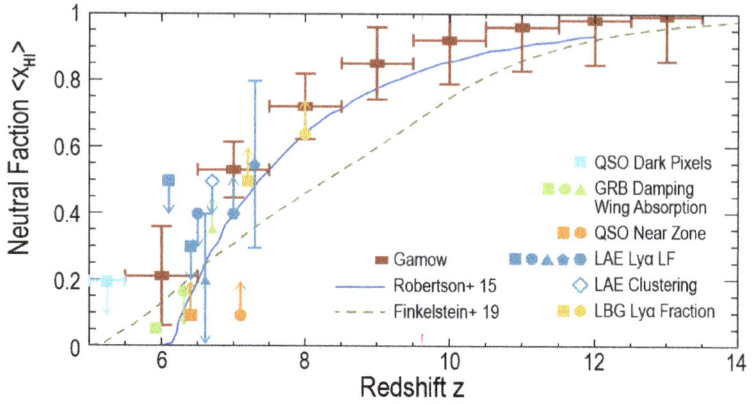

Figure 5. Recent measurements of the neutral fraction <XHI > versus redshift, z, by various techniques, including GRBs (green; see [35]). The forecast Gamow results are shown as red [34]. The Fisher matrix forecast assumes 26 GRBs at z > 5. We assume follow-up spectra with a signal-to-noise ratio (SNR) of 20 at the continuum per R = 3000 resolution element and assume accurate host redshift determinations via metal absorption lines. The results illustrate that Gamow can map the re-ionization history in detail from z∼6 to 14. Predicted re-ionization curves (blue and dashed red lines) illustrate the degree of variance between theoretical models compared to the Fisher forecasting.

4.2. Instrument Design

The lobster-eye X-ray telescope (LEXT) with a wide FoV of ∼1350 sq deg detects GRBs and locates them with arcminute precision. The LEXT utilizes an array of slumped micro-pore optics (MPOs) with 40 μm square pores and a focal length of 30 cm [36]. The focal plane uses heritage MIT-Lincoln Labs CCDs. The required performance is to detect at least 20 z > 6-long GRBs over a 2.5-year prime mission. A rapidly slewing spacecraft autonomously points the photo-z infra-red telescope (PIRT) within 100 s to identify high-redshift (z > 6) GRBs. The photo-z technique takes advantage of the Hydrogen Lyman-σ absorption, which creates a sharp blue-ward drop out. A 30 cm aluminum RC telescope passively cooled to 200 K feeds a dichroic prism beam splitter to place five images onto a single H2RG detector covering the 0.6 to 2.5 μm band [37]. The FoV is 10 arcmin square. This design provides the required 15 μJy (21 mag AB) 5-σ sensitivity in a single 500 s exposure (see [38]). The high redshift science objectives require rapid follow-up observations to provide NIR medium-resolution spectroscopy (R∼3000) to measure the profile of the Ly-σ absorption line and metal absorption lines from the host galaxy. To obtain a sufficient signal-to-noise ratio in a reasonable observing time requires 6 m class or greater telescopes. Ground-based telescopes begin observations within an hour and JWST and the ELT within 2–3 days (Kann et al. 2024, to be submitted). An L2 orbit provides >95% observing efficiency with pointing optimized for these follow-up observatories. A low-bandwidth continuous S-band connection provides real-time alerts to the ground.

4.3. Expected Performance

Gamow's capabilities are also optimized for the identification of EM counterparts to binary neutron star (BNS) and neutron star black hole (NSBH) mergers detected by the A+ generation of GW detectors. Within <200 s of a GW alert, the real-time low-bandwidth link uploads commands to re-point the LEXT at the GW uncertainty region so as to detect and locate to 3 arcmin precision the accompanying short GRB afterglow (Figure 6). Figure 2 from White et al. [31] shows that the measurement predictions are based on the extrapolation of Swift short GRB afterglow detections to the LEXT pass-band and 600 Mpc source distances (the horizon of the LVK A+ GW detectors). Approximately 80% of Swift short GRBs have X-ray afterglow detections. This demonstrates that they will be detectable by Gamow at BNS-appropriate distances. If a transient X-ray source is detected, an autonomous PIRT pointing will refine the position to 1 arcsec precision and use simultaneous five-band photometry to follow the early phase of the merger. The X-ray and optical–NIR multi-band and multi-wavelength capabilities combined with an agile spacecraft will enable a broad swath of time domain astronomy science, very similar to Swift [39]. While waiting for GRBs and GW events, we envision that Gamow will undertake a community-driven comprehensive time domain astronomy program. The Gamow Explorer was proposed in the 2021 NASA MIDEX call. While it was not successful, there are considerations to re-propose for future NASA opportunities.

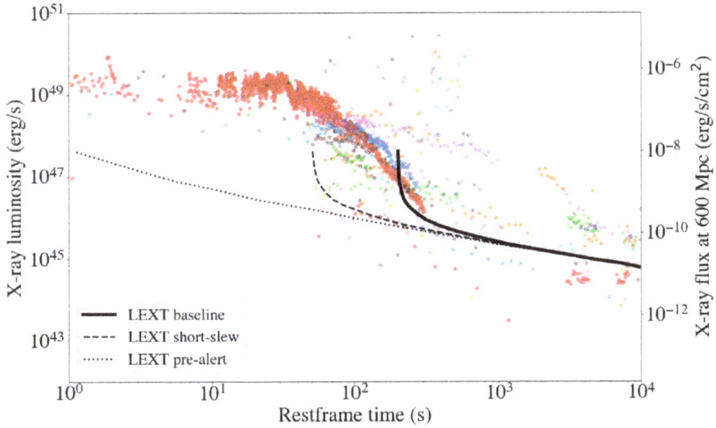

Figure 6. The ability of the Gamow LEXT to detect the afterglows of on-axis BNS and NSBH mergers. X-ray luminosity (left axis) and X-ray flux at 600 Mpc (right axis) for Swift short GRBs at known redshift. The black line is the LEXT sensitivity starting at 0 s (base-line). In practice, it will frequently be earlier than this (short-slew) and, sometimes, before the BNS merger (pre-alert; figure from [31]; see Chan et al. [40], Nitz and Canton [41], Banerjee et al. [42]).

5. HiZ-GUNDAM

5.1. Mission Overview

The high-z gamma-ray bursts for unraveling the dark ages mission (HiZ-GUNDAM) (https://www.isas.jaxa.jp/en/missions/spacecraft/future/hiz-gundam.html) is a future satellite mission, a competitive medium-class mission at ISAS/JAXA, designed to advance time domain astronomy through the observation of high-energy transient phenomena [43]. Two scientific goals are defined: (1) the exploration of the early universe with high-redshift gamma-ray bursts and (2) a contribution to multi-messenger astronomy. These scientific objectives require observational capabilities to detect high-energy transients and to carry out automatic/rapid follow-ups in the near-infrared band. This section provides the description of the current specifications and designs of the satellite and mission payloads, but it is important to note that they are subject to change in subsequent studies and developments.

5.2. Instrument Design

HiZ-GUNDAM has two types of mission payloads: the wide-field X-ray monitor (WFXM) and the near-infrared telescope (NIRT). These payloads are optimized and minimized to achieve the scientific goals [44–48]. We show a schematic view in Figure 7 and basic specification of the WFXM and NIRT in Tables 1 and 2, respectively.

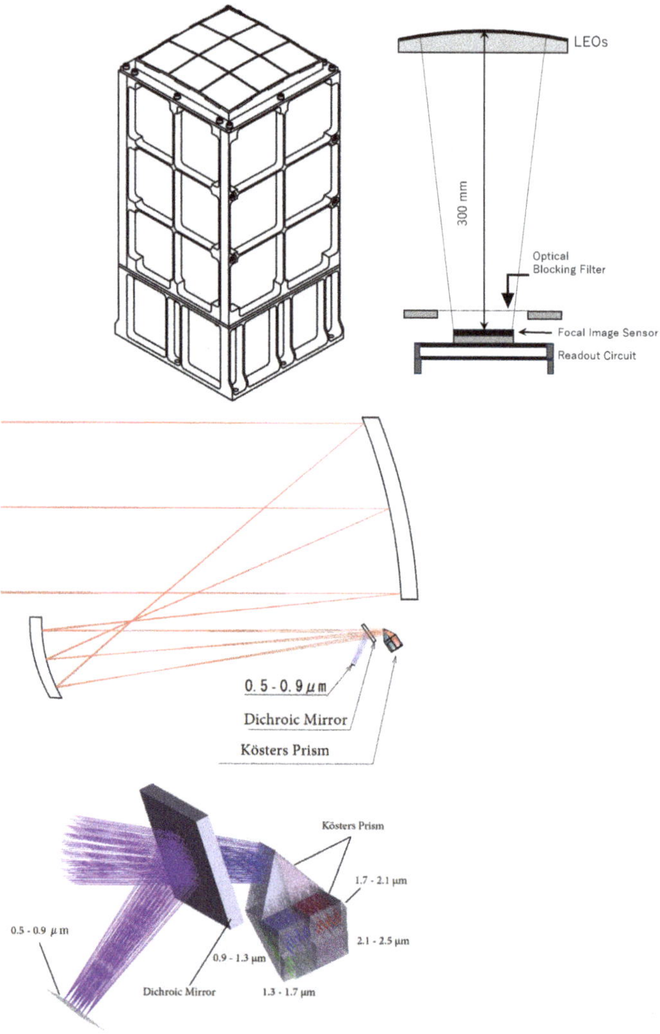

Figure 7. A schematic view of the HiZ-GUNDAM WFXM (**top**) (copyright MEISEI ELECTRIC Co., LTD.). Schematic views of the HiZ-GUNDAM NIRT (**middle** and **bottom**).

The wide-field X-ray monitors (WFXMs) consist of lobster-eye optic arrays and focal imaging sensors (pnCCDs). They are designed to detect high-energy transients within a wide field of view of >0.5 steradian in the 0.5–4 keV energy range. The current design of a single module of the WFXM includes a lobster-eye optic array of 3×3 with a 300 mm focal length and a pnCCD with a pixel size of approximately 100 μm on a 55×55 mm^2 format at the focus. Multiple modules are installed aboard the spacecraft to monitor the wide FoV.

The near-infrared telescope (NIRT) has an aperture size of 30 cm in diameter and simultaneously observes at five wavelength bands between 0.5 and 2.5 μm using a dichroic mirror and Kösters. The NIR telescope is cooled to <200 K by radiative cooling to maintain the best sensitivity up to the 2.5 μm band. The mirrors and structure of the NIRT are primarily made of aluminum alloy to ensure imaging performance remains consistent even if the temperature of the telescope changes in orbit, i.e., an athermal configuration.

Table 1. HiZ-GUNDAM wide-field X-ray monitor.

Item	Specification
Optics	Lobster-Eye
Focal Length	300 mm
Focal Detector	pnCCD
Size of Focal Detector	55×55 mm^2
Energy Band	0.5–4.0 keV
Field of View	>0.5 str (in total)
Sensitivity (100 s)	$\sim 10^{-10}$ erg cm^{-2} s^{-1}
Localization Accuracy	~3 arcmin
Time Resolution	~0.1 s

Table 2. HiZ-GUNDAM near-infrared telescope.

Item	Specification
Telescope Type	Offset/Athermal
Aperture Size	30 cm
Telescope Temp.	<200 K
Focal Detector	HyViSi (Optical)
	HgCdTe (NIR)
Field of View	15×15 arcmin^2
Band	Sensitivity (2 min $\times$ 5)
0.5–0.9 μm	21.4 mag (AB)
0.9–1.3 μm	21.3 mag (AB)
1.3–1.7 μm	21.4 mag (AB)
1.7–2.1 μm	20.8 mag (AB)
2.1–2.5 μm	20.7 mag (AB)

5.3. Expected Performance

In Figure 8, we show schematic views of the HiZ-GUNDAM satellite. The size of the satellite bus is about 1 m^3, except for the solar paddles and mission payloads. For orbital placement, we have chosen a Sun-synchronous polar orbit along the twilight line. This choice is based on the favorable thermal conditions for the NIRT, which is cooled down to <200 K. However, X-ray observation may face disturbances from both the South Atlantic Anomaly and aurora belt at the high-latitude polar region. The nominal operation of HiZ-GUNDAM follows the sequence outlined below:

1. Set the satellite attitude to a solar separation angle of 120 degrees and a forward movement of 50 degrees.
2. Maintain the inertial pointing direction for approximately 560 s, during which HiZ-GUNDAM monitors X-ray transients.
3. After the monitoring time, the satellite slews to the same attitude configuration as described in (1), but with a different pointing direction to prevent thermal radiation from the Earth exposing the NIRT.

This nominal sequence is repeated, and the observation of eight different fields of view is performed in every orbit. The satellite continuously monitors the X-ray sky, except during the maneuver. This optimized sequence facilitates follow-up observations with the NIR telescope. HiZ-GUNDAM can execute this operation for 97% of the GRBs discovered by itself, ensuring more than 10 min of follow-up time in each case.

Figure 8. Schematic view of the HiZ-GUNDAM satellite.

6. LEAP

6.1. Mission Overview

The LargE Area burst Polarimeter (https://sti.usra.edu/in-the-news/leap-mission-study/ (LEAP) is a mission that was proposed in 2021 to NASA's Astrophysics Mission of Opportunity Program. The LEAP instrument is a wide-FoV Compton polarimeter that measures GRB polarization over the energy range from 50 keV to 1 MeV and performs GRB spectroscopy from 20 keV to 5 MeV. If approved (a final decision on the selection is expected during the first quarter of 2024), it will be deployed as an external payload on the International Space Station (ISS) in 2027 for a three-year mission [49,50]. The LEAP science investigation is based on the ability to distinguish between three classes of GRB models [51]. The baseline science investigation requires the observation of 65 GRBs with a minimum detectable polarization (MDP) of 30% or better. Evidence of polarized γ-ray emission in GRBs (>100 keV) has been accumulated in recent years, but the limited sensitivity of these measurements does not yet yield a clear picture of the underlying physics [52–54]. A sensitive and systematic study of GRB polarization, such as the one that will be provided by LEAP, is needed to remedy this situation.

6.2. Detector Design

The LEAP payload (Figure 9) consists of an array of seven independent polarimeter modules, each with a 12 $\times$ 12 array of optically isolated high-Z and low-Z scintillation detectors read out by individual PMTs. Each polarimeter module includes 84 plastic scintillator elements, 58 CsI(Tl) scintillator elements, and two ^{60}Co calibration sources. Each calibration source consists of a small plastic scintillator doped with ^{60}Co, which permits electronic tagging of each decay via the coincident β particle. As a Compton polarimeter, scatter events recorded by the scintillator array within each module are used to measure polarization. Within each module, the arrangement of plastic and CsI(Tl) scintillation detectors is designed to optimize the polarization response. The dominant type of scatter event is one involving only two detector elements, in which incident photons scatter from a low-Z plastic detector element into a high-Z CsI(Tl) element. The distribution of azimuthal scatter angles for these events provides a polarization signature. The total effective area for polarimetry is $\sim$1000 cm^2 at energies above 100 keV. Since the total energy deposit is a sum of the energy deposits in all triggered elements, spectroscopic information is provided by all types of events of any multiplicity (both single events and scatter events). To characterize the GRB parameters, spectroscopic measurements (20–5000 keV) are obtained using all event types (both multiple and single events), with a total effective area that reaches >3000 cm^2 between 50 and 500 keV.

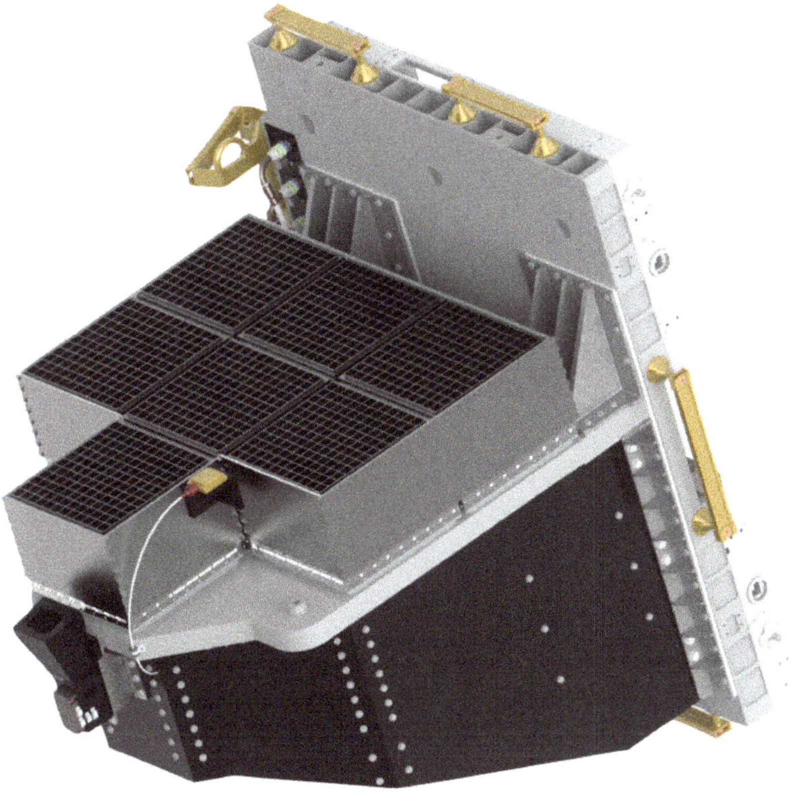

Figure 9. The LEAP payload includes seven independent polarimeter modules. When mounted on the outside of the ISS, the array of modules points toward the local zenith and scans the sky for GRBs.

To accurately reconstruct the source spectrum and polarization, LEAP self-sufficiently determines the source direction using single events from all 420 CsI(Tl) elements, whose relative response provides the source localization. Localization errors of $<5^\circ$ (1σ) are obtained at a rate of about 40 per year. This is sufficiently precise to enable rapid follow-up by many ground-based instruments using the rapid burst response messages that will be generated and distributed (in real time) by LEAP.

6.3. Expected Performance

As a wide-FoV instrument, LEAP maintains some level of polarization sensitivity out to at least 75° off-axis, providing an effective FoV of $\sim 1.5\pi$ sr. Minimal obstructions from the ISS within the FoV maximize sky exposure and minimize photon absorption and scattering effects. However, scattered flux from ISS structures (such as solar panels) must always be considered in the analysis of both the spectrum and the polarization. Response simulations spanning a range in energy, spectral shape, and incidence direction have been used in estimates of instrument performance. Figure 10 shows, for a three-year mission, the number of expected GRBs as a function of MDP. These estimates show that LEAP will attain its requirement of 65 GRBs with $<30\%$ MDP, with a significant margin.

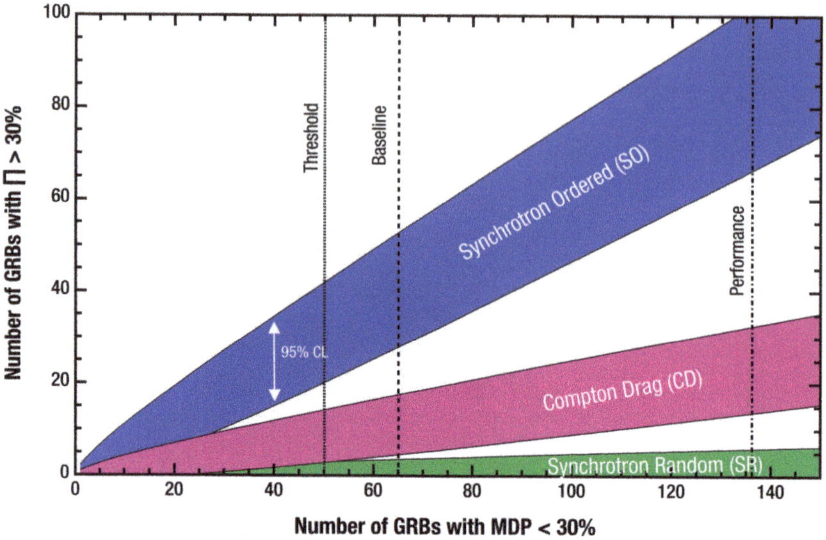

Figure 10. The number of GRBs measured with a polarization degree > 30% versus the number of GRBs with an MDP < 30% provides a convenient way to distinguish between different model classes. For the case shown here, 65 GRB measurements are required to distinguish the three model classes. LEAP will be able to measure a total of 135 GRBs with MDP < 30% (and 23 GRBs with MDP < 10%) during its baseline 32 months of science operations.

7. MoonBEAM

7.1. Mission Overview

The Moon Burst Energetics All-sky Monitor (MoonBEAM) is a proposed gamma-ray mission to observe the entire sky instantaneously for relativistic astrophysical explosions from a cislunar orbit. It is designed to explore the behavior of matter and energy under extreme conditions by observing the prompt emission from GRBs, identifying the conditions capable of launching transient relativistic jets and the origins of high-energy radiation from the relativistic outflows. MoonBEAM provides the essential continuous all-sky gamma-ray observations for time domain and multi-messenger astrophysics by reporting on any prompt emission of a GRB and by providing the critical first alerts to the community for contemporaneous and follow-up observations.

7.2. Detector Design

MoonBEAM achieves instantaneous, all-sky coverage by positioning six gamma-ray detector assemblies at the corners of the spacecraft to minimize blockage (see Figure 11), and by deploying the observatory in an Earth–Moon Lagrange Point 3 cislunar orbit instead of low-Earth orbit (LEO) to reduce the particle background from radiation belts, atmospheric interactions, and planetary occultation from 30% to <<1% of the sky.

Each detector assembly consists of a NaI(Tl)/CsI(Na) phoswich scintillator coupled to flat-panel photomultiplier tubes. The phoswich design allows for both localization improvement and increased effective area for spectroscopy. It is sensitive to 10–5000 keV photons, with an energy resolution better than 12% at 662 keV.

Figure 11. Six scintillating detectors positioned for an instantaneous all-sky field of view, no slewing required. Coupled with a cislunar orbit, MoonBEAM provides an unprecedented all-sky sensitivity that cannot be achieved in low-Earth orbit.

7.3. Expected Performance

MoonBEAM is expected to detect more than 1000 GRBs over 30 months of operation, with an intrinsic localization capability similar to that of the *Fermi* Gamma-ray Burst Monitor. In the absence of a detection, MoonBEAM provides unprecedented sensitive gamma-ray upper limits for any externally detected transients such as mergers seen in gravitational waves and supernovae detected in optical wavelengths, estimated to be in the order of thousands over the operation time of MoonBEAM. Figure 12 shows the 1-second limiting flux sensitivity of MoonBEAM across the entire sky.

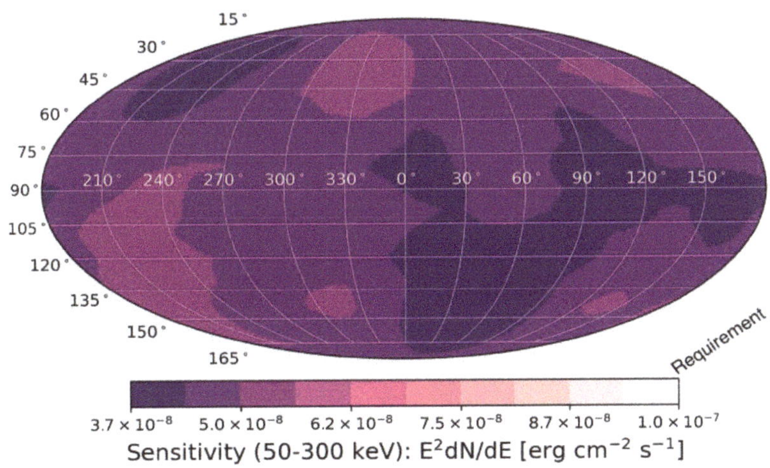

Figure 12. MoonBEAM's 1-second limiting flux sensitivity across the entire sky in any instant, providing the continuous sensitivity needed to study astrophysical jet formation and emissions.

The wide-field, sensitive, and continuous gamma-ray coverage is necessary to advance our current understanding of astrophysical jet formation, structure, and evolution. MoonBEAM achieves sensitivity improvement over current missions in LEO because of the combined advantage of its cislunar orbit and instrument design. It will join the Interplanetary Gamma-ray Burst Timing Network [55] as one of the few missions outside of

LEO with gamma-ray sensitivity and the only one outside of LEO with the capability of on-board transient localization.

8. POLAR-2

8.1. Mission Overview

POLAR-2 (https://www.unige.ch/dpnc/polar-2 is a dedicated GRB polarimeter foreseen to be launched in 2027 towards the China Space Station (CSS). The detector is being developed by an international collaboration consisting of teams from Switzerland, Poland, Germany, and China. The project is currently ready for the production of the flight model after a prototype was successfully tested for physics performance and space qualification in 2023. POLAR-2 is a successor of the POLAR mission, which detected 55 GRBs between October 2016 and April 2017 and performed polarization measurements in the energy range of 50–500 keV of 14 of these [56], as well as of the Crab pulsar [57].

Although the measurement results of POLAR are the most constraining GRB polarization measurements to date, they are only able to constrain the polarization degree to be below ≈40%. Therefore, the results remain consistent with the majority of the existing theoretical predictions [58]. This, along with the hint of an evolution of the polarization with time observed in two of the GRBs [56], indicates the need for a significantly more sensitive detector. For this purpose, the POLAR-2 detector was initiated in 2017, followed by an approval for launch towards the CSS, through a United Nations Office for Outer Space Affairs call in 2019.

8.2. Detector Design

Compared to its predecessor, the POLAR-2 detector will be a factor of four larger in size, thereby employing a total of 6400 plastic scintillator bars. These scintillators are read out in groups of 64, using segmented SiPM arrays connected to their own front-end electronics, as indicated in Figure 13. As the GRB photons enter the detector, they can undergo Compton scattering in the detector array, followed by photo-absorption in a second scintillator. Their azimuthal scattering angle can be constrained using the relative position of the two scintillators in which the photon interacted. As the photons will scatter preferentially perpendicular to their initial polarization, this measurement allows determining the polarization of the incoming photon flux.

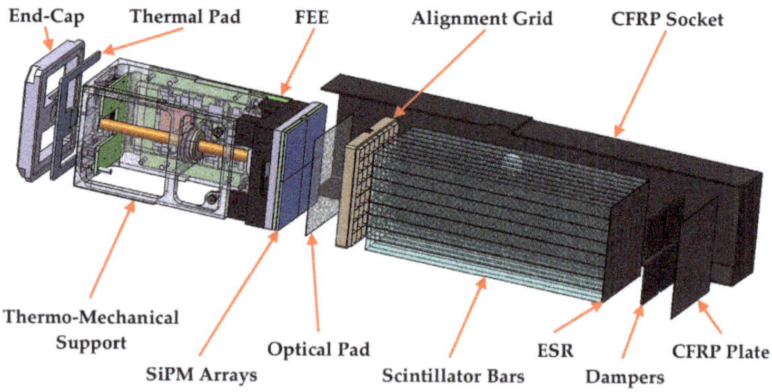

Figure 13. An exploded view of one of the 100 POLAR-2 detector modules that make up the instrument. The detector measures the polarization of the incoming photons when these undergo Compton scattering in the segmented scintillator array. The scintillators are read out by a temperature-controlled Silicon photomultiplier (SiPM) array connected to its own front-end electronics. Taken with permission from [59].

While the increase in detector area provides an increase in effective area bu a factor of four compared to POLAR, further improvements to its design allow for additional improvements in its sensitivity. In particular, the switch from the PMTs as used in POLAR to SiPMs results in a significant increase in sensitivity at lower energies.

8.3. Expected Performance

Thanks to all the design improvements, POLAR-2 is approximately an order of magnitude more sensitive compared to POLAR, as indicated in Figure 14. As a result, POLAR-2 will be able to perform constraining polarization measurements for GRBs with fluences as low as 10^{-6} erg cm^{-2}, while measurements able to constrain the polarization degree below 10% will be possible for about 10–15 GRBs per year.

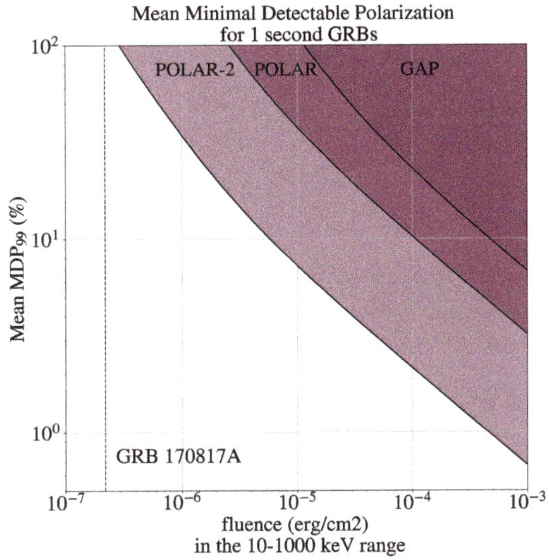

Figure 14. The MDP (99%) for 1 s GRBs as a function of GRB fluence for the POLAR and POLAR-2 instruments. In addition, the sensitivity of the GAP, the first dedicated gamma-ray polarimeter, is indicated. For reference, the fluence of the very weak GRB 170817A is indicated with a dotted line.

The POLAR-2 detector will also contribute to transient alerts thanks to its large effective area, which exceeds 2000 cm^2. This large effective area, combined with continuous observations of half the sky and almost continuous communication to the ground will allow POLAR-2 to send alerts within one minute from the onset of about one GRB every 2 days. As the instrument furthermore has access to a GPU on-board the CSS, studies are currently ongoing on how to optimize GRB spectral and location information within such alerts [60].

9. StarBurst

9.1. Mission Overview

The StarBurst Multimessenger Pioneer is a highly sensitive and wide-field gamma-ray monitor designed to detect the prompt emission of gamma-ray bursts. StarBurst is designed as a SmallSat to be deployed to LEO as a secondary payload using the Evolved Expendable Launch Vehicle Secondary Payload Adapter (ESPA) Grande interface for a nominal 1-year mission starting in 2027 to coincide with LIGO's scheduled fifth observing run. StarBurst will utilize NASA's Tracking and Data Relay Satellites (https://www.nasa.gov/mission/tracking-and-data-relay-satellites/ (TDRS) system to report possible electromagnetic counterparts to gravitational wave mergers with low latency via the GCN system. StarBurst is among the first of NASA's new Pioneer-class missions, intended to perform compelling astrophysics science at a lower cost than missions in the Explorers Program. The Pioneer program provides opportunities for early-to-mid-career researchers to propose innovative experiments and lead space or suborbital science investigations for the first time.

9.2. Instrument Design

StarBurst relies heavily on the heritage of the GBM, Glowbug, and BurstCube instruments, consisting of an array of 12 NaI(TI) scintillator detectors that utilize new, low-mass, and low-voltage SiPMs to cover an energy range from 30 keV to 2 MeV. The StarBurst detectors are arranged to form a half cube, with two detectors mounted side-by-side to form four sides of the cube and four detectors comprising the top, as shown in Figure 15. This configuration maximizes the available active surface area of the instrument given the volume constraints imposed on ESPA-Grande Rideshare Payloads and provides coverage of the entire unocculted sky. Each of the individual StarBurst detectors consists of a 24 cm $\times$ 24 cm $\times$ 1.6 cm NaI(TI) scintillator read out by an array of SiPMs and enclosed in a housing with a thin (1 mm) aluminum window transparent down to 30 keV and a beryllium-copper back shield to provide strong attenuation below $\sim$ 200 keV. The scintillation light is read out by a 2 $\times$ 38 linear array of 6 mm $\times$ 6 mm J-Series SiPMs from ON Semiconductor (formerly SensL) optically coupled through an elastomeric silicone optical pad to a single edge of the NaI(Tl) crystal. Edge readout of the NaI(TI) exploits the planar geometry of the crystal to pipe scintillation light to the SiPM array. Each of the detectors has a dedicated bias voltage supply and Multi-Channel Analyzer (MCA) for independent operation. The MCA is a Commercial-off-the-Shelf (COTS) Bridgeport Instruments slimMorpho with an on-board 20 MHz oscillator and flash ADC.

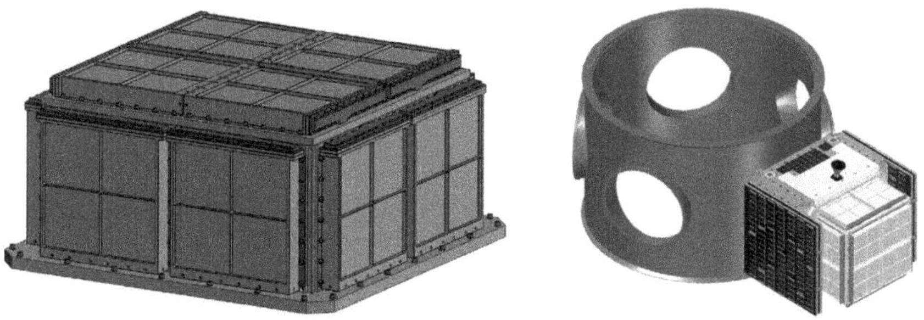

Figure 15. The StarBurst instrument (**left**). The integrated StarBurst observatory shown mounted to an ESPA-Grande ring (**right**).

9.3. Expected Performance

The scientific performance of the StarBurst instrument was assessed through Monte Carlo simulations using *Geometry and Tracking 4* (*GEANT4*) ([61]). The peak StarBurst effective area between 50 and 300 keV, averaged over azimuth, is roughly 3000 cm^2, or roughly 500% that of the GBM (Figure 16. This provides an estimated detection efficiency of >90% for a 64 ms peak flux (photons cm^{-2} s^{-1}) of 0.9 for SGRBs (<2 s) and a 1024 ms peak flux of 0.25 for long GRBs (>2 s). Because StarBurst will have a similar field of view and duty cycle as the GBM, it is estimated that StarBurst will detect 158 SGRBs per year, compared to the 40 and 8.6 SGRBs detected by the GBM and Swift, respectively. StarBurst is estimated to achieve a localization uncertainty within 8° (1σ) for an SGRB with a 1 photon cm^{-2} s^{-1} 64 ms peak flux. For SGRBs with a 64 ms peak flux comparable to GRB 170817, or about 3 photons cm^{-2} s^{-1}, StarBurst is expected to achieve a localization uncertainty of <3° (1 σ). Employing the technique developed in Howell et al. [62], along with the StarBurst detection efficiency and the projected A+ (LHVKI) BNS detection efficiency, provides an estimate joint GW–SGRB detection rate of roughly seven joint GW–SGRB detections per year.

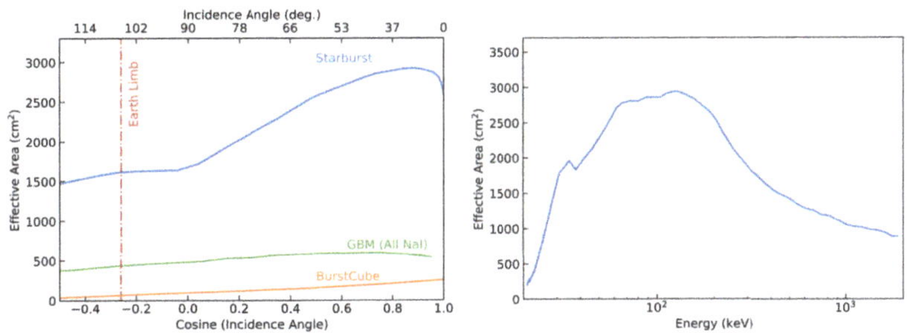

Figure 16. The StarBurst effective area averaged over azimuth, as a function of the angle from the instrument boresight (**left**). The StarBurst effective area as a function of energy (**right**).

10. STROBE-X

10.1. Mission Overview

The Spectroscopic Time-Resolving Observatory for Broadband Energy X-rays (https://strobe-x.org/) (STROBE-X) is a mission proposed to the NASA call for an X-ray probe-class ($1B PI-managed cost cap) mission responding to the recommendations of the 2020 Astrophysics Decadal Survey (Astro2020; see Figure 17). It has a broad range of science goals with a focus on time domain and transient events in the era of multi-wavelength and multi-messenger astronomy. It combines a huge collecting area, high throughput on bright sources, broad energy coverage, and excellent spectral and temporal resolution in a single facility. With its wide field of view, agile spacecraft, and low-latency communications, it would be a critical component of the NASA Time Domain and Multi-Messenger (TDAMM) program.

10.2. Instrument Design

STROBE-X carries three instruments. The Low-Energy Modular Array (LEMA) covers the soft or low-energy band (0.2–12 keV) with an array of lightweight optics (3 m focal length) that concentrate incident photons onto small solid-state detectors with a CCD-level (85–175 eV) energy resolution, a 100 ns time resolution, and low background rates. This technology has been fully developed for NICER and will be scaled up to take advantage of the longer focal length of LEMA, which provides a factor of 8.5 improvement in effective area over NICER with over 1.6 m^2. The High-Energy Modular Array (HEMA) covers the harder or higher energy band (2–30 keV or beyond), with modules of Si drift detectors and micro-pore collimators originally developed for the European LOFT and eXTP mission concepts. HEMA provides a factor of 5.5 improvement in effective area (3.4 m^2) and $\sim$3 in spectral resolution (200–300 eV) over the RXTE/PCA. The Wide-Field Monitor (WFM) comprises a set of coded-aperture cameras operating in the 2–50 keV band, which have a combined instantaneous field of view of 1/3 of the sky, with arcmin localization capability. It will act as a trigger for pointed observations of X-ray transients and will also provide high duty-cycle, high time-resolution, and high spectral-resolution monitoring of the dynamic X-ray sky. The WFM will have 15-times the sensitivity of the RXTE All-Sky Monitor, enabling multi-wavelength and multi-messenger investigations with a large instantaneous field of view, down to a new, order-of-magnitude lower flux regime. On-board processing will detect bursts in real time and provide notifications to the ground, as well as triggering autonomous slews to get the pointed instrument on source in a few minutes.

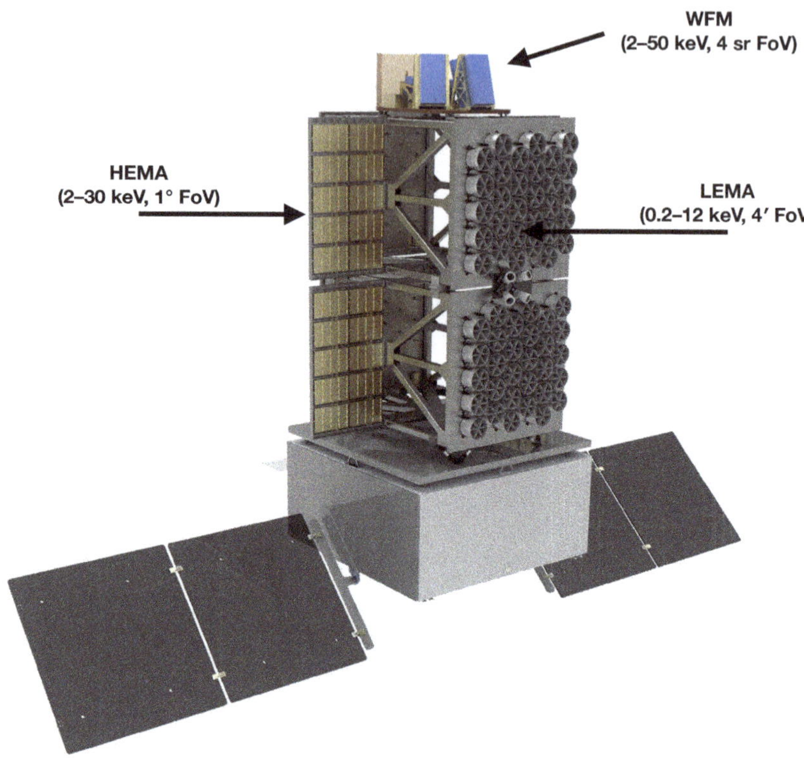

Figure 17. Rendering of the STROBE-X spacecraft, showing the 3 science instruments. LEMA and HEMA are co-aligned narrow FoV instruments, while the WFM has an instantaneous FoV of 1/3 of the sky.

10.3. Expected Performance

STROBE-X will make major advances in several areas of GRB and multi-messenger astronomy (Figure 18). The WFM will detect and localize >5 short GRBs per year and distribute the position accurate to 2 arcmin, brightness, and timing to the ground in <5 min. In some cases, it will measure the redshift directly from the X-ray data (from the location of absorption edges in their spectra). STROBE-X will study the plateau emission of both short and long GRBs and provide unique diagnostics of whether the emission comes from millisecond magnetars or structured jets, revealing the nature of the central engines.

STROBE-X will also be triggered by ground-based GW detections and get the pointed instruments on source within <11.5 min, given access to early afterglows with the tremendous collecting area and CCD-quality spectral resolution of LEMA.

The large grasp and softer response than many GRB missions will give STROBE-X access to unusual GRB phenomena that have not been well studied so far, including X-ray flashes (XRFs) and ultra-long GRBs (ULGRBs). Detecting and localizing >10 XRFs per year will increase the samples of these sources and break the degeneracies in the models for the source mechanisms.

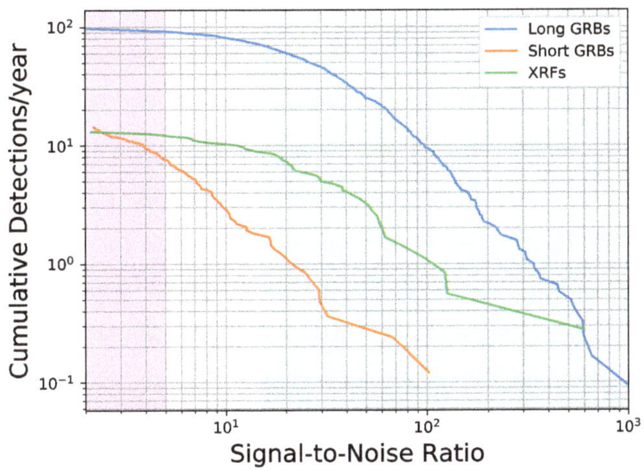

Figure 18. The expected cumulative on-board detection rate of canonical gamma-ray bursts (GRBs) and X-ray flashes (XRFs) by the *STROBE-X*/WFM as estimated by folding observed GRB and XRF spectra through the WFM responses and accounting for the effective field-of-view of the WFM. It will detect ∼100 long-duration GRBs, ∼7 short-duration GRBs, and ∼12 XRFs on-board per year. The onboard detection rate of long GRBs exceeds that of *Swift*/BAT, while the short GRB detection rate is comparable. A unique capability is the down-link of event data to the ground for the WFM, enabling sub-threshold searches to double the number of short GRB detections (purple shading). The detection rate of XRFs exceeds that of previous instruments and is a particular science focus for the WFM. The blue shading shows the region of the signal-to-noise ratio, where high-fidelity spectroscopy can be performed in the prompt X-ray for these sources.

11. SVOM

11.1. Mission Overview

The Space-based Variable astronomical Object Monitor (SVOM) mission is dedicated to gamma-ray burst studies and, in general, to time domain astrophysics, including multi-messenger science. It is optimized to detect and follow-up all types of GRBs, but particularly tailored for high-redshift GRBs thanks to its low-energy triggering threshold of around 4 keV. One of the main goals of SVOM is to produce a complete catalog of GRBs, including their redshift measurements. For this purpose, SVOM will follow an anti-solar pointing strategy and will avoid the galactic plane, in order to facilitate the observations from ground-based observatories and robotic telescopes. That is why SVOM is composed of a space segment, as well as a few ground-based dedicated follow-up facilities. SVOM alerts will be promptly transmitted to the ground through a dedicated network of VHF antennas and through the Chinese Beidou inter-satellite communications link. The goal is for ground observers to receive SVOM alerts less than 30 s after the GRB is detected on board. The SVOM satellite will be launched from China around mid-2024 with an LM 2-C rocket and injected in a low-Earth orbit (h∼600 km) with an inclination of about 30°. It will carry four co-aligned instruments. Two instruments (ECLAIRs and GRM) are sensitive in the hard X-/soft gamma-ray energy range and have a wide FoV, in order to monitor vast regions of the sky and detect gamma-ray transients. Two narrow FoV instruments (MXT and VT) will be used to follow-up and characterize the afterglow emission. The SVOM-integrated payload is shown in Figure 19.

Figure 19. The fully integrated SVOM payload.

11.2. Instrument Design

ECLAIRs is a coded-mask telescope, composed of a 54 × 54 cm^2 pseudo-random coded mask made of a Ti-Ta-Ti sandwich (10/0.6/10 mm) placed 45.8 cm above a pixelated detection plane made of 80 × 80 CdTe crystals (4 × 4 × 1 mm^3). Its FoV is about 2 sr (89°× 89°) wide. ECLAIRs is sensitive in the 4 keV–150 keV energy range, and it comprises on-board software to detect and localize (to better than 13 arcmin) in near-real time the GRBs that appear in its FoV. Once a new transient is detected, ECLAIRs issues an alert and requests the platform to slew so that the error box can be observed by the narrow-field instruments.

ECLAIRs is complemented by the Gamma-ray Monitor (GRM), a set of three 1.5 cm thick NaI scintillators of 16 cm in diameter, each one offset by 120° with respect to each other and with a combined FoV of ∼2.6 sr. The GRM has poor localization capabilities, but it extends the SVOM spectral range up to about 5 MeV and increases the probability of the simultaneous detection of short GRBs and GW alerts.

The Microchannel X-ray Telescope is a light (<42 kg) and compact (focal length of ∼1.15 m) X-ray-focusing telescope; its sensitivity below 1 mCrab makes it the ideal instrument to detect, identify, and localize down to the arcmin level X-ray afterglows of the SVOM GRBs. Its optical design is based on a "lobster-eye" grazing-incidence X-ray optics, inspired by the vision of some crustacean decapods. It is composed of 25 square MPO plates of 40 mm each arranged in a 5 × 5 configuration. Although lobster-eye optics were originally developed for large-FoV telescopes (several tens of square degrees), the MXT optical design is optimized for a (relatively) small FoV of 58 × 58 arcmin^2. The "lobster-eye" technique results in a peculiar point-spread function (PSF), made by a central peak and two cross-arms. The MXT optics is coupled with a focal plane based on a pnCCD sensor, cooled at -65°, sensitive in the 0.2–10 keV energy band. The MXT will localize GRB afterglows to better than the arcminute for the majority of them.

The Visible Telescope (VT) is a Ritchey–Chretien telescope with a 40 cm diameter primary mirror. Its field of view is 26 × 26 arcmin^2 wide, adapted to cover the ECLAIRs error box in most of the cases. It has two channels, a blue one (400–650 nm) and a red one (650–1000 nm), and a sensitivity limit of $M_V = 22.5$ in 300 s, allowing the detection ∼80% of the ECLAIRs GRBs. The main characteristics of the SVOM space segment are summarized in Table 3.

Table 3. Summary of the characteristics of the SVOM space instruments.

	ECLAIRs	GRM	MXT	VT
Energy/Wavelength	4–150 keV	15–5000 keV	0.1–10 keV	650–1000 nm
Field of View	2 sr	2.6 sr (combined)	$58' \times 58'$	$26' \times 26'$
Localization Accuracy	$<12'$	$<20°$	$<2'$	$<1''$
Expected GRBs Year^{-1}	60	90	50	40

The SVOM mission is also provided with a number of dedicated telescopes on the ground. In particular, here, we mention the following:

- The Ground-Based Wide-Angle Cameras (GWACs), a set of 36 optical cameras with a combined FoV of 5400 deg^2, located in Ali (China), whose goal is to catch the prompt optical emission for the ECLAIRs GRBs;
- The Chinese Ground Follow-up Telescope (C-GFT), a robotic 1 m class telescope, with a 21 × 21 arcmin^2 FoV, located in Xinglog (China) and sensitive in the 400-950 nm wavelength range;
- The French Ground Follow-up Telescope (F-GFT, Colibri), a robotic 1 m class telescope, with a 26 × 26 arc min^2 FoV, located in San Pedro Martir (Mexico) and with multi-band photometry capabilities over the 400–1700 nm wavelength range.

Other robotic telescopes will be part of the SVOM follow-up system, but they will not be fully dedicated to SVOM.

11.3. Expected Performance

The understanding of the GRB physics requires observations in the largest spectral domain and in the largest temporal interval, from the possible precursor up to the transition between the prompt and afterglow emissions. Simultaneous observation of the prompt GRB event in the gamma-ray, X-ray, and visible bands, combined with narrow field observations of the afterglow in the X-ray, visible, and near-infrared bands immediately after the beginning of the event will enable a better understanding of the mechanisms at work in such events. New measurements (gravitational wave, gamma-ray polarization, neutrinos) may also become possible in the future, and their impact on the physical understanding of the GRB phenomenon will be maximized if these measurements are made for bursts whose "standard" properties, including the distance, are well measured. The SVOM mission is well adapted to these objectives. Compared to previous missions, it offers simultaneously (i) the capacity to trigger on all types of GRBs (especially on X-ray-rich and ultra-long ones); (ii) an excellent efficiency of the follow-up and the redshift measurement; (iii) a good spectral coverage of the prompt emission by ECLAIRs+GRM, allowing a detailed modeling (for a significant fraction of SVOM GRBs, the GWAC will provide, in addition, a measurement or an upper limit on the prompt optical emission); and (iv) a good temporal and spectral coverage of the prompt and afterglow emission thanks to the MXT, VT, and GFTs.

Concerning multi-messenger astrophysics, SVOM with its ground and space instruments will offer a large and complementary follow-up capability through ToOs. The GWAC with its 5000 sq. deg. coverage can start the observation from the alert reception. The GFTs with their small FoV will confirm GWAC candidates and will be able to perform follow-up for well-localized events. To activate the satellite instruments, we will rely on a specific ToO program to send the observation program using S-band stations. This program guarantees less than 12 h between the alert and the start of space observations (less can be expected for most cases) and can be activated around 20-times per year. From space, the MXT and its 1 sq. deg. FoV will have the possibility to cover a larger sky portion using a specific tiling procedure.

12. THESEUS

12.1. Mission Overview

The Transient High-Energy Sky and Early Universe Surveyor (https://www.isdc.unige.ch/theseus/) (THESEUS) mission concept aims to fully exploit GRBs for investigating the early universe, around the epoch of re-ionization, and substantially advancing multi-messenger astrophysics (see Figure 20). THESEUS is planned to also simultaneously increase the discovery space of high-energy transient phenomena and allow tests of fundamental physics. The core science goals of THESEUS are summarized as follows:

- Investigating the first billion years of the universe through high-redshift GRBs, thus shedding light on the main open issues in modern cosmology, like (i) the population of primordial low-mass and -luminosity galaxies; (ii) the drivers and evolution of cosmic re-ionization; and (iii) the star-formation rate (SFR) and metallicity evolution up to the "cosmic dawn" and across Pop-III stars.
- Providing a substantial advancement of multi-messenger and time domain astrophysics by enabling the identification, accurate localization, and study of (i) electromagnetic counterparts to sources of gravitational waves and neutrinos, which will be routinely detected in the mid-2030s by the second- and third-generation gravitational wave (GW) interferometers and future neutrino detectors and (ii) all kinds of GRBs and most classes of other X-/gamma-ray transient sources.

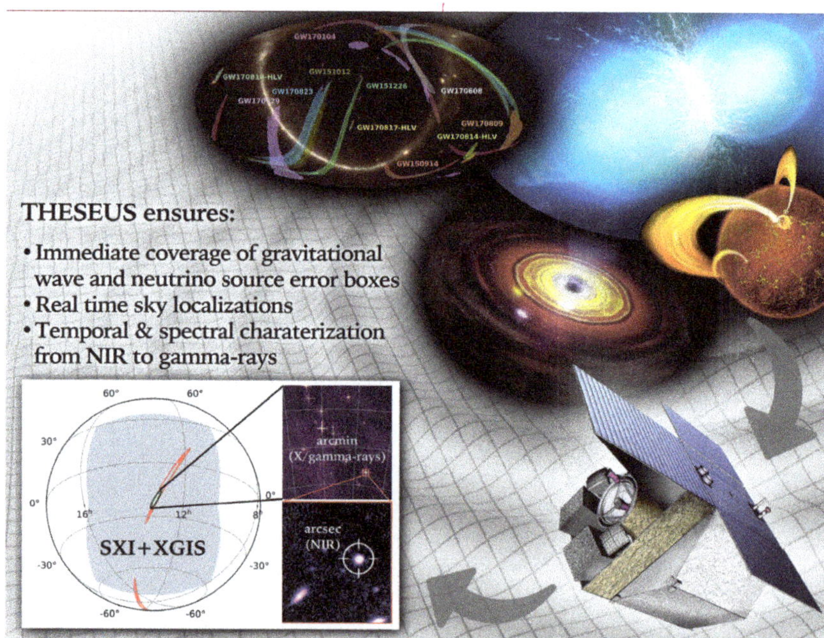

Figure 20. Examples of THESEUS's capabilities for multi-messenger and time domain astrophysics.

The achievement of these scientific objectives will be possible by a mission concept including: (a) a set of innovative wide-field monitors with an unprecedented combination of a broad energy range, sensitivity, FoV, and localization accuracy and (b) an on-board autonomous fast follow-up in the optical/NIR band, arcsec location, and redshift measurement of detected GRB/transients.

THESEUS has been selected twice by the European Space Agency (ESA) for a phase A study (in 2018 and 2022), aiming at demonstrating its technological and programmatic feasibility within the boundaries of a medium-sized mission. THESEUS is currently undergoing its second phase A study, with a further selection step expected in 2026 and eventually a launch planned in 2037. Nominal scientific operations are being planned for four years, but the lack of on-board consumables makes it feasible for the mission to extend operations in space well beyond the nominal lifetime.

12.2. Instruments' Design

Three instruments are planned on-board THESEUS. The Soft X-ray Imager (SXI) uses lobster-eye wide-field ($\sim$0.5 sr) focusing optics to increase the mission's sensitivity to fast transients in the 0.3–5 keV energy band. The use of such optics provides uniform sensitivity across a very large field of view while maintaining arcminute localization accuracy. The X- and Gamma-ray Imaging Spectrometer (XGIS) is a GRB and transient monitor providing an unprecedented combination of an exceptionally wide energy band (2 keV–10 MeV), imaging capabilities, and location accuracy. The latter achieves $<$15 arcmin up to 150 keV over an FoV of 2π. The instrument is also characterized by an energy resolution of a few hundred eV at energies $<$30 keV and a time resolution of a few µs over the whole energy band. Finally, the Infra-Red Telescope (IRT) is mainly conceived to detect, identify, and measure the redshift of GRB afterglows detected by the SXI and the XGIS, especially those at high redshifts ($z > 6$). The IRT is a 70 cm Korsch telescope, optimized for an off-axis line of sight (LoS) of 0.884 deg. The optical design will implement two separated FoVs, one for photometry with a minimal size of 15 $\times$ 15 arcmin (potentially extendable to 17 $\times$ 20 arcmin) and one for spectroscopy of 2 $\times$ 2 arcmin. On the photometric field of view, the IRT will be able to acquire images using five different filters (I, Z, Y, J, and H) and, on the spectroscopic field of view, will provide moderate resolution (R$\sim$400) slit-less spectroscopy in the 0.8–1.6 µm range.

12.3. Expected Performance

THESEUS has two main core science objectives: the characterization of the physics of the high-redshift universe and the electromagnetic characterization of gravitational wave transients (above all, the neutron star binary mergers). Long and short GRBs will be exploited for this scope. The first objective is instrumental to the understanding of the emergence of the first structures in the universe, including the mass function of the high-redshift galaxies and the metal enrichment in the early stages of the universe (at the epoch of the re-ionization). The second objective has transformational potential, building upon the first-ever gravitational wave multi-messenger detection thus far, i.e., GW170817. THESEUS's observations and discoveries in this field are expected to unveil the nature of ultra-dense matter, pin down the physics of relativistic jets, and help us understand the nucleosynthesis of heavy elements (in turn, probing fundamental aspects of general relativity and cosmology). We show in Figure 21 the total number of GRBs expected to be detected during the nominal lifetime of the mission (corresponding to 3.45 yrs of scientific operations) compared to the total number of GRBs discovered from 2005 up to 2020. The transformational capabilities of THESEUS in these respects can well be appreciated from this figure.

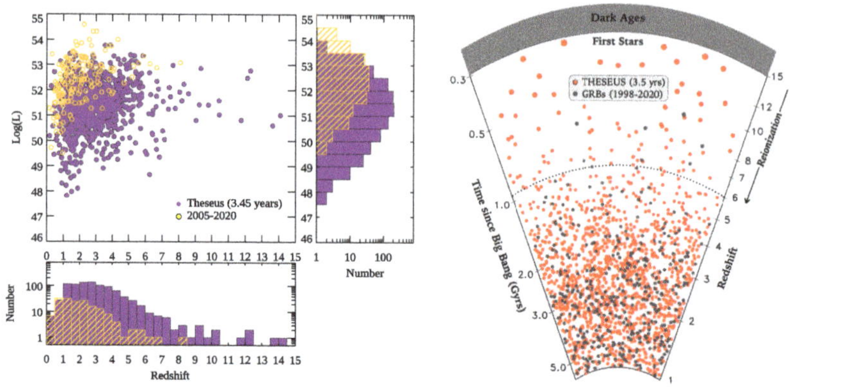

Figure 21. (**Left**) Distribution of long GRBs with redshift determination in the peak isotropic luminosity versus redshift plane now (yellow points and hatched histogram) and after the nominal operation life of THESEUS (purple points and full histogram). (**Right**) A different version of the left-side figure where the GRBs discovered by THESEUS and those detected up to 2020 are displayed in a cone representing the cosmic evolution.

13. Conclusions

In this paper, we provided an overview of a number of missions, either close to beginning scientific operations or being planned for the coming/far future, that are expected to dramatically widen our capability of detecting and characterizing bright impulsive transient events of astrophysical interest, such as (but not limited to) GRBs. The list of missions is not meant to be exhaustive, especially in view of the recent fast-growing interest of the international community in the fields of time domain and multi-messenger astrophysics, which is driving many parallel developments in several countries around the globe. The missions presented here have been or are being conceived based on largely overlapping science objectives, which have, as main celestial targets, transient and fast variable sources (down to time scales as short as fractions of a second). Catching and studying these objects typically requires the simultaneous availability of large-FoV X-/γ-ray instruments efficiently (i.e., with high duty cycles) monitoring the high-energy sky and either narrower field instruments providing higher sensitivity measurements in complementary energy domains (from space, as well as from the ground) or advanced detector technologies paving the way to poorly explored regions of the relevant parameter space (e.g., high-energy polarimetry, simultaneous energy–time-resolved spectroscopy, etc.). Given the largely different programmatic states of all summarized missions and the dynamic evolution of the design of at least those missions that are still in the early design and programmatic stage, providing an exhaustive comparison of their performance capabilities in different scientific fields is hardly achievable with sufficient confidence and, thus, beyond the scope of the present paper.

At the time of writing, the EP is the first of the described mission to come on-line, as the launch was successfully executed on 2024 January 9 and information on the first results is expected to be publicly available soon. SVOM is planned to be the second in line, with a launch planned for June 2024. As described in the previous sections, both missions are likely to boost the number of GRBs, as well as other high-energy transients, to be discovered, characterized, and possibly followed-up in different energy domains during the next (at least) $\sim$5 years. The lifetime of both the EP and SVOM will significantly overlap with the scientific runs of the current generation of GW detectors, such as LIGO and VIRGO, possibly providing the discovery of electromagnetic counterparts of merging binary systems. A similar conclusion applies to the case of StarBurst, the launch of which is currently planned in 2027, and the 1-year expected duration of science operations shall match the LIGO's scheduled so-called fifth observing run. In 2027, also the installation of POLAR-2 onto the CSS is being planned, and this should bring to the community a much deeper insight into the polarization properties of GRBs, extending the outcomes of previous investigations in this domain from its predecessor, POLAR. Less certain, at present, is the path to the beginning of operations for a few more missions presented, the eXTP, Gamow, HiZ-GUNDAM, LEAP, and MoonBEAM. The advanced design state of all these missions could possibly bring them to space between the late 2020s and the early 2030s, making them largely complementary and uniquely valuable for discoveries in the field of time domain and multi-messenger astrophysics to the EP, SVOM, and StarBurst. STROBE-X and THESEUS are being planned for scientific operations no earlier than the mid- to late-2030s. STROBE-X is competing with several other candidate missions in both the X-ray and IR domain within the NASA 2023 Probe mission call (https://explorers.larc.nasa.gov/2023APPROBE/, while THESEUS is competing against two further candidate missions for a launch opportunity in ~2037 within the context of the ESA's seventh call for medium-sized missions (https://www.esa.int/Science_Exploration/Space_Science/Final_three_for_ESA_s_next_medium_science_mission). Although the possible launch dates of STROBE-X and THESEUS are projected relatively far in the future, their timeline could interestingly match the planned observational runs from the third generation of GW detectors, such as the Einstein Telescope (ET) (see, e.g., [63,64] and the references therein) and the Cosmic Explorer (CE) (see, e.g., [65] and the references therein). The enhanced sensitivity of these instruments would greatly increase the number of possibly detected electromagnetic counterparts of GW sources, reaching up to a few tens per year, as reported, e.g., in the case of THESEUS [66].

Author Contributions: E.B. coordinated the overall work; W.Y. and H.S. provided information on the Einstein Probe; E.B., S.Z., M.F., M.H. and A.S. provided information on the eXTP; N.W., S.G., T.-C.C., M.S., W.B., C.K. and A.v.d.H. provided information on Gamow; D.Y., A.D. and H.M. provided information on HiZ-GUNDAM; M.L.M., J.G., C.W.-H. and P.V. provided information on LEAP; C.M.H., C.W.-H., A.G. and P.J. provided information on MoonBEAM; M.K., N.P. and N.D.A. provided information on POLAR-2; D.K. and J.E.G. provided information on StarBurst; P.R., C.F., P.S.R. and T.M. provided information on STROBE-X; D.G., B.C. and J.W. provided information on SVOM; E.B., L.A., D.G., P.O. and A.S. provided information on THESEUS. All authors have read and agreed to the published version of the manuscript.

Funding: L.A. acknowledges support from the Italian Ministry of University and Research through grant PRIN MIUR 2020-2020KB33TP METE and the INAF grant program 2022.

Data Availability Statement: This paper contains information about different missions either being planned for the future or starting nominal operations soon. Further details on all missions considered can be found on the corresponding websites (if applicable) or in complementary material in the literature.

Acknowledgments: We thank all referees for their constructive comments and suggestions.

Conflicts of Interest: The authors declare no conflicts of interest.

References

1. Klebesadel, R.W.; Strong, I.B.; Olson, R.A. Observations of Gamma-ray Bursts of Cosmic Origin. *ApJL* **1973**, *182*, L85. [CrossRef]
2. Vigliano, A.; Longo, F. Gamma-ray Bursts: 50 Years and Counting! *Universe* **2024**, *10*, 57. [CrossRef]
3. Aptekar, R.L.; Frederiks, D.D.; Golenetskii, S.V.; Ilynskii, V.N.; Mazets, E.P.; Panov, V.N.; Sokolova, Z.J.; Terekhov, M.M.; Sheshin, L.O.; Cline, T.L.; et al. Konus-W Gamma-ray Burst Experiment for the GGS Wind Spacecraft. *SSRv* **1995**, *71*, 265–272. [CrossRef]
4. Tsvetkova, A.; Frederiks, D.; Svinkin, D.; Aptekar, R.; Cline, T.L.; Golenetskii, S.; Hurley, K.; Lysenko, A.; Ridnaia, A.; Ulanov, M. The Konus-Wind Catalog of Gamma-ray Bursts with Known Redshifts. II. Waiting-Mode Bursts Simultaneously Detected by Swift/BAT. *ApJ* **2021**, *908*, 83. [CrossRef]
5. Harmon, B.A.; Fishman, G.J.; Wilson, C.A.; Paciesas, W.S.; Zhang, S.N.; Finger, M.H.; Koshut, T.M.; McCollough, M.L.; Robinson, C.R.; Rubin, B.C. The Burst and Transient Source Experiment Earth Occultation Technique. *ApJS* **2002**, *138*, 149–183. [CrossRef]
6. Band, D.; Matteson, J.; Ford, L.; Schaefer, B.; Palmer, D.; Teegarden, B.; Cline, T.; Briggs, M.; Paciesas, W.; Pendleton, G.; et al. BATSE Observations of Gamma-ray Burst Spectra. I. Spectral Diversity. *ApJ* **1993**, *413*, 281. [CrossRef]
7. Kaneko, Y.; Preece, R.D.; Briggs, M.S.; Paciesas, W.S.; Meegan, C.A.; Band, D.L. The Complete Spectral Catalog of Bright BATSE Gamma-ray Bursts. *ApJS* **2006**, *166*, 298–340. [CrossRef]

8. Frontera, F.; Guidorzi, C.; Montanari, E.; Rossi, F.; Costa, E.; Feroci, M.; Calura, F.; Rapisarda, M.; Amati, L.; Carturan, D.; et al. The Gamma-ray Burst Catalog Obtained with the Gamma-ray Burst Monitor Aboard BeppoSAX. *ApJS* **2009**, *180*, 192–223. [CrossRef]
9. Boella, G.; Butler, R.C.; Perola, G.C.; Piro, L.; Scarsi, L.; Bleeker, J.A.M. BeppoSAX, the wide band mission for X-ray astronomy. *Astron. Astrophys. Suppl. Ser.* **1997**, *122*, 299–307. [CrossRef]
10. Demianski, M.; Piedipalumbo, E.; Sawant, D.; Amati, L. Cosmology with gamma-ray bursts. I. The Hubble diagram through the calibrated $E_{p,I}$-E_{iso} correlation. *A&A* **2017**, *598*, A112. [CrossRef]
11. Abdalla, E.; Abellán, G.F.; Aboubrahim, A.; Agnello, A.; Akarsu, Ö.; Akrami, Y.; Alestas, G.; Aloni, D.; Amendola, L.; Anchordoqui, L.A.; et al. Cosmology intertwined: A review of the particle physics, astrophysics, and cosmology associated with the cosmological tensions and anomalies. *J. High Energy Astrophys.* **2022**, *34*, 49–211. [CrossRef]
12. Abbott, B.P.; Abbott, R.; Abbott, T.D.; Acernese, F.; Ackley, K.; Adams, C.; Adams, T.; Addesso, P.; Adhikari, R.X.; Adya, V.B.; et al. Gravitational Waves and Gamma-rays from a Binary Neutron Star Merger: GW170817 and GRB 170817A. *ApJL* **2017**, *848*, L13. [CrossRef]
13. Troja, E.; Fryer, C.L.; O'Connor, B.; Ryan, G.; Dichiara, S.; Kumar, A.; Ito, N.; Gupta, R.; Wollaeger, R.T.; Norris, J.P.; et al. A nearby long gamma-ray burst from a merger of compact objects. *Nature* **2022**, *612*, 228–231. [CrossRef] [PubMed]
14. Dichiara, S.; Tsang, D.; Troja, E.; Neill, D.; Norris, J.P.; Yang, Y.H. A Luminous Precursor in the Extremely Bright GRB 230307A. *ApJL* **2023**, *954*, L29. [CrossRef]
15. Tavani, M.; Barbiellini, G.; Argan, A.; Boffelli, F.; Bulgarelli, A.; Caraveo, P.; Cattaneo, P.W.; Chen, A.W.; Cocco, V.; Costa, E.; et al. The AGILE Mission. *A&A* **2009**, *502*, 995–1013. [CrossRef]
16. Singh, K.P.; Tandon, S.N.; Agrawal, P.C.; Antia, H.M.; Manchanda, R.K.; Yadav, J.S.; Seetha, S.; Ramadevi, M.C.; Rao, A.R.; Bhattacharya, D.; et al. ASTROSAT mission. In Proceedings of the Space Telescopes and Instrumentation 2014: Ultraviolet to Gamma ray, Montréal, QC, Canada, 29 July 2014; Society of Photo-Optical Instrumentation Engineers (SPIE) Conference Series; Takahashi, T., den Herder, J.W.A., Bautz, M., Eds.; SPIE: Bellingham, DC, USA, 2014; Volume 9144, p. 91441S. [CrossRef]
17. Atwood, W.B.; Abdo, A.A.; Ackermann, M.; Althouse, W.; Anderson, B.; Axelsson, M.; Baldini, L.; Ballet, J.; Band, D.L.; Barbiellini, G.; et al. The Large Area Telescope on the Fermi Gamma-ray Space Telescope Mission. *ApJ* **2009**, *697*, 1071–1102. [CrossRef]
18. Winkler, C. INTEGRAL: Overview and Mission Concept. *ApJS* **1994**, *92*, 327. [CrossRef]
19. Gehrels, N.; Chincarini, G.; Giommi, P.; Mason, K.O.; Nousek, J.A.; Wells, A.A.; White, N.E.; Barthelmy, S.D.; Burrows, D.N.; Cominsky, L.R.; et al. The Swift Gamma-ray Burst Mission. *ApJ* **2004**, *611*, 1005–1020. [CrossRef]
20. Yuan, W.; Zhang, C.; Chen, Y.; Ling, Z. The Einstein Probe Mission. In *Handbook of X-ray and Gamma-ray Astrophysics*; Springer: Singapore, 2022; p. 86. [CrossRef]
21. Santangelo, A.; Zhang, S.N.; Feroci, M.; Hernanz, M.; Lu, F.; Xu, Y. The Enhanced X-ray Timing and Polarimetry Mission: eXTP. In *Handbook of X-ray and Gamma-ray Astrophysics*; Bambi, C., Ed.; Springer: Singapore, 2023; p. 154. [CrossRef]
22. Rachevski, A.; Zampa, G.; Zampa, N.; Campana, R.; Evangelista, Y.; Giacomini, G.; Picciotto, A.; Bellutti, P.; Feroci, M.; Labanti, C.; et al. Large-area linear Silicon Drift Detector design for X-ray experiments. *J. Instrum.* **2014**, *9*, P07014. [CrossRef]
23. Mineo, T.; Fraser, G.W.; Martindale, A.; Feldman, C.; Campana, R.; Cusumano, G.; Feroci, M. Effects of capillary reflection in the performance of the collimator of the Large Area Detector on board LOFT. *Exp. Astron.* **2014**, *37*, 69–84. [CrossRef]
24. Soffitta, P.; Baldini, L.; Baumgartner, W.; Bellazzini, R.; Bongiorno, S.D.; Bucciantini, N.; Costa, E.; Dovčiak, M.; Ehlert, S.; Kaaret, P.E.; et al. The Imaging X-ray polarimetry explorer (IXPE) at last! In Proceedings of the UV, X-ray, and Gamma-ray Space Instrumentation for Astronomy XXIII, San Diego, CA, USA, 5 October 2023; Society of Photo-Optical Instrumentation Engineers (SPIE) Conference Series; Siegmund, O.H., Hoadley, K., Eds.; SPIE: Bellingham, DC, USA, 2023; Volume 12678, p. 1267803. [CrossRef]
25. Zhang, B.; Mészáros, P. Gamma-ray Burst Beaming: A Universal Configuration with a Standard Energy Reservoir? *ApJ* **2002**, *571*, 876–879. [CrossRef]
26. Ghirlanda, G.; Nava, L.; Ghisellini, G.; Firmani, C. Confirming the γ-ray burst spectral-energy correlations in the era of multiple time breaks. *A&A* **2007**, *466*, 127–136. [CrossRef]
27. Amati, L.; Frontera, F.; Vietri, M.; in't Zand, J.J.M.; Soffitta, P.; Costa, E.; Del Sordo, S.; Pian, E.; Piro, L.; Antonelli, L.A.; et al. Discovery of a Transient Absorption Edge in the X-ray Spectrum of GRB 990705. *Science* **2000**, *290*, 953–955. [CrossRef] [PubMed]
28. Gehrels, N.; Ramirez-Ruiz, E.; Fox, D.B. Gamma-ray Bursts in the Swift Era. *Ann. Rev. A&A* **2009**, *47*, 567–617. [CrossRef]
29. Pélangeon, A.; Atteia, J.L.; Nakagawa, Y.E.; Hurley, K.; Yoshida, A.; Vanderspek, R.; Suzuki, M.; Kawai, N.; Pizzichini, G.; Boër, M.; et al. Intrinsic properties of a complete sample of HETE-2 gamma-ray bursts. A measure of the GRB rate in the Local Universe. *A&A* **2008**, *491*, 157–171. [CrossRef]
30. in't Zand, J.J.M.; Bozzo, E.; Qu, J.; Li, X.D.; Amati, L.; Chen, Y.; Donnarumma, I.; Doroshenko, V.; Drake, S.A.; Hernanz, M.; et al. Observatory science with eXTP. *Sci. China Physics, Mech. Astron.* **2019**, *62*, 29506. [CrossRef]
31. White, N.E.; Bauer, F.E.; Baumgartner, W.; Bautz, M.; Berger, E.; Cenko, B.; Chang, T.C.; Falcone, A.; Fausey, H.; Feldman, C.; et al. The Gamow Explorer: A Gamma-ray Burst Observatory to study the high redshift universe and enable multi-messenger astrophysics. In Proceedings of the UV, X-ray, and Gamma-ray Space Instrumentation for Astronomy XXII, San Diego, CA, USA, 24 August 2021; Society of Photo-Optical Instrumentation Engineers (SPIE) Conference Series; Siegmund, O.H., Ed.; SPIE: Bellingham, DC, USA, 2021; Volume 11821, p. 1182109. [CrossRef]
32. Salvaterra, R. High redshift Gamma-ray Bursts. *J. High Energy Astrophys.* **2015**, *7*, 35–43. [CrossRef]

33. Fryer, C.L.; Lien, A.Y.; Fruchter, A.; Ghirlanda, G.; Hartmann, D.; Salvaterra, R.; Upton Sanderbeck, P.R.; Johnson, J.L. Properties of High-redshift Gamma-ray Bursts. *ApJ* **2022**, *929*, 111. [CrossRef]
34. Lidz, A.; Chang, T.C.; Mas-Ribas, L.; Sun, G. Future Constraints on the Reionization History and the Ionizing Sources from Gamma-ray Burst Afterglows. *ApJ* **2021**, *917*, 58. [CrossRef]
35. Ota, K.; Iye, M.; Kashikawa, N.; Konno, A.; Nakata, F.; Totani, T.; Kobayashi, M.A.R.; Fudamoto, Y.; Seko, A.; Toshikawa, J.; et al. A New Constraint on Reionization from the Evolution of the Lyα Luminosity Function at z $\sim$6–7 Probed by a Deep Census of z = 7.0 Lyα-Emitter Candidates to 0.3 L. *Astrophys. J.* **2017**, *844*, 85. [CrossRef]
36. Feldman, C.; O'Brien, P.; White, N.; Baumgartner, W.; Thomas, N.; Lodge, A.; Bautz, M.; Hinrichsen, E. LEXT: A lobster eye optic for Gamow. In Proceedings of the Optics for EUV, X-ray, and Gamma-ray Astronomy X, San Diego, CA, USA, 23 August 2021; Society of Photo-Optical Instrumentation Engineers (SPIE) Conference Series; O'Dell, S.L., Gaskin, J.A., Pareschi, G., Eds.; SPIE: Bellingham, DC, USA, 2021; Volume 11822, p. 118221D. [CrossRef]
37. Seiffert, M.; Balady, A.; Chang, T.C.; Dyer, R.; Fausey, H.; Guiriec, S.; Hart, M.; Morris, R.O.; Rodriguez, J.I.; Roming, P.; et al. The Photo-z Infrared Telescope (PIRT)—A space instrument for rapid follow up of high-redshift gamma-ray bursts and electromagnetic counterparts to gravitational wave events. In Proceedings of the UV/Optical/IR Space Telescopes and Instruments: Innovative Technologies and Concepts X, San Diego, CA, USA, 20 August 2021; Society of Photo-Optical Instrumentation Engineers (SPIE) Conference Series; Barto, A.A., Breckinridge, J.B., Stahl, H.P., Eds.; SPIE: Bellingham, DC, USA, 2021; Volume 11819, p. 1181906. [CrossRef]
38. Fausey, H.M.; van der Horst, A.J.; White, N.E.; Seiffert, M.; Willems, P.; Young, E.T.; Kann, D.A.; Ghirlanda, G.; Salvaterra, R.; Tanvir, N.R.; et al. Photometric redshift estimation for gamma-ray bursts from the early Universe. *MNRAS* **2023**, *526*, 4599–4612. [CrossRef]
39. Gehrels, N.; Cannizzo, J.K. How Swift is redefining time domain astronomy. *J. High Energy Astrophys.* **2015**, *7*, 2–11. [CrossRef]
40. Chan, M.L.; Messenger, C.; Heng, I.S.; Hendry, M. Binary neutron star mergers and third generation detectors: Localization and early warning. *Phys. Rev. D* **2018**, *97*, 123014. [CrossRef]
41. Nitz, A.H.; Canton, T.D. Pre-merger Localization of Compact-binary Mergers with Third-generation Observatories. *Astrophys. J. Lett.* **2021**, *917*, L27. [CrossRef]
42. Banerjee, B.; Oganesyan, G.; Branchesi, M.; Dupletsa, U.; Aharonian, F.; Brighenti, F.; Goncharov, B.; Harms, J.; Mapelli, M.; Ronchini, S.; et al. Pre-merger alert to detect prompt emission in very-high-energy gamma-rays from binary neutron star mergers: Einstein Telescope and Cherenkov Telescope Array synergy. *A&A* **2023**, *678*, A126. [CrossRef]
43. Yonetoku, D.; Mihara, T.; Doi, A.; Sakamoto, T.; Tsumura, K.; Ioka, K.; Amaya, Y.; Arimoto, M.; Enoto, T.; Fujii, T.; et al. High-redshift gamma-ray burst for unraveling the Dark Ages Mission: HiZ-GUNDAM. In Proceedings of the Space Telescopes and Instrumentation 2020: Ultraviolet to Gamma ray, San Diego, CA, USA, 20 August 2021; den Herder, J.W.A., Nikzad, S., Nakazawa, K., Eds.; International Society for Optics and Photonics, SPIE: Bellingham, DC, USA, 2020; Volume 11444, p. 114442Z. [CrossRef]
44. Li, J.; Sakamoto, T.; Serino, M.; Yonetoku, D.; Sawano, T.; Mitsuish, I.; Mihara, T. X-ray performance and simulation study of lobster eye optics. In Proceedings of the Space Telescopes and Instrumentation 2020: Ultraviolet to Gamma ray, Online, 13 December 2020; den Herder, J.W.A., Nikzad, S., Nakazawa, K., Eds.; International Society for Optics and Photonics, SPIE: Bellingham, DC, USA, 2020, Volume 11444, p. 114447C. [CrossRef]
45. Ogino, N.; Arimoto, M.; Sawano, T.; Yonetoku, D.; Goto, H.; Hyeongsoon, W.; Hiraga, J.; Sakamoto, T.; Sei, K.; Yatsu, Y. Development of a fast readout system of a CMOS image sensor for the time-domain astronomy. In Proceedings of the Space Telescopes and Instrumentation 2020: Ultraviolet to Gamma ray, online 13 December 2020; Society of Photo-Optical Instrumentation Engineers (SPIE) Conference Series; den Herder, J.W.A., Nikzad, S., Nakazawa, K., Eds.; SPIE: Bellingham, DC, USA, 2020; Volume 11444, p. 114445L. [CrossRef]
46. Sawano, T.; Yonetoku, D.; Arimoto, M.; Li, J.; Mihara, T.; Ogino, N.; Sakamoto, T.; Serino, M. A detection algorithm for faint sources based on 1-d projection for a lobster-eye X-ray imaging system. In Proceedings of the Space Telescopes and Instrumentation 2020: Ultraviolet to Gamma ray, online, 13 December 2020; Society of Photo-Optical Instrumentation Engineers (SPIE) Conference Series; den Herder, J.W.A., Nikzad, S., Nakazawa, K., Eds.; SPIE: Bellingham, DC, USA, 2020; Volume 11444, p. 114445K. [CrossRef]
47. Goto, H.; Yonetoku, D.; Ogino, N.; Takahashi, S.; Sato, T.; Mukai, K.; Arimoto, M.; Sawano, T.; Mihara, T.; Sakamoto, T.; et al. Development of a method for aligning lobster eye optics onboard HiZ-GUNDAM with visible light and shape measurements. In Proceedings of the Society of Photo-Optical Instrumentation Engineers (SPIE) Conference Series, Montréal, QC, Canada, 31 August 2022; den Herder, J.W.A., Nikzad, S., Nakazawa, K., Eds.; Society of Photo-Optical Instrumentation Engineers (SPIE): Bellingham, DC, USA, 2022; Volume 12181, p. 121815J. [CrossRef]
48. Tsumura, K.; Yonetoku, D.; Kawabata, K.; Matsuura, S.; Noda, H.; Urata, Y.; Niino, Y.; Sano, K.; Ohashi, A.; Doi, A.; et al. Development of an optical and near-infrared telescope onboard the HiZ-GUNDAM mission. In Proceedings of the Space Telescopes and Instrumentation 2020: Optical, Infrared, and Millimeter Wave, online, 13 December 2020; Society of Photo-Optical Instrumentation Engineers (SPIE) Conference Series; Lystrup, M., Perrin, M.D., Eds.; SPIE: Bellingham, DC, USA, 2020; Volume 11443, p. 114430R. [CrossRef]

49. McConnell, M.L.; Baring, M.; Bloser, P.F.; Briggs, M.; Ertley, C.; Fletcher, G.; Gaskin, J.; Gelmis, K.; Goldstein, A.; Grove, J.E.; et al. The LargE Area burst Polarimeter (LEAP)—A NASA mission of opportunity for the ISS. In Proceedings of the UV, X-ray, and Gamma-ray Space Instrumentation for Astronomy XXII, San Diego, CA, USA, 1–5 August 2021; p. 35. [CrossRef]
50. Oñate-Melecio, K.; Ertley, C.; McConnell, M.L.; Legere, J.; Bloser, P.F.; Briggs, M.; Gaskin, J.; Goldstein, A.; Grove, J.E.; Hui, M.; et al. Evaluation of a prototype detector for the LargE Area burst Polarimeter (LEAP). In Proceedings of the UV, X-ray, and Gamma-ray Space Instrumentation for Astronomy XXII, San Diego, CA, USA, 1–5 August 2021; p. 36. [CrossRef]
51. Toma, K.; Sakamoto, T.; Zhang, B.; Hill, J.E.; McConnell, M.L.; Bloser, P.F.; Yamazaki, R.; Ioka, K.; Nakamura, T. Statistical Properties of Gamma-ray Burst Polarization. *Astrophys. J.* **2009**, *698*, 1042–1053. [CrossRef]
52. McConnell, M.L. High energy polarimetry of prompt GRB emission. *New Astron. Rev.* **2017**, *76*, 1–21. [CrossRef]
53. Tatischeff, V.; McConnell, M.L.; Laurent, P. *Astronomical Polarisation from the Infrared to Gamma Rays*; Astrophysics and Space Science Library: Toledo, Spain, 2019; pp. 109–146. [CrossRef]
54. Chattopadhyay, T. Hard X-ray polarimetry—An overview of the method, science drivers, and recent findings. *J. Astrophys. Astron.* **2021**, *42*, 106. [CrossRef]
55. Hurley, K.; Pal'shin, V.D.; Aptekar, R.L.; Golenetskii, S.V.; Frederiks, D.D.; Mazets, E.P.; Svinkin, D.S.; Briggs, M.S.; Connaughton, V.; Meegan, C.; et al. The Interplanetary Network Supplement to the Fermi GBM Catalog of Cosmic Gamma-ray Bursts. *ApJS* **2013**, *207*, 39. [CrossRef]
56. Kole, M.; De Angelis, N.; Berlato, F.; Burgess, J.M.; Gauvin, N.; Greiner, J.; Hajdas, W.; Li, H.C.; Li, Z.H.; Pollo, A.; et al. The POLAR gamma-ray burst polarization catalog. *A&A* **2020**, *644*, A124. [CrossRef]
57. Li, H. Gamma-ray Polarimetry of the Crab Pulsar Observed by POLAR. In Proceedings of the 44th COSPAR Scientific Assembly, Online, 16–24 July 2022; Volume 44, p. 1847.
58. Gill, R.; Kole, M.; Granot, J. GRB Polarization: A Unique Probe of GRB Physics. *Galaxies* **2021**, *9*, 82. [CrossRef]
59. De Angelis, N. Development of the Next Generation Space-Based Compton Polarimeter and Energy Resolved Polarization Analysis of Gamma-ray Bursts Prompt Emission. Ph.D. Thesis, University of Geneva, Geneva, Switzerland, 2023. [CrossRef]
60. Kole, M.; Koziol, G.; Droz, D. HAGRID–High Accuracy GRB Rapid Inference with Deep learning. *arXiv* **2023**, arXiv:2309.01493. https://doi.org/10.48550/arXiv.2309.01493.
61. Agostinelli, S.; Allison, J.; Amako, K.; Apostolakis, J.; Araujo, H.; Arce, P.; Asai, M.; Axen, D.; Banerjee, S.; Barrand, G.; et al. Geant4—A simulation toolkit. *Nucl. Instruments Methods Phys. Res. Sect. Accel. Spectrometers Detect. Assoc. Equip.* **2003**, *506*, 250–303. [CrossRef]
62. Howell, E.J.; Ackley, K.; Rowlinson, A.; Coward, D. Joint gravitational wave-gamma-ray burst detection rates in the aftermath of GW170817. *MNRAS* **2019**, *485*, 1435–1447. [CrossRef]
63. Punturo, M.; Abernathy, M.; Acernese, F.; Allen, B.; Andersson, N.; Arun, K.; Barone, F.; Barr, B.; Barsuglia, M.; Beker, M.; et al. The Einstein Telescope: A third-generation gravitational wave observatory. *Class. Quantum Gravity* **2010**, *27*, 194002. [CrossRef]
64. Maggiore, M.; Broeck, C.V.D.; Bartolo, N.; Belgacem, E.; Bertacca, D.; Bizouard, M.A.; Branchesi, M.; Clesse, S.; Foffa, S.; García-Bellido, J.; et al. Science case for the Einstein telescope. *J. Cosmol. Astropart. Phys.* **2020**, *2020*, 050. [CrossRef]
65. Reitze, D.; Adhikari, R.X.; Ballmer, S.; Barish, B.; Barsotti, L.; Billingsley, G.; Brown, D.A.; Chen, Y.; Coyne, D.; Eisenstein, R.; et al. Cosmic Explorer: The U.S. Contribution to Gravitational-Wave Astronomy beyond LIGO. *arXiv* **2019**, arXiv:1907.04833. https://doi.org/10.48550/arXiv.1907.04833.
66. Stratta, G.; Amati, L.; Branchesi, M.; Ciolfi, R.; Tanvir, N.; Bozzo, E.; Götz, D.; O'Brien, P.; Santangelo, A. Breakthrough Multi-Messenger Astrophysics with the THESEUS Space Mission. *Galaxies* **2022**, *10*, 60. [CrossRef]

MDPI
St. Alban-Anlage 66
4052 Basel
Switzerland
www.mdpi.com

Universe Editorial Office
E-mail: universe@mdpi.com
www.mdpi.com/journal/universe

www.ingramcontent.com/pod-product-compliance
Lightning Source LLC
Chambersburg PA
CBHW061830210926
51850CB00006B/849